Earth and Environmental Sciences Library

Earth and Environmental Sciences Library (EESL) is a multidisciplinary book series focusing on innovative approaches and solid reviews to strengthen the role of the Earth and Environmental Sciences communities, while also providing sound guidance for stakeholders, decision-makers, policymakers, international organizations, and NGOs.

Topics of interest include oceanography, the marine environment, atmospheric sciences, hydrology and soil sciences, geophysics and geology, agriculture, environmental pollution, remote sensing, climate change, water resources, and natural resources management. In pursuit of these topics, the Earth Sciences and Environmental Sciences communities are invited to share their knowledge and expertise in the form of edited books, monographs, and conference proceedings.

Gabor Korvin

Statistical Rock Physics

Springer

Gabor Korvin
Earth Sciences
King Fahd University of Petroleum
and Minerals
Dhahran, Saudi Arabia

ISSN 2730-6674 ISSN 2730-6682 (electronic)
Earth and Environmental Sciences Library
ISBN 978-3-031-46699-1 ISBN 978-3-031-46700-4 (eBook)
https://doi.org/10.1007/978-3-031-46700-4

This Springer imprint is published by the registered company Springer Nature Switzerland AG
The registered company address is: Gewerbestrasse 11, 6330 Cham, Switzerland

Paper in this product is recyclable.

Preface

The book is a systematic and comprehensive treatment of the powerful stochastic models and computational tools that have emerged in *Rock Physics* in the last 30 years. In its spirit, scope and style, it is a sequel to my previous book, G. Korvin, "*Fractal Models in the Earth Sciences*" (Amsterdam: Elsevier, 1992), that was well received at the time it appeared, and has introduced generations of Earth Scientists to modeling Earth formations and processes by *fractals*. Two years ago, I was asked by the editors of Springer's *Encyclopedia of Mathematical Geosciences* to contribute a few chapters to this important reference book, and when working on one of them, "*Statistical Rock Physics*", I realized that this new discipline has reached such a maturity in recent years that it would deserve a self-consistent, full-fledged, monographic treatment. I am grateful to Profs. Fritz Agterberg and Daya Sagar for recommending my book for publication, and to *Springer Nature,* especially to Ms. Anette Buettner, Ms. Hemavathy Manivannan, and Ms. Deepa Varadharajan, for espousing this idea and for helping me through the arduous process of book writing, editing and processing.

Statistical Rock Physics is a part of *Rock Physics* (*Petrophysics*). Its concepts, methods and techniques have been borrowed from *Stochastic and Integral Geometry* and *Statistical Physics.* In this new science, one describes the *interior geometry* of rocks; derives their *effective* physical properties based on their random composition and the random arrangement of their constituents; and builds models to simulate the past geological processes that had formed the rock. Most techniques that will be discussed belong to the toolbox of the "*Digital Rock Physics*" (DRP) technology of *Core Analysis.* My aim has been to help readers to understand the claims and published results of these powerful new methods, and—most importantly—to creatively apply stochastic geometry and statistical physics to their own research tasks. My own understanding of these tools had matured by incorporating them, and teaching them for more than two decades in a heuristically simplified but mathematically precise form, in my *Geophysical Inversion*, *Solid Earth Geophysics*, *Reservoir Characterization*, *Well Logging* and other courses at the *King Fahd University of Petroleum and Minerals* (Dhahran, Kingdom of Saudi Arabia).

In all the chapters and sections of the book, the underlying mathematics will be discussed in full, numerical solutions will be emphasized, a full set of references will be given and theory will go hand-in-hand with practical applications to hydraulic permeability, electric conduction, rock failure, Nuclear Magnetic Relaxation (NMR), mechanics of random grain packings, compaction of shale and various other phenomena.

There are eleven self-consistent chapters, each devoted to an independent part of *Statistical Rock Physics*, each containing practical case histories, a full set of references and tabulated lists of the relevant rock physics applications.

Chapter 1 *Random Functions and Random Fields, Autocorrelation Functions* is a brief survey of random functions and fields. It discusses the two-point and multipoint correlation functions (Sects. 1.1–1.3), basic concepts used throughout the book. Section 1.4 describes how to estimate specific surface area and mean grain size from the 2-point correlation function.

Chapter 2 *Models of Tortuosity* is devoted to the wonderful world of tortuous pathways inside the pore space of rocks, available for fluids, electric current, heat, sound waves, diffunding particles, and the developing cracks around harder grains of the rock. The effect of tortuosity on permeability and other effective properties is treated in details.

The *Internal Topology of Rocks* is dealt with in Chap. 3. We acquaint the reader with the modern mathematical tools for studying the topology of pore space. In Sect. 3.1, *Minkowski functionals* are introduced through the paradigm of Mark Kac (1966) "Can one hear the shape of a drum?". *Euler characteristics*, a powerful descriptor of connectivity is treated in 3.2. The elegant and sophisticated technique, *Persistent Homology*, is built up in a series of easy-to-follow steps in 3.3. These abstract constructions are illustrated with practical examples, from diverse fields of Petrophysics.

The common way to compute hydraulic and electric flow in porous media is to substitute the pore structure by a network consisting of nodes (for pores) and bonds (for the interconnecting throats). Different ways of constructing such models, and computation of their hydraulic and electric conductivity, are reviewed in Chap. 4 *Random Network Models* (Sect. 4.1). Section 4.2 introduces more than a dozen measures to characterize the randomness and connectedness of such models. Section 4.3 treats three recently discovered phenomena arising in large-scale networks: the emergence of connectivity and appearance of a giant component in evolving random networks; the scale-free distribution of the nodes' coordination numbers; and the networks' "small-world" property.

Chapter 5 *Menger Sponge Models* treats Menger Sponge (in 3-D) and Sierpiński Carpet (in 2-D) which are the most useful fractal models of pore space. Sections 5. 1–5.2 introduce these models and their combinatorial and topological properties; random and multifractal generalizations and the permeability of these models are discussed in Sect. 5.3. Appendix (5.4) introduces a useful measure, *lacunarity*, to distinguish between textures of the same fractal dimension but different geometry.

Chapter 6 *Coordination Number of Grains* reviews the combinatorial geometry of granular packings in 2-D and 3-D, and the average coordination number (number of contacts per grain) ⟨Z⟩, for different cases. The mechanical stability of random packs requires the satisfaction of a set of linear equations, and the consistency of these equations determines the possible values of ⟨Z⟩. The coordination number will be also determined in the presence of friction and adhesive forces. Recent attempts to find an *Equation of State* (EOS) between packing fraction and mean coordination number are also discussed.

In Chap. 7 *The Shape of Pebbles, Grains and Pores*, we start with the different classification schemes and numerical formulae used to describe the shape of grains, pebbles and rock fragments, then acquaint the reader with the evolution equations leading to the characteristic ovoid shape of beach pebbles. Further topics include the number of stable and unstable points of equilibrium of a pebble or rock fragment, dependence on particle shape of the settling velocity of sediments in water, and the surface roughness and surface fractal dimension of irregular particles.

Chapter 8 *Entropy and Rock Physics* defines entropy as measure of information in Sect. 8.1 through a classic problem, "how to keep the forecaster honest?" (Good 1954). The section discusses "rock entropy" and its role in acoustic scattering, the entropy of the Markov Chains of sedimentary successions and the changes of entropy as an earthquake prediction tool in seismically active regions. Section 8.2 derives *maximum entropy* formalism and demonstrates its use in three worked-out examples: maximum entropy (ME) pore shape inversion; an *ab initio* derivation of Athy's (1930) compaction law; and the distribution of contact forces in stressed granular material. A mathematical Appendix studies the relation between the Hurst exponent and the entropy increase in self-affine random processes.

Chapter 9 *Effective Properties of Rocks* is about the computation of the macroscopic properties of a composite material based upon the relative fractions and specific properties of its constituent parts. The history of *Effective Medium Theory* (EMT), and the mathematical treatment of its different versions (*Maxwell's Approximation*, the *Self Consistent Method*, and *Differential Effective Method*) are given in Sect. 9.1. Section 9.1a applies EMT to find the effective conductivity of a network of random conductors (Kirkpatrick 1973). *Differential Effective Medium* (DEM) approximation (Bruggeman 1935) is treated in Sect. 9.2, in the context of electric conductivity and elasticity. In 9.3., we discuss *variational inequalities* (Hashin and Shtrikman 1963a, b), and a *generalized mean value theorem* (Korvin 1978, 1982) for effective rock properties, and present an example (Korvin 2012) for the anisotropic case. Section 9.4 reviews attempts to theoretically derive *Archie's law* (1942) $\sigma_{rock} = \sigma_{fluid}\Phi^{m}$ from basic principles, and to determine the *cementation exponent m.*

Chapter 10 *Markov Random Fields and Random Walks* has two main topics. Markov Random Fields are spatial generalizations of *Markov Chains* in time, and are mainly used in geophysical inverse problems and geologic image processing. Discrete-time and continuous-time Random Walk models are among the most popular petrophysical simulation tools. Three important applications are discussed in detail: anomalous diffusion; the NMR response of fluids in porous rocks; and an attempt (of Ioannidis et al. 1997) to derive the formation factor *F* in Archie's equation.

Mathematical details (derivation of the Gibbs Distribution; proof of the Hammersley-Clifford Theorem; Polya's Theorem and its proof; and partial differential equations governing Continuous Time Random Walks) are discussed in appendices.

Chapter 11, *Thermodynamic Algorithms*, is the largest, and theoretically most demanding part of the book. Four important computational tools are reviewed, all of which are based on thermodynamic principles. *Simulated Annealing* (Sect. 11.1) is presented through two applications: *Reconstructing 3D pore space from 2D sections*, and *multiphase fluid distribution in porous rocks*. The *Lattice Boltzman Method* (LBM), a promising new tool of Digital Rock Physics, is treated in Sect. 11.2, including its theoretical derivation, computational aspects and generalizations. The applications include simulation of *multiphase flow*, and *Knudsen flow in tight shales*. The powerful method of RNG (*Renormalization Group*, Sect. 11.3) stems from the *Theory of Phase Transitions*. It is applied to find the *critical point of failure*, and *time of failure*, of stressed rocks; to describe the *erosion of tilted landscapes*, and for *permeability upscaling*. Section 11.4 is devoted to *discrete scale invariance*, and its use for *earthquake prediction* and *rock fragmentation*.

When writing the book, I have kept in mind the needs of researchers and graduate students, especially those contemplating writing their thesis on modern petrophysics. I hope it will be useful both as a handbook for research, and as a source of supplementary material for advanced courses in Petrophysics and Computational Geology/Geophysics. Its expected readership includes Earth Scientists, Geophysicists, Geologists, Petrophysicists, Petroleum Engineers, Applied Mathematicians, and Computer Scientists.

Though there are copious references to base mathematics, most of the treatment will be understandable without looking them up, assuming a reasonable grasp of and working skills in Undergraduate Algebra and Analysis. Some topics, however, such as *Homology* in Chap. 8, *Complex Analysis* in Sect. 11.4, and usually the mathematics relegated to appendices, will require more strenuous efforts. Notation is standard, both "log" and "ln" will refer to natural logarithm, unless otherwise noted.

I have deliberately abstained from mentioning, listing, let alone recommending, commercially available software products or commercial rock analysis services. I have tried to provide as complete lists of references to scientific papers as I could, depending on their availability on the Internet. Absence of a particular work only reflects my ignorance, and is not a value judgment. In the interest of my expected readership, young researchers and students, I preferred to use such references which are freely downloadable in pdf.

Boris Rybak (1923–2003), Professor of Animal Physiology at Université de Caen, used to close his directorial introductions to the booklets in the "*Discours de la Méthode*" series of the Gauthier-Villars Publisher (Paris/Bruxelles/Montréal, 1970s) with the sentence: "*Toute acte de l'homme, donc toute œuvre—créations et inventions—est autobiographique*" ("Every human act, thus every work-be it a creation or an invention-is autobiographical", transl. by my son, Adam Korvin).

This is certainly true in the case of my two books. While my "*Fractal Models*" book (1992) bears witness to my "salad days" (Antony and Cleopatra 1:5), the first half of my career, in Hungary and Australia; its present sequel is intimately connected with the second 30 years, more than two decades of Professorship at the KFUPM University, Saudi Arabia, followed by almost a decade spent in happy seclusion in Hejce, a small village in the mountains of NE Hungary, where this book had been prepared.

It happened during my stay in Saudi Arabia that my ideas about "Statistical Rock Physics" started to crystallize (Korvin 2016). I had been invited as mathematical consultant to many projects of two outstanding scientists, Prof. Klavdia Oleschko (soil physicist, UNAM, Mexico), and Dr. Abdulazeez Abdulraheem (petrophysicist, KFUPM, Saudi Arabia). I gratefully thank both of them for giving me access to their data, discussing with me many of these ideas and helping me to confront mathematics with ground truth.

Hejce and Budapest, Hungary Gabor Korvin
August/September of 2023

References

Archie, G.E., 1942. The electrical resistivity log as an aid in determining some reservoir characteristics. Petroleum Transactions of AIME 146: 54–62.

Athy, L.I., 1930. Compaction and oil migration. Bull. Am. Ass. Petrol. Geol. 14: 25–35.

Bruggeman, D.A.G., 1935. Berechnung verschiedener physikalischer Konstanten von heterogenen Substanzen. I. Dielektrizitätskonstanten und Leitfähigkeiten der Mischkörper aus isotropen Substanzen. Annalen der Physik 416(7): 636–664.

Good, I.J., 1954. Uncertainty and Business Decisions. Liverpool: Liverpool University Press.

Hashin, Z., and S. Shtrikman, 1963a. A variational approach to the elastic behavior of multiphase materials. Journal of the Mechanics and Physics of Solids 11: 127–140.

Hashin, Z., and Shtrikman, S., 1963b. Conductivity of polycrystals. Phys. Rev. 130(1): 129–133.

Ioannidis, M.A., Kwiecen, M.J. and Chatzis, I., 1997. Electrical conductivity and percolation aspects of statistically homogeneous porous media. Transport in Porous Media 29(1): 61–83.

Kac, M., 1966. Can one hear the shape of a drum? Amer. Math. Monthly 73:1–24.

Kirkpatrick, S., 1973. Percolation and conduction. Reviews of Modern Physics 45(4): 574–588.

Korvin, G., 1978. The hierarchy of velocity formulae: Generalized mean value theorems. Acta Geod. Geoph. et Mont. Acad. Sci. Hung. 13(1-2): 211–222.

Korvin Gabor, 1982. Axiomatic characterization of the general mixture rule. Geoexploration 19(4): 267–276.

Korvin, G., 1992. Fractal Models in the Earth Sciences. Amsterdam: Elsevier.

Korvin, G., 2012. Bounds for the resistivity anisotropy in thinly-laminated sand-shale. Petrophysics 53(1): 14–21.

Korvin, G., 2016. Permeability from Microscopy: Review of a Dream. Arabian J. of Science & Engineering 41(6): 2045–2065.

Korvin, G., 2021. Statistical Rock Physics. In: Daya Sagar, B., Cheng, Q., McKinley, J., Agterberg, F. (Eds) Encyclopedia of Mathematical Geosciences. Encyclopedia of Earth Sciences Series. Cham: Springer.

Contents

Chapter 1
Random Functions and Random Fields, Autocorrelation Functions

Chapter Highlights The brief survey of *random functions and fields*, and properties of the *two-point and multi-point correlation functions* (Sects. 1.1–1.3), introduces basic concepts used throughout the book. Section 1.4. describes how to estimate specific surface area and mean grain size from the 2-point correlation function.

1.1 Random Functions along a Line, Random Fields, and their Autocorrelation Functions

In 1976, the British statistician George Box penned the famous line, "All models are wrong, some are useful." The 1-, 2-, and 3-point *correlation functions*, which have found an amazingly wide range of applications in all possible branches of petrophysics (from "acoustic emissions" to "weathered limestones", Table 1.1) are the rock-physicist's most useful tools for characterizing random pore spaces and predicting effective rock properties. These functions have been borrowed from the *Theory of* (1-D) *Stochastic Processes*, and (2- and 3-D) *Random Fields* (Chernov 1960; Rytov et al. 1987).

A *random function* (also called *Stochastic Process*), $\{f(x)\}_\alpha$ is a *family of functions* depending on a random parameter α, where the independent variable x varies along some line. A given $f(x)$ picked at random from among all possible $\{f(x)\}_\alpha$ is a *realization.* At any fixed point on the line x_1, $f(x_1)$ is a *random value*, it attains different values y with the probabilities $Prob\big[f(x_1) < y\big] = F(x_1, y)$. The following *expected values* (taken with respect to α, i.e. over all realizations of $\{f(x)\}_\alpha$) are often enough to characterize a random function: $\langle f(x_1)\rangle$, $\big\langle f^2(x_1)\big\rangle$, $\langle f(x_1)\cdot f(x_2)\rangle$, termed *mean value*, *mean square value* and *autocorrelation function.* Throughout the book, the expected value of a random variable r will be denoted by $\langle r\rangle$ or $\overline{r}$.

A random function is *translation invariant* if its statistical properties do not change with respect to a shift along the line; in particular, if $\langle f(x)\rangle = \langle f(0)\rangle$; $\big\langle f^2(x)\big\rangle = \big\langle f^2(0)\big\rangle$;

$$\langle f(x_1)\cdot f(x_2)\rangle = \langle f(0)\cdot f(x_2 - x_1)\rangle := R_{ff}(|x_1 - x_2|) \tag{1.1}$$

G. Korvin, *Statistical Rock Physics*, Earth and Environmental Sciences Library,
https://doi.org/10.1007/978-3-031-46700-4_1

Table 1.1 Some uses of the autocorrelation function in rock physics

Topic	Authors
Acoustic emissions	Lokajicek et al. (2004), Prasad et al. (2009)
Aggregates	Kjems and Freltoft (1991)
Amorph systems	Ciccariello et al. (1981)
Anisotropy	Whitehouse (1974), Erslev and Ge (1990), Odgaard et al. (1990), Derbyshire et al. (1992), Pfleiderer and Halls (1993), Swan and Garratt (1995), Odgaard et al. (1997), Millard et al. (1998), Ikeda et al. (2000), Rotter et al. (2001), Saha and Wehrli (2004), Yokoyama and Nakashima (2005), Wald et al. (2005, 2007), Wehrli et al. (2006), Xiang et al. (2007), Chiang et al. (2006, 2009), Varga and Zysset (2009), Faessel and Jeulin (2010), Singh et al. (2012), Svitelman and Dinariev (2013), Jiao and Chawla (2014a, b), Monfared and Ulm (2016)
Asphalt	Velasquez et al. (2010)
Bacteria in soil	Naoise et al. (2003)
Bulk modulus	Torquato and Lado (1986)
Capillary displacement	Hazlett (1995)
Carbonates	Okabe (2004)
Cement	Han and Kim (2018), Kim et al. (2019a, b, 2020)
Chalk	Bekri et al. (2000), Talukdar and Torsaeter (2002), Talukdar et al. (2002b, c)
Chemical index of alteration	Xiao et al. (2010)
Climatic predictability	Rehman and El-Gebeily (2009)
Coal	Radlinski et al. (2001)
Coda attenuation	Di Martino et al. (2021)
Concrete	Chung and Han (2013), Roubin et al. (2015)
Conductivity	Torquato and Lado (1986), Ioannidis et al. (1997), Liang et al. (2000), Moctezuma-Berthier et al. (2002), Li et al. (2006), Stenzel et al. (2016), Neumann et al. (2020)
Cracked media	Ponte Castañeda and Willis (1995)
Crystallography	Ikeda et al. (2000), Heilbronner (2010)
Dielectricity	Hilfer (1991), Haslund et al. (1994)
Diffusion	Prager (1963a, b), Doi (1976), Ho and Strieder (1979), Torquato (1986a), Rubinstein and Torquato (1988), Tassopoulos and Rosner (1992), Valiullin and Skirda (2001), Kainourgiakis et al. (2002a, b), Howard et al. (2020), Westin et al. (2002), František and Mukamel (2005), Yokoyama and Nakashima (2005), Nakashima et al. (2008)
Earthquakes	Kagan and Knopoff (1980)
Effective rock properties	Beran (1968), Wang and Pan (2008)

(continued)

Table 1.1 (continued)

Topic	Authors
Elasticity	Corson (1974c), Kröner (1977), Berryman (1985a), Beran et al. (1996), Lin and Garmestani (2000), Arns et al. (2002), Gatt et al. (2005), Cai et al. (2019), Zerhouni et al. (2021)
Fabric analysis	Lin (1982), Heilbronner-Panozzo (1992), Heilbronner and Barrett (2014)
Fatigue analysis	Khristenko et al. (2020)
Flow	Doi (1976), Thompson et al. (1987), Adler et al. (1990), Hilfer (1992, 1993), Bentz and Martys (1994), Hazlett (1997), Coles et al. (1998), Kainourgiakis et al. (2002a, b), Alim et al. (2017)
Fluid displacement	Prodanovic et al. (2007)
Formation factor	Adler et al. (1992), Ioannidis et al. (1995)
Fractures	Renard et al. (2018)
Fuel cells	Wargo et al. (2012), Cecen et al. (2014)
Gas shale	Hubler et al. (2017)
Glaciology	Mätzler (2002)
Grain size	Barnard et al. (2007), Buscombe (2008), Buscombe and Masselink (2008), Warrick et al. (2009), Buscombe et al. (2010), Cheng and Liu (2015)
Hypsometry	Pérez et al. (2009)
Image correlation spectroscopy	Wiseman et al. (2002), Costantino et al. (2005), Robertson and George (2012)
Image fusion	Gerke et al. (2015)
inelasticity	Garmestani et al. (1998, 2001)
Joint Roughness Coefficient	Gravanis and Pantelidis (2019)
Magnetic anisotropy	Rees (1965), Schultz-Krutisch and Heller (1985), Pfleiderer and Halls (1990, 1993)
Microstructure informatics	Torquato (1986b), Berryman and Milton (1988), Boger et al. (1992), Biswal et al. (1999), Mason and Brent (1999), Adams et al. (2002), Radlinski et al. (2002), Torquato (2002a), Appoloni et al. (2004), Saheli et al. (2004), Gokhale and Singh (2005), Gokhale et al. (2005), Kumar et al. (2006), Fullwood et al. (2008), Yin et al. (2008), Velasquez et al. (2010), Kalidindi et al. (2011), Baniassadi et al. (2011b), Niezgoda et al. (2011), Jiao et al. (2013), Wildenschild and Sheppard (2013), Xu et al. (2013), Emami (2014), Hyman et al. (2015), Li et al. (2018a, b), Chen et al. (2019)
Microtomography	Fusseis et al. (2014)
Orientation coherence	Adams et al. (1987)
Partial saturation	Toms et al. (2007), Toms-Stewart et al. (2009)
Percolation	Torquato et al. (1988), Cinlar and Torquato (1995), Ioannidis et al. (1997), Chung and Han (2013)

(continued)

Table 1.1 (continued)

Topic	Authors
Permeability	Berryman and Milton (1985), Berryman and Blair (1986, 1987), Rubinstein and Keller (1987), Blair et al. (1993, 1996), Coker et al. (1996), Ioannidis et al. (1996), Chatzis et al. (1997), Liang et al. (1999, 2000), Cobaleda (2000), Keehm et al. (2003), Torabi et al. (2008), Bernard et al. (2020), Röding et al. (2020), Neumann et al. (2020)
Permittivity	Brown (1955)
Pore structure	Rink and Schopper (1976), Ioannidis and Lange (1998), Sun and Koch (1998), Anovitz and Cole (2015), Xiao (2021), Hyman and Winter (2014), Demberg et al. (2018)
Reconstruction	Joshi (1974), Quiblier (1984), Adler et al. (1992), Deutsch (1992), Yao et al. (1993), Deutsch and Journel (1994), Fernandes (1994), Ouenes et al. (1994), Roberts (1997), Levitz (1998), Liang et al. (1998), Yeong and Torquato (1998a,b), Biswal and Hilfer (1999), Cule and Torquato (1999), Manwart and Hilfer (1999), Kainourgiakis et al. (2000, 2002a,b), Manwart et al. (2000), Saccomano et al. (2001), Sheehan and Torquato (2001), Hidajat et al. (2002), Øren and Bakke (2002), Talukdar and Torsaeter (2002), Talukdar et al. (2002a, b, c), Torquato and Haslach (2002), Jin et al. (2003), Okabe (2004), Koutsourelakis (2006), Koutsourelakis and Deodatis (2006), Kumar et al. (2006), Zhao et al. (2007), Fullwood et al. (2008), Jiao et al. (2007, 2008, 2009), Politis et al. (2008), Capek et al. (2009), Patelli and Schuëller (2009, 2012), Chung and Han (2010), Torquato (2010), Hajizadeh et al. (2011), Tahmasebi and Sahimi (2012), Xu et al. (2013), Bodla et al. (2014), Feng et al. (2014), Guo et al. (2014), Hyman and Winter (2014), Jiao and Chawla (2014a), Pant et al. (2014), Ballani and Stoyan (2015), Chen and Jiao (2015), Gerke and Karsanina (2015), Karsanina et al. (2015), Papakostas et al. (2015), Chen et al. (2016), Hasanabadi et al. (2016), Havelka et al. (2016), Karimpouli and Pejman (2016), Pant (2016), Mandzhieva (2017), Mosser et al. (2017), Feng et al. (2018), Karsanina and Gerke (2018), Li et al. (2018a, b), Lemmens et al. (2019), Coiffier et al. (2020), Feng et al. (2020), Gravey and Mariethoz (2020), Cherkasov et al. (2021), Kim et al. (2021), Zheng and Zhang (2022)
Reservoir rocks	Ehrlich et al. (1984), Wu et al. (2006)
Scattering	Debye and Bueche (1949), Guinier et al. (1955), Debye et al. (1957), Frisch and Stillinger (1963), Glatter (1980), Glatter and Kratky (1982), Mildner and Hall (1986), Berk (1987, 1991), Lindner and Zemb (1991), Radlinski et al. (1999, 2004), Radlinski (2006)
Sedimentary cycles	Herman et al. (2009), Xiao et al. (2013)
Shale	Gerke et al. (2015)
Soil	Gerke et al. (2012), de Lima Moraes et al. (2017), Karsanina et al. (2021)

(continued)

Table 1.1 (continued)

Topic	Authors
Specific heat	Steel (1968)
Specific surface	Berryman (1987, 1998), Rabbani et al. (2014)
Stiffness tensor	Torquato (1997, 1998)
Strain	Shimamoto and Ikeda (1976), Fry (1979), Hanna and Fry (1979), Crespi (1986), Bay et al. (1999), Viggiani et al. (2004), Louis et al. (2007), Thissen and Brandon (2015), Tudisco et al. (2015, 2017), Renard et al. (2018), McBeck et al. (2018)
Suspensions	Torquato and Lado (1991)
Texture	Jeulin (2000)
Thermal conductivity	Corson (1974d), Wang et al. (2007), Baniassadi et al. (2011a)
Transport	Prager (1963a, b), Hilfer (1993), Roberts and Teubner (1995), Levitz (1998), Bekri et al. (2000), Hidajat et al. (2002), Kainourgiakis et al. (2000, 2002a, b), Scholz et al. (2015)
Tuff	Berge et al. (1997)
Two-phase materials	Corson (1974a,b), Torquato (1986b), Torquato and Stell (1982, 1983a, b, 1985), Torquato and Lu (1993), Markov (1995), Lin and Garmestani (2000), Saheli et al. (2004), Koutsourelakis and Deodatis (2006), Belvin et al. (2009
Viscoelasticity	Deng et al. (2012)
Viscoplasticity	Adams et al. (1970)
Viscous flow	Prager (1961, 1963a, b)
Wave propagation	Stanke (1986)
Weathering of limestone	Rozenbaum et al. (2007)

Assume that $\langle f(x)\rangle = \langle f(0)\rangle = 0$ (1.2) (the process has *zero mean value*). The function R_{ff} is the *autocorrelation function* (*ACF*) of $f(x)$. It is an *even function*, $R(\xi) = R(-\xi)$, by translational invariance and the Cauchy–Schwarz inequality it assumes its maximum at $\xi = 0$, where

$$R_{ff}(0) = \langle f^2(x)\rangle \tag{1.2}$$

The *normalized autocorrelation function* is

$$\rho_{ff}(\xi) = R_{ff}(\xi)/\langle f^2(x)\rangle. \tag{1.3}$$

If the argument of $\{f(\cdot)\}_\alpha$ is *time*, t, the “*translational invariance* of $\{f(t)\}_\alpha$” with respect to time shifts is called *stationarity*, and we speak about “*stationary time series*”. For such a stationary time series its *autocorrelation function* is a powerful *periodicity indicator*, and a tool for predicting future from observed past values (Levinson 1947; Wiener 1949). The autocorrelation function of a *stationary random function* $x(t)$ is computed as

$$R_{xx}(\tau) = \langle x(t)x(t+\tau)\rangle = \lim|_{T\to\infty}\frac{1}{2T}\int_{-T}^{T} x(t)x(t+\tau)dt, \tag{1.4}$$

where τ is called the *lag*. If $x(t)$ is *sampled*, writing $t = n \cdot \Delta t$; $\tau = k \cdot \Delta t$; $x_n = x(n \cdot \Delta t)$, and approximating the integral with summation, we have, apart from the factor Δt:

$$R_{xx}(k) = \langle x_n x_{n+k}\rangle = \lim|_{N\to\infty}\frac{1}{2N+1}\sum_{n=-N}^{N} x_n x_{n+k} \tag{1.5}$$

In Eqs. (1.4 and 1.5) we assume that the function $x(t)$, and the sequence $\{\ldots, x_{-1}, x_0, x_1, \ldots\}$, are of *finite average energy*, in the sense that

$$R_{xx}(0) = \langle x^2(t)\rangle = \lim|_{T\to\infty}\frac{1}{2T}\int_{-T}^{T} x^2(t)dt < \infty,$$

$$R_{xx}(0) = \langle x_n^2\rangle = \lim|_{N\to\infty}\frac{1}{2N+1}\sum_{n=-N}^{N} x_n^2 < \infty.$$

We use two kinds of averaging in these equations: $\langle\cdot\rangle$ and $\lim|_{T\to\infty}\frac{1}{2T}\int_{-T}^{T}\cdots dt$ (or, numerically, $\lim|_{N\to\infty}\frac{1}{2N+1}\sum_{n=-N}^{N}\cdots$) their assumed equivalence for almost all realizations follows from the *ergodicity* of the process what we always assume.

In practice, (1.5) is only computed for limited values of lags $k = 0, \pm1, \pm2, \ldots, \pm M$ and the number of samples, N, is also limited. To get a reliable estimate of the average value in Eq. (1.5), the rule of thumb is that M (the *maximal lag*) cannot be larger than about 15% of the total number of data. Some of my Readers might have already encountered the *ACF* in *Seismic Processing* ("*deconvolution*") or in geostatistics ("*Kriging*", in the disguise of "*semivariance*").

In some cases the ACF has a particularly simple form:

- The ACF of a *completely random* (*uncorrelated*) *noise* of zero mean is the *Dirac delta function* $\delta(t)$;
- The ACF of a *periodic process* is periodic with the same period;
- The *Campbell-Rice Theorem* (Rice 1944) states that the *central part of the ACF of a series of randomly placed wavelets of the same shape and with random amplitudes is the same as the ACF of a single wavelet*, apart from a constant multiplying factor;
- A famous stationary process is the *random telegraph wave* $x(t)$, defined by three properties: $x(t)$ might assume two different values,

$$x(t) = +1 \text{ or } x(t) = -1 \tag{1.6a}$$

$$x(0) = \begin{cases} +1 \text{ } \textit{with probability } 0.5 \\ -1 \text{ } \textit{with probability } 0.5 \end{cases} \tag{1.6b}$$

the number of changes between the values ± 1 in the interval $[t_1, t_2], t_1 \leq t_2,$ $|t_2 - t_1| = \tau$, follows a *Poisson distribution* with density λ, that is

$$Prob(there\ are\ N\ changes\ in\ [t_1, t_2]) = e^{-\lambda\tau}\frac{(\lambda\tau)^N}{N!} \tag{1.6c}$$

Then $R_{xx}(\tau) = \langle x(t)x(t+\tau)\rangle = (+1) \cdot Prob(an\ even\ number\ of\ changes\ in$ $[t, t+\tau]) + (-1) \cdot Prob(an\ odd\ number\ of\ changes\ in\ [t, t+\tau]) =$

$$\begin{aligned} &= e^{-\lambda\tau}\sum_{k=0}^{\infty}\frac{(-1)^{2k}(\lambda\tau)^{2k}}{(2k)!} + e^{-\lambda\tau}\sum_{k=0}^{\infty}\frac{(-1)^{2k+1}(\lambda\tau)^{2k+1}}{(2k+1)!} \\ &= e^{-\lambda\tau}\sum_{k=0}^{\infty}\frac{(-\lambda\tau)^k}{(k)!} = e^{-2\lambda\tau} \end{aligned} \tag{1.7}$$

that is, the ACF is *exponentially decaying.*

Following are the most important properties of the (1-D) ACF:

(i) The ACF is *symmetrical*,

$$R_{xx}(\tau) = R_{xx}(-\tau) \tag{1.8a}$$

(ii) Its central value is the largest in absolute value: *If* $\tau \neq 0$ *then*

$$|R_{xx}(\tau)| \leq R_{xx}(0) \tag{1.8b}$$

(iii) If $x(t) = A \cdot \sin(\omega t + \varphi)$ then

$$R_{xx}(\tau) = \frac{A^2}{2}\cos(\omega\tau) \tag{1.8c}$$

independently of the phase φ;

(iv) More generally, the ACF of a periodic function is also periodic, with the same period.

(v) *Campbell-Rice Theorem*: Let $s(t)$ be a *wavelet* of finite energy and zero *dc (direct current) component,* that is

$$\int_{-\infty}^{\infty} s^2(t)dt < \infty \tag{1.8d.1}$$

$$\int_{-\infty}^{\infty} s(t)dt = 0 \tag{1.8d.2}$$

and let

$$R(\tau) = \int_{-\infty}^{\infty} s(t)s(t+\tau)dt \tag{1.8d.3}$$

denote the ACF of $s(t)$. Consider a function $x(t)$ which is the superposition of a random number of delayed and weighted replicas of this wavelet:

$$x(t) = \sum_n r_n s(t - t_n), \tag{1.8d.4}$$

where the r_n values are *random amplitudes*, the t_n values are *random arrival times* of the signals. Suppose the r_n values are normally distributed with zero mean, r^2 scatter, and they are uncorrelated, that is $\langle r_i r_j \rangle = r^2 \delta_{ij}$ (where $\delta_{ij} = \begin{cases} 1 \text{ } if \text{ } i = j \\ 0 \text{ } if \text{ } i \neq j \end{cases}$ is the *Kronecker delta symbol*).

If the t_n values are Poisson distributed with density λ, then

$$R_{xx}(\tau) = const \cdot R_{ss}(\tau) \tag{1.8d.5}$$

where the constant in (1.8d5) is λr^2.

(vi) By the *Wiener-Khinchine Theorem*, the Fourier Transform $\mathcal{F}R_{xx}(\tau)$ of the ACF of $x(t)$ is always *non-negative*, and it is equal to the *power spectrum*:

$$\mathcal{F}R_{xx}(\tau) = 2\int_{0}^{\infty} R_{xx}(\tau)e^{-i\omega\tau}\, d\tau = \left| \int_{-\infty}^{\infty} x(t)e^{-i\omega t}\, dt \right|^2 = |\mathcal{F}x(t)|^2 \equiv W(\omega) \tag{1.8d.6}$$

where $i = \sqrt{-1}$; $\omega = 2\pi f$, $W(\omega) \geq 0$ is the *power spectrum.*

(vii) The *arcsine law of correlation* (Tomoda 1956; Van Vleck and Middleton 1966): Let $f(t)$ be a Gaussian process with zero mean, and $\varphi(t)$ its *clipped* version, that is

$$\varphi(t) = \operatorname{sign} f(t) = \begin{cases} +1 \text{ } if \text{ } f(t) \geq 0 \\ -1 \text{ } if \text{ } f(t) < 0 \end{cases} \tag{1.9a}$$

Then $R_{ff}(\tau) = \sin\left[\frac{\pi}{2} \cdot R_{\varphi\varphi}(\tau)\right]$ or

$$R_{\varphi\varphi}(\tau) = \frac{2}{\pi}\arcsin\left[R_{ff}(\tau)\right] \tag{1.9b, c}$$

In several tasks of *Geo-scientific time-series analysis* we use the *Autocorrelation Matrix* (ACM), which is built up from values of the ACF. In *Seismic Processing*, e.g., this matrix has a basic role in *deconvolution, predictive deconvolution, signal estimation, optimal filtering*, etc. Let us use the notation $R(n) = R_{xx}(n \cdot \Delta t)$, let m be a positive integer (in seismic and geostatistical practice $m \gg 1$, it is between 20 and 150). The ACM is defined as

$$R = \begin{bmatrix} R(0) & R(1) & R(2) & \cdots R(m-1) \\ R(1) & R(0) & R(1) & \cdots R(m-2) \\ R(2) & R(1) & R(0) & \cdots R(m-3) \\ \vdots & & & \\ R(m-1) & R(m-2) & R(m-3) & \cdots \quad R(0) \end{bmatrix} \tag{1.10a}$$

The ACM R has some very special properties:

(i) The matrix is symmetrical, namely its elements r_{ij} satisfy

$$r_{ij} = r_{ji} \tag{1.10b}$$

(ii) It has a further symmetry, namely its elements satisfy

$$r_{ij} = R(|i-j|) \tag{1.10c}$$

Such matrices are called *Toeplitz matrices* and they can be inverted with a very fast technique, called *Levinson algorithm* (Levinson 1947; the algorithm has been further improved by Durbin 1960).

(iii) The ACM is *positive semidefinite*, that is its eigenvalues are greater than or equal to zero. In spite of this very favorable property, the ACM is usually severely *ill-conditioned* (Treitel and Wang 1976; Korvin 1978; O'Dowd 1990, 1991).

Recall that a symmetric real matrix $\boldsymbol{M} = \left[m_{ij}\right]_{i,j=0}^{N}$, $m_{ij} = m_{ji}$ is *positive semidefinite* if for any real vector $\boldsymbol{v} = \{v_0, v_1, \ldots, v_N\}$ one has

$$\boldsymbol{vMv}^T = \sum_{i=0}^{N}\sum_{j=0}^{N} m_{ij}v_iv_j \geq 0. \tag{1.11}$$

where the superscript T denotes "transposed". Consider now the ACM $\boldsymbol{R} = \left[r_{ij}\right]_{i,j=0}^{N}$ of the time series $\{x_n\}$. Apply the vector $\boldsymbol{v} = \{v_0, v_1, \ldots, v_N\}$ as digital convolution filter (Korvin 2022) to $\{x_n\}$ and denote the filtered series by $\{y_n\} = \{x_n\} * \{v_0, v_1, \ldots, v_N\}$ that is

$$y_n = \sum_{k=0}^{N} v_k x_{n-k} \tag{1.12}$$

Obviously the filtered series has non-negative expected energy, that is

$$\begin{aligned} 0 \le \langle y_n^2 \rangle &= \left\langle \sum_{k=0}^{N} \sum_{l=0}^{N} v_k v_l x_{n-k} x_{n-l} \right\rangle = \sum_{k=0}^{N} \sum_{l=0}^{N} v_k v_l \langle x_{n-k} x_{n-l} \rangle \\ &= \sum_{k=0}^{N} \sum_{l=0}^{N} v_k v_l r_{|k-l|} = \sum_{k=0}^{N} \sum_{l=0}^{N} v_k v_l r_{kl} \end{aligned} \tag{1.13}$$

As this is true for any real vector $\{v_0, v_1, \ldots, v_N\}$, we have proved the *positive semidefiniteness* of the ACM $\boldsymbol{R}$. Actually, as the length, $(N + 1)$ of the vector $\{v_0, v_1, \ldots, v_N\}$ is arbitrary, we have proved somewhat more than intended: we have shown that the ACM $\boldsymbol{R}$ does not only meet the condition (1.11) of *positive semidefiniteness*, but also another condition of "*corner-semi-definite*" matrices (McMillan 1955; Torquato 2002a, b, 2006), namely that all of its left-upper-corner sub-matrices are positive semidefinite.

A *random field* is *homogeneous* if its statistical properties are invariant with respect to a shift, that is if $\langle f(\boldsymbol{x}) \rangle$ and $\langle f^2(\boldsymbol{x}) \rangle$ are constant, and the autocorrelation function only depends on the vector $\boldsymbol{x} - \mathbf{y}$:

$$\langle f(\boldsymbol{x}) f(\mathbf{y}) \rangle := R_{ff}(\boldsymbol{x}, \mathbf{y}) \equiv R_{ff}(\boldsymbol{x} - \mathbf{y}) \tag{1.14}$$

If the statistical properties of the field are also invariant with respect to *rotations* and *reflections*, we speak about a *homogeneous and isotropic random field*. In such a field the autocorrelation is a function of the magnitude of $\mathbf{x} - y$: $\langle f(\boldsymbol{x}) f(y) \rangle = R_{ff}(|\boldsymbol{x} - \mathbf{y}|)$. The autocorrelation function

$$R(r) = a^2 \exp(-r/r_0) \tag{1.15}$$

where $r = \sqrt{\left[(x_1 - y_1)^2 + (x_2 - y_2)^2 + (z_1 - z_2)^2\right]}$ frequently arises in such a 3-dimensional isotropic field. Here, r_0 is the *correlation distance*, for this value of r the autocorrelation function decreases to $1/e$ times its value at $r = 0$.

1.2 Autocorrelation Functions of Binary-Valued Random Fields

For any random homogeneous isotropic medium consisting of m phases (e.g. $m = 3$, the phases are grains, brine-filled-, and oil-filled pores) we define the following *characteristic function* for the different phases (Beran 1968; Baniassadi et al. 2014):

$$\chi^{(phase\, i)}(\boldsymbol{x}) = \begin{cases} 1 \; if\; \boldsymbol{x}\; in\, phase\; i \\ 0 \quad otherwise \end{cases} \tag{1.16}$$

The volume fraction of phase i ($i = 1, 2, \ldots, m$) is denoted by Φ_i, $\Phi_i = \langle \chi^{(phase\, i)}(\boldsymbol{x}) \rangle = Prob(x \in phase\; i)$. The *n-point correlation functions* are then defined as

$$C_n(\boldsymbol{x}_1, \boldsymbol{x}_2, \ldots, \boldsymbol{x}_n) = \langle \chi^{(\alpha_1)}(\boldsymbol{x}_1) \cdot \chi^{(\alpha_2)}(\boldsymbol{x}_2) \cdots \chi^{(\alpha_n)}(\boldsymbol{x}_n) \rangle \tag{1.17}$$

for all possible m^n combinations of phases. For the description of porous rocks we generally use two phases, "pore" (p), and "grain" (g), and there are four kinds of *two-point correlations*: C_2^{pp}, C_2^{pg}, C_2^{gp}, C_2^{gg}, where e.g., $C_2^{pp}(\boldsymbol{x}, \boldsymbol{y}) = \langle \chi^{(p)}(\boldsymbol{x}) \cdot \chi^{(p)}(\boldsymbol{y}) \rangle = Prob\left[(\boldsymbol{x}\; in\, pore) and\, (\boldsymbol{y}\; in\, pore)\right]$, etc. It can be proved that only one out of the four posible two-point correlation functions for a two-phase microstructure is independent and the three others are connected by the following system of equations (Frisch and Stillinger 1963):

$$C_2^{pp}(\boldsymbol{x}, \boldsymbol{y}) = C_2^{gg}(\boldsymbol{x}, \boldsymbol{y}) + C_2^{pg}(\boldsymbol{x}, \boldsymbol{y}) + C_2^{gp}(\boldsymbol{x}, \boldsymbol{y}) = 1 \tag{1.18a}$$

$$C_2^{pg}(\boldsymbol{x}, \boldsymbol{y}) = C_2^{gp}(\boldsymbol{x}, \boldsymbol{y}) \tag{1.18b}$$

$$C_2^{pp}(\boldsymbol{x}, \boldsymbol{y}) + C_2^{pg}(\boldsymbol{x}, \boldsymbol{y}) = \varphi_p = \Phi \tag{1.18c}$$

The expected value $\langle \chi^{(p)}(\boldsymbol{x}) \rangle = \varphi_p$ is usually denoted by Φ, and called *porosity*. In some studies (Torquato 2002a; Jiao et al. 2009; Pant 2016; Röding et al. 2020; Cherkasov et al. 2021) they also consider the case when $\boldsymbol{x}$ and/or $\boldsymbol{y}$ are on the *grain/ pore boundary surface* (s), and characterize the pore structure with the *three independent two-point correlations* C_{pp}, C_{sp}, C_{ss}. Surface correlation functions describe the information about the random interface between phases, and they are not *probability functions* because the probability of finding a point on the interface in a continuum medium would be zero.

Following are the most important properties of the two-point correlation functions of a two-phase (pore and grain) 3-D medium (Shepp 1963; Matheron 1993; Markov 1995; Ioannidis et al. 1996; Chatzis et al. 1997; Torquato 2002a, b, 2006; Bodla et al. 2014; Emami 2014; Feng et al. 2014). We shall sometimes use the vector $\vec{\boldsymbol{r}} = \boldsymbol{y} - \boldsymbol{x}$; $r = |\vec{r}|$; the *autocorrelation function* (ACF) $R_2(\vec{\boldsymbol{r}})$ is the normalized form of the two-point *correlation function* $C_2(\vec{\boldsymbol{r}})$:

$$R_2(\vec{\boldsymbol{r}}) = \frac{C_2(\vec{\boldsymbol{r}}) - \Phi^2}{\Phi - \Phi^2} \tag{1.19}$$

(i) $C_2(r = 0) = \Phi$ (1.20)

(ii) $\lim_{r\to\infty} C_2(r) = \Phi^2$ in the lack of *long range order* or *periodicity* (1.21)

(iii) $-1 \le R_2(r) \le 1$ (1.22)

(iv) The two-point correlation function for *phase 2* (grains) is related to the same function for phase 1 (pores) as:

$$C_2^{(2)}(r) = C_2^{(1)}(r) - 2\varphi_1 + 1 \tag{1.23a}$$

and, consequently, the autocovariance function for phase 1 is equal to that for phase 2,

$$R_2^{(pore)}(r) \equiv R_2^{(grain)}(r) \tag{1.23b}$$

(v) The 3-dimensional Fourier transform of the autocovariance $R_2(\boldsymbol{r})$, denoted by $\tilde{R}(\boldsymbol{k})$, is *real* and *nonnegative* for all wave vectors,

$$\tilde{R}(\boldsymbol{k}) = \int_{\mathbb{R}^3} R_2(\boldsymbol{r})e^{-i\boldsymbol{k}\cdot\boldsymbol{r}}d\boldsymbol{r} \ge 0 \text{ for all } \boldsymbol{k} \in \mathbb{R}^3 \tag{1.24}$$

(vi) Equivalently with (1.24), $R_2(\boldsymbol{r})$ is *positive semidefinite* in the sense that for any finite number m of points $\boldsymbol{r}_1, \boldsymbol{r}_2, \ldots, \boldsymbol{r}_m$ in $\mathbb{R}^3$ and arbitrary real numbers

$$a_1, a_2, \ldots, a_m, \sum_{i=1}^{m}\sum_{j=1}^{m} a_i a_j R_2(\boldsymbol{r}_i - \boldsymbol{r}_j) \ge 0. \tag{1.25}$$

(vii) Both two-point correlation functions satisfy the inequality

$$0 \le C_2^{(i)}(r) \le \varphi_i \text{ for } i = pore, grain \text{ and for all } r \tag{1.26}$$

(viii) For the autocovariance function the corresponding inequality is

$$-\min\left(\varphi_{1,}^2 \varphi_{1,}^2\right) \le R_2(r) \le \varphi_1\varphi_2 \tag{1.27}$$

(ix) *Triangular inequality* (Shepp 1963; Matheron 1993; Torquato 2006): For any three vectors $\boldsymbol{r}, \boldsymbol{s}, \boldsymbol{t} \in \mathbb{R}^3$ such that $\boldsymbol{r} = \boldsymbol{t} - \boldsymbol{s}$ we have

$$R_2^{(i)}(\boldsymbol{r}) \ge R_2^{(i)}(\boldsymbol{s}) + R_2^{(i)}(\boldsymbol{t}) - \varphi_i \tag{1.28}$$

To prove this inequality, introduce a new random variable

$$Y^{(i)}(\boldsymbol{x}) = 2\chi^{(phase\ i)}(\boldsymbol{x}) - 1 = \begin{cases} 1 & \text{if } \boldsymbol{x} \text{ in phase } i \\ -1 & \text{otherwise} \end{cases} \tag{1.29}$$

We have $\langle Y^{(i)} \rangle = 2\varphi_i - 1$. For any three points $\boldsymbol{x}_1, \boldsymbol{x}_2, \boldsymbol{x}_3$ the sum $Y^{(i)}(\boldsymbol{x}_1) - Y^{(i)}(\boldsymbol{x}_2) + Y^{(i)}(\boldsymbol{x}_3)$ is an odd number (−3, −1, 1, or 3) so that

$$\left\langle \left[Y^{(i)}(\boldsymbol{x}_1) - Y^{(i)}(\boldsymbol{x}_2) + Y^{(i)}(\boldsymbol{x}_3) \right]^2 \right\rangle \geq 1 \tag{1.30}$$

Using $\langle Y^{(i)}(\boldsymbol{x}_k) Y^{(i)}(\boldsymbol{x}_l) \rangle = 4C_2^{(i)}(\boldsymbol{x}_k - \boldsymbol{x}_l) - 4\varphi_i + 1$, a little algebra yields (1.28).

(x) Torquato (2006) generalized the triangular inequality and proved that

$$\sum_{i=1}^{m} \sum_{j=1}^{m} \varepsilon_i \varepsilon_j R_2(\boldsymbol{r}_i - \boldsymbol{r}_j) \geq 1 \text{ where } \varepsilon_i = \pm 1 \text{ for } i = 1, 2, \ldots, m \text{ and } m \text{ odd} \tag{1.31}$$

(xi) The derivative of $C_2^{(i)}(\boldsymbol{r})$ at $r = 0$ is strictly negative:

$$\left. \frac{dC_2^{(i)}}{dr} \right|_{r=0} = \left. \frac{d\chi}{dr} \right|_{r=0} < 0 \text{ for all } 0 < \varphi_i < 1 \tag{1.32}$$

It follows from (1.32) that the Gaussian ACF

$$f(r) = \exp\left[-\left(\frac{r}{a} \right)^2 \right] \tag{1.33}$$

or the "*stretched exponentials*"

$$f(r) = \exp\left[-\left(\frac{r}{a} \right)^\lambda \right], \lambda > 1 \tag{1.34}$$

cannot be autocorrelation functions of a two-phase isotropic, homogeneous random medium because their slopes at $r = 0$ are zero.

(xii) $C_2^{(i)}(r)$ is convex at the origin (Markov 1995; Torquato 2006):

$$\left. \frac{d^2}{dr^2} C_2^{(i)} \right|_{r=0} = \left. \frac{d^2}{dr^2} \chi \right|_{r=0} \geq 0 \tag{1.35}$$

The properties of *anisotropic correlation functions* are outside the scope of the Chapter, the Reader is referred to the references compiled in Table 1.1. It has been claimed (Swan and Garratt 1995; Saccomano et al. 2001; Naoise et al. 2003; de Lima Moraes 2017) that for analysing anisotropy the *semivariogram* works better than the ACF. By definition (Jensen et al. 1997; Saccomano et al. 2001), the *semivariogram* of a random field $Z(\boldsymbol{x})$ is defined as

$$\gamma(\boldsymbol{x}, \boldsymbol{y}) = \frac{1}{2}\left\langle [Z(\boldsymbol{x}) - Z(\boldsymbol{y})]^2 \right\rangle \tag{1.36}$$

If the field $Z(\boldsymbol{x})$ is homogeneous, by (1.36) $\gamma(\boldsymbol{x}, \boldsymbol{y})$ and the two-point correlation $\langle Z(\boldsymbol{x})Z(\boldsymbol{y})\rangle$ are linearly related:

$$\langle Z(\boldsymbol{x})Z(\boldsymbol{y})\rangle = \left\langle Z^2(\boldsymbol{x})\right\rangle - \gamma(\boldsymbol{x}, \boldsymbol{y}) \tag{1.37}$$

In *Geostatistics* (Jensen et al. 1997) γ is estimated, in case of isotropy, as

$$\gamma(h) = \frac{1}{2N(h)} \sum_{|x_i - y_j| = h} \left[Z(\boldsymbol{x}_i) - Z(\boldsymbol{y}_j)\right]^2 \tag{1.38}$$

where $N(h)$ is the number of all pairs of points $\{\boldsymbol{x}_i, \boldsymbol{y}_j\}$ such that $|\boldsymbol{x}_i - \boldsymbol{y}_j| = h$.

1.3 Three-point Correlations and Other Correlation Measures for Binary-Valued Random Fields

The classic papers by researchers of *material characterization* reveal the frustration these scientists had felt at the inadequacy of the *two-pont correlation function*. Already Patterson, as early as in (1939, 1944) realized that different point patterns could lead to the same 2-point CF, and he coined the word "*homometric*" for such structures. In *Crystallography*, this difficulty has been referred to as the *phase problem*. In the context of *Random Binary Media*, the phenomenon that different microstructures may have the same correlation function is named *microstructural degeneracy*. Jio et al. (2009) explains degeneracy as follows: A two-phase medium in the N^d *voxels* (in 2-D, "*pixels*") of a d-dimensional cube of linear size $N\varepsilon$, where ε is the side of a voxel, can be represented by a d-dimensional (0,1)-matrix $\boldsymbol{I}$. In 2 dimensions, e.g.,

$$\boldsymbol{I} = \begin{bmatrix} I_{11} & I_{12} & \cdots & I_{1N} \\ I_{21} & I_{22} & \cdots & I_{2N} \\ \vdots & \vdots & \vdots & \vdots \\ I_{N1} & I_{N2} & \cdots & I_{NN} \end{bmatrix} \tag{1.39}$$

where the elements $I_{ij}(i, j = 1, 2, \ldots, N)$ take the value $I_{ij} = 1\, in\, pore, I_{ij} = 0\, in\, grain$.

To find the elements of $\boldsymbol{I}$ from the values of the 2-point correlation function $C_2(r)$ (the so-called "*reconstruction problem*"), one has to solve the following set of nonlinear algebraic equations for every value of r such that $r \leq N/2$ and r^2 is an integer (*op. cit.*: Supporting Information, Eqs. S1–S3):

$$\sum_{(j_1,\ldots,j_d)\in\Omega}\left[\sum_{i_1=1}^{N}\cdots\sum_{i_d=1}^{N} I_{i_1\cdots i_d}\cdot I_{(i_1+j_1)\cdots(i_d+j_d)}\right]-\omega\cdot N^d\cdot C_2(r)=0 \tag{1.40}$$

where

$$\Omega=\left\{j_1,\ldots,j_d | j_1^2+j_2^2+\cdots+j_d^2=r^2, r\le N/2\right\} \tag{1.41}$$

and ω is the number of elements of Ω. Since the number of equations ($N/2$) is less than the number of unknowns (N^d), the system (1.40) cannot be solved uniqely.

Gommes et al. (2012a, b) attacked the question "What is the number of microstructures compatible with a specified correlation function?" with tools of *Information Theory* and provided a method to estimate the *information content* of any two-point correlation function.

Theoretically (Baniassadi et al. 2014), the use of n-point correlations (with $n\to\infty$) would reveal more and more morphological details, and the structures would be uniquely defined by infinite-order correlation functions. It has been by now well established that $C_2(\boldsymbol{x},\boldsymbol{y}) = \langle\chi(\boldsymbol{x})\chi(\boldsymbol{y})\rangle$ does not provide an accurate and full description of the medium (Yeong and Torquato 1998a, b; Sheehan and Torquato 2001; Jiao et al. 2007, 2008). Zerhouni et al. (2021), e.g. write, "from the perspective of the effective elastic properties, the two-point correlation functions alone do not carry more information than the volume fractions, which is consistent with what is observed numerically here and confirms that the two-point correlation functions are not relevant descriptors of the microstructure". According to Chen et al. (2019), "two-point statistics alone might not be sufficient to represent certain complex microstructures".

To predict long-range connectivity, several new *morphological descriptors* have been proposed (Okabe 2004), such as *the lineal path length distribution*, which is the probability of finding a line segment of length z completely in either void or solid phase (Lu and Torquato 1992a, b, c; Coker and Torquato 1995; Coker et al. 1996; Yeong and Torquato 1998b; Manwart et al., 2000; Hidajat et al. 2002; Talukdar and Torsaeter 2002), or the pore *chord length distribution* (Serra 1992; Torquato and Lu 1993; Levitz 1998; Roberts and Torquato 1999; Kainourgiakis et al. 2000). These new descriptors have been found to yield better characterization of the system than the one- and two-point correlation functions because they contain information about *connectedness* and *long-range behavior*. Computationally, however, they are very demanding.

Torquato (2002, 2002a) introduced the following correlation functions for determining the *effective properties* of randomly heterogeneous materials:

(1) *N-point correlation functions* ($N = 2, 3$)
(2) *Surface correlation functions*
(3) *Lineal-path function*
(4) *Chord-length density function*
(5) *Pore-size functions*
(6) *Percolation and cluster functions*

(7) *Nearest-neighbor functions.*

(1) *Three-point correlation functions*

The *three-point* (*pore–pore*) *correlation function* (Berryman 1985a, b, 1988; Torquato 2002a, b; Velasquez et al. 2010) is defined similarly to the *two-point function* described above:

$$C_3^{(ppp)}(\boldsymbol{x}_1, \boldsymbol{x}_2, \boldsymbol{x}_3) = \left\langle \chi^{(p)}(\boldsymbol{x}_1) \cdot \chi^{(p)}(\boldsymbol{x}_2) \cdot \chi^{(p)}(\boldsymbol{x}_3) \right\rangle \tag{1.42}$$

where, as before,
$\chi^{(p)}(\boldsymbol{x}) = \begin{cases} 1 \ \textit{if } \boldsymbol{x} \textit{ in pore} \\ 0 \ \textit{otherwise} \end{cases}$. In other words, it is the probability that all three points are in the pore space. Similarly to the two-point case, in homogeneous and isotropic media, the three-point correlation function only depends on the distances between the three points:

$$C_3^{(ppp)}(\vec{x}_1, \vec{x}_2, \vec{x}_3) = C_3^{(ppp)}(r_{12}, r_{13}),\ r_{ij} = \left|\vec{x}_j - \vec{x}_i\right| \tag{1.43}$$

As three points always lie on the same plane, we can represent the contours of $C_3^{(ppp)}(r_{12}, r_{13}) = C_3(r_1, r_2)$ on the (r_1, r_2)–plane. By easy geometry:

(i) the maximum of the 3-point correlation occurs at the origin of the contour plot when three points occupy the same position, and

$$C_3^{(ppp)}(\vec{x}_1, \vec{x}_1, \vec{x}_1) = C_1^{(p)}(\vec{x}_1) = \Phi \tag{1.44a}$$

(ii) $C_3^{(ppp)}(\vec{x}_1, \vec{x}_1, \vec{x}_3) = C_3^{(ppp)}(0, r_{13}) = C_2^{(pp)}(r_{13})$ (1.44b)

and

(iii) $C_3^{(ppp)}(\vec{x}_1, \vec{x}_2, \vec{x}_1) = C_2^{(pp)}(r_{12}, 0) = C_2^{(pp)}(r_{12})$ (1.44c)

Consequently, a 3-point CF determines the lower order, 2-point and 1-pont correlation functions.

As, because of the *Wiener-Khinchin Theorem*, 2-point CFs can be easily computed using FFT (*Fast Fourier Transform*, Cooley and Tukey 1965) the question naturally arises whether the inverse is true: can we build up the (computationally very demanding) n-point CF from lower-order correlations? This was discussed in Baniassadi et al. (2012, 2014) who proved that no, theoretically we cannot, but there exists an approximate solution, where *weight functions* W_i^n are used to connect subsets of $(n-1)$-point correlation functions to estimate the full set of n-point correlation functions, in the form (for $n \geq 5$):

$C_n(x_1, x_2, \ldots, x_n)$

$$= \sum_{i=1}^{n} \left(W_i^n \cdot \frac{\prod_{i=1}^{\binom{n-1}{n-2}} C_{n-1}(x_i, \ldots, x_{n-1})}{\prod_{i=1}^{\binom{n-1}{n-3}} C_{n-2}(x_i, \ldots, x_{n-2})} \cdot \frac{\prod_{i=1}^{\binom{n-1}{n-4}} C_{n-3}(x_i, \ldots, x_{n-3})}{\prod_{i=1}^{\binom{n-1}{n-5}} C_{n-4}(x_i, \ldots, x_{n-4})} \right) \tag{1.45}$$

(The optimal weights W_i^n are computed in Baniassadi et al. 2014).

Another, very smart idea to avoid the computation of the n-point (in particular, 3-point) CF was put forward in Chen et al. (2019), who developed a set of "*novel hierarchical statistical microstructural descriptors*" instead of C_n, called *n-point polytope functions* P_n. The authors claim that "these novel polytope functions successively include higher-order n-point statistics of the features of interest in the microstructure in a concise, expressive, explainable, and universal manner; and can be directly computed from multi-modal imaging data including x-ray tomographic radiographs, optical/SEM/TEM micrographs, and EBSD color maps".

The polytope function P_n is defined as: $P_n(r)$ *is the probability that all n vertices of a randomly selected regular n-point polytope with edge length r lies in the phase of interest.*

To compute P_n in 2-D we use *n-point regular polygons*, (in the limit $n \to \infty$ the shape converges to a circle and we quantify its size by its radius r). For media in the 3-D space, we use such polyhedra whose edges are of the same length, i.e. the five *Platonic solids* (regular polyhedra: tetrahedron, octahedron, dodecahedron, icosahedron, and cube) and the thirteen *Archimedean solids* (the *semi-regular polyhedra*, see Torquato and Jiao 2009).

For $n = 2$, the *polytope function* is the same as the 2-point CF: $P_2(r) \equiv C_2(\mathrm{r})$. For any value of n the polytope shrinks to a single point as

$$r \to 0, \text{ and } P_n(0) = \varphi \tag{1.46a}$$

which is the volume fraction of the phase of interest. When $r \to \infty$, the n vertices get so far from each other that they can be considered independent, so that

$$\lim_{r\to\infty} P_n(r) = \varphi^n \tag{1.46b}$$

The asymptotics (1.1. 46a, b) suggest to introduce a *rescaled polytope function*

$$f_n(r) = \frac{P_n(r) - \varphi^n}{\varphi - \varphi^n} \tag{1.47a}$$

for $n > 1$, and

$$f_0(r = 0) = 1; f_0(r \to \infty) = 0 \tag{1.47b, c}$$

We can ask, with Jiao et al. (2009), "how much additional useful information do $C_3(\boldsymbol{x},\boldsymbol{y},z)$ and $C_4(\boldsymbol{x},\boldsymbol{y},z,\boldsymbol{q})$ contain over and above C_2?" Their answer is disappointing: "triangular and tetrahedral statistics do not significantly increase information content over and above pair statistics for textures possessing no long-range order … [a] higher order n-point correlation function such as three-point correlation function or four-point correlation function comes at a considerably higher computational complexity, but with little improvement in long range characterization". The missing information what low-order correlation functions would never provide is about *long-range features*, most importantly about *phase connectivity* throughout the sample (Torquato and Stell 1983b; Capek et al. 2009).

(2) *Surface correlation functions*

The *two-point surface correlation functions* express the probability that either both points $\boldsymbol{x}_1, \boldsymbol{x}_2$ lie on the random interface between pores and grains, or one of them lies on the boundary, the other is in a pore or a grain. These CFs look promising for problems of viscous flow, trapping and diffusion problems (Doi 1976; Rubinstein and Keller 1987; Rubinstein and Torquato 1988).

Denote by $\mathcal{V}_1$ the random pore space, by $\mathcal{V}_2$ the complementary volume occupied by grains, define the *indicator function* $\chi(\boldsymbol{x})$ as usually by $\chi^{(i)}(\boldsymbol{x}) = \begin{cases} 1 & \text{if } x \in \mathcal{V}_i \\ 0 & \text{otherwise} \end{cases}$, $i = 1, 2$. The indicator function for the interface is defined as

$$\mathcal{M}(\boldsymbol{x}) = \left|\nabla\chi^{(i)}(\boldsymbol{x})\right| = \left|\nabla\chi^{(2)}(\boldsymbol{x})\right| \tag{1.48}$$

$\mathcal{M}(\boldsymbol{x})$ is a *generalized function* (such like *Dirac's delta*), it is 1 on the interface, 0 elsewhere.

The 1-point surface correlation function is the *specific surface* $s(\boldsymbol{x})$ (interface area per unit volume, see Sect. 1.4.1) at point $\mathbf{x}$:

$$s(\boldsymbol{x}) = \langle\mathcal{M}(\boldsymbol{x})\rangle \tag{1.49}$$

It is non-negative but cannot be interpreted as a "probability measure" of the interface, because a random point would lie with zero probability on the boundary. In case of statistical homogeneity $s(\boldsymbol{x}) = const = s_{spec}$. There are two independent *two-point surface correlation* (SC) functions (s surface, p pore):

$$F_{sp}(\boldsymbol{x},\boldsymbol{y}) = \left\langle\mathcal{M}(\boldsymbol{x})\cdot\chi^{(p)}(\boldsymbol{y})\right\rangle \tag{1.50}$$

and

$$F_{ss}(\boldsymbol{x},\boldsymbol{y}) = \langle\mathcal{M}(\boldsymbol{x})\cdot\mathcal{M}(\boldsymbol{y})\rangle \tag{1.51}$$

In case of translational invariance they only depend on the vector $\boldsymbol{x}-\boldsymbol{y}$, in case of isotropy only on the distance $|\boldsymbol{x}-\boldsymbol{y}| = r$. In the limit $r \to \infty$, for isotropic binary media without long-range order we have

$$\lim_{r \to \infty} F_{ss}(r) = s_{spec}^2 \,;\ \lim_{r \to \infty} F_{sp}(r) = s_{spec} \cdot \Phi \tag{1.52a, b}$$

Many-point surface correlation functions are defined similarly to Eqs. (1.50–1.51), for example for m points on interface, $n - m$ points inside pores

$$\begin{aligned} & \boldsymbol{F}_{ss\cdots spp\cdots p}(\boldsymbol{x}_1, \boldsymbol{x}_2, \ldots, \boldsymbol{x}_m, \boldsymbol{x}_{m+1}, \ldots, \boldsymbol{x}_n) \\ & = \left\langle \left[\prod_{i=1}^{m} \mathcal{M}(\boldsymbol{x}_i) \right] \cdot \left[\prod_{j=m+1}^{n} \chi^{(p)}(\boldsymbol{x}_j) \right] \right\rangle \end{aligned} \tag{1.53}$$

(3) *Lineal-path function*

The *lineal-path function* (or *lineal path length distribution*) is the probability $L(z)$ of finding a line segment of length z completely in one of the phases, it is used as a two-point *morphological descriptor* by Lu and Torquato (1992a, b, c), Coker and Torquato (1995), Coker et al. (1996), Yeong and Torquato (1998b), Manwart et al. (2000), Hidajat et al. (2002), Talukdar and Torsaeter (2002). Algorithms for its calculation from digitized images of porous media are found in Coker and Torquato (1995). The lineal-path function also arises in *Stochastic Geometry*, in the form $\varphi_i\left(1 - L^{(i)}(r)\right)$, where it is called *linear contact distribution function* (Chiu et al. 2013).

The function contains useful information on *phase connectivity*, but only along lineal paths. It monotonically decreases with r from

$$L^{(i)}(0) = \varphi_i \text{ to } L^{(i)}(\infty) = 0 \tag{1.54}$$

The largest value $r_{max} = \max\left\{ r | L^{(i)}(r) \neq 0 \right\}$ for which it is not zero gives the longest posible lineal path in the phase of interest. In isotropic materials one obviously has, for any $r \geq 0$, that

$$L^{(1)}(r) + L^{(2)}(r) + L^{(12)}(r) = 1 \tag{1.55}$$

where $L^{(12)}(r)$ is the probability that a rigid line of length r, randomly thrown on the medium, lands in such a position that its endpoints are in different phases.

(4) *Chord-length density function*

The *chord-length probability density function* or *chord-length distribution function* $p^{(i)}(r)$ (Matheron 1975; Torquato and Lu 1993) is based on the lineal-path function $L^{(i)}(r)$. *Chords* are the segments between intersections of a random line with the interface between the two phases. In isotropic binary media, $p^{(i)}(r)dr$ is the probability of finding a chord of length between r and $r + dr$ in the phase of interest i,

$$\int_0^\infty p^{(i)}(r)dr = 1 \tag{1.56}$$

The mean chord length for phase i is:

$$\left\langle l_{chord}^{(i)} \right\rangle = \int_0^\infty rp^{(i)}(r)\,dr = \begin{cases} \frac{2\varphi_i}{s_{spec}} & d = 1 \\ \frac{\pi\varphi_i}{s_{sec}} & d = 2 \\ \frac{4\varphi_i}{s_{sec}} & d = 3 \end{cases} \tag{1.57}$$

where d is dimension of the space (Underwood 1973). The chord-length probability density function is related to the second derivative of the *lineal-path function* (Torquato and Lu 1993):

$$p^{(i)}(r) = \frac{\left\langle l_{chord}^{(i)} \right\rangle}{\varphi_i} \cdot \frac{d^2 L^{(i)}(r)}{r^2} \tag{1.58}$$

The function has been applied to study *gas diffusion* and *radiative transport* in porous media (Ho and Strieder 1979; Tassopoulos and Rosner 1992), and fluid flow through sedimentary rocks (Thompson et al. 1987).

(5) *Pore-size functions*

The *pore-size probability density function* $P^{(i)}(r)$, first applied by Prager (1963a,b) for diffusion and viscous flow in porous rocks, can be applied to three-dimensional isotropic binary or multiphase systems. $P^{(i)}(r)dr$ is the probability that a randomly chosen point in the phase of interest i lies at a distance between r and $r + dr$ from the nearest point on the interface of two phases. The function $P^{(i)}(r)$ satisfies

$$\int_0^\infty P^{(i)}(r)dr = 1 \tag{1.59}$$

has its maximum value at $r = 0$, where

$$P^{(i)}(0) = s_{spec}/\varphi_i \tag{1.60a}$$

and

$$\lim_{r\to\infty} = 0 \tag{1.60b}$$

The *cumulative distribution function* corresponding to the pdf $P^{(i)}(r)$ is

$$F^{(i)}(\delta) = \int_{\delta}^{\infty} P^{(i)}(r)dr \tag{1.61}$$

it gives the probability that a sphere of radius r lies entirely in the phase of interest. It is the 3-D version of the *lineal function*, and also yields information on the connectedness in the phase of interest. It is related to the *spherical contact distribution function* $1 - F^{(i)}(\delta)$ used in *Stochastic Geometry* (Chiu et al. 2013).

(6) *Cluster functions*

A *cluster* of one of the phases i is a *topologically connected part* of the phase, where any two points $\boldsymbol{x}, \boldsymbol{y}$ can be connected with a continuous curve without passing through any other phases. Clustering has a significant effect on percolation and mechanical properties, and a *two-point cluster function* has been introduced for its quantification (in the context of *Percolation Theory*, by Torquato et al. 1988).

Suppose that the total volume occupied by phase $\mathcal{V}^{(i)}$ is the union of mutually disjoint clusters $\left\{\mathcal{C}_j^{(i)}\right\}$: $\mathcal{V}^{(i)} = \bigcup_j \mathcal{C}_j^{(i)}, \mathcal{C}_k^{(i)} \cap \mathcal{C}_k^{(i)} = \emptyset$ (1.61). The *two-point cluster function* $C_2^{(i)}(\boldsymbol{x}, \boldsymbol{y})$ is defined similarly to the *two-point correlation function*, but only in connected subsets of the given phase: it is the probability of finding the two points in the same cluster. Mathematically, let

$$I_k^{(i)}(\boldsymbol{x}) = \begin{cases} 1 \; if \, \boldsymbol{x} \in \mathcal{C}_k^{(i)} \\ 0 \; otherwise \end{cases} \tag{1.62a}$$

then

$$C_2^{(i)}(\boldsymbol{x}, \boldsymbol{y}) = \sum_k \left\langle I_k^{(i)}(\boldsymbol{x}) I_k^{(i)}(\boldsymbol{y}) \right\rangle \tag{1.62b}$$

In case of homogeneity, $C_2^{(i)}(\boldsymbol{x}, \boldsymbol{y}) = C_2^{(i)}(\boldsymbol{x} - \boldsymbol{y})$, if the material is also isotropic, then $C_2^{(i)}(\boldsymbol{x}, \boldsymbol{y}) = C_2^{(i)}(|\boldsymbol{x} - \boldsymbol{y}|) = C_2^{(i)}(r)$.

The *cluster function*

$$C_2^{(i)}(r) \text{ satisfies } C_2^{(i)}(0) = \varphi_i \tag{1.63a}$$

and (in the absence of *longe-range order*),

$$\lim_{r \to \infty} C_2^{(i)}(r) = 0 \tag{1.63b}$$

Observe that if all clusters $\left\{\mathcal{C}_j^{(i)}\right\}$ are *convex*, the lineal-path functions and the cluster functions agree. This is no longer true if there are *concave* clusters in the decomposition (1.61).

According to Jiao et al. (2009), "a superior two-point signature of random textures is the two-point cluster function $C_2(r)$, which is sensitive to topological connectedness information. ... $C_2(r)$ contains not only appreciably more information than S_2, but more information than a variety of other two-point quantities, including the surface–surface correlation function, the surface–void correlation function, the pore-size function, lineal-path function, and the chord-length density function". However, there are some difficulties.The two-point cluster function from 2D sectional information might be very different from that obtained from 3-D data (Kim et al. 2021). Pant (2016) mentions the "cumbersome" computations involved in determining this function: "While computation of all the other correlation functions requires a simple scan of the digital image, computation of two-point cluster function requires identification of different clusters (Jiao et al. 2009). This process is extremely time consuming, and becomes extremely cumbersome when the cluster identification has to be repeated millions of times for reconstruction. Due to this very reason, only a few studies actively employ two-point cluster functions for reconstruction purposes."

(7) *Nearest-neighbor functions*

The *nearest-neighbor functions* are defined for such special structures which consist of small spherical particles of equal size distributed randomly in another, continuous material of different physical properties. For such structures the *nearest-neighbor distribution function* $H_P(r)$ was first used by Hertz (as early as 1909). $H_P(r)dr$ gives the probability that the center of an arbitrary embedded particle lies at a distance between r and $r + dr$ from the center of the nearest particle (Torquato et al. 1990). Another, similar function is $H_V(r)$, called *void nearest-neighbor probability density function* (introduced, in the *physical theory of liquids*, by Reiss et al. 1959). The probability that an arbitrary point in the system lies at a distance between r and $r + dr$ from the center of the nearest particle is given by $H_V(r)dr$. The associated *cumulative distribution functions* are called *exclusion probabilities*, and denoted by $E_P(\delta)$ and $E_V(\delta)$:

$$E_P(\delta) = 1 - \int_0^\delta H_P(r)dr \tag{1.64}$$

$$E_V(\delta) = 1 - \int_0^\delta H_V(r)dr \tag{1.65}$$

$E_P(\delta)$ is the probability of finding a spherical region of radius δ, such that (a) it is centered at some arbitrary particle's midpoint, (b) and is empty of other particle centers. Similarly, $E_V(\delta)$ is the probability of finding a spherical region of radius δ centered at some *arbitrary point* in the system, which is empty of other particle centers. If all spheres have radius *R*, and they are randomly distributed in the system, $E_V(\delta)$ determines the fraction of space where a further sphere of radius

$\delta - R$ can be inserted into the system. These probability functions are discussed in Vezzetti (1975), Ziff (1977), Truskett et al. (1998). They were generalized to "nearest-surface functions" by Lu and Torquato (1992c) who introduced four new functions $h_P(r), h_v(r), e_P(\delta), e_V(\delta)$ in analogy to $H_P(r), H_v(r), E_P(\delta), E_V(\delta)$ based on distances to the *surface* of the nearest particle instead of its center. Korvin et al. (2013) used the *entropy of the shortest distance of random points to the nearest surface* to improve the detection of pore-boundaries in 2-D images.

(8) *Other correlation schemes*

Out of the several innovative ideas to compute the ACF (autocorrelation function) using nonstandard ways of correlation, here we can only discusss three: (a) the *co-occurrence correlation measures*; (b) *Moran's I index*; and (c) the *local variance* (LV).

Co-occurrence correlation measures

The idea of *co-occurrence currelation functions* (CCFs) stems from the theory of the *gray-level co-occurrence matrix* (GLCM), used in *image analysis* (Haralick et al. 1973). Their use for purposes of reconstructing binary media is considered in (Feng et al. 2018; Kim et al. 2021). Haralick et al. (1973) introduced 28 possible expressions to classify *gray-level* textures, the following four of them have been found potential candidates for *binary-media reconstruction*:

Uniformity:

$$f_1(r) = \sum_i \sum_j P_{ij}^2(r) \tag{1.66}$$

Entropy:

$$f_2(r) = -\sum_i \sum_j P_{ij}(r) \log P_{ij}(r) \tag{1.67}$$

Contrast:

$$f_3(r) = \sum_i \sum_j (i-j)^2 P_{ij}(r) \tag{1.68}$$

Homogeneity:

$$f_4(r) = \sum_i \sum_j \frac{1}{1+(i-j)^2} P_{ij}(r) \tag{1.69}$$

In Eqs. (1.66–1.69) the subscripts $i, j = 1, 2$ refer to the two phases in the binary image; the elements P_{ij} of the *co-occurrence matrix* count how often pairs of pixels with a specific value occur in the image at distance r (we asume isotropy). The properties of these functions are discussed in Haralick et al. (1973), Feng et al. (2018).

Moran's I index

Formally, this index is defined as (Moran 1950; Anselin 1995; Pérez-Peña et al. 2009):

$$I = \left(\frac{N}{\sum_i \sum_j w_{ij}} \right) \cdot \left[\frac{\sum_i \sum_j w_{ij}(x_i - \mu)(x_j - \mu)}{\sum_i (x_i - \mu)^2} \right] \tag{1.70}$$

where N is the total number of observing the values of some property at *pairs of locations* (ξ_i, ξ_j); x_i, x_j are the measured values; $\mu = \langle x_i \rangle$; and the weight

$$w_{ij} = 1/|\xi_i - \xi_j|, w_{ii} = 0 \tag{1.71}$$

is reciprocal of the distance between the two points. Moran's index can be used to decide whether or not to reject the *null hypothesis* (according to which the data are randomly distributed, and there is no correlation). For this, we compute a standardized *Z score* with a p value (Anselin 1995):

$$Z = \frac{I - E(I)}{S_{E(I)}} \tag{1.72}$$

where I is the empirical value calculated from the sample, $E(I)$ is the theoretical mean value of I for a random distribution, and $S_{E(I)}$ is its theoretical standard deviation. The p value is the probability that the null hypothesis has falsely been rejected. In *Geography* (Anselin 1995) a popular use of the index is to study the *degree of clustering of a spatial variable*, because the value of I changes between $(-1, +1)$, in such a way that

$$I = \begin{cases} -1 & \textit{the variable is perfectly dispersed} \\ \approx 0 & \textit{the variable is randomly dispersed} \\ +1 & \textit{the variable is perfectly clustered together} \end{cases} \tag{1.73}$$

The *local variance* (LV)

Saccomano et al. (2001) introduced a new, *integral measure* of the similarity between the values $Z(\boldsymbol{x})$ and $Z(\boldsymbol{y})$ measured at two locations separated by a distance not larger than h (i.e. for $|x - y| \leq h$). Define first the *dispersion variance* of $Z(\boldsymbol{x})$ in a volume V (Samper and Carrera 1990):

$$\sigma_V^2 = \left\langle \frac{1}{V} \int_V [Z(\boldsymbol{x}) - Z_V]^2 d\boldsymbol{x} \right\rangle \tag{1.74}$$

where Z_V is the mean value of $Z(\boldsymbol{x})$ in the volume V:

$$Z_V = \frac{1}{V} \int_V Z(\boldsymbol{y}) d\boldsymbol{y} \tag{1.75}$$

The *dispersion variance* computed in a ball of diameter h, denoted by $\sigma_V^2(h)$, is the new correlation measure, Saccomano et al. (2001) call it LV (*Local Variance*). If we recall the definition of *semivariance* ($\gamma(\boldsymbol{x}, \boldsymbol{y}) = \frac{1}{2}\left\langle [Z(\boldsymbol{x}) - Z(\boldsymbol{y})]^2 \right\rangle$ Eq. 1.1.36), we immediately see that it is related to σ_V^2. Defining $\overline{\gamma}_{VV}$ the mean value of the semivariance within the volume V, as

$$\overline{\gamma}_{VV} = \int_V \int_V \gamma(\boldsymbol{x} - \boldsymbol{y}) d\boldsymbol{x} d\boldsymbol{y} \tag{1.76}$$

it can be proved (Samper and Carrera 1990; Saccomano et al. 2001) that

$$\sigma_V^2 = \overline{\gamma}_{VV} \tag{1.77}$$

An unbiased *local estimator* of $\sigma_{V_i}^2(h)$ within a ball of diameter h around $\boldsymbol{x}_i$ is

$$\sigma_{V_i}^2(h) = \frac{1}{m_i} \sum_{j \in I_i} \left[Z(\boldsymbol{x}_j) - \langle Z(\boldsymbol{x}_i) \rangle \right]^2 \tag{1.78}$$

where $I_i = \{ j \| \boldsymbol{x}_i - \boldsymbol{x}_j | \leq h/2 \}$; $(m_i + 1)$ is the number of points $\boldsymbol{x}_j$ satisfying this condition; and

$$\langle Z(\boldsymbol{x}_i) \rangle = \frac{1}{m_i + 1} \sum_{j \in I_i} Z(\boldsymbol{x}_j) \tag{1.79}$$

is the average measured value in the ball. The *global estimator* of $\sigma_V^2(h)$ is the weighted average of the $\sigma_{V_i}^2(h)$ values:

$$LV(h) = \frac{1}{\sum m_i} \cdot \sum m_i \sigma_{V_i}^2(h) \tag{1.80}$$

1.4 Estimating Specific Surface Area and Mean Grain Size from the 2-point Correlation Function

1.4.1 Specific Surface, and its Role in Permeability Estimation

By the Kozeny-Carman ("KC") equation (Carman 1956) the permeability of porous sedimentary rocks is given by

$$k = \frac{1}{b}\Phi^3 \frac{1}{S_{spec}^2} \frac{1}{\tau^2} \tag{1.81}$$

where b is a *shape factor* of order one, $\Phi \in [0, 1]$ is porosity, S_{spec} is *specific surface* defined as *total surface area per unit bulk volume*, τ is *tortuosity*. Combining $\frac{1}{b\tau^2}$ to a single constant C, the KC law is expressed as

$$k = C\frac{\Phi^3}{S_{spec}^2} \tag{1.82}$$

Sedimentologists use three different concepts of specific surface (Timur 1968), namely:

S_{spec} = surface area per unit bulk volume of the rock; S_0 = surface area per unit volume of solid material; S_p = surface area per unit volume of pore space.

For porosity Φ and spherical grains of the same radius r the three measures of specific surface are interrelated as:

$$S_{spec} = (1 - \Phi)S_0;\ S_{spec} = \Phi S_p;\ S_p = \frac{1 - \Phi}{\Phi} S_0 \qquad \text{(1.83a, b, c)}$$

To prove this, take a unit bulk volume of the rock, $(1 - \Phi)$ part of it consists of solid material which contains $n = (1 - \Phi)/\left(\frac{4r^3\pi}{3}\right)$ grains, that is the total surface in unit bulk volume is

$$S_{spec} = n \cdot 4r^2\pi = \frac{3(1 - \Phi)}{r} \tag{1.84}$$

A unit grain volume contains $N = 1 : \frac{4r^3\pi}{3}$ grains, which have a total surface area

$$S_0 = N \cdot 4r^2\pi = \frac{3}{r} \tag{1.85}$$

and from Eqs. (1.84–1.85) one has $S_{spec} = (1 - \Phi)S_0$ as in (1.83a). Now, take a bulk volume V_b of rock in such a way that its pore space occupies a unit total volume, that is $V_b\Phi = 1$, whence $V_b = \frac{1}{\Phi}$. The total surface associated with this unit pore volume is $S_p = V_b S_{spec} = \frac{S_{spec}}{\Phi}$, that is $S_{spec} = \Phi S_p$ as in (1.83b). The third equation in Eqs. (1.83) follows from the first two.

Correspondingly, the KC equation can be put in three forms (which are equivalent only for convex grains of equal shape and size):

$$k = C\frac{\Phi^3}{S_{spec}^2} \tag{1.86a}$$

$$k = C\frac{\Phi^3}{(1-\Phi)^2 S_0^2} \tag{1.86b}$$

$$k = C\frac{\Phi}{S_p^2} \tag{1.86c}$$

where

$$C = \frac{1}{b\tau^2} \tag{1.86d}$$

The estimation of τ (tortuosity) from digital rock images will be discussed in Chap. 2.

The specific surface S_p figures in the following important geometrical theorem:

Theorem (Debye et al. 1957). *Select in a random 2-D section of an isotropic porous rock two randomly placed points A and B a distance r apart, where r is small related to the correlation distance. Then the probability that A and B are in different media* (i.e. one is in a pore, the other in a grain) *is given by*

$$Prob(A\&B\,are\,in\,different\,media) = S_p\frac{r}{2} \tag{1.87}$$

where $Prob(X)$ is the probability of the event X.

The proof (Debye et al. *op. cit.*), is geometric: Consider a finite volume of rock, with total pore-surface area $S_{area}(\Pi)$, total pore volume $V(\Pi)$, let Π denote the total pore space, Γ the space occupied by grains. Suppose A is in Π, B in Γ and $\overline{AB} = r << 1$. Let the segment AB move in such a way that it follows the pore-surface S while keeping A and B always in different media. Then point A must always be some distance $h < r$ away from the surface S. If the distance of A from S is between h and $h + dh$, the total pore volume where A can be is $S_{area}(\Pi)dh$. For any fixed position of A, the direction of the AB segment must be within a certain *solid angle* to assure that it crosses S. Using the well known equation for the surface area of the *spherical cap*, the probability that the AB radial segment intersects S is $\frac{S_{area}(\Pi)dh}{V(\Pi)} \cdot \frac{2\pi r(r-h)}{4\pi r^2} = S_p \cdot \frac{2\pi r(r-h)}{4\pi r^2}$ and the probability that A and B are in different media is $Prob(A\&B\,are\,in\,different\,media) = 2S_p\int_0^r \frac{r-h}{2r}dh = \frac{S_p r}{2}$, as stated (The factor "1/2" arises because A and B are interchangeable). As pointed out in Korvin (2016), the beauty of this theorem is that it can be used for *MC* (*Monte Carlo*) *estimation* of the specific surface S_p.

Berryman and Blair (1986) made more precise this classic geometric result and demonstrated that the *specific surface area of the pore boundaries can be estimated from the slope of the ACF* at $\rho = 0$ as

$$S_p = -4\frac{d}{d\rho}R(\rho)\bigg|_{\rho=0} \tag{1.88}$$

an equation which, when substituted into Kozeny-Carman's Eq. (1.86c), and assuming reasonable default values for the shape factor b and tortuosity τ in (1.86d), can be used to predict permeability *from the autocorrelation analysis of the microscopic image.*

Equation (1.88) holds for d = 3 dimension only. More precisely (Berryman 1987), we have

$$\frac{dS_2^{(i)}}{dr} = \begin{cases} -S_p/2 \; for\, d = 1 \\ -S_p/\pi \; for\, d = 2 \\ -S_p/4 \; for\, d = 3 \end{cases} \tag{1.89}$$

If the characteristic function $\chi(\boldsymbol{x}) = \begin{cases} 1 \; if\; x \in \Pi \\ 0 \; if\; x \in \Gamma \end{cases}$ of the pore space has an exponential ACF with *correlation distance* ρ_0, i.e. if

$$R(\rho) = \exp\left[-\frac{\rho}{\rho_0}\right], \tag{1.90}$$

then Eq. 1.88 gives

$$S_{spec} = \frac{4}{\rho_0} \tag{1.91}$$

Using this in the KC Eq. (1.81) gives

$$k = \text{const} \cdot \Phi^3 \rho_0^2 \tag{1.92}$$

which is not only *dimensionally correct* (because permeability has the dimension *length*^{2}) but also *physically reasonable*, because on the right-hand-side (RHS) the "ρ_0" factor is of the order of pore size which has a similarly huge *dynamic range* in sedimentary rocks as the permeability k itself has.

Berryman and Blair's paper (1986) has attracted many follow-ups which attempted, with varying success, to predict permeability from the analysis of thin sections. Ioannidis et al. (1996) observed that the exponential ACF, $R(r) = \exp\left(-\left|\frac{r}{r_0}\right|\right)$, though frequently reported for ordinary black-and-white photographs, is not applicable for microscopic images of sedimentary rock, where a *stretched exponential* function (Johnston 2006) $R_{\lambda,n}(u)$ better fits the experimentally obtained ACF. This ACF is defined as

$$R_{\lambda,n}(u) = \exp\left[-\left|\frac{u}{\lambda}\right|^n\right] \tag{1.93}$$

where λ is the *correlation distance*, n is a positive real exponent. The derivative of this ACF at $u = 0$ is

$$\frac{d}{du}R_{\lambda,n}(u) = -\frac{n}{\lambda^n}u^{n-1}_{u=0} \tag{1.94}$$

which, except for $n = 1$, cannot express specific surface area when substituted to Eq. (1.88) because for $n < 1$ the RHS is divergent, while for $n > 1$ the RHS is zero. To avoid this difficulty, Ioannidis et al. (*op. cit.*) introduced an *average correlation distance*

$$I_S = \int_0^\infty R_{\lambda,n}(u)du = \int_0^\infty \exp\left[-\left(\frac{u}{\lambda}\right)^n\right]du \tag{1.95}$$

With the substitution $x = \left(\frac{u}{\lambda}\right)^n$ and recalling the definition of the Gamma function $\Gamma(z) = \int_0^\infty t^{z-1}e^{-t}dt$ we find

$$I_S = \int_0^\infty \exp\left[-\left(\frac{u}{\lambda}\right)^n\right]du = \frac{\lambda}{n}\Gamma\left(\frac{1}{n}\right) \tag{1.96}$$

using the tabulated integral No. 3.478 in Gradshteyn and Ryzhik (1980). If for a given rock type n does not vary in a wide range, the changes in the expression $\frac{1}{n}\Gamma\left(\frac{1}{n}\right)$ are not too significant and the scaling rule $k = const \cdot \Phi^3\rho_0^2$ (Eq. 1.92) obeyed in case of the exponential ACF, $R(\rho) = \exp\left[-\frac{\rho}{\rho_0}\right]$, would still hold, but more generally, as

$$k \propto \Phi^\alpha I_s^\beta \tag{1.97}$$

(Ioannidis et al. 1996: *Abstract*), or:

$$\ln k = a + b\ln\Phi + c\ln I_S \tag{1.98}$$

(*ibid*.: Eq. 6; Cobaleda 2000: Eq. 2.)

Ioannidis et al. (1966) used *backscatter SEM* images of thin sections from 15 Canadian rock samples of different lithology, and found the parameters λ, n of the stretched exponential I_s (in Eq. 1.96), and the following fitting parameters to

Eq. (1.98): $a = 9.3252$; $b = 5.750$; $c = 1.572$. There were only two cases when the fit did not work satisfactorily, (a) *samples with low porosity but high permeability* (*Gilwood sandstone* with $\Phi = 0.134$, $k = 412$ mD, possible reason for the discrepancy: *small fractures* which did not affect porosity but increased permeability), and (b) *samples with high porosity and low permeability* (*Viking sandstone* with $\Phi = 0.197$, $k = 6.5$ mD, possible reason: *diagenetic clay* blocking the fluid paths, see Korvin 1992b).

Another characteristic length based on the two-point correlation function has beeen used in the upper bounds for the permeability of random binary media (Berryman and Milton 1985; Coker et al. 1996):

$$k \leq \frac{2\lambda_B^2}{3(1-\Phi)^2} \tag{1.99}$$

where k is permeability, Φ is porosity, the characteristic length is given as:

$$\lambda_B = \left\{ \int_0^\infty \left[S_2^{(p)}(r) - \Phi^2 \right] r dr \right\}^{1/2} \tag{1.100}$$

Ioannidis et al. (1996) proposed the following expression for the *correlation length* in random binary media, which can also be used as characteristic length:

$$\overline{\lambda} = \int_0^\infty R_x^{(i)}(r) dr \tag{1.101}$$

where

$$R_x^{(i)}(r) = \frac{S_2^{(i)} - \Phi_i^2}{\Phi_i - \Phi_i^2} \tag{1.102a}$$

In terms of the ACF, the characteristic length is (Talukdar and Torsaeter 2002):

$$\overline{\lambda} = \int_0^\infty \left(\Phi_i - \Phi_i^2\right) S_2^{(i)}(r) dr \tag{1.102b}$$

The interface area can be calculated from the slope of the two-point correlation function at $r = 0$. Using a linear approximation for slope estimation, the specific surface area is obtained as

$$s = \beta \cdot [S_2(0) - S_2(1)] \tag{1.103}$$

where the value of β is 4 and 6 for 2-D and 3-D images, respectively (Yeong and Torquato 1998a).

Rabbani et al. (2014) defined a *specific perimeter* P_s as

$$P_s = \frac{Object\ perimeter}{Object\ surface\ area} \tag{1.104}$$

and asked if there may be a relationship between *specific perimeter* and *specific surface* of a porous material. If the grains are perfect spheres, the ratio between the *specific surface* and *normalized specific perimeter* $P_s^{(norm)}$ is

$$C_\theta = \frac{4\pi r^2/(4/3)\pi r^3}{2\pi r/\pi r^2} = C_\theta = \frac{3}{2} \tag{1.105}$$

where C_θ is a *geometric coefficient* which can be used for estimating the specific surface area (S) of rock grains from 2-D images. The *Geometric Coefficient* C_θ has been found to be about 1.35 for various types of rocks which gave the following empirical rule:

$$S = C_\theta \cdot P_s^{(norm)} \approx 1.35 P_s^{(norm)} \tag{1.106}$$

1.4.2 Mean Grain Size from the 2-Point Correlation Function

There have been many attempts (Debye and Bueche 1949; Debye et al. 1957; Lin 1982; Berryman 1985a, b; Heilbronner-Panozzo 1992; Mätzler 2002; Torquato 2002a, b; Buscombe et al. 2010; Bodla et al. 2014; Thissen and Brandon 2015) to extract grain-size information from the normalized ACF of the image,

$$R_{II}(\boldsymbol{v}) = \frac{\langle [I(\boldsymbol{u}) - \langle I \rangle] \cdot [I(\boldsymbol{u}+\boldsymbol{v}) - \langle I \rangle] \rangle}{\langle [I(\boldsymbol{u}) - \langle I \rangle]^2 \rangle} \tag{1.107}$$

but with limited success, especially for dense aggregates (Thissen and Brandon 2015). To understand the cause of this difficulty (Bernard et al. 2020: Appendix) consider the *inverse image* I_{inv} associated with a binary image I, defined as

$$I_{inv}(\boldsymbol{x}) = \begin{cases} 0 \ if \ I(\boldsymbol{x}) = 1 \\ 1 \ if \ I(\boldsymbol{x}) = 0 \end{cases} = 1 - I(\boldsymbol{x}) \tag{1.108}$$

Using (1.107–1.108) a little algebra shows that

$$R_{I_{inv}I_{inv}}(\boldsymbol{v}) \equiv R_{II}(\boldsymbol{v}) \tag{1.109}$$

that is the medium of interest and the complementary, background, medium are characterized with the same ACF, and consequently, the very same *correlation length*. (See also the equation, $R_2^{(pore)}(r) \equiv R_2^{(grain)}(r)$, 1.1.23b). Thus, the correlation length cannot be used for grain-size estimation.

We encounter a similar difficulty when trying to determine *strain anisotropy* (Fry 1979) from the ACF analysis of such a sample, where the originally spherical particles, initially embedded at points of a regular lattice into a soft material, had undergone anisotropic stress during which the *particles' shape* has become elipsoidal (*Ramsey effect*) and the *interparticle distances* have also become distorted (*Fry effect*). To avoid this difficulty, Thissen and Brandon (2015) recommend that "it is especially important distinguishing between [short range] self-correlation and [long range] neighbor-correlation".

In the exponential autocorrelation model (Debye and Bueche 1949; Debye et al. 1957; Mätzler 2002),

$$\rho(r) = \exp\left(-\frac{r}{R_C}\right) \tag{1.110}$$

the characteristic distance for the decay is R_C. For *identical non-overlapping spheres* R_C is the radius of the spheres, that is

$$R_C \approx -\left(\frac{d\rho(r=0)}{dr}\right)^{-1} \tag{1.111}$$

For *fully penetrable spheres* Berryman (1985a, b) found

$$\rho(r) = \exp[-\lambda v_2(r)] \tag{1.112}$$

where λ is the *number density of spheres*, and $v_2(r)$ is the normalized volume of the *union of two spheres* of radius R which are separated by a distance r:

$$v_2(r) = \frac{V_2(r)}{R^3} = \begin{cases} \frac{4\pi}{3}\left(1 + \frac{3}{4}x - \frac{x^3}{16}\right) & \textit{for } x \le 2 \\ \frac{8\pi}{3} & \textit{for } x > 2 \end{cases} \tag{1.113}$$

with $x = r/R$.

A related idea is to estimate grain size using another characteristic point in the *short-lag part of the ACF* (Heilbronner-Panozzo 1992; Buscombe et al. 2010), namely, the "*half-height lag*" r^*, which is the lag where ρ first drops to 0.5. According to Heilbronner-Panozzo (1992) the average grain radius is

$$R \approx 1.5r^* \tag{1.114}$$

Buscombe et al. (2010) determined

$$\langle R\rangle \approx \pi r^* \tag{1.115}$$

from images of unconsolidated sediments. Torquato (2002a, b: 79) started out from the theoretical ACF of a single sphere

$$\rho(r) = \begin{cases} 1 - \frac{3}{4}\frac{r}{R} + \frac{1}{16}\left(\frac{r}{R}\right)^3 & if \ \ r \le R \\ 0 & for \ r > R \end{cases} \tag{1.116}$$

Again, letting r^* denote "half-height lag", for which $\rho(r^*) = 0.5$, this ACF yields

$$\langle R\rangle \approx 1.44\, r^* \tag{1.117}$$

which is very close to the value in (1.114).

According to Thissen and Brandon (2015), "all of these approximations work poorly for dense aggregates", and they proposed an ACF consisting of two terms

$$\rho(r) = f(r, R_S)\cdot\left[1 - \frac{3}{4}\frac{r}{R_S} + \frac{1}{16}\left(\frac{r}{R_S}\right)^3\right] + f_N\, \exp\left[-\left(\frac{r}{R_N}\right)^\beta\right]\sin\left(\pi\frac{r}{R_N}\right) \tag{1.118}$$

where

$$f(r, R_S) = \begin{cases} 1 \ if \ \ r \le 2R_S \\ 0 \ if \ \ r > 2R_S \end{cases} \tag{1.119}$$

and R_S, f_N, R_N, β are unknown parameters, among which R_S represents the *average grain radius*. (For details of this model, see *op. cit.*).

Bodla et al. (2014) noted that for isotropic materials without long-range order,

$$\rho(r = 0) = \Phi; \ \lim_{r\to\infty} \rho(r) = \Phi^2 \tag{1.120}$$

In certain cases, we see an *oscillatory behavior* of $\rho(r)$ around the asymptotic value Φ^2. In such a case, denote by r_g the location of the first minimum; if there is no such oscillation let r_g denote the location where the ACF can be considered asymptotic. According to these authors the *average particle diameter* is

$$\langle D_p\rangle = \begin{cases} \frac{2r_g}{1.3} & for\ \rho(r)\ with\ distinct\ minimum \\ r_g & for\ \rho(r)\ without\ distinct\ minimum \end{cases} \tag{1.121}$$

As for the *effective pore size*, according to Dullien (1979) it is

$$D_{eff} = \frac{4\Phi}{s} \tag{1.122}$$

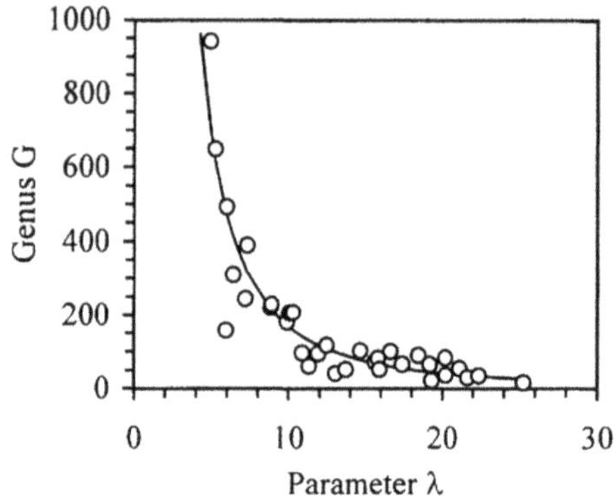

Fig. 1.1 Dependence of the genus on correlation length λ. (After Ioannidis and Lange 1998: Fig. 3; with kind permission of the WIT PRESS)

while (Anovitz and Cole 2015: Eq. 73) claim

$$D_{eff} = \frac{8\Phi(1-\Phi)}{s} \tag{1.123}$$

In both equations, s is the *specific surface area.*

Ioannidis and Lange (1998) used stretched exponential $\rho(u) = \exp\left[-\left({}^{u}/_{\lambda}\right)^{n}\right]$ to approximate the ACF and found a surprising and wonderful result: the *genus per unit volume* G_V, which is a *topological measure of connectedness* (see Sect. 3.2), regularly decreased with increasing λ, as shown in Fig. 1.1.

References

Adams BL, Canova GR, Molinari A (1970) A statistical formulation of viscoplastic behavior in heterogeneous polycrystals. Textures Microstruct 11:57–71

Adams BL, Morris PR, Wang TT, Willden KS, Wright SI (1987) Description of orientation coherence in polycrystalline materials. Acta Metall 35(12):2935–2946

Adams BL, Lyon MB, Kalidindi SR, Garmestani H (2002) Spectral integration of microstructure and design. Mater Sci Forum 408–412:493–498

Adler PM, Jacquin CG, Quiblier JA (1990) Flow in simulated porous media. Int J Multiphase Flow 16:691–702

Adler PM, Jacquin CG, Thovert JF (1992) The formation factor of reconstructed porous media. Water Resour Res 28(6):1571–1576

Alim K, Parsa S, Weitz DA, Brenner MP (2017) Local pore size correlations determine flow distributions in porous media. Phys Rev Letts 119:144501

Anovitz LM, Cole DR (2015) Characterization and analysis of porosity and pore structures. Rev Mineral Geochem 80(1):61–164

Anselin L (1995) Local indicators of spatial association—LISA. Geogr Anal 27:93–115

Appoloni CR, Fernandes CP, de Mello Innocentinic MD, Macedo Á (2004) Ceramic foams porous microstructure characterization by X-ray microtomography. Mater Res 7(4):557–564

Arns CH, Knackstedt MA, Pinczewski WV, Garboczi EJ (2002) Computation of linear elastic properties from microtomographic images: methodology and agreement between theory and experiment. Geophysics 67(5):1396–1405

Ballani F, Stoyan D (2015) Reconstruction of random heterogeneous media. J Microsc 258(3):173–178

Baniassadi M, Addiego F, Laachachi A, Ahzi S, Garmestani H, Hassouna F, Ruch D (2011a) Using SAXS approach to estimate thermal conductivity of polystyrene/zirconia nanocomposite by exploiting strong contrast technique. Acta Mater 59(7):2742–2748

Baniassadi M, Garmestani H, Li DS, Ahzi S, Khaleel M, Sun X (2011b) Three-phase solid oxide fuel cell anode microstructure realization using two-point correlation functions. Acta Mater 59:30–43

Baniassadi M, Ahzi S, Garmestani H, Ruch D, Remond Y (2012) New approximate solution for N-point correlation functions for heterogeneous materials. J Mech Phys Solids 60(1):104–119

Baniassadi M, Safdari M, Garmestani H, Ahzi S, Geubelle PH, Remond Y (2014) An optimum approximation of n-point correlation functions of random heterogeneous material systems. J Chem Phys 140:074905

Barnard P, Rubin D, Harney J, Mustain N (2007) Field test of an autocorrelation technique for determining grain size using a digital camera. Sed Geol 201:180–195

Bay BK, Smith TS, Fyhrie DP, Saad M (1999) Digital volume correlation: three-dimensional strain mapping using X-ray tomography. Exp Mech 39(3):217–226

Bekri S, Xu K, Yousefian F, Adler PM, Thovert JF, Muller J, Iden K, Psyllos A, Stubos AK, Ioannidis MA (2000) Pore geometry and transport properties in North Sea chalk. J Petrol Sci Eng 25:107–134

Belvin A, Burrell R, Gokhale A, Thadhani N, Garmestani H (2009) Application of two-point probability distribution functions to predict properties of heterogeneous two-phase materials. Mater Charact 60:1055–1062

Bentz DP, Martys NS (1994) Hydraulic radius and transport in reconstructed model three-dimensional porous media. Transp Porous Media 17:221–238

Beran MJ (1968) Statistical continuum theories. Interscience Publishers, New York

Beran MJ, Mason TA, Adams BL, Olsen T (1996) Bounding elastic constants of an orthotropic polycrystal using measurements of the microstructure. J Mech Phys Solids 44(9):1543–1563

Berge PA, Berryman JG, Blair SC, Pana C (1997) Scalar properties of transversely isotropic tuff from images of orthogonal cross sections. UCRL-JC- Lawrence Livermore Natl. Lab., Livermore, Calif

Berk NF (1987) Scattering properties of a model bicontinuous structure with a well defined length scale. Phys Rev Lett 58(25):2718

Berk NF (1991) Scattering properties of the leveled-wave model of random morphologies. Phys Rev A 44:5069–5079

Bernard D, Léonard F, Plougonven E, Bruno G (2020) On the use of autocorrelation functions, permeability tensors, and computed tomography to characterise the anisotropy of diesel particulate filter materials. Phil Mag 100(22):2802–2835

Berryman JG (1985a) Variational bounds on elastic constants for the penetrable sphere model. J Phys D: Appl Phys 18:585–597

Berryman JG (1985b) Measurement of spatial correlation functions using image processing techniques. J Appl Phys 57:2374–2384

Berryman JG (1987) Relationship between specific surface area and spatial correlation functions for anisotropic porous media. J Math Phys 28:244–245

Berryman JG (1988) Interpolating and integrating 3-point correlation-functions on a lattice. J Comp Phys 75:86–102

Berryman JG (1998) Planar spatial correlations, anisotropy, and specific surface area of stationary random porous media. J Appl Phys 83(3):1685–1693

Berryman JG, Blair SC (1986) Use of digital image analysis to estimate fluid permeability of porous materials: Application of two-point correlation functions. J Appl Phys 60(6):1930

Berryman JG, Blair SC (1987) Kozeny-Carman relations and image processing methods for estimating Darcy's constant. J Appl Phys 62(6):2221

Berryman JG, Milton GW (1985) Normalization constraint for variational bounds on fluid permeability. J Chem Phys 83(2):754

Berryman JG, Milton GW (1988) Microgeometry of random composites and porous media. J Phys D: Appl Phys 21:87–94

Biswal B, Hilfer R (1999) Microstructure analysis of reconstructed porous media. Phys A 266:307–311

Biswal B, Manwart C, Hilfer R, Bakke S, Øren PE (1999) Quantitative analysis of experimental and synthetic microstructures for sedimentary rock. Phys A 273(3–4):452–475

Blair SC, Berge PA, Berryman JG (1993) Two-point correlation functions to characterize microgeometry and estimate permeabilities of synthetic and natural sandstones. Technical Report, Livermore, CA (United States): Lawrence Livermore National Lab. (LLNL)

Blair SC, Berge PA, Berryman JG (1996) Using two-point correlation functions to characterize microgeometry and estimate permeabilities of sandstones and porous glass. J Geophys Res: Solid Earth 101:20359–20375

Bodla KK, Garimella SV, Murthy JY (2014) 3D reconstruction and design of porous media from thin sections. Int J Heat Mass Transf 73:250–264

Boger F, Feder J, Jøssang T, Hilfer R (1992) Microstructural sensitivity of local porosity distributions. Phys A 187(1–2):55–70

Box GEP (1976) Science and statistics. J Am Stat Assoc 71(356):791–799

Brown WF Jr (1955) Solid mixture permittivities. J Chem Phys 23(8):1514–1517

Buscombe D (2008) Estimation of grain-size distributions and associated parameters from digital images of sediment. Sed Geol 210:1–10

Buscombe D, Masselink G (2008) Grain-size information from the statistical properties of digital images of sediment. Sedimentology 56:421–438

Buscombe D, Rubin DM, Warrick JA (2010) A universal approximation of grain size from images of noncohesive sediment. J Geophys Res 115(F2):F02015.1–F02015.17

Cai X, Brenner R, Peralta L, Olivier C, Gouttenoire P-J, Chappard C, Grimal Q (2019) Homogenization of cortical bone reveals that the organization and shape of pores marginally affect elasticity. J R Soc Interface 16(151):20180911

Capek P, Hejtmanek V, Brabec L, Zikanova A, Kocirik M (2009) Stochastic reconstruction of particulate media using simulated annealing: improving pore connectivity. Transp Porous Media 76(2):179–198

Carman PC (1956) Flow of gases through porous media. Butterworth, London

Cecen A, Fast T, Kumbur EC, Kalidindi SR (2014) A data-driven approach to establishing microstructure-property relationships in porous transport layers of polymer electrolyte fuel cells. J Power Sources 245:144–153

Chatzis I, Jewlal DM, Ioannidis MA (1997) Core sample permeability estimation using statistical image analysis. Paper SCA-9723

Chen S, Li H, Jiao Y (2015) Dynamic reconstruction of heterogeneous materials and microstructure evolution. Phys Rev E 92(2):023301

Chen PE, Xu W, Chawla N, Ren Y (2019) Novel hierarchical correlation functions for quantitative representation of complex heterogeneous materials and microstructural evolution. arXiv:1905.06979

Chen S, Kirubanandham A, Chawla N, Jiao Y (2016) Stochastic multi-scale reconstruction of 3D microstructure consisting of polycrystalline grains and second-phase particles from 2D micrographs. Metall Mater Trans A 47:1–11

Cheng Z, Liu H (2015) Digital grain-size analysis based on autocorrelation algorithm. Sed Geol 327:21–31

Cherkasov A, Ananev A, Karsanina M, Khlyupin A, Gerke K (2021) Adaptive phase-retrieval stochastic reconstruction with correlation functions: three-dimensional images from two-dimensional cuts. Phys Rev E 104(3):035304

Chernov LA (1960) Wave propagation in a random medium. McGraw Hill, New York

Chiang MYM, Wang X, Landis FA, Dunkers J, Snyder CR (2006) Quantifying the directional parameter of structural anisotropy in porous media. Tissue Eng 12:1597–1606

Chiang MYM, Landis FA, Wang X, Smith JR, Cicerone MT, Dunkers J, Luo Y (2009) Local thickness and anisotropy approaches to characterize pore size distribution of three-dimensional porous networks. Tissue Engineering - Part C: Methods 15:65–76

Chiu SN, Stoyan D, Kendall WS, Mecke J (2013) Stochastic geometry and its applications. Wiley, Hoboken, NJ

Chung S-Y, Han T-S (2010) Reconstruction of random two-phase polycrystalline solids using low-order probability functions and evaluation of mechanical behavior. Comput Mater Sci 49:705–719

Chung S-Y, Han T-S (2013) Correlation between low-order probability distribution functions and percolation of porous concrete. Mag Concr Res 65:448–460

Ciccariello S, Cocco G, Benedetti A, Enzo S (1981) Correlation functions of amorphous multiphase systems. Phys Rev B 23(12):6474–6485

Cinlar E, Torquato S (1995) Exact determination of the two-point cluster function for one-dimensional continuum percolation. J Stat Phys 78(3–4):827–839

Cobaleda GE (2000) A study of image autocorrelation as a method to characterize porosity and estimate permeability. Cienc Tecnol Futur 2(1):69–81

Coiffier G, Renard P, Lefebvre S (2020) 3d geological image synthesis from 2d examples using generative adversarial networks. Front Water 2:30

Coker DA, Torquato S (1995) Extraction of morphological quantities from a digitized medium. J Appl Phys 77(12):6087–6099

Coker DA, Torquato S, Dunsmuir JH (1996) Morphology and physical properties of Fontainebleau sandstone via a tomographic analysis. J Geophys Res B: Solid Earth 101(8):17497–17506

Coles ME, Hazlett RD, Muegge EL, Jones KW, Andrews B, Dowd B, Siddons P, Peskin A, Spanne P, Soll W (1998) Developments in synchrotron X-ray microtomography with applications to flow in porous media. SPE Reservoir Eval Eng 1:288–296

Cooley JW, Tukey JW (1965) An algorithm for the machine calculation of complex Fourier series. Math Comput 19(90):297–301

Corson P (1974a) Correlation functions for predicting properties of heterogeneous materials: I. Experimental measurements of spatial correlation functions in multiphase solids. J Appl Phys 45:3159–3164

Corson P (1974b) Correlation functions for predicting properties of heterogeneous materials, II. Empirical construction of spatial correlation functions for two-phase solids. J Appl Phys 45:3165–3170

Corson PB (1974c) Correlation-functions for predicting properties of heterogeneous materials. 3. Effective elastic-moduli of 2-phase solids. J Appl Phys 45:3171–3179

Corson PB (1974d) Correlation-functions for predicting properties of heterogeneous materials. 4. Effective thermal-conductivity of 2-phase solids. J Appl Phys 45:3180–3182

Costantino S, Comeau JW, Kolin DL, Wiseman PW (2005) Accuracy and dynamic range of spatial image correlation and cross-correlation spectroscopy. Biophys J 89(2):1251–1260

Crespi JM (1986) Some guidelines for the practical application of Fry's method of strain analysis. J Struct Geol 8(7):799–808

Cule D, Torquato S (1999) Generating random media from limited microstructural information via stochastic optimization. J Appl Phys 86(6):3428–3437

Debye P, Bueche AM (1949) Scattering by an inhomogeneous solid. J Appl Phys 20(6):518–525

Debye P, Anderson HR Jr, Brumberger H (1957) Scattering by an inhomogeneous solid. II: the correlation function and its application. J Appl Phys 28(6):679–683

de Moraes AGL, Francelin MR, de Carvalho M Jr, Pereira MG, Thomazini A, Gonçalves CE, Schaefer R (2017) Environmental correlation and spatial autocorrelation of soil properties in Keller Peninsula, Maritime Antarctica. Rev Bras Cienc Solo 41:e0170021

Demberg K, Laun FB, Bertleff M, Bachert P, Kuder TA (2018) Experimental determination of pore shapes using phase retrieval from q-space NMR diffraction. Phys Rev E 97(5–1):052412

Deng H, Liu Y, Gai DH, Dikin DA, Putz KW, Chen W, Brinson LC, Burkhart C, Jiang B, Papakonstantopoulos GJ (2012) Utilizing real and statistically reconstructed microstructures for the viscoelastic modeling of polymer nanocomposites. Compos Sci and Tech 72:1725–1732
Derbyshire E, Unwin DJ, Fang XM, Langford M (1992) The Fourier frequency-domain representation of sediment fabric anisotropy. Comput Geosci 18:63–73
Deutsch CV (1992) Annealing techniques applied to reservoir modeling and the integration of geological and engineering (Well Test) data. PhD Dissertation, Stanford University, Stanford CA
Deutsch CV, Journel AG (1994) The application of simulated annealing to stochastic reservoir modeling. SPE Adv Technol Ser 2(02):222–227
Di Martino MDP, De Siena L, Healy D, Vialle S (2021) Petro-mineralogical controls on coda attenuation in volcanic rock samples. Geophys J Int 226:1858–1872
Doi M (1976) A new variational approach to the diffusion and the flow problem in porous media. J Phys Soc Jpn 40:567–572
Dullien FAL (1979) Porous media; fluid transport and pore structure. Academic Press, San Diego, CA
Durbin J (1960) The fitting of time series models. Rev Inst Int Stat 28:233–243
Ehrlich R, Kennedy SK, Crabtree SJ, Cannon RL (1984) Petrographic image analysis; I. Analysis of reservoir pore complexes. J Sediment Res 54(4):1365–1378
Emami A (2014) Design optimization of heterogeneous microstructured materials. MSc. in Mechanical Engineering Thesis, Purdue University, Indianapolis, Indiana
Erslev EA, Ge H (1990) Least-squares center-to-center and mean object ellipse fabric analysis. J Struct Geol 12:1047–1059
Faessel M, Jeulin D (2010) Segmentation of 3D microtomographic images of granular materials with the stochastic watershed. J Microsc 239:17–31
Feng JW, Li CF, Cen S, Owen DRJ (2014) Statistical reconstruction of two-phase random media. Comput Struct 137:78–92
Feng J, He X, Qing L, Li Y (2018) Reconstruction of three-dimensional heterogeneous media from a single two-dimensional section via co-occurrence correlation function. Comput Mater Sci 144:181–192
Feng J, Teng Q, Li B, He X, Chen H, Li Y (2020) An end-to-end three-dimensional reconstruction framework of porous media from a single two-dimensional image based on deep learning. Comput Methods Appl Mech Eng 368:113043
Fernandes CP (1994) Caracterização morfotopologica de espaços porosos: reconstituição multiescala e simulação de processos de invasão de fluidos não-molhantes. PhD Thesis. Federal University of Santa Catarina, Florianopolis, Brazil
František Š, Mukamel S (2005) Multipoint correlation functions for continuous-time random walk models of anomalous diffusion. Phys Rev E 72:031108
Frisch HL, Stillinger FH (1963) Contribution to the statistical geometric basis of radiation scattering. J Chem Phys 38:2200–2207
Fry N (1979) Random point distributions and strain measurement in rocks. Tectonophysics 60(1):89–105
Fullwood DT, Niezgoda SR, Kalidindi SR (2008) Microstructure reconstructions from 2-point statistics using phase-recovery algorithms. Acta Mater 56:942
Fusseis F, Xiao X, Schrank C, De Carlo F (2014) A brief guide to synchrotron radiation-based microtomography in (structural) geology and rock mechanics. J Struct Geol 65:1–16
Garmestani H, Lin S, Adams BL (1998) Statistical continuum theory for inelastic behavior of two-phase medium. Int J Plast 14:719–731
Garmestani H, Lin S, Adams BL, Ahzi S (2001) Statistical continuum theory for large plastic deformation of polycrystalline materials. J Mech Phys Solids 49(3):589–607
Gatt J-M, Monerie Y, Laux D, Baron D (2005) Elastic behavior of porous ceramics: application to nuclear fuel materials. J Nucl Mater 336(2):145–155

Gerke KM, Karsanina MV, Skvortsova EB (2012) Description and reconstruction of the soil pore space using correlation functions. Eurasian Soil Sci 45(9):962–973
Gerke KM, Karsanina MV (2015) Improving stochastic reconstructions by weighting correlation functions in an objective function. Europhys Lett 111:56002
Gerke KM, Karsanina MV, Mallants D (2015) Universal stochastic multiscale image fusion: an example application for shale rock. Sci Rep 5:1
Glatter O (1980) Evaluation of small-angle scattering data from lamellar and cylindrical particles by the indirect transformation method. J Appl Crystallogr 13:577–584
Glatter O, Kratky O (1982) Small angle x-ray scattering. Academic Press, London, UK
Gokhale A, Singh H (2005) Computer simulations of realistic microstructures using statistical two-point correlation functions. Microsc Microanal 11:1632–1633
Gokhale AM, Tewari A, Garmestani H (2005) Constraints on microstructural two-point correlation functions. Scripta Mater 53:989–993
Gommes C, Jiao Y, Torquato S (2012a) Density of states for a specified correlation function and the energy landscape. Phys Rev Lett 108(8):080601
Gommes CJ, Jiao Y, Torquato S (2012b) Microstructural degeneracy associated with a two-point correlation function and its information content. Phys Rev E 85:051140
Gradshteyn IS, Ryzhik IM (1980) Table of integrals, series, and products. 4th English ed, Jeffrey A, Zwillinger D (eds). Academic Press, New York
Gravanis E, Pantelidis L (2019) Determining of the Joint Roughness Coefficient (JRC) of rock discontinuities based on the theory of random fields. Geosciences (7):295
Gravey M, Mariethoz G (2020) Quicksampling v1. 0: a robust and simplified pixel-based multiple-point simulation approach. Geosci Model Dev 13:2611–2630
Guinier A, Fournet G, Walker CB, Yudowitch CB (1955) Small-angle scattering of x-rays. Wiley, New York
Guo E-Y, Chawla N, Jing T, Torquato S, Jiao Y (2014) Accurate modeling and reconstruction of three-dimensional percolating filamentary microstructures from two-dimensional micrographs via dilation-erosion method. Mater Charact 89:33–42
Hajizadeh A, Safekordi A, Farhadpour FA (2011) A multiple-point statistics algorithm for 3d pore space reconstruction from 2d images. Adv Water Resour 34:1256
Han T-S, Kim J-S (2018) Alternative parameter to characterize pore distribution in cement paste, and its relation to mechanical properties. In: Computational modelling of concrete structures: proceedings of the conference on computational modelling of concrete and concrete structures (EURO-C 2018). CRC Press, Boca Raton FL, p 311
Hanna SS, Fry N (1979) A comparison of methods of strain determination in rocks from southwest Dyfed (Pembrokeshire) and adjacent areas. J Struct Geol 1:155–162
Haralick R, Shanmugam K, Dinstein I (1973) Textural features for image classification. IEEE Trans Syst Man Cybern 6:610–621
Hasanabadi A, Baniassadi M, Abrinia K, Safdari M, Garmestani H (2016) Efficient three-phase reconstruction of heterogeneous material from 2d cross-sections via phase-recovery algorithm J. Microsc 264:384
Haslund E, Hansen BD, Hilfer R, Nøst B (1994) Measurement of local porosities and dielectric dispersion for a water-saturated porous medium. J Appl Phys 76:5473–5480
Havelka J, Kučerová A, Sýkora J (2016) Compression and reconstruction of random microstructures using accelerated lineal path function. Comput Mater Sci 122:102–117
Hazlett RD (1995) Simulation of capillary-dominated displacements in microtomographic images of reservoir rocks. Transp Porous Media 20:21–35
Hazlett RD (1997) Statistical characterization and stochastic modeling of pore networks in relation to fluid flow. Math Geol 29:801–822
Heilbronner-Panozzo R (1992) The autocorrelation function: an image processing tool for fabric analysis. Tectonophysics 212(3–4):351–370
Heilbronner R (2010) Mapping texture domains in quartzite microstructures. J Geol Soc India 75(1):160–170

Heilbronner R, Barrett S (2014) Image analysis in earth sciences. Microstructures and textures of earth materials. Springer, Berlin
Herman EK, Toran L, White WB (2009) Quantifying the place of karst aquifers in the groundwater to surface water continuum: a time series analysis study of storm behavior in Pennsylvania water resources. J Hydrol 376:307–317
Hertz P (1909) Über den gegenseitigen durchschnittlichen Abstand von Punkten, die mit bekannter mittlerer Dichte im Raume angeordnet sind. Math Ann 67:387–398
Hidajat I, Rastogi A, Singh M, Mohanty KK (2002) Transport properties of porous media reconstructed from thin-sections. SPE J 7:40–48
Hilfer R (1991) Geometric and dielectric characterization of porous media. Phys Rev B 44(1):60
Hilfer R (1992) Local-porosity theory for flow in porous media. Phys Rev B 45:7115
Hilfer R (1993) Local porosity theory for electrical and hydrodynamical transport through porous media. Phys A 194(1):406–414
Ho FG, Strieder W (1979) Asymptotic expansion of the porous medium, effective diffusion coefficient in the Knudsen number. J Chem Phys 70:5635–5639
Howard MP, Lequieu J, Delaney KT, Ganesan V, Fredrickson GH, Truskett TM (2020) Connecting solute diffusion to morphology in triblock copolymer membranes. Macromolecules 53(7):2336–2343
Hubler MH, Gelb J, Ulm F-J (2017) Microtexture analysis of gas shale by XRM imaging. J Nanomech Micromech 7(3):04017005
Hyman JD, Winter CL (2014) Stochastic generation of explicit pore structures by thresholding Gaussian random fields. J Comput Phys 277:16–31
Hyman JD, Guadagnini A, Winter CL (2015) Statistical scaling of geometric characteristics in stochastically generated pore microstructures. Comput Geosci 19:845–854
Ikeda S, Nakano T, Nakashima Y (2000) Three-dimensional study on the interconnection and shape of crystals in a graphic granite by X-ray CT and image analysis. Min Mag 64:945–959
Ioannidis MA, Lange E (1998) Microgeometry and topology of statistically homogeneous porous media. Trans Ecol Environ 17:223–230
Ioannidis MA, Kwiecien M, Chatzis I (1995) Computer generation and application of 3D model porous media: from pore-level geostatistics to the estimation of formation factor. In: Petroleum computer conference, Houston, TX, Paper SPE 30201, pp 185–194
Ioannidis MA, Kwiecien MJ, Chatzis I (1996) Statistical analysis of the porous microstructure as a method for estimating reservoir permeability. J Petr Sci Eng 16:251–261
Ioannidis MA, Kwiecien MJ, Chatzis I (1997) Electrical conductivity and percolation aspects of statistically homogeneous porous media. Transp Porous Media 29:61–83
Jensen JL, Lake LW, Corbett PWM, Goggin DJ (1997) Statistics for petroleum engineers and geoscientists. Prentice Hall PTR, New Jersey
Jeulin D (2000) Random texture models for material structures. Stat Comput 10:121–132
Jiao Y, Chawla N (2014a) Three dimensional modeling of complex heterogeneous materials via statistical microstructural descriptors. Integr Mater Manuf Innov 3(1):1–19
Jiao Y, Chawla N (2014b) Modeling and characterizing anisotropic inclusion orientation in heterogeneous material via directional cluster functions and stochastic microstructure reconstruction. J Appl Phys 115(9):093511
Jiao Y, Stillinger FH, Torquato S (2007) Modeling heterogeneous materials via two-point correlation functions: Basic principles. Phys Rev E 76:031110
Jiao Y, Stillinger FH, Torquato S (2008) Modeling heterogeneous materials via two-point correlation functions. II. Algorithmic details and applications. Phys Rev E 77(3):031135
Jiao Y, Stillinger FH, Torquato S (2009) A superior descriptor of random textures and its predictive capacity. Proc Natl Acad Sci 106(42):17634–17639
Jiao Y, Padilla E, Chawla N (2013) Modeling and predicting microstructure evolution in lead/tin alloy via correlation functions and stochastic material reconstruction. Acta Mater 61(9):3370–3377

Jin G, Patzek TW, Silin DB (2003) Physics-based reconstruction of sedimentary rocks. In: SPE western regional/AAPG pacific section joint meeting (Society of Petroleum Engineers, 2003). Paper Number: SPE-83587-MS
Johnston C (2006) Stretched exponential relaxation arising from a continuous sum of exponential decays. Phys Rev B 74:184430 (+ Erratum, in Phys Rev B 77:179901)
Joshi M. (1974) A class of stochastic models for porous media. PhD thesis, Univ. of Kansas, Lawrence
Kagan YY, Knopoff L (1980) Spatial distribution of earthquakes: the two-point correlation function. Geophys J R Astr Soc 62:303–320
Kainourgiakis ME, Kikkinides ES, Steriotis TA, Stubos AK, Tzevelekos KP, Kanellopoulos NK (2000) Structural and transport properties of alumina porous membranes from process-based and statistical reconstruction techniques. J Colloid Interface Sci 231:158–167
Kainourgiakis ME, Steriotis TA, Kikkinides ES, Romanos G, Stubos AK (2002a) Adsorption and diffusion in nanoporous materials from stochastic and process-based reconstruction techniques. Colloids Surf a: Physicochem Eng Asp 206:321–334
Kainourgiakis ME, Kikkinides ES, Stubos AK (2002b) Diffusion and flow in porous domains constructed using process-based and stochastic techniques. J Porous Mater 9(2):141–154
Kalidindi SR, Niezgoda SR, Salem AA (2011) Microstructure informatics using higher-order statistics and efficient data-mining protocols. JOM J Miner Metals Mater Soc 63(4):34–41
Sadegh K, Tahmasebi P (2016) Conditional reconstruction: an alternative strategy in digital rock physics. Geophysics 81(4):D465–D477
Karsanina MV, Gerke KM (2018) Hierarchical optimization: fast and robust multiscale stochastic reconstructions with rescaled correlation functions. Phys Rev Lett 121:265501
Karsanina MV, Gerke KM, Skvortsova EB, Mallants D (2015) Universal spatial correlation functions for describing and reconstructing soil microstructure. PLoS ONE 10(5):e0126515
Karsanina MV, Lavrukhin EV, Fomin DS, Yudina AV, Abrosimov KN, Gerke KM (2021) Compressing soil structural information into parameterized correlation functions. Eur J Soil Sci 72(2):561–577
Keehm Y, Mukerji T, Prasad M, Nur A (2003) Permeability prediction from thin sections using the Lattice-Boltzmann flow simulation. In: 2003 SEG annual meeting. OnePetro
Khristenko U, Constantinescu A, Tallec PL, Oden JT, Wohlmuth B (2020) A statistical framework for generating microstructures of two-phase random materials: application to fatigue analysis. Multiscale Model Simul 18(1):21–43
Kim J-S, Kim J, Han T-S (2019a) Microstructure characterization of cement paste from micro-CT and correlations with mechanical properties evaluated from virtual and real experiments. Mater Charact 155:109807
Kim J-S, Kim J, Kang J, Han T-S (2019b) Construction of virtual interfacial transition zone (ITZ) samples of hydrated cement paste using extended stochastic optimization. Cem Concr Compos 102:84–93
Kim J-S, Kim S-Y, Han T-S (2020) Sensitivity and uncertainty estimation of cement paste properties to microstructural characteristics using FOSM method. Const Build Mater 242:118–159
Kim S-Y, Lee JH, Kim JH, Han T-S (2021) Comparison of microstructure characterization methods by two-point correlation functions and reconstruction of 3D microstructures using 2D TEM images with high degree of phase clustering. Mater Charact 172:110876
Kjems JK, Freltoft T (1991) Neutron and X-ray scattering from aggregates. In: Pynn R, Skjeltorp A (eds) Scaling phenomena in disordered systems. Springer, Berlin, pp 133–140
Korvin G (1978) Some notes on a problem of Treitel and Wang. Geophys Trans 21(8):53–58
Korvin G (1992b) A percolation model for the permeability of kaolinite-bearing sandstone. Geophys Trans 37(2–3):177–209
Korvin G (2016) Permeability from microscopy: review of a dream. Arab J Sci Eng 41(6):2045–2065
Korvin G (2022) Digital filters. In: Daya Sagar BS, Cheng Q, McKinley J, Agterberg F (eds) Encyclopedia of mathematical geosciences. Encyclopedia of earth sciences series. Springer, Cham

Korvin G, Sterligov B, Oleschko K, Cherkasov S (2013) Entropy of shortest distance (ESD) as pore detector and pore-shape classifier. Entropy 15(6):2384–2397
Koutsourelakis P (2006) Probabilistic characterization and simulation of multiphase random media. Prob Eng Mech 21:227–234
Koutsourelakis PS, Deodatis G (2006) Simulation of multidimensional binary random fields with application to modeling of two-phase random media. J Eng Mech/ASCE 132
Kröner E (1977) Bounds for effective elastic moduli of disordered materials. J Mech Phys Solid 25(2):137–155
Kumar H, Briant CL, Curtin WA (2006) Using microstructure reconstruction to model mechanical behavior in complex microstructures. Mech Mater 38:818–832
Lemmens L, Rogiers B, Jacques D, Huysmans M, Swennen R, Urai JL, Desbois G, Laloy E (2019) Nested multiresolution hierarchical simulated annealing algorithm for porous media reconstruction. Phys Rev E 100:053316
Levinson N (1947) The Wiener RMS error criterion in filter design and prediction. J Math Phys 25:261–278
Levitz P (1998) Off-lattice reconstruction of porous media: critical evaluation, geometrical confinement and molecular transport. Adv Colloid Interface Sci 77:71–106
Li DS, Saheli G, Khaleel M, Garmestani H (2006) Quantitative prediction of effective conductivity in anisotropic heterogeneous media using two-point correlation functions. Comput Mater Sci 38(1):45–50
Li H, Singh S, Chawla N, Jiao Y (2018a) Direct extraction of spatial correlation functions from limited x-ray tomography data for microstructural quantification. Mater Charact 140:265–274
Li H, Chen PE, Jiao Y (2018b) Accurate reconstruction of porous materials via stochastic fusion of limited bimodal microstructural data. Transp Porous Media 125:5–22
Liang ZR, Fernandes CP, Magnani FS, Philippi PC (1998) A reconstruction technique for three-dimensional porous media using image analysis and Fourier transforms. J Petrol Sci Eng 21:273–283
Liang ZR, Philippi PC, Fernandes CP, Magnani FS (1999) Prediction of permeability from the skeleton of three-dimensional pore structure. SPE Reservoir Eval Eng 2:161–168
Liang Z, Ioannidis MA, Chatzis I (2000) Permeability and electrical conductivity of porous media from 3D stochastic replicas of the microstructure. Chem Eng Sci 55(22):5247–5262
Lin C (1982) Microgeometry I: autocorrelation and rock microstructure. J Int Assoc Math Geol 14:343–360
Lin S, Garmestani H (2000) Statistical continuum mechanics analysis of an elastic two-isotropic-phase composite material. Compos B Eng 31(1):39–46
Lindner P, Zemb T (1991) Neutron, x-ray and light scattering. Elsevier, Amsterdam
Lokajicek T, Rudajev V, Vilhelm J (2004) Possibilities of forecasting of rock sample total fracture by application of correlation analysis of acoustic emission events series. In: DGZfP-proceedings BB 90-CD lecture 1
Louis L, Wong TF, Baud P (2007) Imaging strain localization by X-ray radiography and digital image correlation: deformation bands in Rothbach sandstone. J Struct Geol 29(1):129–140
Lu BL, Torquato S (1992a) Lineal-path function for random heterogeneous materials. Phys Rev A 45:922–929
Lu BL, Torquato S (1992b) Lineal-path function for random heterogeneous materials. 2. Effect of polydispersivity. Phys Rev A 45:7292–7301
Lu BL, Torquato S (1992c) Nearest-surface distribution functions for polydispersed particle systems. Phys Rev A 45:5530–5544
Mandzhieva R (2017) Introduction to digital core analysis: 3D reconstruction, numerical flow simulations and pore network modeling. MSc thesis, Norwegian University of Science and Technology, Trondheim, Norway
Manwart C, Hilfer R (1999) Reconstruction of random media using Monte Carlo methods. Phys Rev E 59:5596–5599

Manwart C, Torquato S, Hilfer R (2000) Stochastic reconstruction of sandstones. Phys Rev E 62:893–899
Markov K (1995) On the “triangular” inequality in the theory of two-phase random media. Technical report, Annuaire L’Universite de Sofia, Faculte de Mathematiques et Informatique
Mason Thomas A, Adams BL (1999) Use of microstructural statistics in predicting polycrystalline material properties. Metall Mater Trans A 30:969–979
Matheron G (1975) Random sets and integral geometry. Wiley, New York
Matheron G (1993) Une conjecture sur covariance d’un ensemble aleatoire. Cahiers De Géostatistique 107:107–113
Mätzler C (2002) Relation between grain-size and correlation length of snow. J Glaciol 48(162):461–466
McBeck J, Kobchenko M, Hall SA, Tudisco E, Cordonnier B, Meakin P, Renard F (2018) Investigating the onset of strain localization within anisotropic shale using digital volume correlation of time-resolved X-ray microtomography images. J Geophys Res: Solid Earth 123:7509–7528
McMillan B (1955) History of a problem. J Soc Ind Appl Math 3(3):119–128
Mildner DFR, Hall PL (1986) Small-angle scattering from porous solids with fractal geometry. J Phys D Appl Phys 19:1535–1545
Millard J, Augat P, Link TM, Kothari M, Newitt DC, Genant HK, Majumdar S (1998) Power spectral analysis of vertebral trabecular bone structure from radiographs: orientation dependence and correlation with bone mineral density and mechanical properties. Calcif Tissue Int 63:482–489
Moctezuma-Berthier A, Vizika O, Adler PM (2002) Macroscopic conductivity of vugular porous media. Transp Porous Media 49(3):313–332
Monfared S, Ulm FJ (2016) A molecular informed poroelastic model for organic-rich, naturally occurring porous geocomposites. J Mech Phys Solids 88:186–203
Moran PAP (1950) Notes on continuous stochastic phenomena. Biometrika 37(1–2):17–23
Mosser L, Dubrule O, Blunt MJ (2017) Reconstruction of three-dimensional porous media using generative adversarial neural networks. Phys Rev E 96:043309
Nakashima Y, Kamiya S, Nakano T (2008) Diffusion ellipsoids of anisotropic porous rocks calculated by X-ray computed tomography-based random walk simulations. Water Resour Res 44:W12435
Nunan N, Wu K, Young IM, Crawford JW, Ritz K (2003) Spatial distribution of bacterial communities and their relationships with the micro-architecture of soil. FEMS Microbiol Ecol 44:203–215
Neumann M, Stenzel O, Willot F, Holzer L, Schmidt V (2020) Quantifying the influence of microstructure on effective conductivity and permeability: virtual materials testing. Int J Solids Struct 184:211–220
Niezgoda SR, Yabansu YC, Kalidindi SR (2011) Understanding and visualizing microstructure and microstructure variance as a stochastic process. Acta Mater 59(16):6387–6400
Odgaard A, Jensen EB, Gundersen HJG (1990) Estimation of structural anisotropy based on volume orientation. A new concept. J Microsc 157:149–162
Odgaard A, Kabel J, van Rietbergen B, Dalstra M, Huiskes R (1997) Fabric and elastic principal directions of cancellous bone are closely related. J Biomech 30:487–495
O’Dowd RJ (1990) Ill-conditioning and pre-whitening in seismic deconvolution. Geophys J Int 101:489–491
O’Dowd RJ (1991) The Wiener-Levinson algorithm and ill-conditioned normal equations. Geophys J Int 106:399–406
Okabe H (2004) Pore-scale modelling of carbonates. PhD thesis, University of London
Øren P-E, Bakke S (2002) Process based reconstruction of sandstones and prediction of transport properties. Transp Porous Media 46:311
Ouenes A, Bhagavan S, Bunge PH, Travis BJ (1994) Application of simulated annealing and other global optimization methods to reservoir description: myths and realities. In: Proceedings of the SPE 69th annual technical conference and exhibition held in New Orleans, LA, USA, Paper 28415

Pant LM (2016) Stochastic characterization and reconstruction of porous media. PhD thesis, University of Alberta

Pant LM, Mitra SK, Secanell M (2014) Stochastic reconstruction using multiple correlation functions with different-phase-neighbor-based pixel selection. Phys Rev E 90:023306

Papakostas GA, Nolan JW, Vordos N, Gkika D, Kainourgiakis ME, Mitropoulos AC (2015) On 3D reconstruction of porous media by using spatial correlation functions. J Eng Sci Technol Rev 8(4):78–83

Patelli E, Schuëller GI (2009) On optimization techniques to reconstruct microstructures of random heterogeneous media. Comput Mater Sci 45:536–549

Patelli E, Schuëller GI (2012) Computational optimization strategies for the simulation of random media and components. Comput Optim Appl 53:903–931

Patterson AL (1939) Homometric structures. Nature 143(3631):939–940

Patterson AL (1944) Ambiguities in the X-ray analysis of crystal structures. Phys Rev 65(5–6):195

Pérez-Peña JV, Azañón JM, Booth-Rea G, Azor A, Delgado J (2009) Differentiating geology and tectonics using a spatial autocorrelation technique for the hypsometric integral. J Geophys Res 114:F02018

Pfleiderer S, Halls HC (1990) Magnetic susceptibility anisotropy of rocks saturated with ferrofluid: a new method to study pore fabric? Phys Earth Planet Inter 65(1–2):158–164

Pfleiderer S, Halls HC (1993) Magnetic pore fabric analysis: verification through image autocorrelation. J Geophys Res 98(B3):4311–4316

Politis MG, Kikkinides ES, Kainourgiakis ME, Stubos AK (2008) A hybrid process-based and stochastic reconstruction method of porous media. Micropor Mesopor Mat 110:92–99

Ponte Castañeda P, Willis J (1995) The effect of spatial distribution on the effective behavior of composite materials and cracked media. J Mech Phys Solids 43(12):1919–1951

Prager S (1961) Viscous flow through porous media. Phys Fluids 4:1477–1482

Prager S (1963a) Interphase transfer in stationary two-phase media. Chem Eng Sci 18(4):227–231

Prager S (1963b) Diffusion and viscous flow in concentrated suspensions. Physica 29:129–139

Prasad M, Mukerji T, Reinstaedtler M, Arnold W (2009) Acoustic signatures, impedance microstructure, textural scales, and anisotropy of kerogen-rich shale. In: SPE annual technical conference and exhibition. OnePetro

Prodanovic M, Lindquist WB, Seright RS (2007) 3D image-based characterization of fluid displacement in a Berea core. Adv Water Resour 30:214–226

Quiblier JA (1984) A new three-dimensional modeling technique for studying porous media. J Colloid Interface Sci 98:84–102

Rabbani Arash, Saeid Jamshidi, and Saeed Salehi, 2014. Determination of specific surface of rock grains by 2d imaging. *Journal of Geological Research*: Article ID 945387.

Radlinski AP (2006) Small-angle neutron scattering and rock microstructure. Rev Mineral Geochem 63:363–397

Radlinski AP, Radlinska EZ, Agamalian M, Wignall GD, Lindner P, Randl OG (1999) Fractal geometry of rocks. Phys Rev Lett 82:3078–3081

Radlinski AP, Mastalerz M, Hinde AL, Hainbuchner M, Rauch H, Baron M, Lin J-S, Fan L, Thiyagarajan P (2001) Non-invasive measurements of pore size distribution in coal pellets using x-ray and neutron techniques. In: International coalbed methane symposium 2001 proceedings, May 14–18, University of Alabama, Tuscaloosa, Alabama, USA, pp 163–175

Radlinski AP, Ioannidis MA, Hinde AL, Hainbuchner M, Baron M, Rauch H, Kline SR (2002) Multiscale characterization of reservoir rock microstructure: combining small-angle neutron scattering and image analysis. SCA2002-35, pp 1–12

Radlinski A, Ioannidis M, Hinde A, Hainbuchner M, Baron M, Rauch H, Kline S (2004) Angstrom-to-millimeter characterization of sedimentary rock microstructure. J Colloid Interface Sci 274:607–612

Rees AI (1965) The use of anisotropy of magnetic susceptibility in the estimation of sedimentary fabric. Sedimentology 4:257–271

Rehman S, El-Gebeily M (2009) A study of Saudi climatic parameters using climatic predictability indices. Chaos Solitons Fractals 41(3):1055–1069

Reiss H, Frisch HL, Lebowitz JL (1959) Statistical mechanics of rigid spheres. J Chem Phys 31:369–380

Renard F, McBeck J, Cordonnier B, Zheng X, Kandula N, Sanchez JR, Kobchenko M, Noiriel C, Zhu W, Meakin P, Fusseis F, Dysthe DK (2018) Dynamic in situ three-dimensional imaging and digital volume correlation analysis to quantify strain localization and fracture coalescence in sandstone. Pure Appl Geophys 176:1083–1115

Rice SO (1944) Mathematical analysis of random noise. Bell Syst Tech J 23:282–332. Reprinted in:Wax N (ed) (1954) Selected papers on noise and stochastic processes. Dover, New York, pp 133–194

Rink M, Schopper JR (1976) Pore structure and physical properties of porous sedimentary rocks. Pure Appl Geophys 114:273–284

Roberts AP (1997) Statistical reconstruction of three-dimensional porous media from two-dimensional images. Phys Rev E 56(3):3203

Roberts AP, Teubner M (1995) Transport properties of heterogeneous materials derived from Gaussian random fields: bounds and simulation. Phys Rev E 51:4141–4154

Roberts AP, Torquato S (1999) Chord-distribution functions of three-dimensional random media: approximate first-passage times of Gaussian processes. Phys Rev E 59:4953–4963

Robertson C, George SC (2012) Theory and practical recommendations for autocorrelation-based image correlation spectroscopy. J Biomed Opt 17(8):080801–080801

Röding M, Ma Z, Torquato S (2020) Predicting permeability via statistical learning on higher-order microstructural information. Sci Rep 10(1):15239

Rotter M, Berg A, Langenberger H, Grampp S, Imhof H, Moser E (2001) Autocorrelation analysis of bone structure. J Magn Reson Imaging 14:87–93

Roubin E, Colliat J-B, Benkemoun N (2015) Meso-scale modeling of concrete: a morphological description based on excursion sets of random fields. Comput Mater Sci 102:183–195

Rozenbaum O, Le Trong E, Rouet J-L, Bruand A (2007) 2D-image analysis: a complementary tool for characterizing quarry and weathered building limestones. J Cult Herit 8:151–159

Rubinstein J, Keller JB (1987) Lower bounds on permeability. Phys Fluids 30:2919–2921

Rubinstein J, Torquato S (1988) Diffusion-controlled reactions: Mathematical formulation, variational principles, and rigorous bounds. J Chem Phys 88:6372–6380

Rytov SM, Kravtsov YA, Tatarskii VI (1987) Principles of statistical radiophysics 1. Elements of random process theory. Springer, Berlin, Heidelberg

Saccomano AF, Savioli GB, Bidner MS (2001) Stochastic modeling of rock heterogeneities applying new autocorrelation estimators and Simulated Annealing. SPE: 69654

Saha PK, Wehrli FW (2004) A robust method for measuring trabecular bone orientation anisotropy at in vivo resolution using tensor scale. Pattern Recogn 37:1935–1944

Samper Calvete F, Carrera Ramírez J (1990) Geoestadística: aplicaciones a la hidrología subterránea. Centro Internacional de Métodos Numéricos en Ingeniería, Barcelona, Spain

Saheli G, Garmestani H, Adams BL (2004) Microstructure design of a two phase composite using two-point correlation functions. J Comput Aided Mater Des 11:103–115

Scholz C, Wirner F, Klatt MA, Hirneise D, Schröder-Turk GE, Mecke K, Bechinger C (2015) Direct relations between morphology and transport in Boolean models. Phys Rev E 92(4):043023

Schultz-Krutisch T, Heller F (1985) Measurement of magnetic susceptibility anisotropy in Buntsandstein deposits from southern Germany. J Geophys 56:51–58

Serra J (1992) Image analysis and mathematical morphology. Academic Press, Orlando FL

Sheehan N, Torquato S (2001) Generating microstructures with specified correlation functions. J Appl Phys 89:53–60

Shepp LA (1963) On positive-definite functions associated with certain stochastic processes. Technical report, Bell Laboratories

Shimamoto T, Ikeda Y (1976) A simple algebraic method for strain estimation from deformed ellipsoidal objects. 1. Basic theory. Tectonophysics 36(4):315–337

Singh SS, Williams JJ, Jiao Y, Chawla N (2012) Modeling anisotropic multiphase heterogeneous materials via directional correlation functions: simulations and experimental verification. Metall Mater Trans A 43:4470–4474

Stanke F (1986) Spatial autocorrelation functions for calculations of effective propagation constants in polycrystalline materials. J Acoust Soc Am 80:1479–1485

Steel G (1968) The two-point correlation function and the specific heat. Phys Lett A 27:550–551

Stenzel O, Pecho O, Holzer L, Neumann M, Schmidt V (2016) Predicting effective conductivities based on geometric microstructure characteristics. AIChE J 62(5):1834–1843

Sun H, Koch M (1998) Fractal generation of surface area of porous media. Stoch Hydrol Hydraul 12:83–96

Svitelman V, Dinariev O (2013) Geostatistical approach to the anisotropy analysis of 3D rock microtomographic models. Comput Geosci 57:116–123

Swan AR, Garratt JA (1995) Image analysis of petrographic textures and fabrics using semivariance. Mineral Mag 59(395):189–196

Tahmasebi P, Sahimi M (2012) Reconstruction of three-dimensional porous media using a single thin section. Phys Rev E 85:066709

Talukdar MS, Torsaeter O (2002) Reconstruction of chalk pore networks from 2D backscatter electron micrographs using a simulated annealing technique. J Petrol Sci Eng 33:265–282

Talukdar MS, Torsaeter O, Ioannidis MA (2002a) Stochastic reconstruction of particulate media from two-dimensional images. J Colloid Interface Sci 248:419–428

Talukdar MS, Torsaeter O, Ioannidis MA, Howard JJ (2002b) Stochastic reconstruction of chalk from 2D images. Transp Porous Media 48:101–123

Talukdar MS, Torsaeter O, Ioannidis MA, Howard JJ (2002c) Stochastic reconstruction, 3D characterization and network modeling of chalk. J Petrol Sci Eng 35:1–21

Tassopoulos M, Rosner DE (1992) Simulation of vapor diffusion in anisotropic particulate deposits. Chem Eng Sci 47:421–443

Thissen CJ, Brandon MT (2015) An autocorrelation method for three-dimensional strain analysis. J Struct Geol 81:135–154

Thompson AH, Katz AJ, Krohn CE (1987) The microgeometry and transport properties of sedimentary rock. Adv Phys 36:625–694

Timur A (1968) An investigation of permeability, porosity, and residual water saturation relationships for sandstone reservoirs. Log Anal 9(4):8–17

Tomoda Y (1956) A simple method for calculating the correlation coefficients. J Phys Earth 4:67–70

Toms J, Müller TM, Gurevich B (2007) Seismic attenuation in porous rocks with random patchy saturation. Geophys Prospect 55:671–678

Toms-Stewart J, Müller TM, Gurevich B, Paterson L (2009) Statistical characterization of gas-patch distributions in partially saturated rocks. Geophysics 74(2):WA51–WA64

Torabi A, Fossen H, Alaei B (2008) Application of spatial correlation functions in permeability estimation of deformation bands in porous rocks. J Geophys Res: Solid Earth 113:B08208

Torquato S (1986a) Interfacial surface statistics arising in diffusion and flow problems in porous media. J Chem Phys 85(8):4622–4628

Torquato S (1986b) Microstructure characterization and bulk properties of disordered two-phase media. J Stat Phys 45:843–873

Torquato S (1997) Effective stiffness tensor of composite media—I. Exact series expansions. J Mech Phys Solids 45(9):1421–1448

Torquato S (1998) Effective stiffness tensor of composite media: II. Applications to isotropic dispersions. J Mech Phys Solids 46(8):1411–1440

Torquato S (2002) Random heterogeneous materials: microstructure and macroscopic properties. Springer, New York

Torquato S (2002a) Statistical description of microstructures. Annu Rev Mater Sci 32:77–111

Torquato S (2006) Necessary conditions on realizable two-point correlation functions of random media. Ind Eng Chem Res 45(21):6923–6928

Torquato S (2010) Optimal design of heterogeneous materials. Annu Rev Mater Res 40:101–129

Torquato S, Haslach HW Jr (2002) Random heterogeneous materials: microstructure and macroscopic properties. Appl Mech Rev 55(4):B62–B63
Torquato S, Lado F (1986) Effective properties of two-phase disordered composite media: II. Evaluation of bounds on the conductivity and bulk modulus of dispersions of impenetrable spheres. Phys Rev B 33(9):6428–6435
Torquato S, Lado F (1991) Trapping constant, thermal conductivity, and the microstructure of suspensions of oriented spheroids. J Chem Phys 94(6):4453–4462
Torquato S, Lu B (1993) Chord-length distribution function for 2-phase random-media. Phys Rev E 4:2950–2953
Torquato S, Stell G (1982) Microstructure of two-phase random media. I. The n-point probability functions. J Chem Phys 77(4):2071–2077
Torquato S, Stell G (1983a) Microstructure of two-phase random media. II. The Mayer-Montroll and Kirkwood-Salsburg hierarchies. J Chem Phys 78:3262–3272
Torquato S, Stell G (1983b) Microstructure of two-phase random media. III. The n-point matrix probability functions for fully penetrable spheres. J Chem Phys 79(3):1505–1510
Torquato S, Stell G (1985) Microstructure of two-phase random media. V. The n-point matrix probability functions for impenetrable spheres. J Chem Phys 82(2):980–987
Torquato S, Jiao Y (2009) Dense packings of the Platonic and Archimedean solids. Nature 460(7257):876
Torquato S, Beasley J, Chiew Y (1988) Two point cluster function for continuum continuum percolation. J Chem Phys 88:6540–6547
Torquato S, Lu B, Rubinstein J (1990) Nearest-neighbor distribution functions in many-body systems. Phys Rev A 41:2059–2075
Treitel S, Wang RJ (1976) The determination of digital Wiener filters from an ill-conditioned system of normal equations. Geophys Prospect 24:317–327
Truskett TM, Torquato S, Debenedetti PG (1998) Density fluctuations in many-body systems. Phys Rev E 58:7369–7380
Tudisco E, Hall SA, Charalampidou EM, Kardjilov N, Hilger A, Sone H (2015) Full-field measurements of strain localisation in sandstone by neutron tomography and 3D-volumetric digital image correlation. Phys Procedia 69:509–515
Tudisco E, Andò E, Cailletaud R, Hall SA (2017) TomoWarp2: a local digital volume correlation code. SoftwareX 6:267–270
Underwood EE (1973) Quantitative stereology for microstructural analysis. In: McCall JL, Mueller WM (eds) Microstructural analysis. Springer, Boston, MA, pp 35–66
Valiullin R, Skirda V (2001) Time dependent self-diffusion coefficient of molecules in porous media. J Chem Phys 114(1):452–458
Van Vleck H, Middleton D (1966) The spectrum of clipped noise. Proc IEEE 54(1):2–19
Varga P, Zysset PK (2009) Sampling sphere orientation distribution: an efficient method to quantify trabecular bone fabric on grayscale images. Med Image Anal 13:530–541
Velasquez R, Marasteanu MO, Labuz JF (2010) Microstructure characterization of asphalt mixtures with 2-and 3-point correlation functions. Road Mater Pavement Des 11:251–272
Vezzetti DJ (1975) A new derivation of some fluctuation theorems in statistical mechanics. J Math Phys 16:31–33
Viggiani G, Lenoir N, Bésuelle P, Di Michiel M, Marello S, Desrues J, Kretzschmer M (2004) X-ray microtomography for studying localized deformation in fine-grained geomaterials under triaxial compression. Comptes Rendus Mécanique 332(10):819–826
Wald M, Vasilic B, Saha PK, Wehrli FW (2005) Study of trabecular bone microstructure using spatial autocorrelation analysis. Proc SPIE 5746:291–302
Wald MJ, Vasilic B, Saha PK, Wehrli FW (2007) Spatial autocorrelation and mean intercept length analysis of trabecular bone anisotropy applied to in vivo magnetic resonance imaging. Med Phys 34:1110–1120
Wang M, Pan N (2008) Predictions of effective physical properties of complex multiphase materials. Mater Sci Eng R 63:1–30

Wang MR, Pan N, Wang JK, Chen SY (2007) Mesoscopic simulations of phase distribution effects on the effective thermal conductivity of microgranular porous media. J Colloid Interface Sci 311:562–570

Wargo EA, Hanna AC, Een A, Kalidindi SR, Kumbur EC (2012) Selection of representative volume elements for pore-scale analysis of transport in fuel cell materials. J Power Sources 197:168–179

Warrick JADM, Rubin P, Ruggiero JN, Harney AED, Buscombe D (2009) Cobble Cam: Grain-size measurements of sand to Boulder from digital photographs and autocorrelation analyses. Earth Surf Proc Land 34:1811–1821

Wehrli FW, Kwon Song H, Saha PK, Wright AC (2006) Quantitative MRI for the assessment of bone structure and function. In: NMR in biomedicine special issue: "NMR of the Musculoskeletal System", pp 731–764

Westin CF, Maier SE, Mamata H, Nabavi A, Jolesz FA, Kikinis R (2002) Processing and visualization for diffusion tensor MRI. Med Image Anal 6:93–108

Wiener N (1949) Extrapolation, interpolation, and smoothing of stationary time series with engineering applications. MIT Press, Cambridge, and Wiley, New York

Whitehouse WJ (1974) The quantitative morphology of anisotropic trabecular bone. J Microsc 101:153–168

Wildenschild D, Sheppard AP (2013) X-ray imaging and analysis techniques for quantifying pore-scale structure and processes in subsurface porous medium systems. Adv Water Resour 51:217–246

Wiseman PW, Capani F, Squier JA, Martone ME (2002) Counting dendritic spines in brain tissue slices by image correlation spectroscopy analysis. J Microsc 205:177–186

Wu K, Van Dijke MI, Couples GD, Jiang Z, Ma J, Sorbie KS, Crawford J, Young I, Zhang X (2006) 3D stochastic modelling of heterogeneous porous media–applications to reservoir rocks. Transp Porous Media 65:443–467

Xiang Y, Yingling VR, Malique R, Yang Li C, Schaffler MB, Raphan T (2007) Comparative assessment of bone mass and structure using texture-based and histomorphometric analyses. Bone 40:544–552

Xiao G, Yang K, Jia H, Tao Z, Xu M, Dong B, Liu L (2021) A new method of central axis extracting for pore network modeling in rock engineering. Geofluids 2021: Article ID 1971622

Xiao S, Liu W, Li A, Yang S, Lai Z (2010) Pervasive autocorrelation of the chemical index of alteration in sedimentary profiles and its palaeoenvironmental implications. Sedimentology 57:670–676

Xiao S, Li R, Chen M (2013) Detecting sedimentary cycles using autocorrelation of grain size. Sci Rep 3:1653

Xu Z, Teng Q, He X, Li Z (2013) A reconstruction method for three-dimensional pore space using multiple-point geology statistic based on statistical pattern recognition and microstructure characterization. Int J Numer Anal Meth Geomech 37(1):97–110

Yao J, Frykman P, Kalaydjian F, Thovert JF, Adler PM (1993) High-order moments of the phase function for real and reconstructed model porous media: A comparison. J Colloid Interface Sci 156:478–490

Yeong CLY, Torquato S (1998a) Reconstructing random media. Phys Rev E 57:495–506

Yeong CLY, Torquato S (1998b) Reconstructing random media. II. Three-dimensional media from two-dimensional cuts. Phys Rev E 58:224–233

Yin XL, Chen W, To A, McVeigh C, Liu WK (2008) Statistical volume element method for predicting micro structure-constitutive property relations. Comput Method Appl Mech Eng 197:3516–3529

Yokoyama T, Nakashima S (2005) Diffusivity anisotropy in a rhyolite and its relation to pore structure. Eng Geol 80:328–335

Zerhouni O, Brisard S, Danas K (2021) Quantifying the effect of two-point correlations on the effective elasticity of specific classes of random porous materials with and without connectivity. Int J Eng Sci 166:103520

Zhao X, Yao J, Yi Y (2007) A new stochastic method of reconstructing porous media. Transp Porous Media 69(1):1–11

Zheng Q, Zhang D (2022) Digital rock reconstruction with user-defined properties using Conditional Generative Adversarial Networks. Transp Porous Med 144:255–281

Ziff RM (1977) On the bulk distribution functions and fluctuation theorems J. Math Phys 18:1825–1831

Chapter 2
Models of Tortuosity

Chapter Highlights This short Chapter is devoted to the wonderful world of *tortuous pathways* inside the pore space of rocks, available for fluids, electric current, heat, sound waves, diffunding particles, and even to the "windings and turnings" of developing cracks around harder grains of the rock. The effect of tortuosity on permeability and other effective properties is treated in details.

2.1 Tortuous Ways of Petrophysics

According to Fu et al. (2021), "tortuosity is a parameter, which is not only used to explicitly describe the sinuosity and complexity of internal percolation paths but also to quantify transport properties of porous media (Clennell 1997; Barrande et al. 2007; Ghanbarian et al. 2013)". And the authors immediately add: "unlike other standard microstructural properties, the concept of tortuosity is vague with multiple definitions and various evaluation methods introduced in different contexts."

Sinuosity (same as *tortuosity,* more precisely *geometrical tortuosity*) is only one of the possible measures to characterize a *curve's complexity,* we find many other measures in *Animal Behaviour Science*, where (Miller et al. 2011), "tortuosity of an animal's path may be defined in the most general sense as departure of the path from straightness and is closely related to the animal's behavior. For example, movement toward a destination is straighter and hence less tortuous than foraging behavior where the animal is searching locally". Miller et al. (*op. cit.*) recommend several measures to quantify tortuosity; (i) the *fractal dimension* (FD) of the path, this shows *how space-filling* is the path; (ii) the *minimum convex polygon area* (MCP), for a sequence of positions $\{P_0, P_1, \ldots, P_n\}$ this is the area of the minimal convex polygon $\mathcal{P}_{\min}$ containing all points P_i. If the movement takes place within a limited confined area $Area_{total}$, then the normalized value $Area(\mathcal{P}_{\min})/Area_{total}$ measures how efficiently the available place is utilized during movement; (iii) The *straightness index* (StI) is a measure of the "orientation efficiency of the animal as it moves toward a goal", letting $dist(P, Q)$ denote the distance between two points, for the path $\{P_0, P_1, \ldots, P_n\}$,

G. Korvin, *Statistical Rock Physics*, Earth and Environmental Sciences Library,
https://doi.org/10.1007/978-3-031-46700-4_2

$$StI = \frac{dist(P_0, P_n)}{dist(P_0, P_1) + dist(P_1, P_2) + \cdots + dist(P_{n-1}, P_n)}, \tag{2.1}$$

it is inverse to *tortuosity*, the smaller *StI*, the more tortuous the route (Batschelet 1981); (iv) *angular dispersion* (AD), introduced by Estevez and Christman (2006) is a measure of the *concentration of the turning angles* around the mean angle of movement. *AD* ranges between 0 and 1 with 0 indicating no concentration, so that the animal turns occur at random in any direction, and 1 indicating no variation in angles, so that the animal moves in a straight line without changes in direction. If $\{\theta_i\}_{i=1}^{n-1}$ is the set of turning angles associated with the *n* locations in the observed path $\{P_1, \ldots, P_{n-1}\}$, the angular dispersion is $AD = \frac{1}{n-1}\sqrt{C^2 + S^2}$ where $C = \sum \cos\theta_i$, $s = \sum \cos\theta_i$

$$AD = \frac{1}{n-1}\sqrt{\left(\sum_{i=1}^{n-1} \cos\theta_i\right)^2 + \left(\sum_{i=1}^{n-1} \sin\theta_i\right)^2} \tag{2.2}$$

(Fisher 1993); (v) In the study of the movement of animals, the *number of crossings* (NC) of the path with itself is also a useful indicator. As the physically meaningful paths of particles in the pores of a rock are *self-avoiding*, this measure of tortuosity seems to be less relevant in petrophysics. The *i*th segment of a path $\overline{P_{i-1}P_i}$ crosses the *j*th segment $\overline{P_{j-1}P_j}$ if and only if there exist two numbers $\lambda, \mu \in (0, 1)$ such that

$$\lambda P_{i-1} + (1 - \lambda)P_i = \mu P_{j-1} + (1 - \mu)P_j \tag{2.3}$$

The number of crossings, *NC*, is the number of pairs of line segments $\left\{\overline{P_{i-1}P_i}; \overline{P_{j-1}P_j}\right\}$ in the path $\{P_0, P_1, \ldots, P_n\}$ that meet this condition. As *NC* would increase with path length, a meaningful indicator of tortuosity is its normalized form, $\frac{NC}{unit\ path\ length}$.

Geographers measure the tortuosity of rivers with *sinuosity*, which is the ratio of the curvilinear length of the curve and the Euclidean distance between the end points of the curve. In other words, it is the "actual path length" divided by the "shortest path length" of a curve. The value starts from 1 (for a straight line) and can be arbitrarily large.

In actual studies of rivers, this is is approximated with the *sinuosity index SI*, defined as

$$SI = \frac{channel\ length}{downvalley\ length}, \tag{2.4}$$

where the *downvalley path* in the direction of *maximum downslope* is not necessarily perfectly straight. Rivers are classified in terms of this index as (Mueller 1968): $1 \leq SI < 1.05$ *almost straight*; $1.05 \leq SI < 1.25$ *winding*; $1.25 \leq SI < 1.50$ *twisty*; $1.50 \leq SI$ *meandering*.

A single *meander* consisting of two inverted half circles (⌒ and ◡)has a sinuosity

$$S = \pi/2 \approx 1.5708 \tag{2.5}$$

independently of the radius r. Continue this construction over a long straight line segment $[0, 2R]$, where $2R = N \cdot (2r), N \gg 1$. The total meandering path from 0 to $2R$ contains N replicas of the elementary meanders, that is its total length is $N \cdot 2r\pi$. The *sinuosity* is

$$S = \frac{N \cdot 2r\pi}{2R} = \frac{\left(\frac{R}{r}\right) \cdot 2r\pi}{2R} = \pi. \tag{2.6}$$

Stølum (1996) claimed that the average sinuosity for rivers is $S = \pi$, and this would indicate that "river meandering is a *Self-Organization Process*", but later studies have proved (Grime 2015) that, over a large number of rivers, $\langle S \rangle < 2$.

In this Chapter we shall use the concept of *geometrical tortuosity* τ_g (later on simply denoted by τ), defined as the ratio of the effective length of flow paths to the straight-line distance L in the flow direction (Adler 1992; Clennell 1997; Ghanbarian et al. 2013). It is given by

$$\tau_g = L_g/L, \tag{2.7}$$

where L_g is the shortest path in the direction of flow passing completely in pores, and L is the length of the direct path. Note that this is the reciprocal of *straightness index* (*StI*, see Eq. 2.1), and the same as *sinuosity* S. (Korvin 2016) defined tortuosity independently of flow direction, as the *expected ratio of hydraulic path length to Euclidean length between two randomly selected points, both in pore space.* Some petrophysical problems where tortuosity is utilized are compiled in Table 2.1.

The simplest rock models of tortuous flow paths (Shen and Chen 2007) consisted of a collection of *randomly oriented capillaries* cutting through the solid frame. Different capillary models yielded $\tau = \sqrt{2}$ (Petersen 1958) or $\tau = \sqrt{3}$ (Bhatia 1985; Dykhuizen and Casey 1989). However, measured values of τ were either smaller or larger than these two constants, and it has been reported that tortuosity correlates with porosity Φ (Archie 1942; van Brakel and Heertjes 1974). As (Boudreau 1996) noted, any physically meaningful tortuosity versus porosity relation $\tau = \tau(\Phi)$ should satisfy the following three requirements:

$$1^{\text{st}} \text{ requirement} \quad \tau = \tau(\Phi) \geq 1 \tag{2.8a}$$

$$2^{\text{nd}} \text{ requirement} \quad \lim_{\Phi \to 1} \tau(\Phi) = 1 \tag{2.8b}$$

$$3^{\text{rd}} \text{ requirement} \quad \tau(\Phi) \to \infty \text{ if and only if } \Phi \to 0. \tag{2.8c}$$

Table 2.1 Some uses of tortuosity of relevance to rock physics

Topic	References
Adsorption	Sing and Williams (2004)
Animal movement	Estevez and Christman (2006), Miller et al. (2011)
Anisotropy	Alam et al. (2006), Backeberg et al. (2017), Dias et al. (2006), Kim et al. (1987), Ohkubo (2008), Zhang et al. (2003)
Building materials	Hu et al. (2012), Nikitsin and Backiel-Brzozowska (2013)
Capillary phenomena	Bhatia (1985), Nikitsin and Backiel-Brzozowska (2013)
Carbonate reservoir rocks	Azar et al. (2008)
Catalyst	Beekman (1990)
Cement	Cao et al. (2019)
Chemical weathering	Navarre-Sitchler et al. (2009)
Concrete	Georgali and Tsakiridis (2005), Lu (1997)
Conductivity (electric)	Cai et al. (2017), De La Rue and Tobias 1959), Ioannidis et al. (1997), Klinkenberg (1951), Moctezuma-Berthier et al. (2002), Mualem and Friedman (1991), Niya and Selvadurai (2018), Schwartz et al. (1995), Stenzel et al. (2016), Weerts et al. (1999, 2001)
Contaminant diffusion	Boving and Grathwohl (2001)
Convective heat transfer	Guo and Zhao (2005)
Corrugated Pore Structure Model	Salmas and Androutsopoulos (2001)
Diagenesis	Berner (1980)
Dielectric constant	Bruggeman (1935)
Diffusion	Andrews and Bennett (1981), Bear (1972), Bhatia (1985), Bini et al. (2019), Boudreau (1996), Bruggemann (1935), du Plessis et al. (2010), Epstein (1989), Garrouch et al. (2001), Grathwohl (2012), Iversen and Jørgensen (1993), Kim and Chen (2006), Kim and Torquato (1992), Klinkenberg (1951), Kreamer et al. (1988), Krom and Berner (1980), Li and Gregory (1974), Mackie and Meares (1955), Malek and Coppens (2003), Manheim and Waterman 1974), Masís-Meléndez et al. (2014), Moldrup et al. (2001), Ostadi et al. (2010), Perkins and Johnston (1963), Petersen (1958), Rayleigh (1892), Shen and Chen (2007), Sweerts et al. (1991), Ullman and Aller (1982), Van Brakel and Heertjes (1974), Weerts et al. (2001), Weissberg (1963), Zalc et al. (2004)
Diffusion of electrolytes	Mackie and Meares (1955)
Diffusive transport	Armatas (2006)
Dissolution of calcite	Archer et al. (1989)
Electrochemistry	Tjaden et al. (2018)
Electrodes	Chung et al. (2013)
Electro-osmosis	Coelho et al. (1996)
Filtration efficiency	Griffiths et al. (2020)

(continued)

Table 2.1 (continued)

Topic	References
Flow (hydraulic)	Abbas (2011), Adler (1992), Adler et al. (1990), Allen and Sun (2017), Carman (1937, 1956), David (1993), De Lima and Niwas (2000), Duda et al. (2011), Epstein (1989), Koponen (1998), Koponen et al. (1996), Kutay et al. (2006), Lemaitre and Adler (1990), Luo et al. (2014), Matyka et al. (2008), Nabovati and Sousa (2007)
Flow in a rock joint	Belem et al. (2000), Lomize (1951), Louis (1969, 1974)
Fluid transport	Bultreys et al. (2016), Hu et al. (2012)
Gas diffusion	Martínez et al. (2008), Millington (1959)
Gas flow	Degruyter et al. (2010), Hosticka et al. (1998)
Gas transport	He et al. (2013)
Heat conductivity	Bruggeman (1935), Wiegmann and Zemitis (2006)
Heat flow	Montes et al. (2007), Li and Chyu (2003), Olives and Mauran (2001)
Hydraulic conductivity	Hunt (2006), Masís-Meléndez et al. (2014), Vervoort and Cattle (2003)
Ion diffusion	Low (1981)
Mass transfer	Li and Chyu (2003), Satterfield (1970)
Membrane's distillation performance	Karanikola et al. (2017)
Molecular diffusion	Sweerts et al. (1991)
Mortar	Schwartz et al. (1995)
Mud	Boudreau and Meysman (2006
Multiphase flow	Herring et al. (2018)
NMR	Yang et al. (2019)
Percolation	Ghanbarian et al. (2023), Ioannidis et al. (1997), Korvin (1992a, b), Sahimi (1994), Shanti et al. (2014)
Permeability	Ahmadi et al. (2011), Allen and Sun (2014), Backeberg et al. (2017), Berg (2014), Borujeni et al. (2013), Carman (1939), de Lima (1995), Fu et al. (2020), Guan et al. (2019), Javadpour (2009), Jin et al. (2004), Keehm and Mukerji (2004), Koponen et al. (1997, 1998), Korvin (1992a, b), Kozeny (1927), Moctezuma-Berthier et al. (2004), Moldrup et al. (2001), Mota et al. (2001), Niya and Selvadurai (2017, 2018), Peng et al. (2014), Piller et al. (2009), Srisutthiyakorn and Mavko (2017), Sun and Wong (2018), Sun et al. (2011), Yu and Cheng (2002
Permeability under stress	Takahashi et al. (2012)
Pharmacy	Wu et al. (2006)
Porosity	Ghanbarian et al. (2013)
Pressure drop measurements	Comiti and Renaud (1989)
Relative permeability	Keehm and Mukerji (2004)

(continued)

Table 2.1 (continued)

Topic	References
Resistivity (electric)	Archie (1942), Bruggeman (1935), Li et al. (2017), Liu and Kitanidis (2013), Montes et al. (2007), Sun and Wong (2018), Winsauer et al. (1952), Wyllie and Rose (1950), Wyllie and Spangler (1952)
Secondary porosity	Perez-Rosales and Luna (2004, 2005)
Solute diffusion	Allaire et al. (2002), Dykhuizen and Casey (1989), Hunt and Skinner (2008), Keller et al. (2015)
Sulfate reduction	Goldhaber et al. (1977), Jorgensen (1978)
Surface diffusion	Ho and Strieder (1981)
Thermal conductivity	Thovert et al. (1990)
Transport	Adler (1992), Akanni et al. (1987), Allen and Sun (2017), Bear and Bachmat (2012), Bird (2002), Bodla et al. (2010), Brodkey and Hershey (2003), Civan (2011), Coleman and Vassilicos (2008), Fredrich et al. (2006), Mauret and Renaud (1997), Nakashima and Watanabe (2002), Neale and Nader (1973), Ning et al. (2016), Tomadakis and Sotirchos (1983)
Ultrasound	Berryman (1981), Bofiglio and Pompoli (2007), Zielinski (2012)
Viscous and thermal characteristic lengths	Fellah et al. (2013)
Water retention	Weerts et al. (1999)
Water saturation	de Lima (1995)

Hundreds of functional forms $\tau(\Phi)$ have been proposed, and the research is still going on. (According to Allen and Sun 2017, "the development of tortuosity-porosity trends remains to be an insightful topic of research".) Table 2.2 (after Boudreau 1996; Shen and Chen 2007; Kong et al. 2015; Allen and Sun 2017; Sobieski and Lipiński 2017; Hamad et al. 2018) shows some of these formulae in historical order.

We must repeat here the warning of Sobieski and Lipiński (2017), who published a similar Table, "in practice, it is impossible to indicate which equation in the Table should be used in a specific case". Obviously, those equations which have adjustable free parameters provide a more reasonable fit.

It is a disturbing feature of Table 2.2 that some of the equations describe *tortuosity* τ in terms of porosity Φ (such as $\tau = \left(1 + \Phi^{-1}\right)/2$, Eq. 2.26), while some others relate *squared tortuosity* to porosity (as e.g. $\tau^2 = 1 - \ln \Phi/2$, Eq. 2.27). Frequently, the same relation is referred to in technical literature in two different forms. Fu et al. (2021) write about the Bruggeman relation: "The Bruggeman correlation is one of the most commonly used porosity-tortuosity relationships, which is expressed as an exponential equation, given by $\tau^2 = \Phi^{1-\alpha}$", while the same relation is introduced in (Hutzenlaub et al. 2013) as "In the past, the most common approach to calculate tortuosity was to apply the Bruggeman equation and thus express the tortuosity τ as a function of the porosity ε: $\tau = \varepsilon^{1-\alpha}$, α is defined as the Bruggeman exponent".

Table 2.2 History of the proposed porosity-tortuosity relations

Year	Author	Model	Equation	Eq. no. and notes
1881	Maxwell	Array of spheres in 3D	$\tau = 1.5 - 0.5\Phi = 1 + \frac{1}{2}(1 - \Phi)$	(2.9)
1892	Rayleigh	Obstacles in rectangular order	$\tau = 2 - \Phi$	(2.10)
1935	Bruggeman		$\tau^2 = \Phi^{1-\alpha}$	(2.11) α empirical constant, $\alpha \approx 1.5$ for equally sized spheres
1937	Carman	Granular bed	$\tau = 0.67\Phi^{-1}$	(2.12)
1942	Archie		$\tau^2 = (\Phi F)^n$, more generally $\tau^2 = \left(A\Phi^{1-m}\right)^n$	(2.13a, b) F formation factor, A, m, n free parameters
1950	Wyllie		$\tau^2 = \Phi F$	(2.14) F formation factor
1955	Mackie and Meares	Diffusion of electrolytes in membrane	$\tau = \left(\frac{2-\Phi}{\Phi}\right)^2$	(2.15)
1958	Petersen	Sand	$\tau = 1 + 52.46P\varphi\left(1 - \Phi^{0.0135}\right)$	(2.16) P shape factor, φ sphericity, Φ porosity
1959	Millington	Monodisperse sphere packings	$\tau^2 = \Phi^{-1/3}$	(2.17)
1960	Amyx et al.		$\tau = (F_R\Phi)^x$	(2.18)F_R formation factor, $0.5 \le x \le 1.5$
1963	Weissberg	Bed of uniform spheres	$\tau = 1 - \frac{1}{2}\ln\Phi$; more generally $\tau = 1 - C\ln\Phi$	(2.19a, b) C free parameter
1965	Petersen	Randomly oriented capillaries	$\tau = \sqrt{2} \approx 1.44$	(2.20)
1972	Bear	Capillary model for a porous medium	$\tau = \Phi^{-3/2}$; more generally $\tau = \Phi^{-C}$	(2.21a, b) C free parameter

(continued)

Table 2.2 (continued)

Year	Author	Model	Equation	Eq. no. and notes
1973	Neale and Nader	Homogeneous isotropic spheres	$\tau = (3 - \Phi)/2$	(2.22)
1974	van Brakel and Heertjes	Monodisperse sphere packings	$\tau^2 = \Phi^{-1/3}$	(2.23)
1974	van Brakel and Heertjes	Overlapping spheres	$\tau^2 = 1 - ln\Phi/2$	(2.24)
1979	Lerman	Sands, mud	$\tau^2 = \left(A\Phi^{1-m}\right)^n$	(2.25) A, m, n free parameters
1981	Berryman	Fluid-saturated porous media	$\tau = \left(1 + \Phi^{-1}\right)/2$	(2.26)
1981	Ho and Strieder	Overlapping spheres	$\tau^2 = 1 - ln\Phi/2$	(2.27)
1981	Low	Ion diffusion in clay	$\tau^2 = \Phi + B(1 - \Phi)$	(2.28) B free parameter
1982	Ullman and Aller	Sands, mud	$\tau^2 = \left(A\Phi^{1-m}\right)^n$	Same as (2.25) A, m, n free parameters
1983	Tomadakis and Sotirchos	Overlapping cylinders	$\tau^2 = 1 - ln\Phi$	(2.29)
1985	Bhatia	Randomly oriented capillaries	$\tau = \sqrt{3} \approx 1.73$	(2.30)
1987	Kim et al.	Isotropic porous system	$\tau = \Phi^{-0.4}$	$0 < \Phi < 0.5$ (2.31)
1987	Akanni et al.	Overlapping spheres	$\tau^2 = 1 - ln\Phi/2$	(2.32)
1989	Comiti and Renaud	Parallelepiped particles	$\tau = 1 + p \cdot ln(1/\Phi)$	(2.33) p shape factor
1989	Dykhuizen and Casey	Randomly oriented capillaries	$\tau = \sqrt{3} \approx 1.73$	(2.34)
1990	Beekman	Heterogeneous catalyst	$\tau^2 = \frac{\Phi}{\left(1-(1-\Phi)^{\frac{1}{3}}\right)}$	(2.35)
1991	Du Plessis and Masliyah	Idealized granular pore structure	$\tau = \frac{\Phi}{1-(1-\Phi)^{2/3}}$	(2.36)

(continued)

Table 2.2 (continued)

Year	Author	Model	Equation	Eq. no. and notes
1993	Iversen and Jørgensen		$\tau^2 = 1 + q(1 - \Phi)$	(2.37) q depends on type of porous media
1993	Iversen and Jørgensen	Soils, catalysts	$\tau^2 = \Phi + B(1 - \Phi)$	(2.38) B free parameter
1993	Iversen and Jørgensen	Sandy marine sediments	$\tau = \sqrt{1 + 2(1 - \Phi)}$	$0.4 < \Phi < 0.9$ (2.39)
1996	Boudreau	Fine-grained unlithified sediments	$\tau^2 = \left[1 + \frac{32}{9\pi}(1 - \Phi)\right]^2 \approx (2 - \Phi)^2$	(2.40)
1996	Boudreau	Fine-grained unlithified sediments	$\tau^2 = 1 - \ln(\Phi^2)$; more generally $\tau^2 = 1 - C \ln \Phi$	(2.41a, b) C free parameter
1996	Koponen et al.	2D overlapping monosized squares	$\tau = 1 + 0.8(1 - \Phi)$; more generally $\tau = 1 + p(1 - \Phi)$	$0.5 < \Phi < 1$; (2.42a, b) p free parameter
1997	Koponen et al.	2D random overlapping monosized squares	$\tau = 1 + 0.65 \frac{1-\Phi}{(\Phi - \Phi_c)^{0.19}}$	$0.5 < \Phi < 1$; Φ_c is critical porosity (2.43)
1997	Koponen et al.	2D overlapping monosized squares	$\tau = p \cdot \frac{(1-\Phi)}{(\Phi - \Phi_c)^m}$	(2.44) $\Phi_c \approx 0.33$; p, m are free parameters
1998	Koponen	2-dimensional squares	$\tau = 0.8(1 - \Phi) + 1$	(2.45)
1998	Pape et al.	Fractal pore geometry	$\tau = 0.67\Phi^{-1}$	(2.46)
2001	Boving and Grathwohl	Sedimentary rocks	$\tau = \Phi^{-1.2}$	(2.47)
2001	Salmas and Androutsopoulos	Corrugated pore structure model	$\tau_{CPSM} =$ $1 + A\left[\frac{D_{\max.eff} - D_{\min.eff}}{D_{\min}}\right].(N_S - 2)^a$	(2.49) A and a are free parameters. For the other symbols see Eqs. 2.77 and 2.78

(continued)

Table 2.2 (continued)

Year	Author	Model	Equation	Eq. no. and notes
2003	Li and Chyu	Tubular	$\log \tau = -0.151 - 0.5333 \log \Phi$	(2.50)
2004	Yu and Li	2D square particles	$\tau = \left[1 + \frac{\alpha}{2} + \frac{\sqrt{\left(\frac{1}{\alpha}-1\right)^2 + \frac{1}{4}}}{1-\alpha}\right]$	(2.51) $\alpha = \sqrt{1 - \Phi}$
2005	Yu	Fractal scaling of a square shaped grains model	$\tau = \left(\frac{L}{\lambda_{\min}}\right)^{D_T - 1}$	(2.52) D_T is "tortuosity dimension"
2006	Dias et al.	Granular and spherical packings	$\tau = \Phi^{-n}$	(2.53) Porosity range 0.4–0.5
2007	Nabovati and Sousa	Lattice Boltzmann Method	$\tau = -0.519\Phi^3 + 0.879\Phi^2 - 1.1567\Phi + 1.8058$	(2.54)
2008	Matyka et al.	Porous media flow	$(\tau - 1) \propto R\frac{S}{\Phi}$	(2.55) R is hydraulic radius, S specific surface area
2008	Matyka et al.	2D overlapping mono-sized squares	$\tau = 1 - 0.77 \ln \Phi$	(2.56)
2010	Lanfrey et al.	Randomly packed identical particles	$\tau = 1.23\frac{(1-\Phi)^{4/3}}{\xi^2 \Phi}$	(2.57) ξ sphericity, $\xi = 1$ for sphere and $\xi < 1$ for non-spherical particles
2010	Lanfrey et al.	Randomly packed identical particles	$\tau = \left(\frac{1}{\Phi^3} - \frac{1}{\Phi^2}\right)^{4/3}$	(2.58)
2010	du Plessis et al.	Loose porous structure	$\tau = \frac{\Phi}{1-(1-\Phi)^m}$	$\begin{cases} \mathit{for}\ 2D\ m = 1/2 \\ \mathit{for}\ 3D\ m = 2/3 \end{cases}$ (2.59)

(continued)

Table 2.2 (continued)

Year	Author	Model	Equation	Eq. no. and notes
2011	Duda et al.	2D freely overlapping squares	$\tau \propto 1 + (1 - \Phi)^{1/2}$	(2.60)
2011	Pisani	Random, partially overlapping shapes	$\tau = \frac{1}{1-\alpha(1-\Phi)}$	(2.61) α = shape factor
2011	Abbas	Monosize spherical packing	$\tau = 1.5312 - 0.5268\Phi^{0.8984}$	(2.62)
2012	Tang et al.	Cubic particles	$\tau = \frac{3}{4} + \frac{1}{8}\sqrt{1 + \frac{1}{4} \cdot \frac{1-\Phi}{2-\Phi-2\alpha}}$ $+\frac{1}{8}\sqrt{1 + \frac{1-\Phi}{2-\Phi-2\alpha}} + \frac{1}{4}\sqrt{1-\Phi}$	(2.63) $\alpha = \sqrt{1-\Phi}$
2013	Liu and Kitanidis	Isotropic spherical grains	$\tau = \Phi^{0.28} + 0.15$	(2.64)
2013	Nikitsin and Backiel-Brzozowska	Capillary tortuosity	$\tau = \left(\frac{1-0.15}{\Phi-0.15}\right)^{0.4}$	(2.65)
2014	Yong et al.	Voronoi channel geometries	$\tau = \left(0.75\Phi^{-1.0819}\right)^{1/2}$	$0.2 < \Phi < 0.6$ (2.66)
2014	Toray Research Center, Inc.		$\frac{l_e}{l_d} = \tau = \sqrt{\frac{\Phi}{2k}} \cdot \left(\frac{V_p}{S_{BET}}\right)$	(2.67) l_e: Pore pathway, l_d: Sample thickness; V_p: Specific volume of pore, Φ porosity, S_{BET}: Specific surface area by BET method; k (Darcy) air permeability

(continued)

Table 2.2 (continued)

Year	Author	Model	Equation	Eq. no. and notes
2016	Korvin	Fractal geometry	$\left\langle \frac{L_{hydr}}{L} \right\rangle = \tau =$ $\Phi + \frac{(1-\Phi)}{Z}\left(\frac{P_0}{d_0}\right)\left(\frac{\sqrt{A}}{d_0}\right)^{D_{P/A}}$	(2.68) d_0, P_0, A_0 Characteristic *diameter*, characteristic *perimeter*, characteristic *area* of the grains (in the 2D section); Z average number of pores adjacent to a grain (in the 2D section)
2016	Korvin	Method of Moments	$\langle \tau \rangle = \frac{1}{\langle cos\vartheta \rangle}$	(2.69) $\cos\vartheta$ is average direction cosine of the major axes of the pores on the 2D microscopic image
2018	Hamad et al.	Porous media	$\tau = 1 + C_1 P \varphi \left(1 - \Phi^{C_2}\right)$	(2.70) Porosity (Φ), shape factor (φ), sphericity factor (P); C_1,C_2 free parameters

Updated, after Boudreau (1996): Table 1; Shen and Chen (2007): Table 1; Kong et al. (2015): Table 2; Allen and Sun (2017): Table 2; Sobieski and Lipiński (2017): Table 1; Hamad et al. (2018)

Boudreau (1996) notes, "When comparing this [diffusion theoretical] treatment with those in other fields (e.g., engineering and hydrology), some confusion may arise because of differences in the definition of tortuosity and the use of a related term, the *tortuosity factor*. Specifically, Carman (1937) defined the *tortuosity factor* as $(dL/dx)^2$ or θ^2. However, this distinction is not always made, with some authors labeling θ^2 as the *tortuosity*."

To put Boudreau's remark in the proper context, recall the *Fickian theory of solute diffusion* (Shackelford and Moore 2013). The *mass flux* of a chemical species in aqueous solution due to diffusion is described by *Fick's first law*, which for the 1-D case states that

$$J_d = -D_0 \frac{\partial C}{\partial x} \tag{2.71}$$

where J_d is the rate of change in mass of the chemical species per unit cross-sectional area perpendicular to the direction of transport, C is the aqueous-phase concentration of the chemical species, x is position in the direction of transport, and D_0 is the *free-solution diffusion coefficient*. For diffusion in porous media Fick's first law is modified to

$$J_d = -\frac{\theta D_0}{T_m} \cdot \frac{\partial C}{\partial x} \tag{2.72}$$

where θ is volume of water per unit volume of the porous medium, and T_m is the *geometric tortuosity factor*:

$$T_m = \left(\frac{L_{eff}}{L}\right)^2, \tag{2.73}$$

where L is the straight-line distance between two points within a porous medium, and L_{eff} is the length of the actual zig-zag path between the same points.

Gommes et al. (2009: Appendix A) also discuss this problem. They point out that the effective diffusivity D_{eff} of a molecule in a porous solid is given by (Adler 1992; Clennell 1997; Satterfield 1970)

$$\frac{D_{eff}}{D_m} = \frac{\Phi}{\tau_g^2}, \tag{2.74}$$

where D_m is the molecular diffusivity *unhindered by any pore wall*, Φ is porosity, and τ is the *geometric tortuosity* of the pore space. They warn, "In the literature, the tortuosity is sometimes introduced [in Eq. 2.74] with an exponent 1; it has to be noted, however, that in this case, it is merely used as a *tortuosity factor* with no precise geometrical interpretation."

We remind the Reader that the Kozeny-Carman equation (Carman 1956) for the permeability of porous sedimentary rocks, dealt with in Chap. 1,

$$k = \frac{1}{b}\Phi^3 \frac{1}{S_{spec}^2} \frac{1}{\tau^2} \tag{2.75}$$

also contains the *tortuosity factor* $1/\tau^2$, where τ is the *geometrical tortuosity*.

The role of the tortuosity factor in the diffusion Eq. (2.72) and the permeability Eq. (2.75) is to compensate for the elongation in path length due to the irregular bendings of the flow channels. This, however, does not provide full compensation in case of channels in which the cross sectional area varies. In such cases the geometric tortuosity should be corrected taking into account the correlation (Selvadurai 2000; Niya and Selvadurai 2017) between *flow rate* and *cross-sectional area* in a *circular pipe* (3-D) and *a parallel-walled cannel* (2-D), respectively, using the equations (Niya and Selvadurai 2018)

$$\tau_{c,3D} = \int_{s=0}^{s=s_{\max}} \left(\frac{w_s}{d(s)}\right)^4 ds; \; \tau_{c,2D} = \int_{s=0}^{s=s_{\max}} \left(\frac{w_s}{h(s)}\right)^3 ds, \tag{2.76a, b}$$

where integration is performed along the length of the path. In 2-D l_s is sample thickness, $h(s)$ the width of the channel, $s_{\max}$ the length of the tortuous path. In 3-D, $d(s)$ is the diameter of the circular cross section. In both equation, w_s is a characteristic length scale. (Niya and Selvadurai 2017) reported that this correction yields reasonable permeability values. If the inflow and outflow sides of the sample are connected through several parallel paths, Kirchoff's law should be applied.

Salmas and Androutsopoulos (2001) introduced the *corrugated pore structure model* (CPSM) to describe channels with varying cross-section. They neglected the curvature of the paths, and estimated the tortuosity τ_{CPSM} based solely on the distribution of the random cross sections. They found, empirically, that

$$\tau_{CPSM} = 1 + A\left[\frac{D_{\max .eff} - D_{\min .eff}}{D_{mean}}\right].(N_S - 2)^a, \tag{2.77}$$

where A, a are adjustable parameters, the pore diameters D are distributed according to the distribution function $F_V(D)$, $D_{\max .eff}$ and $D_{\min .eff}$ are defined by

$$\int_{D_{\max .eff}}^{D_{\max}} F_V(D)dD = \int_{D_{\min}}^{D_{\min .eff}} F_V(D)dD = 0.025, \tag{2.78}$$

and N_S is a characteristic size parameter.

Xiao et al. (2013) added the effect of surface roughness to that of geometrical tortuosity for computing flow rate through a *rough joint*. According to Lomize (1951), Louis (1969) the flow rate through a rough joint is

$$Q = \frac{1}{[1 + A \cdot (\Delta/2d)]^{1.5}} \cdot \frac{wd^3}{12\mu} \cdot \nabla p, \tag{2.79}$$

where Q is the *volumetric flow rate*; d is the *joint's mean aperture*; $\Delta/2d$ is *relative roughness*, ∇p is *pressure gradient*; w is *joint width*; μ is the fluid's *kinematic viscosity*; A is an empirical constant. Let τ be the *geometric tortuosity* of the joint, and denote by T_s the *surface tortuosity*, a concept proposed by Belem et al. (2000). For a tortuous joint, Xiao et al. (*op. cit.*) generalised Eq. (2.79) as

$$Q = \frac{1}{\tau^2 T_s^2 [1 + A \cdot (\Delta/2d)]^{1.5}} \cdot \frac{wd^3}{12\mu} \cdot \nabla p. \tag{2.80}$$

Here, the surface tortuosity T_s is defined as the ratio of the area of the rough fracture surface and the area A_p of the surface determined by the four corner points, delimiting the measurement zone. Belem et al. (2000) called the surface determined by M_1, M_2, M_3, M_4 the π-*surface*. Generally, the four points M_1, M_2, M_3, M_4 are not co-planar, and the area A_p should be measured on the *best fitting mean plane* belonging to the four points. The surface tortuosity is given by

$$T_s = \frac{A_t}{A_p} = \frac{A_t}{A_n} \cos\phi = R_s \cos\phi \tag{2.81}$$

with $0 < T_s < R_s$; $0 < \cos\phi \leq 1$, ϕ is the angle between the normal of the π-*surface* and the normal of the mean plane of the joint surface.

2.2 Mathematical Models of Tortuosity

Allen and Sun (2017) assure us that "the development of tortuosity-porosity trends remains to be an insightful topic of research". I deeply agree with this statement, and will show the Readers a few techniques borrowed from mathematics and physics that can be used to attack this problem, namely: (a) a random walk model (Boudreau and Meysman 2006); (b) a percolation model (Ghanbarian et al. 2023); (c) a fractal scaling argument (Korvin 2016); (d) the method of moments (Korvin 2016).

2.2.1 Tortuosity from Porosity Using Random Walk Around Mud Flakes

Boudreau and Meysman (2006) offered a simple geometric model for the dependence of tortuosity on porosity in marine muds. The model consists of rows of non-overlapping parallel disks of thickness δ, there is small gap of width ε, where

$\varepsilon \ll \delta$, between the rows. All disks have equal size, with radius

$$a = m\delta \tag{2.82}$$

where m is the radius to thickness ratio. Denote by A_p the total area of pores in a layer of disks and by A_d the total area of solid disks in the same layer. For the model, the *areal porosity*, Φ_a and volumetric porosity, Φ_v are approximately equal (assuming $\varepsilon \ll \delta$):

$$\Phi_a = \frac{A_p}{A_d + A_p} = \frac{\delta A_p}{\delta A_d + \delta A_p} = \Phi_v \approx \Phi. \tag{2.83}$$

Let us now suppose there exists a random walker, small in size compared to a disk that can fit into the small horizontal gaps (the "*ant in the labyrinth*" of de Gennes 1976), which attempts to move through the pores vertically in the direction of the gradient, say upwards. If it founds itself in a pore between disks, it will move up until it reaches a solid disk. Then it will choose a *random horizontal direction* and walks in the gap along the disk surface, until the next pore where it can further move upward. Let the walker start from a random point P in a thin layer separating the disks. With probability Φ the point P is in *free space* (below a pore, *case* 1), with probability $(1 - \Phi)$ it is below a disk (*case* 2). In *case* 1, the expected path length to be taken in order to reach the bottom of the next disk-layer is

$$l_f = \delta \tag{2.84}$$

(because $\varepsilon \ll \delta$). To deal with *case* 2, we shall need a geometrical theorem (Boudreau and Meysman 2006):

Theorem 2.1 *Consider a circle with radius a. The mean distance between all posible points P within the circle and all posible points C on the circumference is given by*

$$\langle \Delta \rangle = \langle dist(C, P) \rangle = \frac{32a}{9\pi} \approx 1.1318a. \tag{2.85}$$

Proof In the polar coordinate system (r, σ), r measured on axis x from $x = 0$, $\measuredangle\sigma$ measured from the x axis anti-clockwise, the equation of the circle is

$$r_c = 2a\cos\sigma, \tag{2.86}$$

where a is the radius of the disk and r_c is a point on the circle. On a random radial line, originating from C, choose a point P inside the circle. Then $dist(C, P) = r$ and its mean value is given by

$$\langle \Delta \rangle = \langle dist(C, P) \rangle = \frac{\int_{-\pi/2}^{\pi/2} \int_0^{2a\cos(\sigma)} r^2 dr d\sigma}{\int_{-\pi/2}^{\pi/2} \int_0^{2a\cos(\sigma)} r dr d\sigma} = \frac{\frac{32a^3}{9}}{\pi a^2} = \frac{32a}{9\pi} \approx 1.1318a \tag{2.87}$$

By this theorem, when our walker emerges from below under a disk (*case* 2), the average path to circumnavigate this obstacle and reach the next interlayer is

$$l_d = \delta\left[1 + \frac{32m}{9\pi}\right], \tag{2.88}$$

where $m\delta = a$. Using Eqs. (2.84) and (2.88), the expected path length for the walker to get one layer higher is

$$\bar{l} = \Phi l_f + (1 - \Phi)l_d = \Phi\delta + (1 - \Phi)\delta[1 + 32m/(9\pi)], \tag{2.89}$$

which yields the tortuosity estimate

$$\tau = \frac{\bar{l}}{l_f} = \frac{\bar{l}}{\delta} = \Phi + (1 - \Phi)[1 + 32m/(9\pi)]. \tag{2.90}$$

The best fit of Eq. (2.90) to the measured tortuosity data in Boudreau (1996) is obtained for $m \approx 1 \pm 0.06\left(R^2 = 0.65\right)$, that is $\tau \approx 1 + \frac{32}{9\pi}(1 - \Phi) \approx 2 - \Phi$ (2.91).

2.2.2 A Percolation Model

In a recent paper, Ghanbarian et al. (2023) advocate *Percolation Theory* (Sahimi 1994; Stauffer and Aharony 1994; Bunde and Havlin 1996; Hunt and Skinner 2008) as a physically meaningful way to estimate the tortuosity of flow paths in porous rocks. Percolation Theory has three basic variants: *bond*- and *site percolation* (for discrete systems), and *continuum percolation* (for more general systems). The pore space is modeled as a 2-D or 3-D regular lattice where *pore bodies* are substituted by vertices ("nodes", "sites"), and *pore throats* become *bonds*. (In practice, distinguishing pore bodies from pore throats on a digitized binary, or gray-scale image is not a trivial task, but such image-processing aspects are outside the scope of this book.)

According to a basic result of Percolation Theory in an *infinite system*, the mean distance between two sites on the same *finite cluster*, occupied by the percolating fluid, called the correlation length χ, scales as

$$\chi = C \cdot |p - p_c|^{-\nu} \tag{2.91}$$

where C is a constant, p is the fraction of sites (or bonds) occupied by liquid, p_c is the percolation threshold. The value of the exponent ν is 4/3 in 2-D and 0.88 in 3-D (Stauffer and Aharony 1994). If the *percolating path* is made up of segments of length ε ($\varepsilon \ll 1$, usually taken as equal to bond length), its length $L(\varepsilon)$ measured in units of ε scales as (Mandelbrot 1983)

$$L(\varepsilon) = L_s^{D_x}\varepsilon^{1-D_x} \tag{2.92}$$

where L_s is system size, and the fractal dimension D_x depends on the type of percolation (*site nontrapping invasion percolation*; *site trapping invasion percolation*; *bond trapping invasion percolation*; *random percolation*; the path can be *shortest path* or *energetically optimal path*, all these concepts are defined and discussed in Cieplak et al. (1996), Porto et al. (1997), Lee et al. (1999), Ghanbarian et al. (2023). At the percolation threshold the correlation length χ diverges, while the step (bond length) remains constant. Instead, we can use an equivalent, alternative convention, where the system size is kept constant and the steps tend to zero as

$$\varepsilon \propto \chi^{-1}. \tag{2.93}$$

This yields the scaling relation

$$L(\chi) \propto L_s^{D_x} \chi^{D_x - 1} \propto L_s^{D_x} \left(|p - p_c|^{-\nu} \right)^{D_x - 1}, \tag{2.94}$$

But χ is mean distance between any two sites in the finite system, $L(\chi)$ is average length of the geometrical path $\langle L_g \rangle$, so the geometrical tortuosity is

$$\tau_g(p) = \frac{L(\chi)}{\chi} \propto |p - p_c|^{\nu - \nu D_x}. \tag{2.95}$$

Ghanbarian et al. (*op. cit.*) referred to Fisher's (1971) method of *finite-size scaling* and, instead of (2.95), proposed the following expression:

$$\tau_g(p) \propto \left| p - p_c + \left(\frac{C}{L_s} \right)^{1/\nu} \right|^{\nu - \nu D_x}. \tag{2.96}$$

In the limit of infinite system size $C \ll L_s$, Eq. (2.96) reduces to Eq. (2.95), for $p = p_c$ it predicts a power-law dependence of the tortuosity on system size. Hunt (2006) used the same Eq. (2.96), for the scale-dependence of the hydraulic conductivity in anisotropic networks.

2.2.3 Tortuosity Versus Porosity from Fractal Theory

Korvin (2016) showed with a simple scaling argument that the hydraulic tortuosity in a 2-D cross section of granular porous sedimentary rocks depends on grain size and grain shape. Let L be the vertical size of the section considered (assuming that hydraulic flow goes from top to bottom); Φ porosity; τ tortuosity (defined as the expected ratio of hydraulic path length to Euclidean length between 2 randomly selected points, always $\tau \geq 1$); d_0, P_0, A_0 characteristic *size* (diameter), characteristic *perimeter*, characteristic *area* of the grains (in the 2-D section); Z average number of pores adjacent to a grain (in the 2-D section). We shall denote by $D_{P/A}$ the exponent in Mandelbrot's *perimeter-area scaling law* (Mandelbrot 1983; Feder 1988; Korvin 1992a):

$$P = P_0 \left(\frac{\sqrt{A}}{d_0} \right)^{D_{P/A}}. \tag{2.97}$$

Theorem 2.2

$$\left\langle \frac{L_{hydr}}{L} \right\rangle = \tau = \Phi + \frac{(1-\Phi)}{Z} \left(\frac{P_0}{d_0} \right) \left(\frac{\sqrt{A}}{d_0} \right)^{D_{P/A}}. \tag{2.98}$$

Note that for $\Phi = 1$ we have $\tau = 1$; for $\Phi = 0$ there are no pores at all, that is $Z = 0$ and consequently $\tau = \infty$ as it should be. (The divergence of tortuosity in low-porosity, low permeability clay-bearing sandstones was noted in Korvin (1992a, b).

Equation (2.98) is proven by a heuristic scaling argument. Along a randomly selected top-to-bottom vertical line of length L by the *De-Lesse principle* (Oleschko 1998) a fraction ΦL of the line would go through pore space. Across these parts of the line the flow goes along straight line segments. The remaining $(1-\Phi)L$ fraction of the vertical line is filled by grains, the fluid path would cross $\frac{(1-\Phi)L}{d_0}$ grains if it could flow along a straight vertical line. But it cannot proceed straight, but every time the flow reaches a grain it changes direction and continues in a "throat" following the curvature of the grain's perimeter. By the definition of *the grain/pore coordination number* Z, the periphery P of a grain is adjacent to Z other grains, so that every individual "detour" adds a length $\left(\frac{P}{Z}\right)$ to the hydraulic path. This *detour* is, by Mandelbrot's rule (2.96) equal to $\left(\frac{P}{Z}\right) = \frac{P_0}{Z}\left(\frac{\sqrt{A}}{d_0}\right)^{D_{P/A}}$. As there are $\frac{(1-\Phi)L}{d_0}$ such *detours*, the total hydraulic length from top to bottom is $L_{hydr} = \Phi L + \frac{(1-\Phi)L}{Z}\left(\frac{P_0}{d_0}\right)\left(\frac{\sqrt{A}}{d_0}\right)^{D_{P/A}}$, what is the same as Eq. (2.98) to be proved.

2.2.4 Tortuosity Estimation with the Method of Moments

The moments of order (p, q) for any 2-D digitized object A having the gray value function $g(x, y)$ are given by

$$m_{p,q} = \iint_A x^p y^q g(x, y)dxdy \tag{2.99}$$

where the integral is over the area of the object (Akhiezer 1965). For a binary image

$$g(x, y) = F(x, y) = \begin{cases} 1 \ if \quad (x.y)\epsilon\Pi \\ 0 \ if \ (x, y) \notin \Pi \end{cases}$$

(where Π is the set of pores in the microscopic image of the rock), that is the moments are defined as $m_{p,q} = \iint_A x^p y^q F(x, y)dxdy$ (2.100). Special cases of Eq. (2.100) are (Teague 1980; Jähne 1997):

area of the object:

$$A = m_{0,0} = \iint\limits_A F(x, y)dxdy; \tag{2.100}$$

center of gravity (x_c, y_c) of the object

$$(x_c, y_c) : x_c = \frac{m_{1,0}}{m_{0,0}}; x_c = \frac{m_{0,1}}{m_{0,0}}. \tag{2.101}$$

If the pore shape is closely elliptical, the *second order moments*

$$m_{2,0} = \iint\limits_A x^2 F(x, y)dxdy; m_{0,2} = \iint\limits_A y^2 F(x, y)dxdy;$$

$$m_{1,1} = \iint\limits_A xy \, F(x, y)dxdy$$

are also needed to characterize the *size*, *shape*, and *direction* of the best-fitting ellipse, as follows:

semimajor axis of the ellipse:

$$a = \left(\frac{m_{2,0} + m_{0.2} + \left[\left(m_{2,0} - m_{0.2}\right)^2 + 4m_{1,1} \right]^{1/2}}{0.5m_{0,0}} \right)^{1/2}; \tag{2.102}$$

semiminor axis of the ellipse:

$$b = \left(\frac{m_{2,0} + m_{0.2} - \left[\left(m_{2,0} - m_{0.2}\right)^2 + 4m_{1,1} \right]^{1/2}}{0.5m_{0,0}} \right)^{1/2}; \tag{2.103}$$

tilt angle of the ellipse:

$$\phi = \tan^{-1}\left(\frac{2m_{1,1}}{m_{2,0} - m_{0.2}} \right) \tag{2.104}$$

(where ϕ is the angle between the x axis and the semimajor axis, and the principal value of the $\tan^{-1}$ function is selected in a way to assure. $-\frac{\pi}{2} \leq \tan^{-1} x \leq \frac{\pi}{2}$);

eccentricity of the ellipse:

$$\varepsilon = \frac{\sqrt{a^2 - b^2}}{a} = \frac{(m_{2.0} - m_{0.2})^2 - 4m_{1.1}^2}{(m_{2.0} + m_{0.2})^2}. \tag{2.105}$$

(Obviously, $0 < \varepsilon < 1$, it is zero for a round object, ≈ 1 for an elongated object.)

To apply the *method of moments* compute for every pore P_i, $(i = 1, 2, \ldots, n)$ the length l_i of its major axis and estimate the angle ϑ_i, $\left(-90^\circ \leq \vartheta_i \leq 90^\circ\right)$ what this axis makes with the horizontal reference direction X. Take two points, P, Q randomly and far from each other in the rectangle $XY = \{0 \leq x \leq X; 0 \leq y \leq Y\}$. Consider the 2D projection of a typical hydraulic path from P to Q consisting of line segments. In most cases the projected line segment would coincide with the longer axis $l_i = 2a_i$ of some pore in the XY plane. The broken line consisting of the segments l_i is the "skeleton" of a chain of touching pores, which are either in the XY plane or immediately below or above it in a thin slab S of thickness $\Delta : S = \left\{(x, y, z) | (x, y) \epsilon XY; -\frac{\Delta}{2} \leq z \leq \frac{\Delta}{2}\right\}$. Because of the local homogeneity of the rock sample, we assume that (i) the probability distributions of the major-axis lengths $\{l_i\}$ and of their directions $\{\vartheta_i\}$ are the same on the XY rectangle as in the slab S. We also assume that (ii) $\{l_i\}$, $\{\vartheta_i\}$ are independent random variables and that (iii) $\langle\vartheta_i\rangle = 0$, $\langle\sin\ \vartheta_i\rangle = 0$, as well as (iv) for $i \neq j$ $\langle\cos\vartheta_i \cos\vartheta_j\rangle = \langle\cos\vartheta\rangle^2$. "Expected value", in this context means *expected value over all statistically equivalent random realizations of the microscopic image.*

Next, we define a *random tortuosity* τ. Let $n \gg 1$, and construct a *random chain of n pores* starting from a fixed point P. We select pores randomly from among the actual pores figuring on the XY section, move them without rotation and any change in their shape into positions where they touch each other and form a continuous hydraulic path from P to some random endpoint what we call Q. A particular pore in the chain has a major axis of length l_i and direction angle ϑ_i, that is the X- resp. Y-projections of the major axis are $\xi_i = l_i \cos\vartheta_i$, $\eta_i = l_i \sin\vartheta_i$, $i = 1, 2, \ldots, n$. The tortuosity of this path is

$$\tau = \left\langle \frac{\sum_{i=}^{n} l_i}{\sqrt{\left(\sum l_i \cos\vartheta_i\right)^2 + \left(\sum l_i \sin\vartheta_i\right)^2}} \right\rangle \tag{2.106}$$

If $n \gg 1$, and we assume ergodicity, the expected (average) tortuosity over all possible chains consisting of n pores is the same as the expectation over all realizations. We get:

$$\tau = \left\langle \frac{\sum_{i=}^{n} l_i}{\sqrt{\left(\sum l_i \cos\vartheta_i\right)^2 + \left(\sum l_i \sin\vartheta_i\right)^2}} \right\rangle. \tag{2.107}$$

A simple and efficient estimation of $\langle\tau\rangle$ can be given as follows, using the notation $l = \langle l_i\rangle$ and referring to the assumptions (i)–(iv). Because of assumption (ii) the variables $\{l_i\}$ and $\{\vartheta_i\}$ are independent, and the expected value of the fraction is well approximated by its expected numerator divided by the expected denominator:

$$\langle\tau\rangle = \left\langle \frac{\sum_{i=}^{n} l_i}{\sqrt{\left(\sum l_i \cos\vartheta_i\right)^2 + \left(\sum l_i \sin\vartheta_i\right)^2}} \right\rangle \approx \frac{\left\langle \sum l_i \right\rangle}{\sqrt{\left\langle \left(\sum l_i \cos\vartheta_i\right)^2 + \left(\sum l_i \sin\vartheta_i\right)^2 \right\rangle}}$$
$$\approx \frac{nl}{l\sqrt{n(n-1)\langle\cos\vartheta\rangle^2}} \approx \frac{1}{\langle\cos\vartheta\rangle}. \tag{2.108}$$

where we used assumptions (i) to (iv) and took the limit $n \to \infty$. This approximate rule can be expressed in words as: *the average tortuosity is the reciprocal of the average direction cosine of the major axes of the pores on the 2D microscopic image.*

$$\langle\tau\rangle \approx \frac{1}{\langle\cos\vartheta\rangle}. \tag{2.109}$$

Equation (2.109) correctly expresses that always $\tau \geq 1$ and it also predicts that tortuosity can be different in the 3 mutually perpendicular directions *X, Y, Z*, which is in accordance with experimental findings (Takahashi et al. 2012) on anisotropic permeability, and is along the lines of the *TCT* (*tensorial connectivity-tortuosity*) concept of (Zhang et al. 2003).

References

Abbas MN (2011) Modeling of porosity equation for water flow through packed bed of monosize spherical packing. J Eng Sustain Dev (JEASD) 15(4):205–226

Adler PM (1992) Porous media geometry and transport. Butterworth-Heinemann, Boston, London, Oxford

Adler PM, Jacquin CG, Quiblier JA (1990) Flow in simulated porous media. Int J Multiph Flow 16(4):691–712

Ahmadi MM, Mohammadi S, Hayati AN (2011) Analytical derivation of tortuosity and permeability of monosized spheres: a volume averaging approach. Phys Rev E 83(2):026312

Akanni KA, Evans JW, Abramson IS (1987) Effective transport coefficients in heterogeneous media. Chem Eng Sci 42:1945–1954

Akhiezer NI (1965) The classical moment problem and some related questions in analysis. Hafner, New York

Alam P, Byholm T, Toivakka M (2006) Calculating tortuosity in quasi-random anisotropic packings. Nord Pulp Pap Res J 21(5):670–675

Allaire SE, Gupta SC, Nieber J, Moncrief JF (2002) Role of macropore continuity and tortuosity on solute transport in soils: 1. Effects of initial and boundary conditions. J Contam Hydrol 58(3–4):299–321

Allen R, Sun S (2014) Investigating the role of tortuosity in the Kozeny-Carman equation. In: International conference on numerical and mathematical modeling of flow and transport in porous media, Dubrovnik, Croatia, 29 September–3 October
Allen R, Sun S (2017) Computing and comparing effective properties for flow and transport in computer-generated porous media. Geofluids 2017. Article ID 4517259
Amyx JW, Bass DM Jr, Whiting RL (1960) Petroleum reservoir engineering. McGraw-Hill Book Co., New York
Andrews D, Bennett A (1981) Measurements of diffusivity near the sediment-water interface with a fine-scale resistivity probe. Geochim Cosmochim Acta 45:2169–2175
Archer D, Emerson S, Reimers C (1989) Dissolution of calcite in deepsea sediments: pH and O_2 electrode results. Geochim Cosmochim Acta 53:2831–2845
Archie GE (1942) The electrical resistivity log as an aid in determining some reservoir characteristics. Trans AIME 146(01):54–62
Armatas GS (2006) Determination of the effects of the pore size and pore connectivity distribution on the pore tortuosity and diffusive transport in model porous networks. Chem Eng Sci 61(14):4662–4675
Azar JH, Javaherian A, Pishvaie MR, Nabi-Bidhendi M (2008) An approach to defining tortuosity and cementation factor in carbonate reservoir rocks. J Petrol Sci Eng 60(2):125–131
Backeberg NR, Iacoviello F, Rittner M, Mitchell TM, Jones AP, Day R, Wheeler J, Shearing PR, Vermeesch P, Striolo A (2017) Quantifying the anisotropy and tortuosity of permeable pathways in clay-rich mudstones using models based on x-ray tomography. Sci Rep 7(1):14838
Barrande M, Bouchet R, Denoyel R (2007) Tortuosity of porous particles. Anal Chem 79(23):9115–9121
Batschelet E (1981) Circular statistics in biology. Academic Press, London
Bear J (1972) Dynamics of fluids in porous media. Dover, New York
Bear J, Bachmat Y (2012) Introduction to modeling of transport phenomena in porous media. Springer, Dordrecht
Beekman JW (1990) Mathematical description of heterogeneous materials. Chem Eng Sci 45:2603–2610
Belem T, Homand-Etienne F, Souley M (2000) Quantitative parameters for rock joint surface roughness. Rock Mech Rock Eng 33:217–242
Berg CF (2014) Permeability description by characteristic length, tortuosity, constriction and porosity. Transp Porous Media 103(3):381–400
Berner RA (1980) Early diagenesis: a theoretical approach. Princeton University Press, Princeton
Berryman JG (1981) Elastic wave propagation in fluid-saturated porous media. J Acoust Soc Am 69(2):416–424
Bhatia SK (1985) Directional autocorrelation and the diffusional tortuosity of capillary porous media. J Catal 93:192–196
Bini F, Pica A, Marinozzi A, Marinozzi F (2019) A 3d model of the effect of tortuosity and constrictivity on the diffusion in mineralized collagen fibril. Sci Rep 9(1):2658
Bird RB (2002) Transport phenomena. Appl Mech Rev 55(1):R1–R4
Bodla KK, Murthy JY, Garimella SV (2010) Microtomography-based simulation of transport through open-cell metal foams. Numer Heat Transf Part a: Appl 58(7):527–544
Bofiglio P, Pompoli F (2007) Frequency dependent tortuosity measurement by means of ultrasonic tests. In: Proceedings of 14th ICSV (International Congress of Sound and Vibration), Cairns
Borujeni AT, Lane N, Thompson K, Tyagi M (2013) Effects of image resolution and numerical resolution on computed permeability of consolidated packing using lb and fem pore-scale simulations. Comput Fluids 88:753–763
Boudreau BP (1996) The diffusive tortuosity of fine-grained unlithified sediments. Geochim Cosmochim Acta 60(16):3139–3142
Boudreau BP, Meysman FJR (2006) Predicted tortuosity of muds. Geology 34(8):693–696
Boving TB, Grathwohl P (2001) Tracer diffusion coefficients in sedimentary rocks, correlation to porosity and hydraulic conductivity. J Contam Hydrol 53(4):85–100

Brodkey RS, Hershey HC (2003) Transport phenomena: a unified aproach. Brodkey Publishing
Bruggeman VD (1935) Berechnung verschiedener physikalischer Konstanten von heterogenen Substanzen. i. Dielektrizitätskonstanten und Leitfähigkeiten der Mischkörper aus isotropen Substanzen. Ann. Phys 416(7):636–664
Bultreys T, De Boever W, Cnudde V (2016) Imaging and image-based fluid transport modeling at the pore scale in geological materials: a practical introduction to the current state-of-the-art. Earth Sci Rev 155:93–128
Bunde A, Havlin S (1996) Percolation I. In: Bunde A, Havlin S (eds) Fractals and disordered systems, 2nd edn. Springer, Berlin, pp 59–114
Cai J, Wei W, Hu X, Wood DA (2017) Electrical conductivity models in saturated porous media: a review. Earth Sci Rev 171:419–433
Cao T, Zhang L, Sun G, Wang C, Zhang Y, Yan N, Xu A (2019) Model for predicting the tortuosity of transport paths in cement-based materials. Materials (Basel) 12(21):3623
Carman PC (1937) Fluid flow through granular beds. Trans Inst Chem Eng 50:150–166
Carman PC (1939) Permeability of saturated sands, soils and clays. J Agric Sci 20:262
Carman PC (1956) Flow of gases through porous media. Butterworths Scientific Publications, London
Chung D-W, Ebner M, Ely DR, Wood V, García RE (2013) Validity of the Bruggeman relation for porous electrodes. Model Simul Mater Sci Eng 21(7):074009
Cieplak M, Maritan A, Banavar JR (1996) Invasion percolation and Eden growth: geometry and universality. Phys Rev Lett 76:3754–3757
Civan F (2011) Porous media transport phenomena. Wiley, Hoboken, New Jersey
Clennell MB (1997) Tortuosity: a guide through the maze. Geol Soc Lond Spec Publ 122(1):299–344
Coelho D, Shapiro M, Thovert J, Adler PM (1996) Electroosmotic phenomena in porous media. J Colloid Interface Sci 181(1):169–190
Coleman S, Vassilicos J (2008) Transport properties of saturated and unsaturated porous fractal materials. Phys Rev Lett 100(3):035504
Comiti J, Renaud M (1989) A new model for determining mean structure parameters of fixed beds from pressure drop measurements: application to beds packed with parallelepipedal particles. Chem Eng Sci 44(7):1539–1545
David C (1993) Geometry of flow paths for fluid transport in rocks. J Geophys Res Solid Earth 98(B7):12267–12278
de Gennes PG (1976) La percolation: un concept unificateur. La Recerche 7:72–82
Degruyter W, Burgisser A, Bachmann O, Malaspinas O (2010) Synchrotron X-ray microtomography and lattice Boltzmann simulations of gas flow through volcanic pumices. Geosphere 6(5):470–481
De La Rue Robert E, Tobias CW (1959) On the conductivity of dispersions. J Electrochem Soc 106(9):827
de Lima OA (1995) Water saturation and permeability from resistivity, dielectric, and porosity logs. Geophysics 60(6):1756–1764
De Lima O, Niwas S (2000) Estimation of hydraulic parameters of shaly sandstone aquifers from geoelectrical measurements. J Hydrol 235(1–2):12–26
Dias R, Teixeira JA, Mota M, Yelshin A (2006) Tortuosity variation in a low density binary particulate bed. Sep Purif Technol 51(2):180–184
Duda A, Koza Z, Matyka M (2011) Hydraulic tortuosity in arbitrary porous media flow. Phys Rev E 84(3):036319
Du Plessis JP, Masliyah JH (1988) Mathematical modelling of flow through consolidated isotropic porous media. Transp Porous Med 3:145–161
Du Plessis E, Woudberg S, Prieur du Plessis J (2010) Pore-scale modelling of diffusion in unconsolidated porous structures. Chem Eng Sci 65(8):2541–2551
Dykhuizen RC, Casey WH (1989) An analysis of solute diffusion in rocks. Geochim Cosmochim Acta 53:2797–2805

Epstein N (1989) On tortuosity and the tortuosity factor in flow and diffusion through porous media. Earth Sci Rev 212(2021):103439
Estevez I, Christman MC (2006) Analysis of the movement and use of space by animals in confinement: the effect of sampling effort. Appl Anim Behav 97:221–240
Feder J (1988) Fractals. Plenum Press, New York
Fellah ZEA, Sadouki M, Fellah M, Mitri FG, Ogam E, Dépollier C (2013) Simultaneous determination of porosity, tortuosity, viscous and thermal characteristic lengths of rigid porous materials. J Appl Phys 114(20):204902
Fisher ME (1971) The theory of critical point singularities. In: Green MS (ed) Critical phenomena: proceedings of the 1970 enrico fermi international school of physics, Varenna, Italy. Course 51. Academic Press, New York, pp 1–99
Fisher NI (1993) Statistical analysis of circular data. Cambridge University Press, Cambridge, UK
Fredrich J, DiGiovanni A, Noble D (2006) Predicting macroscopic transport properties using microscopic image data. J Geophys Res Solid Earth 111:B03201
Fu J, Dong J, Wang Y, Ju Y, Owen DRJ, Li C (2020) Resolution effect: an error correction model for intrinsic permeability of porous media estimated from lattice Boltzmann method. Transp Porous Media 132:627–656
Fu J, Thomas HR, Li C (2021) Tortuosity of porous media: image analysis and physical simulation. Earth Sci Rev 212:103439
Garrouch AA, Ali L, Qasem F (2001) Using diffusion and electrical measurements to assess tortuosity of porous media. Ind Eng Chem Res 40(20):4363–4369
Georgali B, Tsakiridis P (2005) Microstructure of fire-damaged concrete. A case study. Cem Concr Compos 27(2):255–259
Ghanbarian B, Hunt AG, Ewing RP, Sahimi M (2013) Tortuosity in porous media: a critical review. Soil Sci Soc Am J 77(5):1461–1477
Ghanbarian B, Hunt AG, Sahimi M, Ewing RP, Skinner TE (2023) Percolation theory generates a physically based description of tortuosity in saturated and unsaturated porous media. Soil Sci Soc Am J 87(2):1920–1929
Goldhaber MB, Aller RC, Cochran JK, Rosenfeld J, Martens C, Berner RA (1977) Sulfate reduction, diffusion, and bioturbation in Long Island Sound sediments: report of the FOAM Group. Am J Sci 277:193–237
Gommes CJ, Bons A-J, Blacher S, Dunsmuir JH, Tsou AH (2009) Practical methods for measuring the tortuosity of porous materials from binary or gray-tone tomographic reconstructions. AIChE J 55(8):2000–2012
Grathwohl P (2012) Diffusion in natural porous media: contaminant transport, sorption/desorption and dissolution kinetics, vol 1. Springer Science & Business Media, New York
Griffiths IM, Mitevski I, Vujkovac I, Illingworth MR, Stewart PS (2020) The role of tortuosity in filtration efficiency: a general network model for filtration. J Membr Sci 598:117664
Grime J (14 Mar 2015) A meandering tale: the truth about pi and rivers. Alex Bellos's adventures in numberland. The Guardian
Guan KM, Nazarova M, Guo B, Tchelepi H, Kovscek AR, Creux P (2019) Effects of image resolution on sandstone porosity and permeability as obtained from x-ray microscopy. Transp Porous Media 127(1):233–245
Guo Z, Zhao T (2005) A lattice Boltzmann model for convection heat transfer in porous media. Numer Heat Transf Part B 47(2):157–177
Hamad HT, Abbood DW, Mustafa AS (2018) Strength tortuosity-porosity relation in locally types of porous media (experimental model). In: IOP conference series: materials science and engineering, vol 454(1), p 012093
He W, Zou J, Wang B, Vilayurganapathy S, Zhou M, Lin X, Zhang KH, Lin J, Xu P, Dickerson JH (2013) Gas transport in porous electrodes of solid oxide fuel cells: a review on diffusion and diffusivity measurement. J Power Sources 237:64–73
Herring A, Sheppard A, Turner M, Beeching L (2018) Multiphase flows in sandstones. http://www.digitalrocksportal.org/projects/135

Ho F-G, Strieder W (1981) A variational calculation of the effective surface diffusion coefficient and tortuosity. Chem Eng Sci 36:253–258
Hosticka B, Norris P, Brenizer J, Daitch C (1998) Gas flow through aerogels. J Non-Cryst Solids 225:293–297
Hu LB, Savidge C, Rizzo DM, Hayden N, Hagadorn JW, Dewoolkar M (2012) Commonly used porous building materials: geomorphic pore structure and fluid transport. J Mater Civ Eng 25(12):1803–1812
Hunt AG (2006) Scale-dependent hydraulic conductivity in anisotropic media from dimensional cross-over. Hydrogeol J 14:499–507
Hunt AG, Skinner TE (2008) Longitudinal dispersion of solutes in porous media solely by advection. Philos Mag 88:2921–2944
Hutzenlaub T, Asthana A, Becker J, Wheeler DR, Zengerle R, Thiele S (2013) FIB/SEM-based calculation of tortuosity in a porous $LiCoO_2$ cathode for a Li-ion battery. Electrochem Commun 27:77–80
Ioannidis M, Kwiecien M, Chatzis I (1997) Electrical conductivity, percolation aspects of statistically homogeneous porous media. Transp Porous Media 29(1):61–83
Iversen N, Jørgensen BB (1993) Diffusion coefficients of sulfate and methane in marine sediments: influence of porosity. Geochim Cosmochim Acta 57(3):571–578
Jähne B (1997) Digitale bildverarbeitung, 4th edn. Springer, Berlin, Heidelberg
Javadpour F (2009) Nanopores and apparent permeability of gas flow in mudrocks (shales and siltstone). J Can Pet Technol 48(08):16–21
Jin G, Patzek T, Silin D (2004) Direct prediction of the absolute permeability of unconsolidated and consolidated reservoir rock. In: 2003 SPE annual technical conference and exhibition (Houston, Texas, USA): SPE 90084
Jorgensen BB (1978) A comparison of methods for the quantification of bacterial sulfate reduction in coastal marine sediments. II. Calculations from mathematical models. Geomicrobiol J 1:29–47
Karanikola V, Corral AF, Jiang H, Sáez AE, Ela WP, Arnold RG (2017) Effects of membrane structure and operational variables on membrane distillation performance. J Membr Sci 524:87–96
Keehm Y, Mukerji T (2004) Permeability and relative permeability from digital rocks: issues on grid resolution and representative elementary volume. In: SEG technical program expanded abstracts 2004. Society of exploration geophysicists, pp 1654–1657
Keller LM, Hilger A, Manke I (2015) Impact of sand content on solute diffusion in opalinus clay. Appl Clay Sci 112:134–142
Kim AS, Chen H (2006) Diffusive tortuosity factor of solid and soft cake layers: a random walk simulation approach. J Membr Sci 279(1–2):129–139
Kim IC, Torquato S (1992) Diffusion of finite-size Brownian particles in porous media. J Chem Phys 96:1498–1503
Kim J, Ochoa J, Whitaker S (1987) Diffusion in anisotropic porous media. Transp Porous Media 2(4):327–356
Klinkenberg LJ (1951) Analogy between diffusion and electrical conductivity in porous rock. Geol Soc Am Bull 62:559–564
Kong W, Zhang Q, Gao X, Zhang J, Chen D, Su S (2015) A method for predicting the tortuosity of pore phase in solid oxide fuel cells electrode. Int J Electrochem Sci 10:5800–5811
Koponen A (1998) Simulations of fluid flow in porous media by lattice-gas and Lattice-Boltzmann methods. PhD thesis, University of Jyväskyla, Jyväskylä, Finland
Koponen A, Kataja M, Timonen J (1996) Tortuous flow in porous media. Phys Rev E 54(1):406
Koponen A, Kataja M, Timonen J (1997) Permeability and effective porosity of porous media. Phys Rev E 56(3):3319
Koponen A, Kandhai D, Hellen E, Alava M, Hoekstra A, Kataja M, Niskanen K, Sloot P, Timonen J (1998) Permeability of three dimensional random fiber webs. Phys Rev 80(4):716–719
Korvin G (1992a) Fractal models in the earth sciences. Elsevier, Amsterdam, pp 17–33

Korvin G (1992b) A percolation model for the permeability of kaolinite-bearing sandstones. Geophys Trans 37(2–3):177–209
Korvin G (2016) Permeability from microscopy: review of a dream. Arab J Sci Eng 41(6):2045–2065
Kozeny J (1927) Ueber kapillare Leitung des Wassers im Boden. Sitzungsber Akad Wiss, Wien 136(2a):271–306
Kreamer DK, Weeks EP, Thompson GM (1988) A field technique to measure the tortuosity and sorption-affected porosity for gaseous diffusion of materials in the unsaturated zone with experimental results from near Barnwell, South Carolina. Water Resour Res 24(3):331–341
Krom MD, Berner RA (1980) The diffusion coefficients of sulfate, ammonia and phosphorus ions in anoxic marine sediments. Limnol Ocean 25:327–337
Kutay ME, Aydilek AH, Masad E (2006) Laboratory validation of lattice Boltzmann method for modeling pore-scale flow in granular materials. Comput Geotech 33(8):381–395
Lanfrey PY, Kuzeljevic ZV, Dudukovic MP (2010) Tortuosity model for fixed beds randomly packed with identical particles. Chem Eng Sci 65:1891–1896
Lee Y, Andrade JS Jr, Buldyrev SV, Dokholyan NV, Havlin S, King PR, Paul G, Stanley HE (1999) Traveling time and traveling length in critical percolation clusters. Phys Rev E 60:3425–3428
Lemaitre R, Adler PM (1990) Fractal porous media iv: three-dimensional Stokes flow through random media and regular fractals. Transp Porous Media 5(4):325–340
Lerman A (1979) Geochemical processes: water and sediment environments. Wiley, New York
Li PW, Chyu MK (2003) Simulation of the chemical/electrochemical reactions and heat/mass transfer for a tubular SOFC in a stack. J Power Sources 124(2):487–498
Li Y-H, Gregory S (1974) Diffusion of ions in seawater and in deep-sea sediments. Geochim Cosmochim Acta 38:703–714
Li W, Zou C-C, Wang H, Peng C (2017) A model for calculating the formation resistivity factor in low and middle porosity sandstone formations considering the effect of pore geometry. J Pet Sci Eng 152:193–203
Liu Y, Kitanidis PK (2013) Tortuosity and Archie's law. In: Mishra PK, Kuhlman KL (eds) Advances in hydrogeology. Springer, New York, pp 115–126
Lomize GM (1951) Flow in fractured rock. Gosenergoizdat, Moscow
Louis C (1969) A study of groundwater flow in jointed rock and its influence on stability of rock masses. Rock mechanics research report, vol 10. Imperial College of Science and Technology, London
Louis C (1974) Rock hydraulics. In: Müller L (ed) Rock mechanics. Springer, New York, pp 299–387
Low PF (1981) Principles of ion diffusion in clays. Chem Soil Environ 40:31–45. American Society for Agronomy, Special Publication
Lu X (1997) Application of the Nernst-Einstein equation to concrete. Cem Concr Res 27(2):293–302
Luo L, Yu B, Cai J, Zeng X (2014) Numerical simulation of tortuosity for fluid flow in two-dimensional pore fractal models of porous media. Fractals 22(04):1450015
Mackie JS, Meares P (1955) The diffusion of electrolytes in a cation-exchange resin membrane. I. Theoretical. Proc R Soc A: Math, Phys Eng Sci 232(1191):498–509
Malek K, Coppens M-O (2003) Knudsen self- and Fickian diffusion in rough nanoporous media. J Chem Phys 119(5):2801–2811
Mandelbrot BB (1983) The fractal geometry of nature. W.H. Freeman, San Francisco
Manheim FT, Waterman LS (1974) Diffusimetry (diffusion constant estimation) on sediment cores by resistivity probe. In: von der Borch CC, Sclater GC (eds) Initial reports of the deep sea drilling project, US Printing Office, vol 22:663–670
Martínez MJ, Shimpalee S, Van Zee JW (2008) Measurement of MacMullin numbers for PEMFC gas-diffusion media. J Electrochem Soc 156(1):B80
Masís-Meléndez F, Deepagoda TKKC, de Jonge LW, Tuller M, Moldrup P (2014) Gas diffusion-derived tortuosity governs saturated hydraulic conductivity in sandy soils. J Hydrol 512:388–396
Matyka M, Khalili A, Koza Z (2008) Tortuosity-porosity relation in porous media flow. Phys Rev E 78(2):026306

Mauret E, Renaud M (1997) Transport phenomena in multi-particle systems—i. limits of applicability of capillary model in high voidage beds-application to fixed beds of fibers and fluidized beds of spheres. Chem Eng Sci 52(11):1807–1817

Maxwell JC (1881) Treatise on electricity and magnetism, 2nd edn. Clarendon Press, Oxford

Miller C, Christman MC, Estevez I (2011) Movement in a confined space: estimating path tortuosity. Appl Anim Behav Sci 135(1–2):13–23

Millington RJ (1959) Gas diffusion in porous media. Science 130:100–102

Moctezuma-Berthier A, Vizika O, Adler PM (2002) Macroscopic conductivity of vugular porous media. Transp Porous Media 49(3):313–332

Moctezuma-Berthier A, Vizika O, Thovert J, Adler PM (2004) One-and two-phase permeabilities of vugular porous media. Transp Porous Media 56(2):225–244

Moldrup P, Olesen T, Komatsu T, Schjønning P, Rolston D (2001) Tortuosity, diffusivity, and permeability in the soil liquid and gaseous phases. Soil Sci Soc Am J 65(3):613–623

Montes J, Cuevas F, Cintas J (2007) Electrical and thermal tortuosity in powder compacts. Granul Matter 9(6):401–406

Mota M, Teixeira JA, Bowen WR, Yelshin A (2001) Binary spherical particle mixed beds: porosity and permeability relationship measurement. Trans Filt Soc 1(4):101–106

Mualem Y, Friedman S (1991) Theoretical prediction of electrical conductivity in saturated and unsaturated soil. Water Resour Res 27(10):2771–2777

Mueller J (1968) An introduction to the hydraulic and topographic sinuosity indexes. Ann Assoc Am Geogr 58(2):371–385

Nabovati A, Sousa ACM (2007) Fluid flow simulation in random porous media at pore level using the Lattice Boltzmann Method. J Eng Sci Technol 2(3):226–237

Nakashima Y, Watanabe Y (2002) Estimate of transport properties of porous media by microfocus x-ray computed tomography and random walk simulation. Water Resour Res 38(12):1272

Navarre-Sitchler A, Steefel CI, Yang L, Tomutsa L, Brantley SL (2009) Evolution of porosity and diffusivity associated with chemical weathering of a basalt clast. J Geophys Res Earth Surf 114:F02016

Neale GH, Nader WK (1973) Prediction of transport processes in porous media. Am Inst Chem Eng J 19:112–119

Nikitsin VI, Backiel-Brzozowska B (2013) Determination of capillary tortuosity coefficient in calculations of moisture transfer in building materials. Int J Heat Mass Transf 56:30–34

Ning Y, He S, Liu H, Wang H, Qin G (2016) A rigorous upscaling procedure to predict macroscale transport properties of natural gas in shales by coupling molecular dynamics with lattice Boltzmann method. In: SPE annual technical conference and exhibition, Dubai, UAE, September 2016: SPE-181689

Niya RSM, Selvadurai APS (2017) The estimation of permeability of a porous medium with a generalized pore structure by geometry identification. Phys Fluids 29:037101

Niya SR, Selvadurai APS (2018) A statistical correlation between permeability, porosity, tortuosity and conductance. Transp Porous Media 121:741–752

Ohkubo T (2008) Tortuosity based on anisotropic diffusion process in structured plate-like obstacles by Monte Carlo simulation. Transp Porous Media 72(3):339–350

Oleschko K (1998) Delesse principle and statistical fractal sets: 1. Dimensional equivalents. Soil Tillage Res 49:255

Olives R, Mauran S (2001) A highly conductive porous medium for solid–gas reactions: effect of the dispersed phase on the thermal tortuosity. Transp Porous Media 3(2):377–394

Ostadi H, Rama P, Liu Y, Chen R, Zhang X, Jiang K (2010) 3d reconstruction of a gas diffusion layer and a microporous layer. J Membr Sci 351(1–2):69–74

Pape H, Clauser C, Iffland J (1998) Permeability prediction for reservoir sandstones and basement rocks based on fractal pore space geometry. SEG Tech Program Expand Abstr 1998:1032–1035

Peng S, Marone F, Dultz S (2014) Resolution effect in x-ray microcomputed tomography imaging and small pore's contribution to permeability for a Berea sandstone. J Hydrol 510:403–411

Perez-Rosales C, Luna E (2004) Characterization of vuggy fractured media: a practical approach. SPE 91863
Perez-Rosales C, Luna E (2005) Naturally fractured reservoirs: how to estimate secondary porosity. SPE 94525
Perkins TK, Johnston OC (1963) A review of diffusion and dispersion in porous media. Trans AIME 228:70–84
Petersen EE (1958) Diffusion in a pore of varying cross-section. AIChE J 3:343–345
Petersen EE (1965) Chemical reaction analysis. Prentice-Hall, Englewood Cliffs, NJ
Piller M, Schena G, Nolich M, Favretto S, Radaelli F, Rossi E (2009) Analysis of hydraulic permeability in porous media: from high resolution x-ray tomography to direct numerical simulation. Transp Porous Media 80(1):57
Pisani L (2011) Simple expression for the tortuosity of porous media. Transp Porous Media 88(2):193–203
Porto M, Havlin S, Schwarzer S, Bunde A (1997) Optimal path in strong disorder and shortest path in invasion percolation with trapping. Phys Rev Lett 79:4060–4062
Rayleigh L (1892) On the influence of obstacles arranged in rectangular order upon the properties of a medium. Phil Mag 34:481–489
Sahimi M (1994) Applications of percolation theory. Taylor and Francis, London
Salmas CE, Androutsopoulos GP (2001) A novel pore structure tortuosity concept based on nitrogen sorption hysteresis data. Ind Eng Chem Res 40:721–730
Satterfield CN (1970) Mass transfer in heterogeneous catalysis. MIT Press, Cambridge
Schwartz LM, Garboczi EJ, Bentz DP (1995) Interfacial transport in porous media: application to dc electrical conductivity of mortars. J Appl Phys 78(10):5898–5908
Selvadurai APS (2000) Partial differential equations in mechanics, vol 2, The biharmonic equation poisson's equation. Springer, Berlin
Shackelford CD, Moore SM (2013) Fickian diffusion of radionuclides for engineered containment barriers: diffusion coefficients, porosities, and complicating issues. Eng Geol 152(1):133–147
Shanti NO, Chan VW, Stock SR, De Carlo F, Thornton K, Faber KT (2014) X-ray micro-computed tomography and tortuosity calculations of percolating pore networks. Acta Mater 71:126–135
Shen L, Chen Z (2007) Critical review of the impact of tortuosity on diffusion. Chem Eng Sci 62(14):3748–3755
Sing KS, Williams RT (2004) The use of molecular probes for the characterization of nanoporous adsorbents. Part Part Syst Character Measur Des Part Properties Behav Powd Other Disperse Syst 21(2):71–79
Sobieski W, Lipiński S (2017) The analysis of the relations between porosity and tortuosity in granular beds. Tech Sci/univ Warm Maz Olszt 20(1):75–85
Srisutthiyakorn N, Mavko GM (2017) What is the role of tortuosity in the Kozeny-Carman equation? Interpretation 5(1):SB57–SB67
Stauffer D, Aharony A (1994) Introduction to percolation theory, 2nd edn. Taylor and Francis, London
Stenzel O, Pecho O, Holzer L, Neumann M, Schmidt V (2016) Predicting effective conductivities based on geometric microstructure characteristics. AIChE J 62(5):1834–1843
Stølum H-H (1996) River meandering as a self-organization process. Science 271(5256):1710–1713
Sun W, Wong T-F (2018) Prediction of permeability and formation factor of sandstone with hybrid lattice Boltzmann/finite element simulation on microtomographic images. Int J Rock Mech Min Sci 106:269–277
Sun WC, Andrade JE, Rudnicki JW (2011) Multiscale method for characterization of porous microstructures and their impact on macroscopic effective permeability. Int J Numer Methods Eng 88(12):1260–1279
Sweerts J-P, Kelly CA, Rudd JW, Hesslein R, Cappenberg TE (1991) Similarity of whole-sediment molecular diffusion coefficients in freshwater sediments of low and high porosity. Limnol Ocean 36:335–342

Takahashi M, Ahn C, Kato M (2012) Hydraulic stress effect on 3-dimensional pore network and permeability change. In: Quian Q, Zhou Y (eds) Harmonising rock engineering and the environment. 12th ISRM international congress on rock mechanics. Taylor & Francis Group, London, pp 542–546
Tang X-W, Sun Z-F, Cheng GC (2012) Simulation of the relationship between porosity and tortuosity in porous media with cubic particles. Chin Phys B 21(10):100201
Teague MR (1980) Image analysis via the general theory of moments. J Opt Soc Am 70(8):920–930
Thovert JF, Wary F, Adler PM (1990) Thermal conductivity of random media and regular fractals. J Appl Phys 68:3872–3883
Tjaden B, Brett DJ, Shearing PR (2018) Tortuosity in electrochemical devices: a review of calculation approaches. Int Mater Rev 63(2):47–67
Tomadakis MM, Sotirchos SV (1983) Transport properties of random arrays of freely overlapping cylinders with various orientation distributions. J Chem Phys 98:616–626
Ullman WJ, Aller RC (1982) Diffusion coefficients in near shore marine sediments. Limnol Ocean 27:552–556
Van Brakel J, Heertjes P (1974) Analysis of diffusion in macroporous media in terms of porosity, tortuosity and constrictivity factor. Int J Heat Mass Transf 17(9):1093–1103
Vervoort R, Cattle S (2003) Linking hydraulic conductivity and tortuosity parameters to pore space geometry and pore-size distribution. J Hydrol 272(1–4):36–49
Weerts A, Bouten W, Verstraten J (1999) Simultaneous measurement of water retention and electrical conductivity in soils: testing the Mualem-Friedman tortuosity model. Water Resour Res 35(6):1781–1787
Weerts A, Kandhai D, Bouten W, Sloot P (2001) Tortuosity of an unsaturated sandy soil estimated using gas diffusion and bulk soil electrical conductivity. Soil Sci Soc Am J 65(6):1577–1584
Weissberg HL (1963) Effective diffusion coefficient in porous media. J Appl Phys 34(9):2636–2639
Wiegmann A, Zemitis A (2006) EJ-HEAT: a fast explicit jump harmonic averaging solver for the effective heat conductivity of composite materials. Berichte des Fraunhofer-Instituts für Techno- und Wirtschaftsmathematik (ITWM report), 94
Winsauer WO, Shearin H Jr, Masson P, Williams M (1952) Resistivity of brine-saturated sands in relation to pore geometry. AAPG Bull 36(2):253–277
Wu San Y, van Vliet LJ, Frijlink HW, van der Voort Maarschalk K (2006) The determination of relative path length as a measure for tortuosity in compacts using image analysis. Eur J Pharm Sci 28(5):433–440
Wyllie M, Rose WD (1950) Some theoretical considerations related to the quantitative evaluation of the physical characteristics of reservoir rock from electrical log data. J Pet Technol 2(04):105–118
Wyllie M, Spangler M (1952) Application of electrical resistivity measurements to problem of fluid flow in porous media. AAPG Bull 36(2):359–403
Xiao W, Xia C, Wei W, Bian Y (2013) Combined effect of tortuosity and surface roughness on estimation of flow rate through a single rough joint. J Geophys Eng 10(4):045015
Yang K, Li M, Ling NN, May EF, Connolly PR, Esteban L, Clennell MB, Mahmoud M, El-Husseiny A, Adebayo AR, Elsayed MM (2019) Quantitative tortuosity measurements of carbonate rocks using pulsed field gradient NMR. Transp Porous Media 130:847–865
Yong Y, Lou X, Li S, Yang C, Yin X (2014) Direct simulation of the influence of the pore structure on the diffusion process in porous media. Comput Math Appl 67(2):412–423
Yu B, Cheng P (2002) A fractal permeability model for bi-dispersed porous media. Int J Heat Mass Transf 45:2983–2993
Zalc JM, Reyes SC, Iglesia E (2004) The effects of diffusion mechanism and void structure on transport rates and tortuosity factors in complex porous structures. Chem Eng Sci 59:2947

Zhang ZF, Ward AL, Gee GW (2003) A tensorial connectivity-tortuosity concept to describe the unsaturated hydraulic properties of anisotropic soils. Vadose Zone J 2(3):313–321

Zielinski T (2012) Inverse identification and microscopic estimation of parameters for models of sound absorption in porous ceramics. In: Proceedings of international conference on noise and vibration engineering (ISMA2012)/international conference on uncertainty in structural dynamics (USD2012), pp 95–108

Chapter 3
The Internal Topology of Rocks

Chapter Highlights We discuss the modern mathematical tools for studying the *topology* of pore space. In Sect. 3.1. *Minkowski functionals* are introduced through the paradigm of Mark Kac (Kac, Am Math Monthly 73:1–24, 1966) '*Can one hear the shape of a drum?*'. *Euler characteristics*, a powerful tool to study *connectivity* is treated in 3.2. The elegant and sophisticated research tool, *Persistent Homology*, is built up in a series of easy-to-follow steps in 3.3. The abstract constructions are illustrated with practical examples, and there are copious references in the text and in Tables (3.1, 3.4, 3.11) for their use in *Petrophysics.*

3.1 Minkowski Functionals: Integral Geometry of the Pore Space

3.1.1 Background and Mathematics

In a landmark paper '*Can one hear the shape of a drum?*' Mark Kac (1966) established a relation between the geometric descriptors of the shape of a "drum" (more exactly, of a *2-D membrane*) $\mathfrak{D}$ and the sound it makes when it is struck. The sound of the drum depends on its shape, and on the material properties and tautness of its skin. The drum vibrates at discrete frequencies called *normal modes*, the lowest is the *fundamental tone*, higher frequencies are called *overtones.*

The vibrations of the drum $\mathfrak{D}\epsilon\mathbb{R}^2$ are described by the wave equation

$$\frac{\partial^2 v(x,y,t)}{\partial t^2} = c^2\left(\frac{\partial^2 v(x,y,t)}{\partial x^2} + \frac{\partial^2 v(x,y,t)}{\partial y^2}\right) \tag{3.1.1}$$

where $v(x, y, t)$ is the transverse displacement of point (x, y) at time t. If by proper choice of the physical units we make $c^2 = 1$, and (by separation of variables) we set $v(x, y, t) = f(t)u(x, y)$ then $u(x,y)$ satisfies

$$\Delta u(x,y) + \lambda u(x,y) = 0 \text{ if } (x,y)\epsilon\mathfrak{D} \tag{3.1.2a}$$

G. Korvin, *Statistical Rock Physics*, Earth and Environmental Sciences Library,
https://doi.org/10.1007/978-3-031-46700-4_3

and

$$u(x, y) = 0 \text{ if } (x, y) \epsilon \partial \mathfrak{D}. \tag{3.1.2b}$$

As well known (Kac 1966; Protter 1987), there are infinitely many *eigenvalues* $0 < \lambda_1 \leq \lambda_2 \leq \lambda_3 \leq \cdots \leq \lambda_k \leq \cdots (\lim_{k \to \infty} \lambda_k = \infty)$ and their corresponding *eigenfunctions* $u_k(x, y)$ such that

$$\Delta u_k(x, y) + \lambda_k u_k(x, y) = 0 \textit{ in } \mathfrak{D}, \tag{3.1.3a}$$

and

$$u_k(x, y) = 0 \textit{ on the boundary } \partial \mathfrak{D}. \tag{3.1.3b}$$

In mathematical terms, the question of Mark Kac, '*Can one hear the shape of a drum?* asks whether it is possible to tell the shape of the domain $\mathfrak{D}$ from the knowledge of the eigenvalues $\{\lambda_k\}_{k=1}^{\infty}$. In 1911 Weil showed that

$$\lambda_k \sim 4\pi^2 \left(\frac{k}{\omega_n V} \right)^{2/n} \textit{as } k \to \infty, \tag{3.1.4}$$

where $V = vol(\mathfrak{D})$, ω_n is the volume of the unit ball in $\mathbb{R}^n$ ($\omega_n = \pi^{n/2} / \Gamma\left(1 + {}^{n}/{}_{2}\right)$; $\omega_0 = 1, \omega_1 = 2, \omega_2 = \pi, \omega_3 = 4\pi/3$) that is *the asymptotic growth of the eigenvalues* λ_k *determines the volume of the drum* $V(\mathfrak{D})$.

Pleilej (1954) and Kac (1966) established further relations between geometrical properties of $\mathfrak{D}$ and the asymptotic behavior of the *Laplacian*'s spectrum. Pleijel expressed the asymptotics of the sum $\sum_{k=1}^{\infty} e^{-t\lambda_k}$ for finite domains $\mathfrak{D} \epsilon \mathbb{R}^2$ as

$$\sum_{k=1}^{\infty} e^{-t\lambda_k} \sim \frac{A(\mathfrak{D})}{2\pi t} - \frac{U(\mathfrak{D})}{4} \cdot \frac{1}{\sqrt{2\pi t}} (\textit{as } t \downarrow 0), \tag{3.1.5}$$

where, in $\mathbb{R}^2$, $A(\mathfrak{D})$ is the area, $U(\mathfrak{D})$ the boundary length of $\mathfrak{D}$. By Eq. (3.1.5), for any finite drum $\mathfrak{D} \subset \mathbb{R}^2$ one can "hear" (i.e. mathematically express in terms of the eigenfrequencies λ_k) its area and its perimeter. For *simply connected domains* in $\mathbb{R}^2$ Pleilej succeeded to add a third term to the right-hand-side (RHS) of Eq. (3.1.5), namely

$$\sum_{k=1}^{\infty} e^{-t\lambda_k} \sim \frac{A(\mathfrak{D})}{2\pi t} - \frac{U(\mathfrak{D})}{4} \cdot \frac{1}{\sqrt{2\pi t}} + \frac{1}{6} (\textit{as } t \downarrow 0). \tag{3.1.6}$$

Mark Kac (1966) proved Eq. (3.1.5) in a different way, proposed a heuristic derivation for Eq. (3.1.6), and recognized that the factor "1" in the numerator of

"$\frac{1}{6}$" on the RHS of (3.1.6) is a *topological measure* of connectivity (the *Euler-Poincaré characteristic*) which expresses that $\mathfrak{D}$ contains no holes in its interior. He conjectured that in case of a smooth drum $\mathfrak{D} \subset \mathbb{R}^2$ with r smooth non-overlapping holes, H_1 *with* $A(H_1) = A_1, U(H_1) = U_1; \cdots$ *with* $A(H_r) = A_r, U(H_r) = U_r$ in its interior, instead of the term $\frac{1}{6}$ there should appear $\left(\frac{1-r}{6}\right)$ and Eq. (3.1.6) would generalize to

$$\sum_{k=1}^{\infty} e^{-t\lambda_k} \sim \frac{A(\mathfrak{D})}{2\pi t} - \frac{U(\mathfrak{D})}{4} \cdot \frac{1}{\sqrt{2\pi t}} + \frac{1}{6}(1-r)\,(as\ t \downarrow 0) \tag{3.1.7}$$

where $A(\mathfrak{D})$ is the area of the solid part of the drum's skin (without the holes), $U(\mathfrak{D})$ is the total boundary length consisting of the outside perimeter of $\mathfrak{D}$ plus the sum $U_1 + \cdots + U_r$ of the perimeters of the r holes. As in $\mathbb{R}^2$ the *Euler-Poincaré characteristic* $\chi(A)$ of a domain A equals the number of connected components minus the number of holes (Michielsen and de Raedt 2001; Hadwiger 1957), we can rewrite Eq. (3.1.7) to

$$\sum_{k=1}^{\infty} e^{-t\lambda_k} \sim \frac{A(\mathfrak{D})}{2\pi t} - \frac{U(\mathfrak{D})}{4} \cdot \frac{1}{\sqrt{2\pi t}} + \frac{\chi(\mathfrak{D})}{6}\,(t \downarrow 0). \tag{3.1.8}$$

In *Integral Geometry* (Hadwiger 1957; Santaló 1976; Stoyan et al. 1989) the shape of a complex d-dimensional domain is characterized by the so-called *Minkowski functionals* $M_\nu(\mathfrak{D})$, $\nu = 0, 1, \ldots, d$. For a two-dimensional drum $\mathfrak{D}$ its Minkowski functionals are related, in turns, to its *area*, to its *boundary length*, and to its *Euler characteristic* (which is a measure of the connectivity of $\mathfrak{D}$) as:

$$A(\mathfrak{D}) = M_0(\mathfrak{D});\ U(\mathfrak{D}) = 2\pi M_1(\mathfrak{D});\ \chi = \pi M_2(\mathfrak{D}) \tag{3.1.9}$$

Equation (3.1.8) shows that for the vibrations of a 2-dimensional drum there is a connection between the shape-descriptors of the drum and the eigenvalues of the *Laplacian*. Since Laplace equations such as (3.1.3a) govern most physical processes inside the pore space Π of rocks, where the boundary conditions (3.1.3b) are prescribed on the surface $\partial\Pi$ separating pores from solid grains, several attempts have been made to express properties such as *electric conductivity* or *hydraulic permeability* of a porous rock in terms of *Minkowski functionals*. Examples, where Minkowski functionals of the pore space are possibly related to rock properties are compiled in Table 3.1.

Before formally introducing *Minkowski functionals*, we discuss a simple geometrical problem, which shows how these functionals had first arisen in a natural way.

In *Integral Geometry*, the *parallel body* or *parallel set* A_ε *at distance* ε of a structure $A \subset \mathbb{R}^d$ is defined as the set of all points of $\mathbb{R}^d$ with distance less than ε to A, where $\varepsilon \geq 0$ (Mecke 2000: 123–124; Michielsen and De Raedt 2001: 481; Arns et al. 2002a, b: 64). Formally,

Table 3.1. Studies on the connection between the rock's Minkowski Functionals and its physical properties

Property or process	References
Chemical dissolution	Khachkova et al. (2020); Noiriel et al. (2004)
Conductivity	Arns (2002)
Connectivity	Vogel (1997a, b)
Correlation function	Korvin (2016); Torquato and Stell (1983)
Dielectric constant	Torquato (1984)
Diffusion, diffusivity	Epstein (1989); Garboczi (1990)
Elasticity	Arns (2002)
General treatment and mathematics	Arns (2002); Arns et al. (2002b); Hadwiger (1957); Mecke (2000); Michielsen and De Raedt (2001); Santaló (1976); Stoyan et al. (1989); Torquato (2002b)
NMR relaxation	Sen et al. (1990)
Permeability	Alqahtani et al. (2018); Arns (2002); Bakke and Øren (1997); Berg (2014); Costa (2006); Epstein (1989); Garboczi (1990); Korvin (2016); Lehmann et al. (2008); Liu et al. (2017a, b); Mecke and Arns (2005); Nair et al. (2021); Noiriel et al. (2004); Schwartz et al. (1993), Sen et al. (1990); Scholz et al. (2012); Slotte et al. (2020); Sudakov et al. (2019); Sun et al. (2011)
Porosity	Costa (2006); Noiriel et al. (2004)
Relative permeability	Lehmann et al. (2008); McClure et al. (2018)
Resistivity	Slotte et al. (2020)
Rock characterization, modeling	Armstrong et al. (2018); Bakke and Øren (1997); Jiang and Arns (2020); Vogel et al. (2010)
Seismic velocity	Arns et al. (2002a)
Tortuosity	Ghanbarian et al. (2013); Korvin (2016);
Transport	Scholz et al. (2015)

$$A_\varepsilon = \left\{x|x\epsilon\mathbb{R}^d;\, dist(A,x) \leq \varepsilon\right\}, \tag{3.1.10a}$$

where

$$dist(A,x) = \min_y(|x-y|;\, y\epsilon A) \tag{3.1.10b}$$

denotes the minimal *Euclidean distance* and $|x|$ is the standard norm in the Euclidean space $\mathbb{R}^d$. An alternative definition of the parallel body is that

$$A_\varepsilon = \bigcup_{x\epsilon A} B_\varepsilon^{(d)}(x), \tag{3.1.10c}$$

that is, it is the union of all d-dimensional balls $B_\varepsilon^{(d)}(x)$ of radius ε and with centers x inside of A. For $\varepsilon < 0$ the parallel body is defined as

$$A_{\varepsilon<0} = \{x; d(A^c, x) \leq |\varepsilon|\}^c, \tag{3.1.10d}$$

In other words it is the *complement set* of all points with distances less than $|\varepsilon|$ to the complement A^c of A. For $\varepsilon < 0$ the parallel body $A_{\varepsilon<0}$ is given by the set of all points x so that the union of the ε -balls centered at them equals A:

$$\bigcup_{x\epsilon A_{\varepsilon<0}} B_\varepsilon(x) = A. \tag{3.1.10e}$$

Using $\varepsilon > 0$ corresponds to *dilation*, and $\varepsilon < 0$ to the *erosion* of a structure.

If the d-dimensional volume of A , $V^{(d)}(A)$ is known, what will be the volume of the parallel body A_ε $(\varepsilon > 0)$? Consider a few examples:

(i) For $d = 1$ let $L^{(1)}(a)$ be a straight line segment of length a, say, $L^{(1)}(a) = [0, a]$, In this case $V^{(1)}(L^{(1)}(a)) = l(L^{(1)}(a)) = a$ and $B_\varepsilon^{(1)}$ is a segment of length 2ε, $0 \leq \varepsilon$. Then $L_\varepsilon^{(1)}(a) = [-\varepsilon, a+\varepsilon]$, and

$$V^{(1)}\left(L_\varepsilon^{(1)}\right) = a + 2\varepsilon; \tag{3.1.11.i}$$

(ii.a) For $d = 2$ let $D^{(2)}(a)$ be a circular disc of radius a, with $V^{(2)}(D^{(2)}(a)) = A(D^{(2)}(a)) = a^2\pi$, and $B_\varepsilon^{(2)}$ is a disc of radius ε, $0 \leq \varepsilon$. Then $D_\varepsilon^{(2)}(a) = D^{(2)}(a+\varepsilon)$ is a disc of radius $a + \varepsilon$, and

$$V^{(2)}\left(D_\varepsilon^{(2)}(a)\right) = \pi(a+\varepsilon)^2 = \pi a^2 + 2\pi a\varepsilon + \pi\varepsilon^2. \tag{3.1.11.ii.a}$$

(ii.b) As another example in 2D, consider a square $S^{(2)}(a)$ with side a, its parallel set $S_\varepsilon^{(2)}(a)$ is a *rounded square* with sides $a + 2\varepsilon$, whose four corners are rounded off with quarter-circles of radius ε. In this case

$$V^{(2)}\left(S_\varepsilon^{(2)}(a)\right) = a^2 + 4a\varepsilon + \varepsilon^2\pi \tag{3.1.11.ii.b}$$

(iii.a) Similarly to case (ii.a) but in 3D, let $A^{(3)}$ be a ball $B^{(3)}(a)$ of radius a with $V^{(3)}(B^{(3)}(a)) = \frac{4a^3\pi}{3}$, then $B_\varepsilon^{(3)}(a) = B^{(3)}(a+\varepsilon)$ is a ball of radius $a + \varepsilon$, whose volume is

$$V^3\left(B_\varepsilon^{(3)}(a)\right) = \frac{4\pi}{3}\cdot(a+\varepsilon)^3 = \frac{4\pi}{3}a^3 + 4\pi a^2\varepsilon + 4\pi a\varepsilon^2 + \frac{4\pi}{3}\varepsilon^3. \tag{3.1.11.iii.a}$$

(iii.b) A 3D analog of the rounded square $S_\varepsilon^{(2)}(a)$ is the *rounded cube*, $C_\varepsilon^{(3)}(a)$ which is the parallel body at distance ε from the 3D cube $C^{(3)}(a)$. It is a cube with edge-length $a + \varepsilon$, rounded off at its corners by 8 identical spherical sectors of radius ε. By simple calculation

$$V^{(3)}\left(C_\varepsilon^{(3)}(a)\right) = a^3 + 6a^2\varepsilon + 3a\pi\varepsilon^2 + \frac{4\pi}{3}\varepsilon^3. \tag{3.1.11.iii.b}$$

To find what is common in Eqs. (3.1.11.i–3.1.11.iii.b), and to generalize these equations, we need some preliminaries. First, note that all shapes in the examples, i.e. the straight line segment, the circular disc, the square and the spherical ball, are *convex*, in the sense that a collection of points K in the d-dimensional Euclidean space $\mathbb{R}^d$ is called a *convex set* if for every pair of points in K, the entire line segment joining them also lies in K, that is $\lambda \cdot x + (1-\lambda) \cdot y \epsilon K$ whenever $x, y \epsilon K$ and $0 < \lambda < 1$. Single points and the empty set are also considered convex. We shall only consider convex sets that are bounded and closed, they are called *compact sets*. Denote the class of such sets by $\mathbb{K}$.

A transformation $\mathcal{T} : x \to x'; x, x' \epsilon \mathbb{R}^d$ is called *isometry* if it preserves the Euclidean distance between any two points x and y: $|x-y| = |x'-y'| = |\mathcal{T}x - \mathcal{T}y|$.

Functionals defined on $\mathbb{K}$ assign a real value $f(K)$ to every $K \epsilon \mathbb{K}$. In *Integral Geometry* (Stoyan et al. 1989: 25) we usually deal with such functionals which satisfy the following requirements:

Non-negativity:

$$f(K) \geq 0 \textit{ if } K \epsilon \mathbb{K}; \tag{3.1.12.a}$$

Motion invariance: A functional f is motion invariant if

$$f(gK) = f(K) \text{ for } K \epsilon \mathbb{K}, g \epsilon \mathcal{G}, \tag{3.1.12.b}$$

where $\mathcal{G}$ denotes the group of all translations and rotations in $\mathbb{R}^d$;

Monotonicity: If

$$K_1, K_2 \epsilon \mathbb{K}, K_1 \subset K_2, \textit{then } f(K_1) \leq f(K_2); \tag{3.1.12.c}$$

Additivity:

$$f(K_1 \cup K_2) = f(K_1) + f(K_2) - f(K_1 \cap K_2) \textit{ if } K_1, K_2, K_1 \cup K_2 \epsilon \mathbb{K}. \tag{3.1.12.d}$$

(The condition $K_1 \cup K_2 \epsilon \mathbb{K}$ in Eq. 3.1.12.d is necessary, because the union of two convex sets is not always convex, though their intersection is.)

Continuity: $\lim_{n \to \infty} f(K_n) = \text{f(K)}$ whenever $\{K_n\}$ is a sequence of compact sets such that

$$\lim_{l \to \infty} K_l = \text{K}. \tag{3.1.12.e}$$

(In Eq. 3.1.12.e the second limit is meant in the *Hausdorff metric*, Hausdorff 1914).

Examples of functionals satisfying Eqs. (3.1.12.a–3.1.12.e) in $d = 1, 2, 3$ dimensions are the *length* $l(K)$ if $d = 1$; the *area* $A(K)$ and *boundary length* $U(K)$ if $d = 2$;

the *volume* $V(K)$ and surface area $S(K)$ if $d = 3$. A further important functional is the *mean breadth* $B(K)$ of a convex body $K \in \mathbb{R}^n$ which is a measure of its size and is defined as follows. Let S^{n-1} be the surface of the n-dimensional unit sphere, and $\vec{n}$ a random direction in S^{n-1} . The *breadth* $B_{\vec{n}}(K)$ of the body K measured in the direction $\vec{n}$ is the *shortest distance* between a pair of $(n-1)$—dimensional parallel hyperplanes such that they are perpendicular to $\vec{n}$ and the body K is enclosed between the two of them. The *mean breadth* is obtained by taking the average of the direction-dependent breadths, $B(K) = \langle B_{\vec{n}}(K) \rangle$ with respect to $\vec{n}$, supposed to be uniformly distributed over S^{n-1}. In 3D (Stoyan et al. 1989: 25)

$$B(K) = \frac{1}{2\pi} \int_0^{\pi/2} \int_0^{2\pi} l\left(p_{\beta,\lambda} K\right) sin\beta d\lambda d\beta, \tag{3.1.13}$$

where $l\left(p_{\beta,\lambda} K\right)$ is the projected length of K on the line passing through the origin and through the point $(sin\beta cos\lambda, sin\beta sin\lambda, cos\beta)$. Volumes, surface areas and mean breadths for some common shapes in $\mathbb{R}^3$ are compiled in Table 3.2, from (Stoyan et al. 1989: 26: Table 1.1; Hadwiger 1957: 215; Santaló 1976: 226).

We can rewrite Eqs. (3.1.11.i–3.1.11.iii.b) in terms of the functionals l, A, U, V, S, B, as follows:

$$for\ d = 1 : V^{(1)}\left(L_\varepsilon^{(1)}\right) = a + 2\varepsilon = l\left(L^{(1)}\right) + 2\varepsilon \tag{3.1.14.i}$$

$$\begin{aligned} for\ d = 2 : V^{(2)}\left(D_\varepsilon^{(2)}(a)\right) &= \pi a^2 + 2\pi a\varepsilon + \pi\varepsilon^2 \\ &= A\left(D^{(2)}\right) + U\left(D^{(2)}\right)\varepsilon + \pi\varepsilon^2 \end{aligned} \tag{3.1.14.ii.a}$$

$$\begin{aligned} for\ d = 2 : V^{(2)}\left(S_\varepsilon^{(2)}(a)\right) &= a^2 + 4a\varepsilon + \varepsilon^2\pi \\ &= A\left(S^{(2)}\right) + U\left(S^{(2)}\right)\varepsilon + \pi\varepsilon^2 \end{aligned} \tag{3.1.14.ii.b}$$

Table 3.2. Volumes, surface areas and mean breadths for some common shapes in $\mathbb{R}^3$

K	V	S	B
Segment, length *l*	0	0	$l/2$
Disc, radius *r*	0	$2\pi r^2$	$(\pi/2)r$
Square, side *a*	0	$2a^2$	a
Ball, radius *r*	$(4/3)\pi r^3$	$4\pi r^2$	$2r$
Cylinder, radius *r*, height *h*	$\pi r^2 h$	$2\pi r(r+h)$	$(h+\pi r)/2$
Cube, side *a*	a^3	$6a^2$	$(3/2)a$
Rectangular parallelepiped, sides *a*, *b*, *c*	abc	$2(ab+bc+ca)$	$(a+b+c)/2$

Source: Stoyan et al. (1989): 26: Table 1.1; Hadwiger (1957): 215; Santaló (1976): 226

$$\textit{for } d = 3: V^3\left(B_\varepsilon^{(3)}(a)\right) = \frac{4\pi}{3}\cdot(a+\varepsilon)^3 = \frac{4\pi}{3}a^3 + 4\pi a^2\varepsilon + 4\pi a\varepsilon^2 + \frac{4\pi}{3}\varepsilon^3$$
$$= V\left(B^{(3)}\right) + S\left(B^{(3)}\right)\varepsilon + 2\pi B\left(B^{(3)}\right)\varepsilon^2 + \frac{4\pi}{3}\varepsilon^3 \qquad (3.1.14.\text{iii.a})$$

$$\textit{for } d = 3: V^{(3)}\left(C_\varepsilon^{(3)}(a)\right) = a^3 + 6a^2\varepsilon + 3a\pi\varepsilon^2 + \frac{4\pi}{3}\varepsilon^3$$
$$= V\left(C^{3)}\right) + S\left(C^{(3)}\right)\varepsilon + 2\pi B\left(C^{(3)}\right)\varepsilon^2 + \frac{4\pi}{3}\varepsilon^3. \qquad (3.1.14.\text{iii.b})$$

Note that in 2D the Eqs. (3.1.14.ii.a, 3.1.14.ii.b), and in 3D the Eqs. (3.1.14.iii.a, 3.1.14.iii.b), have a common form. This is generally true for convex bodies, under certain restrictions on their shape, as we always have

$$\textit{for } d = 2: V^{(2)}\left(K_\varepsilon^{(2)}\right) = A\left(K^{(2)}\right) + U\left(K^{(2)}\right)\varepsilon + \varepsilon^2\pi \qquad (3.1.15.\text{a})$$

$$\textit{for } d = 3: V^{(3)}\left(K_\varepsilon^{(3)}\right) = V\left(K^{(3)}\right) + S\left(K^{(3)}\right)\varepsilon + 2\pi B\left(K^{(3)}\right)\varepsilon^2 + \frac{4\pi}{3}\varepsilon^3. \qquad (3.1.15.\text{b})$$

Equations (3.1.15.a, 3.1.15.b) are special cases of *Steiner's formula* (Hadwiger 1957) for the d-dimensional volume $V^{(d)}(K_\varepsilon)$ of the parallel body K_ε at a distance ε of a convex body K:

$$V^{(d)}(K_\varepsilon) = \sum_{\nu=0}^{d}\binom{d}{\nu}W_\nu^{(d)}(K)\varepsilon^\nu. \qquad (3.1.16)$$

In Steiner's formula (3.1.16), the coefficients $W_\nu^{(d)}(K)$ are called *Minkowski functionals* and are related to the functionals l, A, U, V, S, B, as follows:

$$\textit{For } d = 1: W_0^{(1)}(K) = l(K),\ W_1^{(1)}(K) = 2; \qquad (3.1.17.\text{a})$$

$$\textit{For } d = 2: W_0^{(2)}(K) = A(K),\ W_1^{(2)}(K) = \frac{1}{2}U(K),\ W_2^{(2)}(K) = \pi; \qquad (3.1.17.\text{b})$$

$$\textit{For } d - 2: W_0^{(3)}(K) - V(K),\ W_1^{(3)}(K) - \frac{1}{3}S(K),$$
$$W_2^{(3)}(K) = \frac{2\pi}{3}B(K),\ W_3^{(3)}(K) = \frac{4\pi}{3}. \qquad (3.1.17.\text{c})$$

The Minkowski functionals satisfy the properties (3.1.12.a–3.1.12.e). Their importance is due to the fact that, by a theorem of Hadwiger (1957), *in* $\mathbb{R}^d$ *every motion-invariant, additive and continuous functional* φ *over* $\mathbb{K}$ *can be expressed as a linear combination of the Minkowski functionals* $W_\nu^{(d)}$ *in the form*

$$\varphi(K) = \sum_{\nu=0}^{d} a_\nu W_\nu^{(d)}(K) \tag{3.1.18}$$

where the real coefficients a_ν are independent of K.

For a compact domain $A \subset \mathbb{R}^d$, with smooth surface ∂A, the Minkowski functionals may be expressed as *curvature integrals*. Let A be such a compact domain with $d-1$ *principal radii of curvature* $r_i (i = 1, \ldots, d-1)$. The functionals $W_\nu(A)$, for $\nu \geq 1$, can be defined by the surface integrals

$$W_{\nu+1}(A) = \frac{1}{(\nu+1)\begin{pmatrix} d \\ \nu+1 \end{pmatrix}} \int_{\partial A} S_\nu\left(\frac{1}{r_1}, \cdots, \frac{1}{r_{d-1}}\right) dS \tag{3.1.19}$$

where S_ν denotes the ν-th *elementary symmetric function* (Macdonald 1995), dS is the $(d-1)$-dimensional surface element. In 3D, for example,

$$W_1 = \frac{1}{3}\int dS,\ W_2 = \frac{1}{3}\int\left[\frac{1}{2}\left(\frac{1}{r_1} + \frac{1}{r_2}\right)\right] dS,\ W_3 = \frac{1}{3}\int \frac{1}{r_1 r_2} dS \tag{3.1.19a}$$

In some works, they normalize the functionals W_ν and use, instead,

$$M_\nu(A) = \frac{\omega_{d-\nu}}{\omega_\nu \omega_d} W_\nu(A),\ \nu = 0 \cdots, d, \tag{3.1.20a}$$

where ω_d is the volume of the d-dimensional unit sphere, $\omega_d = \pi^{d/2}/\Gamma\left(1 + {}^{d}/_{2}\right)$, namely

$\omega_0 = 1, \omega_1 = 2, \omega_2 = \pi, \omega_3 = 4\pi/3$. For configurations $\mathcal{V}$ consisting of unions of *voxels* (unit cubes) $\{\mathcal{V}_i\}$ it is more convenient to use another normalization of the Minkowski functionals by introducing

$$V_\nu(\mathcal{V}) = \frac{W_\nu(\mathcal{V})}{\omega_\nu} \tag{3.1.20b}$$

because this assures that for a unit cube

$$V_\nu(\mathcal{V}_1) = 1. \tag{3.1.20c}$$

The Minkowski functionals V_ν of a sphere of radius r are

$$V_0 = \frac{4}{3}\pi r^3,\ V_1 = \pi r^2,\ V_2 = 2r,\ V_3 = 1. \tag{3.1.20d}$$

In literature, we find four common forms and notations of the Minkowski functionals: $M_\nu, V_\nu, \overline{V}_\nu, W_\nu$, defined as follows ($\omega_\nu = \frac{\pi^{\mu/2}}{\Gamma(1+d/2)}$ is the volume of the ν

dimensional unit ball):

$$V_\nu := \frac{\omega_{d-\nu}}{\omega_d} M_\nu;\ \overline{V}_{d-\nu} := \frac{\omega_{d-\nu}}{\omega_d}\binom{d}{\nu} M_\nu;\ W_\nu := \frac{\omega_\nu \omega_d}{\omega_{d-\nu}} M_\nu \qquad (3.1.20\text{e,f,g})$$

In (Schmalzing et al. 1995: Table 1) these different notations are compiled, for the 3D case, and they are expressed in terms of the geometric quantities V (Volume), S (Surface area), H (Mean curvature), and χ (Euler characteristics).

In $\mathbb{R}^d$ the Minkowski functional $W_\nu^{(d)}$ has the *scaling property* that for a λ times dilated domain λA one has

$$W_\nu^{(d)}(\lambda A) = \lambda^{d-\nu} W_\nu^{(d)}(A). \qquad (3.1.21)$$

A fundamental result of *Integral Geometry* is Hadwiger's (1957) *kinematic* formula, which generalizes Steiner's formula (Eq. 3.1.16). The *kinematic formula* relates to the case when instead of a sphere B_ε, an arbitrary convex body B moves inside another convex body A and when moving, it is also randomly rotated. The formula describes the *expected Minkowski functionals* of the intersection $A \cap B$, expectation being taken over the motions $g = (\vec{r}, \Theta) \epsilon \mathcal{G}$, i.e. translations $\vec{r}$ and rotations Θ of B. The kinematic formula states that

$$\int_{\mathcal{G}} W_\nu(A \cap gB) dg = \sum_{\mu=0}^{\nu} \binom{\nu}{\mu} W_{\nu-\mu}(B) M_\mu(A) \qquad (3.1.22)$$

where the integration $\int dg = \int d\vec{r} \times \int d\Theta$ is the direct product of the integrations over *restricted* translations and all possible orientations. The integration over translations is *restricted* in such a way that the translated B should have nonempty intersection with A.

For *rock-physical applications* the definition of Minkowski functionals and the kinematic formula (3.1.22) must be properly generalized, otherwise we could not use them to describe the geometry of a pack of convex grains, or the pore space inside a rock which consists of several overlapping holes and tunnels. For such purposes, Minkowski functionals should be extended for systems consisting of a *finite number of convex bodies*. Introduce the *convex ring* $\mathcal{R}$, which consists of all subsets $A \subset \mathbb{R}^d$ which are finite unions of convex sets:

$$A = \bigcup_{i=1}^{l} K_i; K_i \epsilon \mathbb{K}, l = 1, 2, \ldots \qquad (3.1.23)$$

If $A_1 \epsilon \mathcal{R}$ and $A_2 \epsilon \mathcal{R}$ then also $A_1 \cup A_2 \epsilon \mathcal{R}$ and $A_1 \cap A_2 \epsilon \mathcal{R}$. In order to extend the Minkowski functionals from $\mathbb{K}$ to $\mathcal{R}$ we recursively introduce the *Euler characteristic* χ , which is defined for a single convex body $K \epsilon \mathbb{K}$ as

$$\chi(K) = \begin{cases} 1, K \neq \emptyset \\ 0, K = \emptyset \end{cases}. \tag{3.1.24}$$

The *Euler characteristic* is an additive, motion invariant functional. For the body $A \epsilon \mathcal{R}$ figuring in Eq. (3.1.23), the additivity of χ yields

$$\chi(A) = \chi\left(\bigcup_{i=1}^{l} K_i\right) = \sum_{i=1}^{l} \chi(K_i)$$
$$- \sum_{i<;j} \chi(K_i \cap K_j) + \cdots + (-1)^{l+1} \chi(K_1 \cap \cdots \cap K_l) \tag{3.1.25}$$

As all sets on the RHS of (3.1.25) are convex, Eq. (3.1.24) is applicable for all terms, and it yields the numerical value of $\chi(A)$.

Recalling that any single point $x \in \mathbb{R}^d$ is by definition a convex set, the *characteristic function* of the set A can be written as $I_A(x) = \chi(A \cap x)$. Then the *volume* of A is given by

$$W_0^{(d)} = \int_{\mathcal{G}} I_A(gx)dg = \int_{\mathcal{G}} \chi(A \cap gx)dg \tag{3.1.26a}$$

where the integration is taken over all motions of $\mathcal{G}$. This suggests the following generalization of the Minkowski functionals for $\mathcal{R}$:

$$W_\nu^{(d)}(A) = \int_{\mathcal{G}} \chi(A \cap gE_\nu)dg \text{ for } \nu = 0, \ldots, d-1, \tag{3.1.26b}$$

and

$$W_d^{(d)}(A) = \omega_d \chi(A), \text{ with } \omega_d = \pi^{d/2}/\Gamma(1 + d/2), \tag{3.1.26c}$$

where in (3.1.26b) E_ν is a ν -dimensional *hyperplane* in $\mathbb{R}^d$. The normalization in (3.1.26c) is chosen to assure that for a d-dimensional ball $B_d(r)$ with radius r, $W_\nu^{(d)}(B_d(r)) = \omega_d r^{d-\nu}$ where ω_d is the volume of the unit ball.

Hadwiger (1957) proved that all Minkowski functionals defined on $\mathcal{R}$ are motion invariant and additive:

$$W_\nu^{(d)}(A) = W_\nu^{(d)}\left(\bigcup_{i=1}^{l} K_i;\right) = \sum_{i=1}^{l} W_\nu^{(d)}(K_i)$$
$$- \sum_{i<j} W_\nu^{(d)}(K_i \cap K_j) + \cdots + (-1)^{l+1} W_\nu^{(d)}(K_1 \cap \cdots \cap K_l), \tag{3.1.27}$$

and that any motion-invariant, additive and continuous functionals φ over the convex ring $\mathcal{R}$ can be represented by a linear combination of the $W_{\nu}^{(d)}$, as in Eq. (3.1.18).

The Euler characteristic $\chi(A)$ is a *topological measure* of A, i.e. it is independent of its size. For $d = 2$, $\chi(A)$ equals the number of the connected components of A minus the number of holes. In three dimensions $\chi(A)$ counts the number of connected components of A minus the number of tunnels in it, plus the number of cavities.

The Euler characteristic of the surface ∂A of an *n*-dimensional body A in *d*-dimensional space is

$$\chi(\partial A) = \chi(A)\big[1 - (-1)^{n}\big], \tag{3.1.28}$$

where $n \leq d$ (Likos et al. 1995). In *geometrical topology* the definition of Euler characteristic stems from Euler's (1758) celebrated *polyhedral-formula*

$$V - E + F = 2, \tag{3.1.29}$$

which states a relation between the number of vertices V, number of edges E and number of faces E on any surface which is *homeomorphic* to a sphere. The *Euler characteristic* is defined as $\chi = V - E + F$, different surfaces have different values of χ. (Korvin 1992:132–133, 302–303)

In rock physics Minkowski functionals are especially promising when combined with the *Boolean model,* a tool of *stochastic geometry* (Arns 2002; Arns et al. 2001; Kellerer 1984; Matheron 1975; Mecke 1998, 2000, 2001; Mecke and Wagner 1991; Mecke and Stoyan 2000; Michielsen and De Raedt 2001). This model simulates complex structures in $d = 2$ or $d = 3$ dimensions by randomly placed, *overlapping* convex shapes (spheres, spheroids, sticks, etc.). Because of the overlaps, the Boolean model is mostly used to generate a random *pore space*, except when studying the *diagenesis* of sedimentary rocks where solid grains could melt into each other, or thick sections in transmission microscopy where several layers of grains produce a kaleidoscopic image consisting of overlapping shapes (Miles 1976).

We only consider here the *Boolean model of pore space*. To assure random locations one takes a large number of points $\{x_i\}$ following *Poisson-distribution* with density ρ in a finite convex domain $\Omega \subset \mathbb{R}^d$, and superimposes around each of them pores of random orientation. The pores can overlap. If $\Delta \subset \Omega$ is a randomly selected subset of Ω, the probability that Δ contains exactly N of these points is $\exp(-\rho|\Delta|) \cdot \frac{(\rho|\Delta|)^N}{N!}$ and if $|\Delta| \gg 1$, then $\rho \approx N/|\Delta|$ ($|\Delta|$ denotes the volume of Δ). For the construction we use a collection of shapes, $\{\Pi_k\}$, $\Pi_k \in \mathcal{R}$ where all Π_k are either convex, or a union of a finite number of convex bodies. In this case, the Minkowski functionals V_ν, $\nu = 0, 1, \cdots, d$ exist, let us denote the average Minkowski functionals for the shapes in $\{\Pi_k\}$ by $\langle V_\nu \rangle = m_\nu$, $\nu = 0, 1, \cdots, d$. We randomly assign to each of the Poisson-distributed points $x_i \in \Omega$ some pore-shape $\Pi_i \in \{\Pi_k\}$, and copy it around x_i to get a pore $\Pi(x_i) \subseteq \Omega$, $x_i \in \Pi(x_i)$. (Those familiar with the concept of *Minkowski addition* will realize that $\Pi(x_i) = x_i \oplus \Pi_i$. The *Minkowski addition* $C = A \oplus B$ of two sets A and B is a set consisting of all points

$\vec{c} = \vec{a} + \vec{b}$ where $\vec{a} \in A$ and $\vec{b} \in B$.) Now, subject the pore $\Pi(x_i)$ to random movements $g_i = (\vec{r}_i, \Theta_i)\epsilon\mathcal{G}$ consisting of rotations and such *restricted* translations for which $g_i\Pi(x_i)$ still contains the point x_i. Any given realization of this construction *for fixed N* is called a *Boolean pore model,* $\Pi_N = \cup_{i=1}^{N} g_i\Pi(x_i)$, denote its Minkowski functionals by $V_\nu^{(N)} = V_\nu(\Pi_N)$. In the limit, when both $|\Omega| \to \infty$ and $N \to \infty$ in such a way that $N/|\Omega| \to \rho$, the *expected values of the normalized* Minkowski functionals $\langle V_\nu(\Pi_N)/N\rangle_{N\to\infty,|\Omega|\to\infty,}\ \nu = 0, 1, \cdots, d$ can be determined in terms of the Minkowski functionals of the *average pore shape* $\langle V_\nu\rangle = m_\nu$, $\nu = 0, 1, \ldots, d$, and the density ρ of the Poisson process, using the kinematic Steiner-Hadwiger formula, Eq. (3.1.22) (Mecke and Wagner 1991; Kellerer 1984):

$$\langle V_0(\Pi_N)\rangle/N \sim \left(1 - e^{-\rho m_0}\right)/\rho \tag{3.1.30a}$$

$$\langle V_1(\Pi_N)\rangle/N \sim m_1 e^{-\rho m_0} \tag{3.1.30b}$$

$$\langle V_2(\Pi_N)\rangle/N \sim \left(m_2 - m_1^2\rho\right)e^{-\rho m_0} \tag{3.1.30c}$$

$$\langle V_3(\Pi_N)\rangle/N \sim \left(m_3 - 3m_1m_2\rho + m_1^3\rho^2\right)e^{-\rho m_0} \tag{3.1.30d}$$

These formulae are sometimes called *Miles Equations*, referring to his early (Miles 1976) paper. If the Π_k are *identical spherical pores* of radius r (*IOS*, or *Identical Overlapping Spheres* model: Arns 2002; Torquato 1984; Torquato and Stell 1983; Weissberg 1963), the average Minkowski functionals of the single pore in Eqs. (3.1.30.a–3.1.30.d) are $m_0 = \frac{4}{3}\pi r^3$, $m_1 = \pi r^2$, $m_2 = 2r$, $m_3 = 1$ (see Eq. 3.1.20d). A Boolean model with *spherical grains*, has also been proposed as a model of consolidated sandstone, it yields a good match to *Fontainebleau sandstone* data (Arns et al. 2002a, b; Martys et al. 1994; Schwartz et al. 1993; Thovert et al. 2001).

In Eqs. (3.1.30.a–3.1.30.d) the numerically estimated means $\langle V_\nu(\Pi_N)\rangle$ are only acceptable, if the volume of the domain, $|\Omega|$ is sufficiently large (Andreeva et al. 2021). The *Representative Elementary Volume* (REV) is the smallest value of $|\Omega|$ for which the estimated average Minkowski functionals do no not vary significantly during further volume increases. Note that the REV depends on the lithology of the core, and it might be different for the various Minkowski functionals. According to Andreeva et al. (2021) the REV should be found as follows.

Take a digital core of cubic shape $C^{(M\times M\times M)}$ which consists of M^3 voxels, $M \gg 1$ where each voxel is of size L and volume L^3. Suppose we compute Minkowski functionals for the *white voxels* (representing *pore space*). Take randomly a sample voxel $v\epsilon C^{(M\times M\times M)}$, let $1 \le k_1 < k_2 < \cdots \le M$ be a sequence of increasing odd integers, and for each k_i consider the cube $C^{(k_i\times k_i\times k_i)} \subset C^{(M\times M\times M)}$ centered at the same selected voxel v. Let $\Pi^{(i)} \subseteq C^{(k_i\times k_i\times k_i)}$ be the set of white voxels in $C^{(k_i\times k_i\times k_i)}$, and compute the average normalized Minkovski functionals $V_\nu(k_i) = \left\langle V_\nu\left(\Pi_N^{(i)}\right)\right\rangle/N$ ($\nu = 0, 1, 2, 3$) figuring in Eqs. (3.1.30.a–3.1.30.d). Repeat this for sufficiently many initial sample voxels v, if the graphs of the functions $V_\nu(k_i)$ tend to some asymptotes when $k_i \rightarrow C_\nu$, we accept the volume $V_\nu = (C_\nu L)^3$ as REV for the Minkowski functional V_ν (see Andreeva et al., 2021: Figs. 2–4).

A *3-dimensional binary image* of a rock consists of elementary lattice-cubes ("voxels") $\{\mathcal{V}_{ijk}\}, i, j, k = 1, \ldots, N$ where each voxel may have only one of the two values, 0 or 1. Voxels with a 1-value (*black voxels* or *full voxels*) usually correspond to solid grains, voxels with a 0-value (*white voxels* or *empty voxels*) correspond to pore space.

Denote by Π the *pore space*, that is the union of all white voxels. In the published algorithms to compute the *normalized Minkowski functionals* of Π, $v_\nu = V_\nu\left(\bigcup_{\mathcal{V}_{ijk} \in \Pi} \mathcal{V}_{ijk}\right)/N^3$, $(\nu = 0, 1, 2, 3)$, each voxel is considered as a union of its *interior*, its 6 open *faces*, 12 open *edges* and 8 open *vertices*. For example, the configuration of three black voxels put together in an L-shape consists of $n_c = 3$ open cubes, $n_f = 16$ open faces, $n_e = 28$ open edges, $n_v = 16$ open vertices (Blasquez and Poiraudeau 2003: Fig. 1).

It can be shown (Hyde et al. 1990; Michielsen and Raedt 2001; Mecke and Seyfried 2002) that for any configuration of voxels the computation of Minkowski functionals can be reduced to counting the (black or white, depending on our interest) *elementary objects*, i.e. the number n_c of open cubes, number n_f of open faces, number n_e of open edges, and number n_v of open vertices. In terms of these numbers:

$$\text{The } \textit{enclosed volume} \text{ is } \quad V = n_c, \tag{3.1.31a}$$

$$\text{the } \textit{surface area} \text{ is } \quad S = -6n_c + 2n_f, \tag{3.1.31b}$$

$$\text{the } \textit{mean breadth} \text{ is } \quad 2B = 3n_c - 2n_f + n_e, \tag{3.1.31c}$$

$$\text{the } \textit{Euler characteristic} \text{ is } \quad \chi = -n_c + n_f - n_e + n_v, \tag{3.1.31d}$$

(as an exercise, check that for 3 voxels put together in an L shape, $V = 3, S = 14, 2B = 5, \chi = 1$, the first three are in voxel-size units).

In case of white voxels, a particularly important functional is the *porosity*, Φ, defined for a cube consisting of altogether L^3 voxels as

$$\Phi = n_c/L^3. \tag{3.1.31e}$$

In 2-dimensional cross-section, if a configuration of *pixels* contains n_s open *squares*, n_e open *edges*, and n_v *vertices*, then the area A, perimeter U; and Euler characteristic χ are

$$A = n_s,\ U = -4n_s + 2n_e,\ \chi = n_s - n_e + n_v. \tag{3.1.32a,b,c}$$

In analogy to Eq. (3.1.31.e), by the *De Lesse principle* (Delesse 1847; Oleschko 1998-1999; Stoyan et al. 1989: 280) the porosity can be estimated from 2D microscopic images of a rock consisting of $L \times L$ pixels as

$$\Phi = n_s/L^2. \tag{3.1.32.d}$$

In order to count open faces, open edges and open vertices in a 3D image one should precisely define when voxels are considered as *connected*. The most common definition is (Michielsen and De Raedt 2001) that two voxels are connected when they have either a face, an edge, or a vertex in common, that is an interior pore voxel is connected to 26 neighbors. In Michielsen and De Raedt (2001) the algorithm for computing Minkowski functionals is stated in terms of black voxels. They start out from an initially *white* background and add black voxels one-by-one to the already existing configuration. Their recursive equations determine how the numbers of open faces, open edges and open vertices change when one adds one black voxel at the position $(i, j.k)$ to the already existing pattern. To solve the equations, all 26 neighbors of the voxels must be examined. The algorithm was implemented in FORTRAN 90 on a single processor, improved versions are given in (Blasquez and Poiraudeau 2003; Legland et al. 2011; Schindelin et al. 2015; Ivonin et al. 2020; Boelens and Tchelepi 2021; MATHWORKS 2022).

Hadwiger's characterization theorem (Hadwiger 1957) asserts that in $\mathbb{R}^d$ every motion-invariant, additive and continuous functional φ over $\mathbb{K}$ can be expressed as a linear combination of the Minkowski functionals $W_\nu^{(d)}$ in the form $\varphi(K) = \sum_{\nu=0}^{d} a_\nu W_\nu^{(d)}(K)$ (see Eq. 3.1.18). Of course there are also *non-additive* functionals that yield useful information about the geometry of a voxelized rock, such as the *m-point covariance function*, the *chord-length-distribution*, the *pore-size-distribution*, etc. (Arns 2002; Torquato 2002a, b; Berchtold 2007; Korvin 2016). To discuss them, we need the concept of *characteristic function.*

Let $\Theta \subset \mathbb{R}^d$ be a *random* set in $\mathbb{R}^d$, denote by I_Θ its *characteristic function,* that is for $\vec{x} \epsilon \mathbb{R}^d$ one has $I_\Theta(\vec{x}) = \begin{cases} 1 \ if\ \vec{x} \in \Theta \\ 0 \ if\ \vec{x} \notin \Theta \end{cases}$. The *m-point covariance function* $C_\Theta^{(m)}(\vec{x}_1, \cdots, \vec{x}_m)$ gives the probability that the m points $\vec{x}_1 \epsilon \mathbb{R}^d, \ldots, \vec{x}_m \epsilon \mathbb{R}^d$ all belong to Θ:

$$C_\Theta^{(m)}(\vec{x}_1, \ldots, \vec{x}_m) = Prob\big[\vec{x}_1 \epsilon \Theta, \ldots, \vec{x}_m \epsilon \Theta\big] = \left\langle \prod_{i=1}^{m} I_\Theta(\vec{x}_i) \right\rangle \tag{3.1.33}$$

The 2-point covariance function $C_{\Theta}^{(2)}(\vec{x}_1, \vec{x}_2)$ is *directly related to the Minkowski functionals*. Namely, we have the following relationship between the 2-point-covariance function of a random set $\Theta \subset \mathbb{R}^d$ and its *specific area* (Berryman and Blair 1986; Stoyan et al. 1989: Eq. 6.2.3. on p. 177; Berchtold 2007: Theorem 2.5.2 on pp. 44, 200–201; Korvin 2016): *Let* $\Theta \subset \mathbb{R}^d$ *be a sufficiently regular, isotropic random set in* $\mathbb{R}^d$*, whose surface area and volume exist, and denote by* $R(r) = C_{\Theta}^{(2)}(\vec{x}_1, \vec{x}_2)$ *where* $\vec{x}_1 \epsilon \mathbb{R}^d, \vec{x}_2 \epsilon \mathbb{R}^d, |\vec{x}_1 - \vec{x}_2| = r$*. Then*:

$$S_{spec}(\Theta) = \frac{S(\Theta)}{V(\Theta)} = -\frac{d \cdot \omega_d}{\omega_{d-1}} \cdot \left.\frac{\partial R(r)}{\partial r}\right|_{r=0}, \tag{3.1.34}$$

where ω_d is volume of the d-dimensional unit ball. Equation (3.1.34) establishes a relationship between two Minkowski functionals (S and V) and the derivative of the 2-point covariance function at $r = 0$.

We have seen that the pore space Π of a sedimentary rock can be simulated using the *Boolean model*, i.e. a large system consisting of the union of $N \gg 1$ randomly placed and rotated, overlapping convex bodies, $\Pi = \bigcup_{i=1}^{N} K_i$. By additivity,

$$\chi(\Pi) = \chi\left(\bigcup_{i=1}^{N} K_i\right) = \sum_{i=1}^{N} \chi(K_i)$$
$$- \sum_{i<j} \chi\left(K_i \cap K_j\right) + \cdots + (-1)^{l+1} \chi(K_1 \cap \cdots \cap K_N)$$

that is the Euler characteristic $\chi(\Pi)$ depends on all possible nth-order multiple intersections of the elementary building blocks $\{K_i\}$, and consequently it bears information on the n-point-correlations as well. By the construction of the Minkowski functionals $W_{\upsilon}^{(d)}(\Pi)$ according to Eqs. (3.1.26a, 3.1.26b) this information is inherited by the Minkowski functionals, but to date, no such relations have been found between Minkowski functionals and *higher-order correlations*. Three-point correlations would be especially important to compute corrections to the *Hashin–Shtrikman variational bounds* for the effective properties of composite materials (see Sect. 9.3.a; Hashin and Shtrikman 1962; Beran 1965; Milton 1982).

In topology, multiple-connected surfaces are characterized by their *genus*, defined as "*the greatest number of distinct, continuous non self-intersecting closed curves which may be drawn on a surface without separating it into distinct regions*" (Flegg 1974; Korvin 1992: 302). The genus of the surface of the sphere is 0, the genus of a one-fold torus is 1, of a two-fold torus 2, of an n-fold torus, n. In lay terms, the genus g is the "number of handles" on the body, it is linearly related to the Euler characteristics of the surface as

$$\chi(\partial A) = 2(1 - g) \tag{3.1.35}$$

(Korvin 1992: 303, Eq. 4.5.44; Michielsen and De Raedt 2001: Eq. 33). A pioneering study of Smalley (1967; see also Korvin 1992: 302, Fig. 4.50.c.) used the *genus* as function of geological time t to describe the structural changes in sandstone during *diagenesis*. In this model, for a single sand particle $g = 0$; if two particles are joined together the genus does not change, but if a third one is attached to both of these, the configuration becomes topologically equivalent to a *torus* and the genus rises to one. When more and more particles are joined together, the genus will steadily rise (sedimentation, sand to sandstone transition). During compaction and diagenesis the pores begin to fill, tangential contacts develop between adjacent grains, and the genus decreases (sandstone to *quartzite* transition).

This continuous growth of sand grains, resulting in their interpenetrating each other, is similar to the *dilation* process of *Mathematical Morphology* and *Integral Geometry* where the dilation $A_E(\varepsilon)$ of a set A by a convex *structural element* E is defined as $A_E = A \oplus E(\varepsilon)$, ε being the characteristic size of the structuring element. If we select for the structuring element a d-dimensional ball of radius ε, and increase the radius until the structure becomes a one-phase medium, we can write this as an operation $A_\varepsilon = A \oplus \varepsilon B(1)$.

In a similar way as we defined the *Minkowski addition* of two sets ($Z = X \oplus Y = \{z \text{ such that } \vec{z} = \vec{x} + \vec{y}, \quad \vec{x} \in X \quad \text{and } y \in Y\}$, the *Minkowski subtraction* $Z = X \ominus Y$ may be defined as consisting of all points $\vec{z}$ such that to each point $\vec{y} \in Y$ one can find the points $\vec{x} \in X, \vec{z} \epsilon Z$ with $\vec{z} = \vec{x} - \vec{y}$. If no such points exist, the subtraction $X \ominus Y = \emptyset$ is empty. One should note that dilation and erosion cannot be interchanged, i.e., that $(X_\varepsilon)_{-\varepsilon} \neq X$ and $(X \ominus Y) \oplus Y \subset X$ is only a subset of X.

If $K \epsilon \mathbb{R}^d$ is *convex*, as a generalization of Eq. (3.1.16) we can express all Minkowski functionals (MFs) of the parallel body K_ε as

$$V_\nu(K_\varepsilon) = \sum_{\mu=0}^{d-\nu} \binom{d-\nu}{\mu} \frac{\omega_{\nu+\mu}}{\omega_\nu} V_{\nu+\mu}(K)\varepsilon^\mu. \tag{3.1.36}$$

Thus, if one knows the MFs $V_\nu(K) = V_\nu\left(K_{\varepsilon,\varepsilon=0}\right)$ of a convex body K, Eq. (3.1.36) provides the MFs $V_\nu(\varepsilon)$ for arbitrary ε , and conversely, from the MFs of the parallel volumes $V_\nu(K_\varepsilon)$ of a convex body K one can infer the MFs $V_\nu(K)$ of the original K. This is no longer true for *non-convex shapes* such as the pore space Π of sedimentary rocks. In this case the family of Minkowski functions $V_\nu(\varepsilon)$ of the parallel body Π_ε contains *more information* about the spatial structure than just the $V_\nu(\Pi) = V_\nu\left(\Pi_{\varepsilon,\varepsilon=0}\right)$ obtained at zero dilation.

3.1.2 Applications

The evolution of the *genus* during *compaction* and *diagenesis* (documented by Smalley 1967) suggests to similarly study the evolution of the *Minkowski functionals* of the pore or grain space of rocks during erosion and dilation. Apparently, this methodology has been first used to discriminate between different cosmological models of the early universe (Mecke et al. 1994).

Examples for such a *dynamic display* of the Minkowski functionals of the *intergranular pore space* are shown in Fig. 3.1, for the cases when (a) the solid part of the rock is approximated by a *Boolean model* built up of cubes, (b) by a *Boolean model* consisting of identical spheres, (c) and by a *Gaussian random field model*, (d) while the fourth example is a real sandstone core. All examples were digitized on a 128^3 lattice. The first graph shows that as the grains grow (> 0, *dilation*) the available relative pore volume shrinks, while for $\varepsilon < 0$ (*erosion*), the grains become smaller and the relative pore volume increases to 1. In the last three graphs the morphological measures $V_\nu(\Phi)$ are shown as functions of the *relative pore volume* (*porosity*) $V = V_0^{(pores)}(\varepsilon)/V_{total}$, with $0 \leq V \leq 1$, instead of in terms of ε.

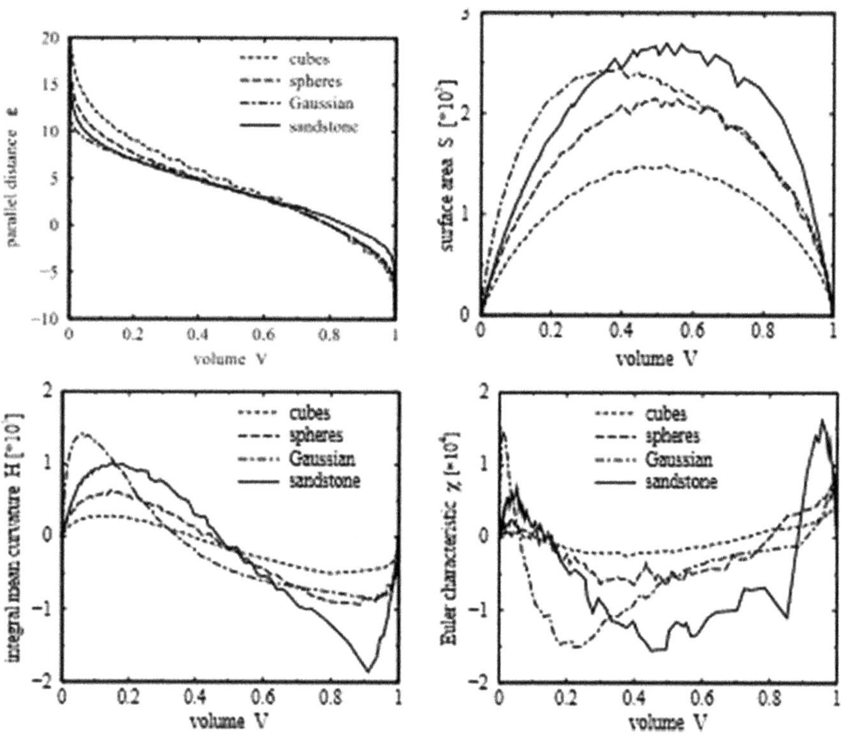

Fig. 3.1 Minkowski functions M_ν (ε) of the parallel body at distance ε for four configurations: **a** Boolean model of overlapping cubes with edge length $L = 16$, **b** spheres of radius $R = 8$, **c** a Gaussian random field, **d** a core of sandstone, all digitized on a 128^3 lattice. (Lengths are in units of pixel size.) The figure on the upper LHS shows the dependence of the relative pore volume $V = V_0(\varepsilon)/V_{total}$ on the parallel distance ε, the others show the development of the surface area S, mean curvature H, and Euler characteristics χ as V changes from 0 to 1. (From Mecke 2000: 124, Fig. 3; with kind permission of Prof. Klaus Mecke, and *Springer Nature*)

Comparing the Minkowski functionals $M_\nu(\varepsilon)$ of the models with those of the real rock, suggests that no simple Boolean models could reproduce the irregular Minkowski functions of the sandstone. Properly constructed Gaussian random fields seem better candidates, but it is difficult to achieve a fair match between the families $M_\nu(\varepsilon)$ for models and for the real rock.

We have seen how to construct a *Boolean model* of the pore space, using a given convex pore body and superimposing its randomly rotated replicas on random points which are Poisson-distributed with density ρ. The *expected* Minkowski functionals $\langle V_0(\Pi_N)\rangle, \ldots, \langle V_3(\Pi_N)\rangle$ were given by the *Miles equations* (3.1.30.a–3.1.30.d).

In rock modeling, or "*reconstruction*" (Arns et al. 2002a, b) we are facing the inverse problem: from a single observed image of a Boolean rock model (or of a real rock) find an equivalent 'pore' ensemble which would generate the same global morphology characterized by the average Minkowski functionals $V_0^{(im)} = \langle V_0(\Pi_N)/N\rangle, \ldots, V_3^{(im)} = \langle V_3(\Pi_N)/N\rangle$ of the image. Counting the number N of distinct pores on the 3D image and assuming $\frac{N}{|\Omega|} = \rho$ where $|\Omega|$ is the size of the 3D image, we find the Poisson density ρ . Using the average MFs $\left\langle V_0^{(im)}(\Pi_N)/N\right\rangle, \ldots, V_3^{(im)} = \langle V_3(\Pi_N)/N\rangle$ of the image we can determine the pore parameters $\{m_0, m_1, m_2, m_3\}$ from Eqs. (3.1.30.a–3.1.30.d), and with the Boolean method we can construct with Monte Carlo technique *infinitely many equivalent models*, all matching the measured MFs $V_i^{(im)}$. This is just another problem, namely there exists a wide range of different Boolean structures leading to the same Minkowski functionals. This *indeterminacy* had been noted by Ohser and Mücklich (2000), and Berchtold (2007).

In some recent studies they construct a rock model with desired MFs without solving Eqs. (3.1.30.a–3.1.30.d), using *Metropolis-Monte Carlo* optimalization (Slotte et al. 2020) or *Deep Neural Networks* (DNNs; Volkhonskiy et al. 2021). Slotte's group (2020) generated the pore space Π with prescribed Minkowski Functionals using the *Metropolis-Monte-Carlo* approach ("*Simulated Annealing*", see Sect. 11.1). They simulated *sedimentation*, that is started out from an initially empty spatial region, and gradually filled it with grains of random size and shape. An "energy" E was assigned to each specific configuration of grains, depending on their surface geometry, and any given realization was created with probability

$$p \propto \exp(-E). \tag{3.1.36}$$

The prescribed values for the Minkowski functionals, $\Phi_0, \sigma_0, \chi_{s0}$ (porosity, specific surface, specific Euler characteristic) were enforced using *Simulated Annealing*, that is, by including a *penalty term*

$$M = \left(\lambda_0(\Phi - \Phi_0)^2 + \lambda_1(\sigma - \sigma_0)^2 + \lambda_2(\chi_s - \chi_{s0})^2\right)/T \tag{3.1.37}$$

in the system energy and slowly reducing the *annealing temperature* T. The λ_n are tunable weights to improve convergence. We note here, that *sedimentation* is only the very first of the multistage physical processes (*sedimentation*, *consolidation*, *compaction* and *cementation*) creating a rock. More realistic models of sandstone formation are discussed in (Bakke and Øren 1997; Øren et al. 1998).

The main applications of the $M_\nu(\varepsilon)$ graphs include finding the REV (*Representative Elementary Volume*) for different calculations on digitized rock samples (Andreeva et al. 2021), and testing the *reconstructions of porous media.* (Mecke 2000; Arns 2002; Arns et al. 2001, 2002a, b; Berchtold 2007; Volkhonskiy et al. 2021). Volkhonskiy's group (2021) reconstructed 3D porous media from 2D slices using **a** *Deep Neural Network* (DNN), and they measured the goodness of reconstruction in terms of the errors in their Minkowski functionals $M_\nu(\varepsilon)$.

As Klaus Mecke noted (2000: 123), "the complicated pore structure of an interconnected three-dimensional network of capillary channels of non-uniform sizes and shapes distinguishes a porous medium from any other solid or planar substrate". Whenever transport (of electric current, contaminants, water, etc.) takes place in this complex network, it is natural to expect that the efficiency of transport would strongly depend on the surficial geometry of the pore space, and that Minkowski Functionals, especially when combined with the Boolean approach, would provide a powerful tool for their modeling and numerical study. If it would be possible to theoretically relate the Minkowski functionals of the sandstone's pore-space to electric conductivity or hydraulic permeability, we could predict these physical properties from the study of a miniature rock cutting, the boldest "dream" petrophysicists have ever had (Korvin 2016).

The effect of the rock's microstructure on its permeability has remained an unsolved problem despite intensive research for almost a century. The role of porosity is better understood (Kozeny 1927; Costa 2006) but the role of the other parameters such as tortuosity, pore constriction, Euler characteristic still have not been satisfactorily studied (Epstein 1989; Scholz et al. 2012; Ghanbarian et al. 2013; Berg 2014). The new paradigm of estimating conductivity and permeability from the *Minkowski functionals* of the pore space has arisen only recently (Mecke 2000; Mecke and Arns 2005; Lehman et al. 2008; Vogel et al. 2010; Scholz et al. 2012, 2015; Liu et al. 2017a, b; Armstrong et al. 2018; Slotte et al 2020; Nair et al. 2021). "The prediction of transport in porous structures from measures of the morphology and topology is an unsolved problem"—wrote Mecke in (2000) (p. 145). The first experiences have not been satisfactory, as we judge from the following quotes:

"We conclude that the permeability and formation factor are not uniquely determined by the Minkowski functionals" (Slotte et al. 2020); "While Minkowski functionals were proposed for the characterization of porous permeability, until now there is no evidence that MFs can reliably characterize the permeability of general porous media" (Nair et al. 2021).

There has been a recent, very reasonable proposal (Ivonin et al. 2020) to include in such predictions the pores' *Betti numbers* separately, in addition to their Euler characteristics: that is the zeroth Betti number associated with the *number of individual pores,* and the first Betti number counting *tunnels* in the sample.

As in other fields of rock physics, *AI* (*Artificial Intelligence*) aficionados claim that all necessary information about permeability and resistivity are already included in the Minkowski Functionals, but the relation is nonlinear and complex, so that only

Deep Neural Networks (DNNs) can mine it out (Al Qahtani et al. 2018; Sudakov et al. 2019; Nair et al. 2021. A *DNN* is a *Neural Network* with several layers: Bengio 2009; Schmidhuber 2015). In a well-documented study by Nair et al. (2021), who used a 9-layered DNN (where the hidden layers had 24,24,48,72,32,4 and 1 nodes), the permeability of computer-generated 2-dimensional porous media was reliably predicted from the *Minkowski tensors* of the pore microstructure. (*Minkowski tensors* are generalizations of the *Minkowski functionals*, also showing their *anisotropy*: Schröder-Turk et al. 2011, 2013).

I do not know of any empirical or theoretical equations stating an explicit dependence of some rock-physical property on *all four* Minkowski Functionals. The Euler characteristic χ of the pore space and its role in determining the *connectivity* of the pore network, and predicting hydraulic permeability, will be further discussed in Sect. 3.2. As for the first two MFs of the pore space (*specific pore volume* and *specific pore surface*), they are frequently grouped together in *empirical formulae* where some effective physical property is related to a combination of porosity Φ and surface-to-volume ratio S/V_p of the pore space, as in case of permeability

$$k = 10^{6.6}\left(\Phi^m V_p/S\right)^{2.1}, \tag{3.1.37}$$

or *NMR-relaxation time*

$$T_1 = 10^{3.08}\left(V_p/S\right)^{0.9}, \tag{3.1.38}$$

(Sen et al. 1990; Mecke 2000:124–125). Here the prefactors $10^{6.6}$ and $10^{3.08}$ take care of the applied physical units and empirical constants, but no geometrical arguments could derive the exponents "m", "2.1″ and "0.9″ in terms of the MFs of the pore space. In the original *Kozeny–Carman equation* (Carman 1936; Korvin 2016) the permeability is given by

$$k = \frac{1}{b}\Phi^3 \cdot \frac{1}{S_{spec}^2} \cdot \frac{1}{\tau_h^2} \tag{3.1.39}$$

where b is a shape factor of order one, $\Phi\epsilon[0, 1]$ is porosity, S_{spec} is *specific surface area* defined as total pore surface area per unit bulk volume of rock, τ_h is *hydraulic tortuosity* (Chap. 2; Korvin 2016). Equation (3.1.39) contains two further geometric factors, b and τ_h which have not been related to the MFs as yet.

A similar problem arises with *Archie's law* (Archie 1942; Arns 2002; Slotte et al. 2020) expressing rock resistivity ρ in terms of *electric tortuosity* τ_e and porosity Φ. (Electric tortuosity is not necessarily the same as τ_h, *hydraulic tortuosity*). If the grains in a porous material are non-conducting and the pores are completely filled with a conducting fluid of resistivity ρ_w (as in a brine-saturated sandstone), by Archie's law the rock's resistivity is proportional to fluid resistivity:

$$\rho = F \cdot \rho_w \tag{3.1.40}$$

where the constant F is the *formation factor*, related to porosity as

$$F = \Phi^{-m}. \tag{3.1.41}$$

The so-called *cementation exponent* "m" may vary between rock types, its default value is $m = 2$. The formation factor can also be expressed as

$$F = \tau_e/\Phi, \tag{3.1.42}$$

from the last two equations one has

$$\tau_e = \Phi^{-(m-1)}. \tag{3.1.43}$$

This equation is a geometrical relation for the pore space. Korvin (2016) derived a similar scaling law between τ_h and Φ using a fractal pore model. A recent review of tortuosity and the connection between τ_e and τ_h is (Ghanbarian et al. 2013).

There have been many attempts to establish Archie's law, and to find the theoretical value of the cementation exponent m. Some notable ideas (for details, see Sect. 9.4) have been *Effective Medium Theory* (Kirkpatrick 1973), and its interesting variant, the *Granular Effective Medium Approximation* (Yonezawa and Cohen 1983), where the conductive fluid is gradually filled with non-conducting grains, simulating sedimentation; *Percolation Theory* (Korvin 1992: 292); and *Fractal Random Walk in the Pore Space* (Roy and Tarafdar 1997). Yonezawa and Cohen (1983) established a *geometrical relation* between the *average coordination number* Z of the pores and the *cementation exponent* m:

$$Z = \frac{2m}{m-1}. \tag{3.1.44}$$

To find the *geometric* meaning of the exponent m in Archie's law, we start out from Korvin's (1982) generalized mean value formula for the conductivity of a 2-component composite:

$$\sigma = \left\{\lambda\sigma_1^t + (1-\lambda)\sigma_2^t\right\}^{1/t}, \tag{3.1.45}$$

where λ, $1-\lambda$ are the respective volume ratios between those parts of the rock which take part, resp. do not take part in the conduction of the DC current, and assume t is a positive parameter. Letting $\sigma_2 \to 0$, the limit is $\sigma = \lambda^{1/t} \cdot \sigma_1$ which formally agrees with Archie's equation, $\sigma = \sigma_1\Phi^m$ if we assume that $\lambda^{1/t} = \Phi^m$ i.e.

$$\lambda = \Phi^{mt}. \tag{3.1.46}$$

As m and t are both positive, $\lambda = \Phi^{mt} < \Phi$, that, is, as the mean value formula (3.1.45) shows, only *a fraction of the available pore space takes part in the DC electric conduction.* This is similar to the hypothesis of Pérez-Rosales (1982) for *hydraulic flow* in porous rocks, according to whom the total porosity consists of a *stagnant part* and a *flowing* part,

$$\Phi = \Phi_s + \Phi_f \tag{3.1.47}$$

in such a way that the fluid particles are only flowing through Φ_f. Pérez-Rosales (1982, see also Korvin et al. 2014: 1131) identified Φ_f as that part of porosity which is associated with the *pore throats*, and hypothesized that

$$\Phi_f = \Phi^m, \tag{3.1.48}$$

thus giving an exciting geometric interpretation of the cementation exponent m. He supported his claim with pore- and throat area measurements on microscopic images. A recent study (Slotte et al. 2020) on the use of Minkowski functionals to predict permeability and resistivity, also noted that only the *connected pore space* is relevant for transport properties, and, consequently, recommends that *isolated pores should be excluded from the calculation of the Minkowski functionals.* These authors however failed to notice any connection between the total volume of such disconnected pores and the cementation exponent.

3.2 Euler Characteristics

In Rock Physics, the main use of the *Euler Characteristic* (EC), or *Euler number* χ is to characterize the connectivity of the pore space. It is a *topological invariant* of the pore space, in the sense that it does not *change* during continuous deformations, such as low-amplitude stretching, compression or twisting which only change the "shape" (i.e. geometry) of the pores but do no not create new pores, destroy old ones, or merge nearby ones together. The Euler characteristic χ can be calculated from the *Betti numbers*, β_i as

$$\chi = \beta_0 - \beta_1 + \beta_2 \tag{3.2.1}$$

where β_0 refers to the number of connected clusters of pores of the system; β_1 is the number of loops formed by chains of adjacent pores; and β_2 counts the number of solid grains totally surrounded by void space. In most cases, such inclusions do not exist and thus usually $\beta_2 = 0$; $\chi = \beta_0 - \beta_1$. As an example, the letter **B** consists of one main connected part ($\beta_0 = 1$) and it has 2 redundant loops ($\beta_1 = 2$) that is its Euler number is $\chi = -1$. As another simple example (Sossa-Azuela et al. 2011), for the Spanish letter **Ñ** one has $\chi(\tilde{\mathrm{N}}) = \beta_0(\tilde{\mathrm{N}}) - \beta_1(\tilde{\mathrm{N}}) + \beta_2(\tilde{\mathrm{N}}) = 2 - 0 + 0 = 2$. If S_1 and S_2 are two structures which are equivalent topologically in the sense that there is a continuous transformation from any one to the other (known as *homeomorphy*), then they have the same Betti numbers, and thus $\chi(S_1) = \chi(S_2)$ (Bobrowski and Skraba 2020).

In some works (Scholz 2014; Wirner 2015; Liu et al., 2017a, b) they also define an *open Euler characteristic* χ_o which does not take into account *enclosed void components* because they do not contribute to flow.

Another topologically invariant property characterizing the connectivity of a surface is its *genus*, *g*. It is the largest number of nonintersecting simple closed curves that can be drawn on the surface without separating it. The genus g can be expressed in terms of the Euler characteristic χ via the relationship

$$\chi = 2 - 2g. \tag{3.2.2}$$

A cube, sphere or a disc has genus zero, a torus has genus one. For 2D or 3D images of rocks, the Euler characteristic is defined for *binary structures*, for example Π = pore space, Γ solid grains; depending on our interest, one is called *foreground*, the other *background*. In the first such studies Π was selected as *foreground*, and χ_Π was computed in order to estimate hydraulic properties (Mecke and Arns 2005; Lehmann et al. 2008). In more recent studies of *multiphase flow* (e.g. Schlüter et al. 2016) one fluid phase (namely *wetting* or *non-wetting*) is considered as *foreground*, while the other fluid phase and the solid phase together constitute the *background*, so that we have two different χ, for the wetting phase χ_w and for the non-wetting phase χ_n. In the binary case, χ_Π and χ_Γ satisfy the following *consistency relation* proved, under mild conditions, by Ohser et al. (2002, 2003). In *n* dimensions:

$$\chi^{(n)}\left(\overline{\Gamma}\right) = (-1)^{n+1}\chi^{(n)}(\Pi) \tag{3.2.3a}$$

or, generally,

$$\chi^{(n)}\left(\overline{X^c}\right) = (-1)^{n+1}\chi^{(n)}(X) \tag{3.2.3b}$$

for such bounded sets X for which both $X \cap B$ and $\overline{X^c \cap B}$ belong to the convex ring $\mathcal{R}$ for all convex bodies B. Here, $\overline{A}$ denotes the *closure* of A , A^c is the complement of A.

If we define the Betti numbers $\beta_0, \beta_1, \beta_2$ both for the solid matrix Γ and the pore space Π, they will satisfy

$$\left.\begin{aligned} \beta_0^\Pi &= 1 + \beta_2^\Gamma \\ \beta_1^\Pi &= \beta_1^\Gamma = g \\ \beta_0^\Gamma &= 1 + \beta_2^\Gamma \end{aligned}\right\}, \tag{3.2.4.a,b,c}$$

implying another *consistency relation* (Sahimi 1985: Eqs. 4.89–4.90)

$$\beta_0^\Pi + \beta_2^\Pi = \beta_0^\Gamma + \beta_2^\Gamma. \tag{3.2.5}$$

The most frequently used equations for computing the Euler characteristic χ are compiled in Table 3.3.

Table 3.3. Equations to compute the Euler number $\chi(X)$

Name of the object X	Eq. #	References	Equation	Symbols
Polytope (n dim.)	3.2.6.a	Ohser et al. (2003)	$\chi^{(n)}(P) = \sum_{k=0}^{n}(-1)^k \#\mathcal{F}^{(k)}(P)$	$\mathcal{F}^{(k)}(P)$ = set of all k-faces of P
Polyhedra (3 dim.)	3.2.6.b	Andrew (2019); Nakahara (2003): Eq. 2.31	$\chi = V - E + F$	$V = \#vertices$, $E = \#\ edges$, $F = \#faces$
Structure partitioned into convex volume elements	3.2.6.c	Vogel and Roth (2001)	$\chi = \#vertices = \#edges + \#faces = \#volumes$	
Network model	3.2.6.d	Vogel (2002)	$\chi = N_n - N_b$	N_n = # of non-isolated nodes, N_b = #bonds
Graph G	3.2.6.e	Huber (2018)	$\chi(G) = V - E$	V = #vertices, E = # edges
Complex network consisting of k- *cliques* (completely connected subgraphs with $k+1$ nodes)	3.2.6.f	Amorim et al. (2019)	$\chi = \sum_k (-1)^{k+1} Cl_k$	Cl_k = #k-cliques
Triangulated surface	3.2.6.g	Konkle et al. (2003)	$\chi = V - E + T$	V = # vertices, E = #edges, T = #triangles
Continuous surfaces	3.2.6.h	Lozano-Durán and Borrell (2016)	$\chi = \frac{1}{2\pi}\iint_\Sigma K d\Sigma$	K = Gaussian curvature Σ = surface
Gauss-Bonnet theorem	3.2.6.i	Chen and Rong (2010)	$\chi = \frac{1}{2\pi}\iint_\Sigma K d\Sigma = \beta_0 - \beta_1 + \beta_2$	$\beta_0, \beta_1 \beta_2$ are Betti numbers

The classic Eq. 3.2.6.b for *polyhedra* can be also used to compute the Euler characteristic $\chi(X)$ if the body $X \in \mathbb{R}^3$ is homeomorphic to some polyhedron P. Then, by the *Euler – Poincaré – Alexandrov Theorem* (Nakahara 2003: Theorem 2.4),

$$\chi(X) = V(P) - E(P) + F(P), \tag{3.2.7}$$

and $\chi(X)$ *is independent of the polyhedron* P *as long as* P *is homeomorphic to* X.

As a simple example, the 2D surface S^2 of the sphere is homeomorphic with the cube for which $(cube) = 8, E(cube) = 12, F(cube) = 6$, so that $\chi\left(S^2\right) = \chi(cube) = 8 - 12 + 6 = +2$.

When modeling pore structures, we can build up complicated configurations by recursively *connecting* two pores with "necks" or "throats". The *connected sum* $X \# Y$ of two surfaces X and Y is a surface obtained by removing a small disc from each of X and Y and then connecting the resulting holes with a cylinder, as in Fig. 3.2.

For any two surfaces X and Y we have the following *Theorem* (Nakahara 2003, Theorem 2.5): *The Euler characteristic of the connected sum X # Y is given by*

$$\chi(X\#Y) = \chi(X) + \chi(Y) - 2 \tag{3.2.8}$$

Indeed, let the polyhedra P_X and P_Y be homeomorphic to X and Y, respectively. With suitable deformation, we achieve that both polyhedra have a small triangular face. Remove these triangles, and connect the two holes with a prism of triangular cross section. The number of vertices remains the same but the number of edges increases by three. As we removed two faces, and created three new ones (the open faces of the prism), the number of faces increases by + 1. Thus the overall change in the Euler numbers is (by Eq. 3.2.7, and the additivity of χ) $0 - 3 + 1 = -2$.

The Euler characteristics is an *extensive property* because its overall value for a domain is the sum of the values for all subdomains within it. Therefore the Euler characteristics depends on the scale over which it is measured. The Betti numbers and the genus are also scale-dependent, they change with the analyzed volume, or with the number of objects (pores, grains, oil ganglia) involved in the analysis. An appropriate normalization of χ and g leads to a more meaningful and representative metrics of connectivity if we have to compare different porous structures.

The most common *normalization* is done by the total volume V of the studied sample, i.e.

$$\chi_V(X) = \frac{\chi(X)}{vol(X)} \tag{3.2.9.a}$$

(more precisely, $\chi^{(n)}(X) = \lim_{r\to\infty} \frac{\mathbb{E}[\chi^{(n)}(X\cap rW)]}{vol(rW)}$, where $vol(\cdot)$ denotes n-dimensional volume, $\mathbb{E}$ is expectation, and W is a window with $vol(W) > 0$, see (Ohser et al. 2003: Eq. 7). χ_V is called *specific Euler characteristic*, or *Euler density*,

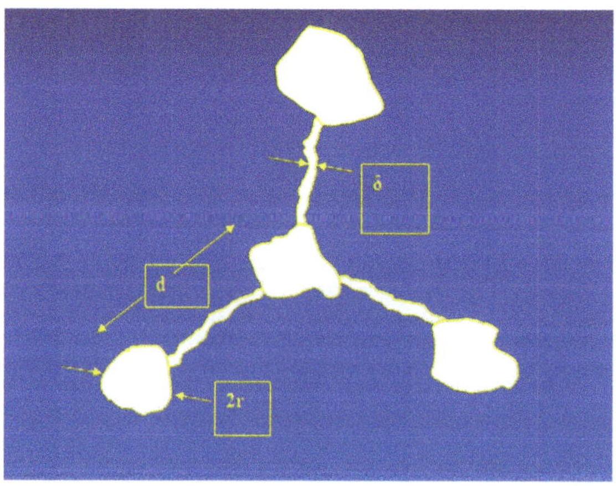

Fig. 3.2 A simple model of rock porosity

this is its most commonly used normalization in rock physics (Gong et al. 2020; Khanamiri and Torsæter 2018; Koestel et al. 2020; Ohser et al. 2003; Okuma et al. 2017; Schlüter et al. 2016; Vogel and Roth 2001; Yang et al. 2019; Zhao 2019).

Specific genus, or *genus density*, is similarly defined,

$$g_V(X) = \frac{g(X)}{vol(X)}, \tag{3.2.9.b}$$

it is used in (Huber 2018; Okuma et al. 2022). If we work with clusters of different size (such as *ganglia of residual oil*), normalization by volume can be replaced by *volume-weighted normalization*:

$$\chi_{V_s} = \frac{\sum n_{cl}\chi_{cl}\overline{V}_{cl}}{\sum n_{cl}\overline{V}_{cl}}, \tag{3.2.9.c}$$

where n_{cl} is the number of those clusters having the Euler characteristic χ_{cl} and $\overline{V}_{cl}$ is their average volume (Ott et al. 2019, 2020).

Another frequently used normalization (Dohnalik and Jarzyna 2015; Scholz 2014; Wirner 2015) is to divide χ by the total *number of objects*, as e.g. by the number of pores:

$$\chi_N\left(\bigcup_{i=1}^{N}\chi(X_i)\right) := \frac{\sum_{i=1}^{N}\chi(X_i)}{N}, \tag{3.2.9.d}$$

assuming that $X_i \cap X_j = \emptyset$ for $\neq j$.

In *multiphase flow studies*, the χ of the pore space *partially saturated* by one of the fluid-phases can be normalized by χ of the *fully saturated* pore space. Herring et al. (2013) normalized the Euler characteristic for those pores filled with an NW (*Non Wetting*) fluid at a given saturation by the characteristic at 100% NW phase saturation:

$$\hat{\chi}_{NW} = \chi_{NW}/\chi_{100\%NW}. \tag{3.2.9.e}$$

(Note that $\chi_{100\%NW}$ is the Euler number of the entire pore space which can be a negative value, so that this normalization might change the sign of χ_{NW} !)

When we study the evolution of χ through a dynamic process, such as *drainage* or *imbibition*, the Euler numbers can be computed *relative to their maximum values* (Vogel 2002), or normalized to the range [0, 1] *using their maximum and minimum values* (Andrew 2019). Amorim and his group (2019), who studied topological phase transitions in complex systems used

$$S_\chi = |log|\langle\chi\rangle|| \tag{3.2.9.f}$$

and called it *Euler entropy*. Soares and co-workers (2017) also used a logarithmic transform "*to facilitate the observation of small variations within a broad range of values*". When the Euler number is negative, they applied a negative sign to the logarithm of the absolute value of the Euler number, that is,

$$S_{\chi}^{*} = sign(\chi) \cdot |log|\chi||. \tag{3.2.9.g}$$

All these normalizations postulate the existence, and knowledge, of a *representative elementary volume.* Bachmat and Bear (1986) defined REV as a volume in which "all averaged geometrical characteristics of the microstructure in a porous material are independent of the size of the volume". Also, the Euler characteristic inside the REVmust be statistically homogenous in order to be meaningfully averaged. What Koestel et al. (2020) state in their soil-physics paper, "in comparison to studies focused on the porosity [of soils], studies investigating REVs for measures of connectivity are scarcer" also holds true for solid rocks. Studies on *the connectivity of the soil pore space* which valuated the Euler density $\chi_V\left(mm^{-3}\right)$ reported that $REV(\chi_V) > 37\,mm \times 37\,mm \times 37\,mm$ (Borges et al. 2018 for topsoil of a *Rhodic Ferralsol*), Vogel et al. (2002) documented a $REV(\chi_V) > 16\,mm \times 16\,mm \times 16\,mm$ for a sample from the B-horizon of a loamy clay *Cambisol.*

Herring et al. (2019) studied the pore space of the *Bentheimer sandstone* and reported $REV\left(\hat{\chi}_{NW}\right) \approx 600 \times 600 \times 600 voxels \approx 42.14\,mm^3$. Ohser et al. (2003) compared the Euler characteristic of autoclaved aerated concrete computed from two models: 320 × 330 × 330 voxels of spacing 31 μm, and 450 × 500 × 230 voxels of 17 μm spacing. Soares et al. (2017) determined χ from cubic subvolumes of 150^3 voxels from the micro-CT images of *carbonate* samples. In an attempt to predict permeability from the Euler characteristic of 3D images, Liu et al. (2017a, b) generated different *quasi-two-dimensional* and 3D models of the following size: the quasi-two-dimensional model's volume was 4000 × 4000 × 10 voxels; the 3D *sphere model* and 3D *stick model* were both of 480 × 480 × 480 voxels size. They also constructed an analogue *Robuglass Model* (a synthetic porous media, imaged with micro-CT) and estimated χ from a domain of size 1000 × 1000 × 572 voxels. As an estimate from below, we quote the opinion (of Gilmanov et al. 2018; Huber 2018) that any attempt to connect topology and macroscopic mechanical properties of 3D porous materials requires a REV size in the order of 100^3 at least.

A single computed Euler characteristics, even if normalized, is in itself not a useful rock-physical property. It has become customary to select subsets $\Pi_d^* \subseteq \Pi$ of the pore space, or $\Gamma_g^* \subseteq \Gamma$ of the ensemble of solid grains, or $\mathcal{V}_t^* \subseteq \mathcal{V}$ of all voxels, which satisfy certain numerical thresholds conditions

$$\mathcal{P} : \Pi_d^* = \{\pi | \pi \epsilon \Pi, \mathcal{P}(\pi) \geq d\}; \Gamma_g^* = \{\gamma | \gamma \epsilon \Gamma, \mathcal{P}(\gamma) \geq g\};$$
$$\mathcal{V}_t^* = \{v | v \epsilon \mathcal{V}, \mathcal{P}(v) \geq t\}$$

then measure the normalized EC over these subsets, and plot $\chi\left(\Pi_d^*\right)$, $\chi\left(\Gamma_g^*\right)$ or $\chi\left(\mathcal{V}_t^*\right)$ *as functions of the threshold parameters* d, g, t.

In hydraulic studies on soils (Vogel 1997a, b, 2000; Vogel and Roth 2001) the EC χ can be measured and plotted as function of the pore size d, starting with the largest pores and successively adding smaller ones. This will result in an increase of pore-connectivity and informs about how important role do the different pore size classes play in conduction. To eliminate pores smaller than a given threshold, we have to borrow the tools of *Mathematical Morphology* (Serra 1982; Vogel 2002). The morphologic procedure called *opening* is based on *morphological erosions* and *dilations* using a *spherical structuring element* $\mathbb{B}^d$, which is a ball of diameter d.

For a 3D binary rock image $X = \Pi \cup \Gamma$ that contains pores Π and grains Γ, the morphologically *eroded set* Π_{eroded} is obtained as

$$\Pi_{eroded} = \left\{x : \mathbb{B}_x^d \subset \Pi\right\} := \mathbb{P} \ominus \mathbb{B}^d \tag{3.2.10}$$

where $\mathbb{B}_x^d$ is the sphere with diameter d centered at point x. Π_{eroded} is the set of all points within Π where the structuring element fits completely into the pore space.

The morphologically *dilated set* is defined as

$$\Pi_{dilated} = \left\{x : \mathbb{B}_x^d \cap \Pi \neq \emptyset\right\} := \Pi \oplus \mathbb{B}^d \tag{3.2.11}$$

The operation "*morphological opening*" combines the operations $\oplus$ and $\ominus$:

$$\Pi_{opened}^d = \left(\Pi \ominus \mathbb{B}^d\right) \oplus \mathbb{B}^d, \tag{3.2.12}$$

it can be proved that *it removes all pores smaller than the structuring element* $\mathrm{B^d}$ *structuring element* $\mathbb{B}^d$. Note from Eq. 3.2.12 that, in general,

$$\left(X \ominus \mathbb{B}^d\right) \oplus \mathbb{B}^d \neq \left(X \oplus \mathbb{B}^d\right) \ominus \mathbb{B}^d \neq X. \tag{3.2.13}$$

Starting with a 3D image recorded with a given voxel-size λ, and applying structuring elements $\mathbb{B}^d$ with increasing diameter $d = \lambda, 2\lambda, 3\lambda, \ldots$ we shall get such realizations Π_{opened}^d of the structure where only pores greater than d are considered. The Euler characteristic $\chi(d)$ of these structures is determined by calculating χ for the different Π_{opened}^d.

Figure 3.3 shows an example for such a representation. The very small pores are disconnected, they form separate clusters, so the volume-averaged EC, $\chi_V(d)$ is negative for $d \ll 10\,mm$. The value $d \approx 10\,mm$ signifies the *percolation threshold*, the set of all pores with diameter $d \geq 10\,mm$ already becomes connective. Interestingly, in this example, continuous paths consisting of very large pores—without including small ones—can be very long. It is tempting to assume that the measured curve (solid line in Fig. 3.3) approximates a theoretical one (dashed line), whose functional form, or the significance of its maximum, are not known as yet.

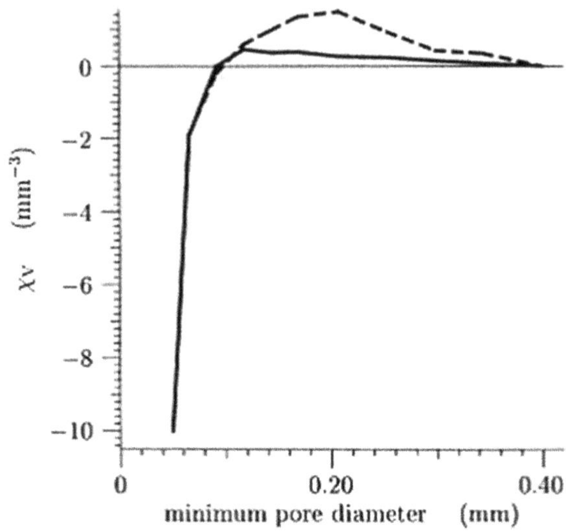

Fig. 3.3 Connectivity function of soil (solid line) and hypothetical connectivity function (dashed line). (From Vogel and Roth 2001: Fig. 5; with kind permission of Prof. Hans-Joerg Vogel, and *Elsevier Publishers*)

In many cases the voxel-values are not simply binary (black and white, i.e. 0 and 1) but they have *grayness* values, $t\epsilon\{0, 1, \cdots, 255\}$, which are sometimes normalized between their minimum and maximum, and linearly mapped to [0,1], so that the smallest value is 0, the largest is 1. The parameter t is variously called *gray scale*, *threshold*, *radio density*, *excursion parameter*, etc. In such cases (Vogel 2002; Wanner et al. 2010; Gilmanov et al. 2018) we use the "*indicator formalism*". If $\mathcal{V}$ is a set of voxels and $f(v)$, $v\epsilon\mathcal{V}$ is a real-valued function defined on the voxels, then we consider the *super-level set* $\mathcal{V}_t^* = \{v|v\epsilon\mathcal{V}, \mathrm{f}(v) \geq t\}$, compute $\chi_t^* \chi\left(\mathcal{V}_t^*\right)$ and plot it as function of the threshold t. An interesting *soil-physical example* was given in (Vogel 2002: Figs. 9, 11).

If we can measure the Euler characteristic of the invaded pore space at different stages S of saturation during an *imbibition* or *draining* process (or during *both*, if we study the *hysteresis* of the process), we can measure χ_S^* at different saturation levels $S\epsilon[0, 1]$ or $S\epsilon[0, 100\%]$. It is convenient to normalize the data relative to $\chi_{S=100\%}^*$ and plot the normalized $\hat{\chi}_S = \chi_S^*/\chi_{S=100\%}^*$ values as function of S. An interesting example is provided by the hysteresis studies of 2-phase flow (Porter et al. 2010; Herring et al. 2015; Schlüter et al. 2016: Fig. 4 and Table 1). The non-wetted EC, $\hat{\chi}_n$ as a function of S_n follows an empirical power law relation of the form

$$\hat{\chi}_n = pS_n^k + \varepsilon \tag{3.2.14}$$

This was first observed by Schlüter's group (2016), who claimed that "the exponent k turns out to be characteristic for the flow regime and independent of the underlying porous medium", an exciting hypothesis worth being followed up.

If the topology of a surface changes as an effect of a *time-dependent external field*, it is a good idea to study its evolution as function of time. Figure 3.4. shows how the Euler characteristic changes with time through simulated *sintering*. The plotted

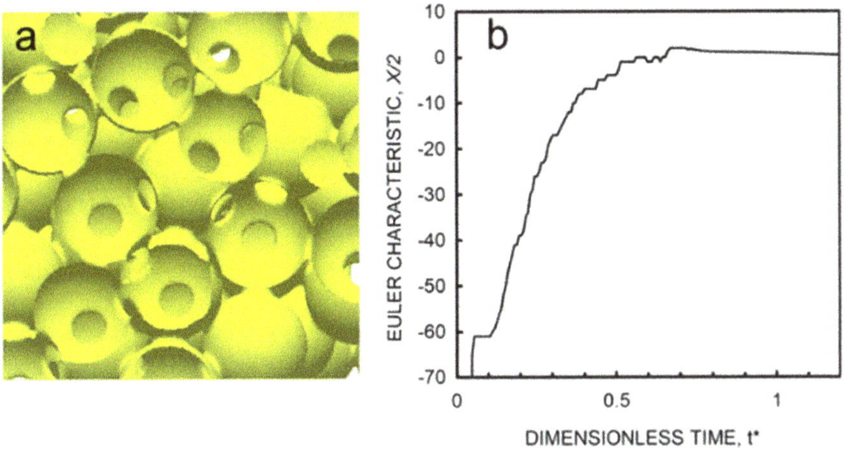

Fig. 3.4 Computer simulation of sintering by bulk diffusion. **a** At the initial stage, **b** half of the Euler characteristic $\chi/2$ as function time. (Taken with thanks from Okuma et al. 2017: Fig. 5; *Creative Commons Attribution 4.0 International License*)

χ /2 is the sum $\chi/2 = \sum_{n=1}^{N}(1-g)$ over all the N pores. Since the sum $\sum g$, is approximately equal to the total number of holes G, $\chi/2 \approx N - G$. The figure shows that as function of dimensionless time t^* the structure gradually changes *from disconnected to fully connected*. The step-wise increase of $\chi/2$ with t^* resembles a *Devil's staircase* (Korvin 1992: 95–100).

In Fig. 3.3. the *Topological Phase Transition* from disconnected to connected (*TPT*) was indicated by the *zero crossing* of the EC. Amorim et al. (2019), who studied *TPT*s in complex systems used $S_\chi = |log|\langle\chi\rangle||$ (see Eq. 3.2.9.f), called in their paper *Euler entropy*. The singularities of this function clearly indicate the *TPT*s. A refined version of this function is $S_\chi^* = sign(\chi) \cdot |log|\chi||$ (Eq. 3.2.9.g), introduced by Soares et al. (2017).

Recently, several empirical relations were proposed between the EC χ (or the *genus* g) and other pore properties. None of them have been theoretically justified. Among others, the EC was related to porosity Φ (Dohnalik and Jarzyna 2015; Aboufoul and Garcia 2017; Andrew 2019; Gong et al. 2020), to the relative density ρ of the solid phase (Okuma et al. 2017), and to the average coordination number Z (Vogel and Roth 2001; Li et al. 2015). Similarly, the genus $g = (\chi - 2)/2$ was related to the correlation length λ (Loannidis and Lang 1998); to average coordination number (Huber 2018), and to the relative density ρ of the solid phase (Okuma et al. 2022). In the *percolation* context Bobrowski and Skraba (2020) theoretically derived χ as function of the *percolation probability p* for *random cubical complexes*. Here, we can only show a few typical examples, and refer the Reader to the original papers.

According to Hadwiger's theorem (Eq. 3.1.18 of Sect. 3.1), the Minkowski functionals $\mathcal{W}_\nu^d$ are linearly independent, and every motion-invariant, additive and continuous functional φ over $\mathbb{R}^d$ can be expressed as their linear combination, $\varphi(K) = \sum_{\nu=0}^{d} a_\nu W_\nu^d(K)$, where $K \epsilon \mathbb{R}^d$ is union of a finite number of convex sets. In light of this theorem one cannot really expect a strict relation between the *two independent Minkowski functionals*, porosity and Euler characteristics. Still, Dohnalik and Jarzyna (2015: Fig. 15) documented an approximate linear relation, and Gong et al. (2020: Fig. 11), who plotted the logarithm of *connectivity density* $\beta = \frac{1-\chi_v}{V}$ against the logarithm of *porosity*, found an approximate power-function relation between the variables.

The group of Okuma plotted the Euler characteristics (2017) and the genus (2022) as functions of relative density ρ to show the structural development of glasses through viscous sintering. They distinguished three stages of sintering: in the *initial phase* $\chi_V(\rho) < 0$, then it decreases up to $\rho \approx 0.74$; in *the intermediate stage* $(0.74 < \rho < 0.96)$ it increases to its maximum; in the *final stage* the trend turns and $\chi_V(\rho)$ goes down to zero. The authors (2017: Fig. 3 and its discussion) claim: "The result is general, and applicable not only to viscous sintering of glasses but also to sintering of crystalline particles."

Some 25 years ago Ioannidis (Ioannidis and Lang 1998; Ioannidis and Chatzis 2000) established a simple linear relation between *normalized genus* g_V and average coordination number $\langle Z \rangle$ for simple *node –bond* network models of the pore space:

$$G_0 = \frac{\langle Z_3 \rangle}{2} - 1, \tag{3.2.15a}$$

where G_0 is the genus per node, $\langle Z_3 \rangle$ the average coordination number of pores with $Z \geq 3$ (only pores with $Z \geq 3$ are "nodes" in a topological sense). Huber (2018) checked and extended this result to pore networks with $Z \geq 2$ as:

$$g_V^*(\overline{z}) = \frac{g(\overline{z})}{N_{REV}} = \frac{\overline{z}}{2} - 1, z \geq 2, \tag{3.2.15b}$$

as well as for more complicated lattice approximations of the pore space. Li et al. (2015: Fig. 3) did simulations on simple *cubic* and *bcc* structures, and obtained similar results: *the dependence of the normalized genus or normalized EC on the average coordination number* $\overline{Z}$ *is linear for* $\overline{Z} \geq 2$*, but the slope of this linear relationship depends on the lattice type.* For a particular lattice structure (*fcc* grid) Vogel and Roth (2001) theoretically derived the slope in this relation:

$$\chi_V = \sqrt{2} \cdot \frac{1 - (1/2) Z_{eff}}{\lambda^3}. \tag{3.2.15c}$$

Here λ is the *grid constant*, Z_{eff} is the *effective coordination number* (for its definition and determination see the cited paper).

Ioannidis and Lang (1998), who approximated the 2D autocorrelation function of *binarized* (0- and 1-valued) *images* of reservoir rocks with a two-parameter (λ, n) stretched exponential model

$$R(u) = \exp\left[-(|u|/\lambda)^n\right], \tag{3.2.16}$$

reported a hyperbolic decrease of the *genus* with correlation distance λ (see Fig. 1. 1).

There are very few theoretical results for the expected value $\langle \chi \rangle$ of the EC on random structures other than the *Boole model* (Eq. 3.1.30.d).

Bobrowski and Skraba (2020) studied percolation on *d-dimensional cubical complexes*. Denote by Q_n^d a *d-dimensional cube* $[0, 1]^d$ (with periodic boundary conditions) and split it to $n = m^d$ equal-sized boxes where m is an integer. Cubical complexes Q are subsets of Q_n^d. Denote by $F_k(Q)$ the number of k-dimensional faces of Q ($k = 0, \ldots, d-1$). The EC of Q can be calculated by Euler's rule $\chi(Q) = \sum_{k=0}^{d} (-1)^k F_k(Q)$(Eq. 3.2.6.a in Table 3.3.).

The *random cubical complex* (percolation cluster) denoted $Q(n, p)$ is generated by taking Q_n^d and declaring each d-dimensional box as either *open* with probability p or closed with probability $1 - p$. $Q(n, p)$ is the union of all *open boxes* together with their lower dimensional faces. Using Eq. 3.2.6.a. and some algebra (Bobrowski and Skraba 2020, Appendix A), the expected EC becomes:

$$\overline{\chi}_Q(p) := \mathbb{E}\{\chi(Q(n, p))\} = n \sum_{k=0}^{d} (-1)^k \binom{d}{k} \left[1 - (1-p)^{2^{d-k}}\right] \tag{3.2.17}$$

Table 3.4 summarizes recent attempts to predict physical properties of porous materials from their Euler characteristics. To date, none of these efforts have resulted in well-established, industry-wide accepted equations, though some follow-ups of the Katz-Thompson (1986) permeability model (Scholz et al. 2012; Scholz 2014; Wirner 2015; Liu et al. 2017a, b; Zhao 2019) seem promising.

I start with the classic Archie Eq. (1942) for the conductivity of brine-filled reservoir rocks

$$\sigma_{rock} = \Phi^m \sigma_w \tag{3.2.18}$$

to indicate the inevitable difficulties when connecting topology with rock physics. In this equation, Φ is porosity, the cementation exponent m is between 1.5 and 2.5. There is ample experimental evidence for the law, but since 1942 nobody could derive it satisfactorily from basic laws of physics. According to Thovert et al. (1990) "It is well verified experimentally though it does not have any theoretical justification". Yonezawa and Cohen (1983) used the *granular effective medium theory* (see Sect. 4.4.b.) to express m as

$$m = \frac{\langle Z \rangle}{\langle Z \rangle - 2} \tag{3.2.19}$$

where $\langle Z \rangle$ is the *average coordination number of the pore network*. As discussed above we have a fair correlation between the *EC* χ and average coordination number

Table 3.4. Applications of different types of Euler characteristics

Application	Reference (name, year, type of Euler characteristic used)
Bone properties	Pothuaud et al. (2002), χ_V; Waldir et al. (2009), χ
CO_2 sequestration	Dahrabou et al. (2016), $\{B_i\}$; Herrring et al. (1995, 2013),$\hat{\chi}_{NW}$
Electric conductivity	Li et al. (2015), $\chi/\chi_{100\%air}$
Flow in porous media	Khanamiri and Torsæter (2018), χ_V ; Ott et al. (2020), χ_{vs} ; Schlüter et al. 2016, χ_V; Scholz (2014), χ_{open}; Wirner (2015), χ_{open}
Granular media	Kondic et al. (2012),$\{B_i\}$
Hydraulic conductivity of asphalt	Aboufoul and Garcia (2017),χ
Mechanical properties of porous materials	Huber (2018), g_V
Permeability	Liu et al. (2017a, b), χ_{open} ; Soares et al. (2017), sign(χ)log($\lvert\chi\rvert$) ; Zhao (2019), χ_V
Phase transitions	Amorim et al. (2019),log($\lvert\chi\rvert$)
Polycrystals	Wanner et al. (2010),χ
Pore space connectivity	Gong et al. (2020), χ_V ; Loannidis and Lang (1998), g_{node}
Pore space in concrete	Ohser et al. (2003), χ_V
Reservoir characterization	Andrew (2019), χ_{minmax} ; Dohnalik and Jarzyna (2015), $\chi_{objects}$; Gilmanov et al. (2018), $\{B_i\}$; Haruzi et al. (2021), χ ; Ott et al. (2019), χ_{vs} ; Yang et al. (2019), χ
Sintering	Okuma et al. (2017), χ_V; Okuma et al. (2022),g_V
Soil physics	Ivonin et al. (2020), χ ; Koestel et al. (2020), χ_V ; Vogel (2002), χ_V ; Vogel and Roth (2001),χ_V; Vogel et al. (2015), χ_V
Turbulence	Sossa-Azuela et al. (2011), g

$\{B_i\}$ = un-normalized Betti numbers; g = un-normalized genus ($\chi = 2-g$); g_V = genus normalized by volume;g_{node} = genus per node;
χ = un-normalized Euler number; χ_{open} = un-normalized Euler number without enclosed void components; χ_V = χ normalized by volume; χ_{vs} = χ scaled by cluster volumes; χ_{vs} (NW = non wetting fluid); $\chi_{objects}$ = χ normalized by number of objects; χ_{minmax} = χ normalized by $\chi_{max} - \chi_{min}$;$\chi/\chi_{100\%air}$

$\langle Z\rangle$ if $\langle Z\rangle \geq 2$, but this cannot be a *universal relation* because *the slope of the approximately linear relationship depends on the lattice type* what we select to model the pore space (Li et al. 2015; Huber 2018). Consequently, we cannot expect *a universal, lattice-independent relation* $m = f(\chi)$.

In hydraulic conduction, the empirical Kozeny-Carman equation (Tiab and Donaldson 2004) is used to estimate the *permeability* of a rock in terms of the *porosity, specific surface* and the *tortuosity* of the pore system:

$$k_{KC} = \left(\frac{1}{2\tau S_0^2}\right)\frac{\Phi^3}{(1-\Phi)^2} \tag{3.2.20}$$

where k_{KC} is the absolute permeability, τ the *tortuosity* (see Chap. 1.23), S_0 is the specific pore surface and Φ is porosity. As k_{KC} is not an additive functional on the pore space, *Hadwiger's Theorem* is not necessarily relevant. Still, because the other two variables on the RHS of Eq. (3.2.20), namely S_0 and Φ are *Minkowski functionals* (Sect. 1.3.1.), it is tempting to assume that the third independent variable of Eq. (3.2.20), *tortuosity*, somehow depends on the *EC* χ , but apparently no such relation has been reported yet. Khanamiri and Torsæter (2018), wrote in a related context: " Supported by a mathematical [i.e. Hadwiger's] theorem, it seemed that χ could be the missing piece to fill the knowledge gaps in the study of two-phase flow in porous media."

Katz and Thompson (1986) used *Percolation Theory* to derive the simple relationship

$$k = \frac{1}{226}\cdot l_c^2\cdot\frac{\sigma}{\sigma_0}, \tag{3.2.21}$$

where: k is absolute permeability, σ is electrical conductivity of the rock, σ_0 is the conductivity of the saturant. As they wrote, " l_c represents a dimension of a very particular subset of pores ... the effective pore size [which] is the smallest pore on the connected path of pores containing the largest pores. We call that effective pore size". The constant $1/226$ corresponds to the assumed pore geometry. They checked the rule (3.2.21) for calculating the permeabilities of some 50 rocks, ranging over 5 orders of magnitude, and found good agreement with measured values.

The Katz-Thomson equation cannot be used unless we know the *electric conductivity* of the rock and the saturant. In the 2010s two PhD students of *Universität Stuttgart*, Christian Scholz and Frank Wirner (Scholz et al. 2012; Scholz 2014; Wirner 2015) found an ingenious way to substitute the σ/σ_0 factor in Eq. (3.2.21) with another function depending on the (2D) Euler characteristic χ (more exactly the "open Euler characteristic χ_o which does not count the enclosed void components because they do not contribute to flow", see Wirner 2015). They observed that the ratio $\sigma(\Phi)/\sigma_0$ as function of porosity Φ satisfies the limit relations

$$\left.\begin{aligned}\lim_{\Phi\to 0}\sigma(\Phi)/\sigma_0 &= \tfrac{0}{\sigma_0} = 0\\ \lim_{\Phi\to 1}\sigma(\Phi)/\sigma_0) &= \tfrac{\sigma_0}{\sigma_0} = 1\end{aligned}\right\} \tag{3.2.22a}$$

and, using only topological arguments (for a 2D system) without assuming any special functional form for $\chi_o = f(\Phi)$, they showed that the function $\frac{1-\chi_o}{N}$, where χ_o is the Euler characteristic of the conducting phase in the (open) pore space and $N \gg 1$ is the number of *grains*, obeys the same asymptotics;

$$\left.\begin{array}{l}\lim\limits_{\Phi\to 0, N\to\infty} \frac{1-\chi_o}{N} = 0 \\ \lim\limits_{\Phi\to 1, N\to\infty} \frac{1-\chi_o}{N} = 1\end{array}\right\}. \tag{3.2.22b}$$

Accordingly, instead of the Katz-Thompson's Eq. (3.2.21), they proposed

$$\frac{k}{cl_c^2} = \left(\frac{1-\chi_0}{N}\right)^\alpha, \tag{3.2.23}$$

where α is a free parameter, and in the 2D case the constant is $c = 1/12$ (Scholz et al. 2012: Eq. 4; Scholz 2014: Eq. 6.1; Wirner 2015: Eq. 6.1). Equation (3.2.23) is now called the *Scholz model* (Liu et al. 2017a, b).

The problem with Eq. 3.2.23 is, as Liu et al. (2017a, b) noted, that "calculating the number of grains N in the context of real reservoir rock is not straightforward". For his *Boolean rock models*, Wirner (2015) had the "attractive but bold idea " (as he wrote on p. 84 of his 2015 Dissertation) to find the *effective number of grains*, $\hat{N}$, from the known relations (Mecke and Wagner 1991) for the expected values of the Minkowski functionals $\langle\Phi\rangle$, $\langle S\rangle$, $\langle\chi\rangle$ of a 2D random Boolean model of linear size L, consisting of N randomly placed grains of area A_0, and surface S_0:

$$\left.\begin{array}{rl}\Phi = & exp\left(-\frac{NA_0}{L^2}\right) \\ \frac{S}{L^2} = & \frac{N}{L^2}S_0 exp\left(-\frac{NA_0}{L^2}\right) \\ \frac{\chi}{L^2} = & -\frac{N}{L^2}\left(1-\frac{S_0^2 N}{4\pi L^2}\right)exp\left(-\frac{NA_0}{L^2}\right)\end{array}\right\} \tag{3.2.24a, b, c}$$

The *effective number of grains*, $\hat{N}$, is defined as that value of N which satisfies the system of equations Eq. 3.2.24 a, b, c. To find it in terms of the Minkowski functionals we first use (3.2.24a) to replace the exponential functions in (3.2.24b,c) by Φ , then express S_0 from (3.2.24b) and substitute it to (3.2.24c). Solving the resulting equation for N, we get

$$N = \hat{N} = \frac{\langle S\rangle^2}{4\pi L^2\langle\Phi\rangle^2} - \frac{\langle\chi\rangle}{\langle\Phi\rangle}. \tag{3.2.25}$$

If the model is inside a square of area L^2, its expected area will be $\langle A\rangle = L^2\cdot\langle\Phi\rangle$, and from Eqs. (3.2.23) and (3.2.25) we get the *Scholz model* in the form

$$\frac{k}{cl_c^2} = \left(\frac{1-\langle\chi\rangle}{\hat{N}}\right)^\alpha \text{ with } \hat{N} = \frac{\langle S\rangle^2}{4\pi\langle A\rangle\langle\Phi\rangle} - \frac{\langle\chi\rangle}{\langle\Phi\rangle} \tag{3.2.26}$$

Liu et al. (2017a, b) also tested Scholz's model in the *q2D* (*quasi-2-dimensional*) case and found a relationship between k/cl_c^2 and $(1-\chi)/N$ that is similar to what was reported by Scholz et al. (2012: Fig. 5.b).

The Katz and Thompson (KT) permeability model (Eq. 3.2.21) can be rewritten, in terms of the *formation factor* $F = \frac{\rho_0}{\rho}$ (ρ *resistivity* of the saturated rock, ρ_0

resistivity of the saturant) as

$$k = \frac{1}{c} \cdot l_c^2 \cdot \frac{\sigma}{\sigma_0} = \frac{1}{c} \cdot \frac{l_c^2}{F} \tag{3.2.27}$$

Near the *percolation threshold* the formation factor scales as

$$\frac{1}{F} = \left[\frac{\Phi(1 - p_c)}{1 - \Phi p_c} \right]^m \tag{3.2.28}$$

(Stauffer and Aharony 1992; Ghanbarian et al. 2014; Daigle 2016), the scaling exponent m is *universal* (independent of the lattice) and, in 3D, $m = 2$. From Eqs. (3.2.27–3.2.28)

$$k = \frac{l_c^2}{c} \left[\frac{\Phi(1 - p_c)}{1 - \Phi p_c} \right]^m . \tag{3.2.29}$$

Zhao (2019) analyzed digitized images of 15 rock samples, determined for each the percolation threshold p_c, the normalized Euler characteristic χ_V, and computed their permeability with the LBM approach (*Lattice Boltzman Machine*; see Sect. 11.2). The percolation threshold was fitted with an equation

$$p_c = a \cdot e^{b\chi_v} + c, \tag{3.2.30}$$

which resulted in another k vs. χ_v relation:

$$k = \frac{r_c^2}{c} \left[\frac{\Phi\left(1 - a \cdot e^{b\chi_v} - c\right)}{1 - \Phi\left(a \cdot e^{b\chi_v} + c\right)} \right]^m . \tag{3.2.31}$$

3.3 Persistent Homology

Persistent Homology (PH) is technically among the most advanced methods in *Statistical Rock Physics* and I must make aware the Readers of its sophisticated mathematics, and of the conceptual difficulties in interpreting *barcodes* and *persistence diagrams*. The literature has also noted this:

– "[The treatment could only] scratch the surface of the math involved, …The mathematics behind the tools discussed in this article constitute several semesters of graduate level mathematics courses" (Munch 2017); "For an extensive and rigorous introduction to the computation of persistent homology, we refer readers to the survey papers" (Aktas et al. 2019); "A complete description … is beyond the scope of this paper" (Kramár et al. 2013), etc.;

– "It is difficult to interpret the parameters of persistent homology themselves and difficult to directly relate the parameters to physical properties" (Suzuki et al. 2021).

Texts introducing the necessary mathematics include Carlsson (2009); Carlsson et al. (2005); Cohen-Steiner et al. (2007, 2010); Delgado-Friedrichs et al. (2014; 2015); Edelsbrunner and Harer (2008, 2010); Edelsbrunner et al. (2002); Ghrist (2008, 2014); Hatcher (2002); Kaczynski et al. (2006); Kong and Rosenfeld (1989); Munch (2017); Munkres (1993); Otter et al. (2017); Patania et al. (2017); Robins (1999); Robins et al. (2011); Vejdemo-Johansson and Skraba (2016); Weinberger (2011); Zomorodian (2005); Zomorodian and Carlsson (2005). For the newcomer, I recommend to start with Munch (2017); Aktas et al. (2019), and Moon et al. (2019).

If you want to read first clearly explained case histories (the best way to approach a new field!), I recommend: Aktas et al. (2019); Feng and Porter (2021); Herring et al. (2019); Moon et al. (2019); Munch (2017); Suzuki et al. (2021); Takahashi et al. (2018); Ushizima et al. (2012). Then get the software (most papers specify the software they used) and *start experimenting with your own data.*

To learn interpretation, read Feng and Porter (2021); Herring et al. (2019); Kramár et al. (2013); Moon et al. (2019; Suzuki et al. (2021); Takahashi et al. (2018); Ushizima et al. (2012). The masterful explanations and geological interpretation of the persistence diagrams for the Bentheimer, Berea and Leopard sandstones (Herring et al. 2019), and the Selma Chalk (Moon et al. 2019) are especially recommended.

To test your understanding, try to construct – on your own—barcodes and persistence diagrams for simple toy models, such as found in Aktas et al. (2019); Khachkova et al. (2020: Fig. 2); Kramár et al. (2013: Figs. 1 and 13); Munch (2017: Fig. 1); Zomorodian and Carlsson (2005: Fig. 1).

Persistent homology (PH) is a mathematical tool of *Computational Topology* (Edelsbrunner and Harer 2010) and *Topological Data Analysis* (Munch 2017; Chazal and Michel 2021) that identifies those *topological features* of data that persist across *multiple scales*, and measures their *length of persistence*.

In this definition, "*data*" means the values of a real function $f : X \to \mathcal{R}$ where $\mathcal{R} = (-\infty, \infty)$ is the set of real numbers, $X \subset \mathbb{R}^1$, or $X \subset \mathbb{R}^2$ or $X \subset \mathbb{R}^3$ is either a finite set of points in $\mathbb{R}^1$, $\mathbb{R}^2$ or $\mathbb{R}^3$; or a set of *pixels* in $\mathbb{R}^2$ or *voxels* in $\mathbb{R}^3$; or vertices, edges and in general *k-cliques* (*complete subgraphs with* $k \leq n$ *vertices*) of some *graph* G_n with n vertices; or a *complex* consisting of *simplices* glued in a certain way (a *0-simplex* is a point, a 1-simplex a line segment, a *2-simplex* a triangle, a *3-simplex* a tetrahedron: Aktas et al. 2019). The simplest example for "data" is associated with pixels, when the set of pixels represents the 2D microscopic image of a porous material whose pore space is Π, solid space (grains) is Γ, and $f(x)$ is defined as $f(x) = \begin{cases} 1 \ if \ x \epsilon \Pi \\ 0 \ if \ x \epsilon \Gamma \end{cases}$.

The "*topological features*" of a d-dimensional object $X \subset \mathbb{R}^d$ are characterized by the Betti numbers $\beta_0.\beta_1, \cdots, \beta_{d-1}$ (Kramár et al. 2013; Munch 2017; Herring et al. 2019). In 1-D, β_0 is the *number of distinct components*; in 2-D, β_0 is the *number of distinct components* of X and β_1 is the *number of independent closed loops* in it; in 3-D, β_0 is the number of *distinct components*, β_1 the *number of independent closed (ring-like) tunnels*, and β_2 is the *number of cavities* (*holes*). Betti numbers for some simple shapes in 1-, 2- and 3D are shown in Table 3.5.

If Betti numbers are defined for *sets* (topological spaces), not for functions, how can we compute them for *data*? To construct Betti numbers for a function $f : X \to \mathcal{R}$ where $X \subset \mathbb{R}^d$ we recall the concepts of *level sets*

$$X_{f=r} = \{x \epsilon X \,|f(x) = r\} = f^{-1}(\{r\}); \tag{3.3.1a}$$

of *sublevel sets* (also called *lower level cuts* or [*lower*] *excursion sets*) (Zomorodian and Carlsson 2005; Bubenik 2009; Delgado-Friedrichs et al. 2014; Lisitsa et al. 2020; see Fig. 3.5),

$$X_{f \le r} = \{x \epsilon X \,|f(x) \le r\} = f^{-1}(-\infty, r]; \tag{3.3.b}$$

Table 3.5. Examples for Betti Numbers

Object	Name	d	β_0	β_1	β_2
•	Point	1	1	–	–
	Circle	2	1	1	–
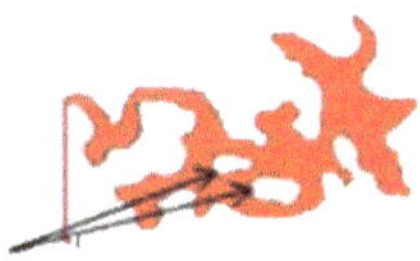	Oil cluster in rock, arrows show holes	2	1	2	–
	Sphere	3	1	0	1
	Torus	3	1	2	1
	Klein bottle	3	1	2	1

Source: Ott et al. (2019): Fig. 7; Munch (2017): Fig. 3

and *superlevel sets* (or *upper level sets*)

$$X_{f\geq r} = \{x\epsilon X\,|f(x) \geq r\} = f^{-1}(r, \infty]. \quad (3.3.c)$$

For a function $f : X \to \mathcal{R}$ where $X \subset \mathbb{R}^d$, the δ -dimensional ($1 \leq \delta \leq d$) Betti numbers $\beta_k^{(\delta)}$, $k = 0, 1, \cdots, \delta - 1$ of the sublevel sets $X_{f\leq r}$ and level sets $X_{f=r}$ can be found by counting:

- for $\delta = 1$ the number of *distinct connected components* ($\beta_0^{(1)}$);
- for $\delta = 2$ the number of *distinct connected components* ($\beta_0^{(2)}$) and the number of *independent closed loops* ($\beta_1^{(2)}$);
- for $\delta = 3$ the number of *distinct connected components* ($\beta_0^{(3)}$), the number of *independent closed ring-like "tunnels"* ($\beta_1^{(3)}$); and the number of *independent holes* ($\beta_2^{(3)}$) *in these sets.*

As these Betti numbers depend on the threshold value r, it is appropriate to denote them by $\beta_k^{(\delta)}(r)$, $k = 0, 1, \cdots, \delta - 1$, $r\epsilon R$, to show this dependence.

An important topological theorem states (Alexandroff and Hilbert 1932; Sahimi 1985; Gilmanov et al. 2018) that two structures $X, Y \subset \mathbb{R}^d$ are *topologically equivalent* (*homeomorphic*) *if and only if all their Betti numbers are equal*:

$$\beta_k^{(\delta)}(X) = \beta_k^{(\delta)}(Y), k = 0, 1, \cdots, \delta - 1; \delta \leq d \quad (3.3.2)$$

For the function in Fig. 3.5 in case of the indicated threshold $r = 2$ the sublevel set

$$X_{f\leq 2} = f^{-1}(-\infty, 2] = (-\infty, 0] \cup [1.5, 2.5) \cup \{4\}$$

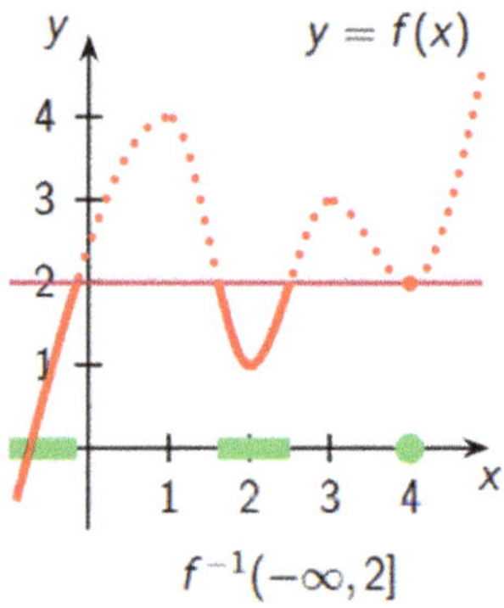

Fig. 3.5 An $X \to \mathcal{R}$, ($X \subseteq \mathcal{R}$) function $f(x)$ and its *level set* ($X_{f=2} = \{0; 1.75; 2.5; 4\}$ where the graph of the function has common points with the line $y = 2$; and the sublevel set (green colored intervals, $X_{f\leq 2} = \{(-\infty, 0]; [1.75, 2.5]; \{4\}\}$ for $r = 2$. (This is Slide #5 of Dr. Peter Bubenik's (2009) ppt presentation, reproduced with his kind permission)

is the union of an infinitely long interval, a finite interval, and a single point. Similarly, by checking the graph in Fig. 3.5 with the thresholds $r = 3, 4$, we find that
$X_{f \leq 3} = f^{-1}(-\infty, 3] = (-\infty, 0.25] \cup [1.75, 4.75)$, and

$$X_{f \leq 4} = f^{-1}(-\infty, 4] = (-\infty, 4.75].$$

This means the sublevel set $X_{f \leq 2}$ consists of three distinct connected parts $(\beta_0^{(1)}(2) = 3)$; $X_{f \leq 3}$ consists of two distinct connected parts $(\beta_0^{(1)}(2) = 2)$, while $X_{f \leq 4}$ is connected $(\beta_0^{(1)}(4) = 1)$.

Observe that $X_{f \leq 2} \subseteq X_{f \leq 3} \subseteq X_{f \leq 4}$, and it is true for any "well-behaved" real-valued functions $f : X \rightarrow \mathcal{R}$ that the sublevel sets $X_{f \leq r} = f^{-1}(-\infty, r]$ satisfy

$$\textit{if} -\infty < s < t < \infty \textit{ then } \mathcal{R}_{f \leq s} \subseteq \mathcal{R}_{f \leq t}. \quad (3.3.3)$$

The monotonicity of inclusions expressed by Eq. (3.3.3) is called *filtering* or *filtration*. For real-valued functions $f : X \rightarrow \mathcal{R}$ the sublevel sets $X_{f \leq r} = f^{-1}(-\infty, r]$ establish such a filtering. It can be proved (*Morse theory*, see Edelsbrunner and Harer 2010) that the *topology* of these sub-level sets (defined by their *Betti numbers*) can only change when r passes a critical value of f (i.e., a *local minimum, saddle point* or *maximum*).

More formally, *filtration* $\mathcal{F}$ *is an indexed family* $\{S_i\}_{i \in \mathcal{I}}$ of objects of a *structure* $\mathcal{S}$ *, with the index i running over some ordered index set* $\mathcal{I}$*, subject to the condition that if* $i \leq j; i, j \in \mathcal{I}$ *then* $S_i \subseteq S_j$,. For example, if $\mathcal{S}$ is a sequence of real numbers, $\mathcal{S} = \{a_1, a_2, a_3, a_4, a_5, a_6, \ldots\}$ and $\mathcal{I}$ is the set of prime numbers $\mathcal{I} = \{1, 2, 3, 5, 7, 11, 13, \ldots\}$, then $\{a_1\} \subset \{a_1, a_2\} \subset \{a_1, a_2, a_3\} \subset \{a_1, a_2, a_3, a_5\} \subset \{a_1, a_2, a_3, a_5, a_7\} \subset \ldots$ is a filtration. Another, geometrical, example for filtration (Fig. 3.6) is when we build up a pentagon-shaped "house" by randomly adding in each step some of the *0-*, *1-* and *2-simplices* from the set {•, |, ▲} (Aktas et al. 2019; a similar example is used in Zomorodian and Carlsson 2005: Fig. 1).

Denote the vertices of the pentagon in Fig. 3.6 counterclockwise as 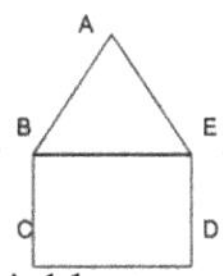

,

and assume that the changes only occur at integer values of the time variable t, e.g.

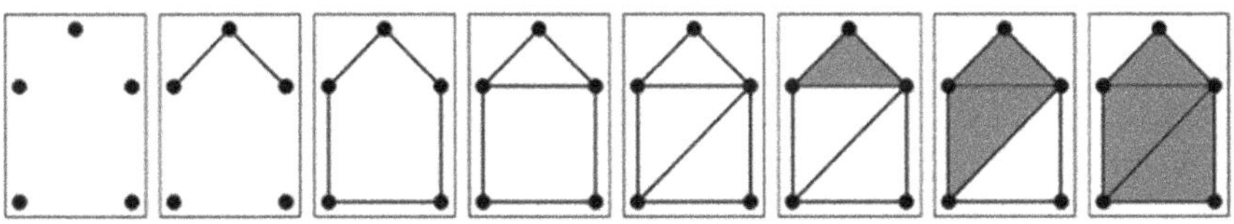

Fig. 3.6 Eight steps of building a pentagonal house-shape from simplicial elements (from left to right, numbered as 0, 1, ..., 7). (Reproduced from Aktas et al. 2019: Fig. 5, without any change; with the kind permission of Dr. Mehmet Aktas, Clinical Assistant Professor; *Creative Commons Attribution 4.0 International License*)

the first shape in Fig. 3.6 retains its form (five separate points $\{A, B, C, D, E\}$) for $t \in [0, 1)$; then at $t = 1$ two more edges, AB and EA are added to it, this shape pertains in the interval $t \in [1, 2)$; at $t = 2$ three more edges are added and there appears the first loop ABCDE, etc. In Table 3.6 we show by *black cells* those time intervals $[i, i+1)$ where the loops are present (we say that they are *born* at the beginning of such an interval, and they *die* at the beginning of the next *white cell*).

Consider a set $X \subseteq \mathbb{R}^d$, let $X_0 = \emptyset$, and consider the consecutive steps of a filtration of X, namely the sequence $\{X_i\}_{i=0,1,2,...}$ where $X_0 = \emptyset \subseteq X_1 \subseteq X_2 \subseteq \cdots$. By definition, the δ-dimensional Betti numbers are all zero for $X_0 = \emptyset$:

$$\beta_k^{(\delta)}(X_0) = 0 \text{ for } k = 0, 1, \cdots, \delta - 1; \delta \leq d \tag{3.3.4}$$

We say that the Betti number $\beta_k^{(\delta)}$ is *persistent* in a subsequence $S = \{X_l, X_{l+1}, \ldots, X_{l+L}\}$ of $X_0 = \emptyset \subseteq X_1 \subseteq X_2 \subseteq \cdots$ *if there exists an integer* $M \geq 1$ *such that*

$$\beta_k^{(\delta)}(X_l) \geq M\,, \beta_k^{(\delta)}(X_{l+1}) \geq M\,, \ldots, \beta_k^{(\delta)}(X_{l+L}) \geq M\,, \tag{3.3.5a}$$

and

At least one of the $\geq$ *relations in* (3.3.5a) *is a strict equality* (3.3.5b), *and, further,* $S = \{X_l, X_{l+1}, \ldots, X_{l+L}\}$ *is maximal in the sense that*

$$\beta_k^{(\delta)}(X_{l-1}) < M\,, \beta_k^{(\delta)}(X_{l+L+1}) < M \tag{3.3.5c}$$

Table 3.6. Persistence of loops in the Intervals $t\epsilon[i, i+1), i = 0, 1, \ldots, 7$ During the filtration shown in Fig. 3.6

Loops (in their order of appearence	Interval							
	[0,1)	[1,2)	[2,3)	[3,4)	[4,5)	[5,6)	[6,7)	[7,∞)
ABCDE			■					
BCDE				■				
ABE				■	■			
BCE					■	■		
CDE					■	■	■	
$\beta_1^{(1)}$ (number of loops	0	0	1	2	3	2	1	0

Source of the data: Aktas et al. (2019): Fig. 5, used with the kind permission of Dr. Mehmet Aktas, Clinical Assistant Professor; *Creative Commons Attribution 4.0 International License*)

In this case we say that the Betti number $\beta_k^{(\delta)}$ "*persists* from l to L", and it is "*born* at l, and *dies* at $l + L + 1$".

In practical terms, this means that

(a) In case of $\beta_0^{(1)}$, $\beta_0^{(2)}$, $\beta_o^{(3)}$: each of $X_l, X_{l+1}, \ldots, X_{l+L}$ consists of at least M *connected parts*, at least one of them consists of exactly M connected parts; and the sequence cannot be extended either to the left or to the right if we want to preserve this property;
(b) In case of $\beta_1^{(2)}$: each of $X_l, X_{l+1}, \ldots, X_{l+L}$ contains at least M *loops*, at least one of them contains exactly M loops; and the sequence cannot be extended either to the left or to the right;
(c) In case of $\beta_1^{(3)}$ each of $X_l, X_{l+1}, \ldots, X_{l+L}$ contains at least M *ring-like tunnels*, and the sequence cannot be extended either to the left or to the right.
(d) In case of $\beta_2^{(3)}$ each of $X_l, X_{l+1}, \ldots, X_{l+L}$ contains at least M *holes* (or "*cavities*"), at least one of them contains exactly M holes; and the sequence cannot be extended either to the left or to the right.

Table 3.6 summarizes that during the filtration shown in Fig. 3.6, in which time intervals $[i, i+1)$ are the loops present (from their *birth* at the beginning of the interval, to their *death* at the beginning of the next one). The Betti number $\beta_1^{(1)}$ is also shown. Table 3.7 shows the development of the Betti numbers $\beta_0^{(1)}$ and $\beta_0^{(2)}$ and of the 2-dimensional Euler characteristics $\chi^{(2)} = \beta_0^{(2)} - \beta_1^{(2)}$ during the same filtration.

According to definitions (3.3.5a, b, c) there are three *persistence ranges* for both Betti numbers $\beta_0^{(1)}$ and $\beta_0^{(2)}$ (number of connected components), namely the intervals $[0, 1), [0, 2)$, and $[0, 8)$. As for $\beta_1^{(2)}$ (number of independent loops), it also has three persistence ranges: $[2, 7), [3, 6)$, and $[4.5)$, see Table 3.8 and Fig. 3.7. In the *Persistence Bar* (PB, see Table 3.8) representation, these ranges are shown by horizontal line segments $[t_{birth}, t_{death}]$; this construction is deftly called "*barcodes*" (Ghrist 2008), because of their resemblance to the small machine-readable labels affixed to retail store items.

Table 3.7. Changes of the 1-dimensional Betti numbers $\beta_0^{(1)}$ (number of connected components) and $\beta_1^{(2)}$ (number of independent loops), and of the Euler characteristics $\chi^{(2)}$ ($\chi^{(2)} = \beta_0^{(2)} - \beta_1^{(2)}$) during the filtration shown in Fig. 3.6

Topology measures	Interval							
	$i = 0$	$i = 1$	$i = 2$	$i = 3$	$i = 4$	$i = 5$	$i = 6$	$i = 7$
	$[0, 1)$	$[1, 2)$	$[2, 3)$	$[3, 4)$	$[4, 5)$	$[5, 6)$	$[6, 7)$	$[7, 8)$
$\beta_0^{(1)} = \beta_0^{(2)}$	5	3	1	1	1	1	1	1
$\beta_1^{(2)}$	0	0	1	2	3	2	1	0
$\chi^{(2)} = \beta_0^{(2)} - \beta_1^{(2)}$	5	3	0	-1	-2	-1	0	1

Source of the data: Aktas et al. 2019: Fig. 5, used with the kind permission of Dr. Mehmet Aktas, Clinical Assistant Professor; *Creative Commons Attribution 4.0 International License*)

Table 3.8. Persistence Bars (PBs) for $\beta_0^{(2)}$ and $\beta_1^{(2)}$ for the filtration shown in Fig. 3.6

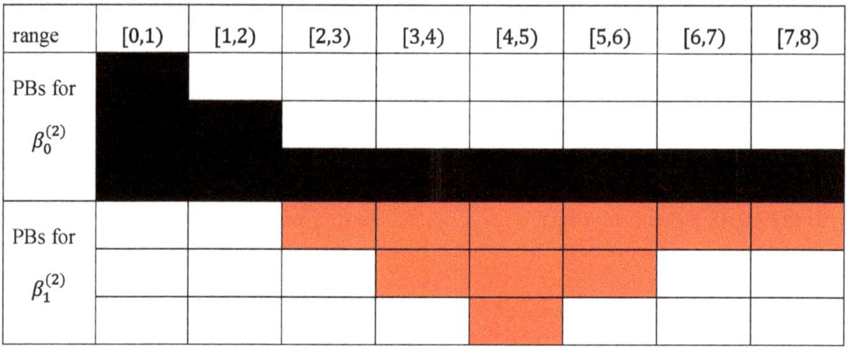

For explanation, see text

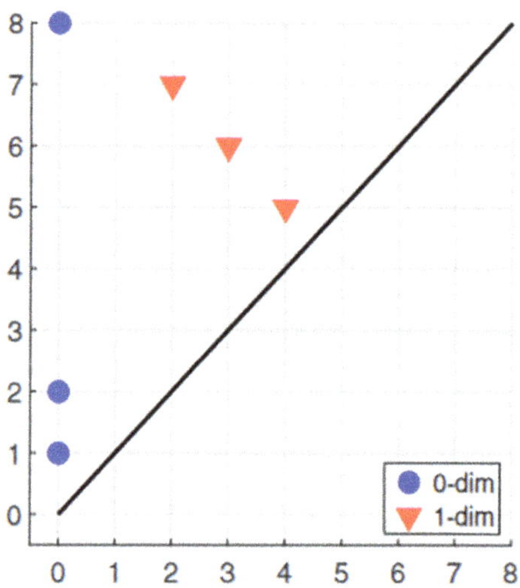

Fig. 3.7 Persistence Diagram (PD) for $\beta_0^{(2)}$ (●) and $\beta_1^{(2)}$ (▼) for the filtration shown in Fig. 3.6. Abscissa: birth, ordinate: death. (Reproduced from Aktas et al. 2019: Fig. 6.b, without any change; with the kind permission of Dr. Mehmet Aktas, Clinical Assistant Professor; *Creative Commons Attribution 4.0 International License*)

Another way of representing the persistence ranges is to plot them as points (x, y) in the rectangular $x = t_{birth}$, $y = t_{death}$ coordinate system (Fig. 3.7). Further examples for this representation are shown in Fig. 3.8.

The Euler characteristics $\chi^{(2)} = \beta_0^{(2)} - \beta_1^{(2)}$ (see Sect. 3.2) is not utilized in PH analysis. However, observe in Table 1.3.2.3 that in this filtration $\chi^{(2)}$ has two zeros, at $i = 2$ and $i = 6$, signaling the appearance of a new "giant" component: at $i = 2$ appears the first large loop, at $i = 6$ the fully filled-out pentagon (Fig. 3.6). Amorim and his group (2019), who studied *Topological Phase Transitions* (TPTs) in complex systems, observed that "under certain conditions, one can identify TPTs via the zeros of the Euler characteristic or by singularities of its logarithm $S_\chi = |log|\langle\chi\rangle||$". In the topological studies of flow in porous media an Euler number approaching 0 indicates that the fluid phase is near the *percolation threshold* of the pore space

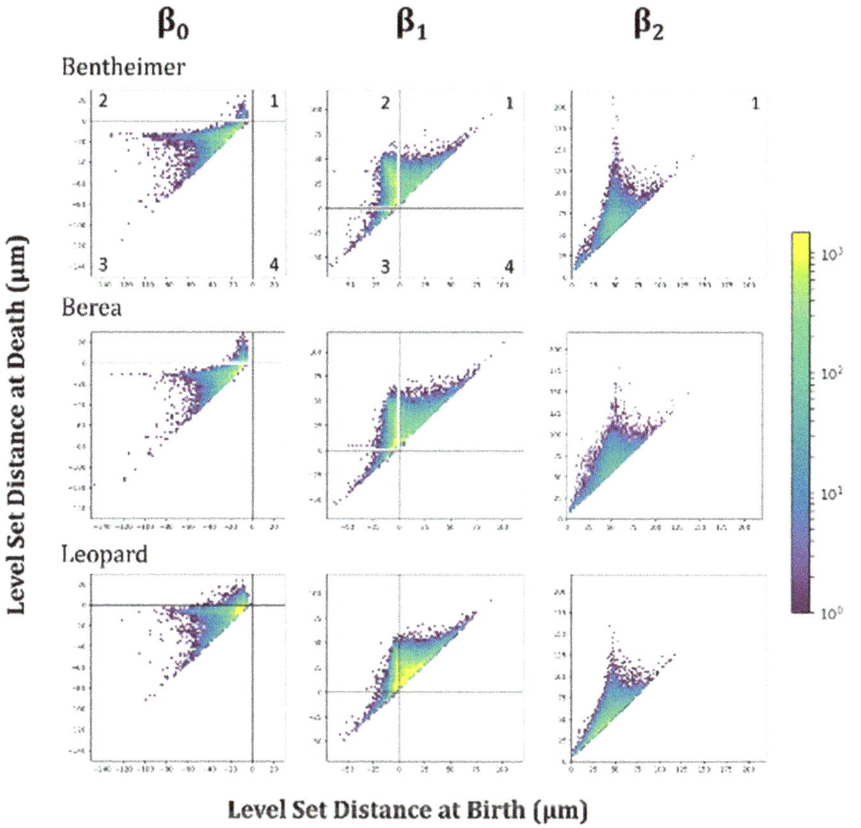

Fig. 3.8 Persistence diagrams for sandstone samples, showing quadrant numbering. Color scale indicates multiplicity. (After Herring et al. 2019: Fig. 5; with the kind permission of Dr. Annalisa Herring, and *Wiley Publishers*)

(Mecke and Arns 2005; Herring et al. 2019). We note that it is even better to use $S^*_\chi = sign(\chi) \cdot |log|\chi||$ (as in Soares et al. 2017), because this clearly indicates whether the change is towards a more connected ($\chi \langle 0 \Rightarrow \chi \rangle 0$), or less connected ($\chi > 0 \Rightarrow \chi < 0$) structure.

In rock-physical applications of *Persistent Homology* (Table 1.3.2.7), the most common way of *filtering pixelized or voxelized binary images* (Ushizima et al. 2012; Herring et al. 2019; Moon et al. 2019) is to use the *sublevel sets* of their *Distance Transform* (DT) or *Signed Distance Transform* (SDT).

Let I be a binary image, denote by Π its pore-space, and by Γ its solid part (grains). The *distance transform* (Fabbri et al. 2008) is a transformation that generates from the 2D or 3D binary image I a map $D(I)$ whose value in each *pixel* (or *voxel*) p is the smallest distance from this pixel to the set Γ:

$$D(p) = \begin{cases} \min\{d(p, q)|q\epsilon\Gamma)\} \textit{ if } p\epsilon\Pi \\ 0 \textit{ if } p\epsilon\Gamma \end{cases} \tag{3.3.6}$$

If the distance $d(p, q)$ is the *Euclidean distance*,

$$(p, q) = \sqrt{(p_x - q_x)^2 + (p_y - q_y)^2} \text{ or}$$
$$(p, q) = \sqrt{(p_x - q_x)^2 + (p_y - q_y)^2 + (p_z - q_z)^2}$$

the transform is called *Euclidean Distance Transform* (EDT).

The *superlevel set* $\Pi^\alpha = \{p\epsilon\Pi|D(p) \geq \alpha\}$ is a subset of the pore space which has an important geometric meaning: a "point" (pixel, voxel) p belongs to the superlevel set Π^α if a ball of radius α, centered at p, does not intersect any grain: $B(x, \alpha)\cap\Gamma = \emptyset$. When, during filtration, we raise the threshold α, we are focusing on increasingly greater pores and wider channels between them (Ushizima et al. 2012; Herring et al. 2019; Moon et al. 2019; Suzuki et al. 2021).

Instead of defining $D(p) = 0$ for pixels inside grains, we can assign *signed distances* to both pixels in pores and pixels in grains, measuring *how far they are from the closest pore/grain boundary* (Korvin et al. 2013) and at the same time indicating their phase. The generally adopted convention is (Herring et al. 2015; Moon et al. 2019) that negative distance values are assigned to the pore phase and positive distances are used for the grain phase:

$$SD(p) = \begin{cases} -\min\{d(p, q)|q\epsilon\Gamma)\} \textit{ if } p\epsilon\Pi \\ \min\{d(p, q)|q\epsilon\Pi)\} \textit{ if } p\epsilon\Gamma \end{cases} \tag{3.3.7}$$

This transform is called *Signed Distance Transform* (SDT), in particular, if we use Euclidean metric, *Signed Euclidean Distance Transform* (SEDT). Geometrically, "a large negative value indicates a large pore size, and a large positive value indicates a large grain size" (Moon et al. 2019).

In *rock-mechanical applications* we study the effect of external force fields on the *grains* (Takahashi et al. 2018). In many cases the forces acting on a pixel (voxel) are encoded as *image brightness* (or *gray scale*) $\theta(p), p\epsilon\Gamma$ where $\theta\epsilon\{0, 1, \ldots, 255\}$ is an integer. In this case, we carry out *superlevel filtration* with respect to the brightness thresholds, meaning that we only keep those pixels whose brightness is above a specified threshold: $\Gamma^\alpha = \{p\epsilon\Gamma|\theta(p) \geq \alpha\}$. If we choose a very high threshold, Γ^α will consist of only a few pixels. As we lower the threshold, more and more pixels will be included. This is indeed a *filtration*, because if $\alpha_1 < \alpha_2 < \alpha_3 < \cdots$, then $\cdots \subseteq \Gamma^{\alpha_3} \subseteq \Gamma^{\alpha_2} \subseteq \Gamma^{\alpha_1}$.

Permanent Homology analysis can also be carried out on *point cloud data* (PCD), and it can detect hidden features such as connected clusters, loops and holes and indicate their dependence on scale. Suppose we are given a finite set of points $\{x_1, x_2, \ldots, x_N\}$ where all N points belong to the same finite geometrical object

$X, X \subset \mathbb{R}^n$. We call $\{x_i\}$ *point cloud data* (PCD), and assume that if X is reasonably smooth, N is large, and the sampling of the points is dense enough, we should be able to estimate the topological invariants of X directly from the PCD. To do this, we must build from the data a *simplicial complex* which preserves the structure of the PCD, and contains the original points as its *vertex set* (Zomorodian and Carlsson 2005; Munch 2017; Aktas et al. 2019; Feng and Porter 2021).

Recall from Geometry (Coxeter 1973) the definition of *simplex* (pl. *simplices* or *simplexes*): *An n-simplex* $\sigma^{(n)}$ (or $\sigma^{(n)}_{v_0,v_1,\ldots,v_n}$) *is the convex hull of* $(n+1)$ *independent points, i.e. the set of all convex combinations* $\lambda_0 v_0 + \lambda_1 v_1 + \cdots + \lambda_n v_n$ *where* $\lambda_0 + \lambda_1 + \cdots + \lambda_i = 1$ *and* $\lambda_i \leq 1$ *for all* $i \in \{0, 1, \ldots, n\}$. *A* 0*-simplex is a point, a* 1*-simplex is a line segment, a* 2*-simplex is a filled triangle, the* 3*-simplex a tetrahedron. The convex hull of any subset* $\{v_{i_1}, \cdots, v_{i_m}\} \subseteq \{v_0, \ldots, v_n\}$, $m < n$ *of the defining points is called an m-face of the n-simplex*; 0*-faces are the vertices* (sing. *vertex*), 1*-faces the edges. The number of m-faces is equal to* $\begin{pmatrix} n+1 \\ m+1 \end{pmatrix}$.

A *simplicial complex* Σ is the union of a finite number of simplexes, but these need to be glued together according to the following rules:

(i) Every face of a simplex in Σ also belongs to Σ. (3.3.8a)
(ii) For any two simplexes σ_1 and σ_2 in Σ, if $\sigma_1 \bigcap \sigma_2 \neq \phi$, then $\sigma_1 \bigcap \sigma_2$ is a common face of both σ_1 and σ_2 (3.3.8b)

(i.e., we can only glue simplexes along their common faces, e.g. 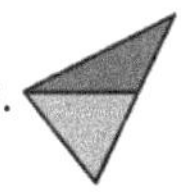 is allowed,

but 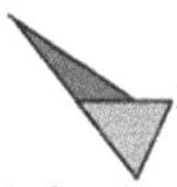 is forbidden).

A family of *simplicial complexes* $\{\Sigma_\delta\}_{\delta=\delta_1,\delta_2,\ldots}$; $\delta_i \in \mathbb{R}$ is called *filtration* if

$$\text{for } \delta_i < \delta_j \text{ we have } \Sigma_{\delta_i} \subseteq \Sigma_{\delta_j}. \tag{3.3.9}$$

We have already had an example of such a filtration in Fig. 3.6.

Common methods for constructing filtered simplicial complexes from point clouds were recently reviewed by Feng and Porter (2021). A popular technique is the Vietoris – Rips algorithm (Vietoris 1927; Hausmann1995; Zomorodian and Carlsson 2005; Edelsbrunner and Harer 2010; Munch 2017; Chazal and Michel 2021; Feng and Porter 2021), which goes as follows:

Let $\mathfrak{X} = \{x_1, x_2, \ldots, x_N\}$ be a point cloud. Given a real number $\varepsilon \geq 0$, we build up the VR (Vietoris–Rips) *simplicial complex* $VR_\varepsilon(\mathfrak{X})$ recursively, as:

(1) *The vertex set of* $VR_\varepsilon(\mathfrak{X})$ *is the point cloud itself,* $\{x_1, x_2, \ldots, x_n\}$.
(2) *For any two different points x, y of* $\mathfrak{X}$ *we include the edge* $\overline{xy}$ *in* $VR_\varepsilon(\mathfrak{X})$ *if their distance is at most:* $d(x, y) \leq \varepsilon$.
(3) *For any integer* $2 \leq k \leq n$ *we include the k-simplex* $\sigma^{(k)}_{\xi_0,\xi_1,\ldots,\xi_k}$ *in* $VR_\varepsilon(\mathfrak{X})$ *if and only if* $\{\xi_0, \xi_1, \ldots, \xi_k\} \subseteq \{v_0, v_1, \ldots, v_n\}$ *and its vertices are pairwise at least as close as* ε, *that is* $d(\xi_0, \xi_1, \ldots, \xi_k)$ *for all pairs* $\xi_i, \xi_j \in \sigma^{(k)}_{\xi_0,\xi_1,\ldots\xi_k}$.

For an increasing sequence of real numbers $0 \leq \varepsilon_1 < \varepsilon_2 < \cdots$ all $VR_{\varepsilon_i}(\mathfrak{X})$ are *simplicial complexes* according to Eqs. (3.3.8a, b), and, by Eq. (3.3.9), the sequence of *simplicial complexes* $VR_{\varepsilon_1}(\mathfrak{X}) \subseteq VR_{\varepsilon_2}(\mathfrak{X}) \subseteq \cdots$ is a *filtration.* Feng and Porter (2021) aptly state that "in this construction, we produce a thickening of a point cloud … thickenings also capture the intuition of blurring an image by squinting at it until the points start to merge".

During this construction, loops and holes appear (are *born*) and then disappear (*die*) and their persistency is a topological characteristics of the point cloud. The *lifespan* of each homological feature can be represented either as an interval (*persistence barcode*, PB) or a point on the *persistence diagram* (PD), as discussed above. Longer bars in PBs and points far away from the diagonal in PDs are considered as essential features of the dataset; very short barcodes or points near the diagonal of the PD are interpreted as mere noise, their elimination is called *simplification* (Edelsbrunner et al. 2002). For example, in the study of Herring et al. (2019), β0, β1 and β2 features with only one voxel persistence were removed from the analysis.

Suppose we subject our rock sample to some physical process (drainage – imbibition cycles, or increasing strain: Kim (2012), Robins et al. (2016), Robinson et al. (2016), Herring et al. (2019), Moon et al. (2019), Suzuki et al. (2021), Thakur et al. (2021); chemical dissolution of the pore space: Khachkova et al. (2020); Lisitsa et al. (2020); Prokhorov et al. (2021); granular crystallization: Saadatfar et al. 2017a, b; etc.) and we want to find the topological changes in the internal rock structure in subsequent stages of the process by comparing the corresponding PDs. In another suggested application of Persistence Diagrams, we might want to decide whether two 3D reconstructions of the same rock from its 2D image (Torquato 2002a, b) have the same topology. For such comparisons, we need to measure the *distance* between PDs. There are two commonly used metrics for the distance between PDs: the *Wasserstein distance* and the *bottleneck distance* (Cohen-Steiner et al. 2007; Munch 2017; Aktas et al. 2019; Takahashi et al. 2018; Khachkova et al. 2020; Lisitsa et al. 2020).

The *Wasserstein distance* is defined (for 2D, Aktas et al. 2019) as: *The r-th Wasserstein distance between two persistence diagrams* PD_1 and PD_2 *is*

$$d_{W_r}(PD_1, PD_2) = \underset{\gamma}{inf} \left(\sum_{x \epsilon PD_1} \|x - \gamma(x)\|_r \right)^{1/r} \tag{3.3.10}$$

where γ *ranges over all one-to-one matchings from* PD_1 *to* PD_2, $r \epsilon \mathbb{R}$, *and*

$$\|p - q\|_r = \left(|p_1 - q_1|^r + |p_2 - q_2|^r\right)^{1/r} \quad for$$
$$\{p = (p_1, p_2), q = (q_1, q_2)\} \epsilon \, \overline{\mathbb{R}}^2. \tag{3.3.11}$$

Here, $\overline{\mathbb{R}}^2$ is the real plane extended with infinity, $\overline{\mathbb{R}}^2 = (\mathbb{R} \cup \{\pm\infty\})^2$, and $|\infty - \infty| := 0$. Usually, one takes $r = 2$. If in Eq. (3.3.11) we let $r \to \infty$, the equation becomes:

$$\|p - q\|_\infty = max(|p_1 - q_1|, |p_2 - q_2|) \text{ for } \{p = (p_1, p_2), q = (q_1, q_2)\} \epsilon \overline{\mathbb{R}}^2, \tag{3.3.12}$$

and we obtain the *bottleneck distance*, defined as:

The bottleneck distance between two persistence diagrams PD_1 and PD_2 is

$$d_B(P, Q) = \underset{\gamma}{inf} \; \underset{x \epsilon PD_1}{sup} \|x - \gamma(x)\|_\infty \tag{3.3.13}$$

where γ *ranges over all one-to-one matchings from* PD_1 *to* PD_2 *and* $\|p - q\|_\infty$ *is given by* Eq. (3.3.12).

Lisitsa et al. (2020), who used PH to study changes in pore-space topology during *chemical dissolution* of rocks, documented that the application of the bottleneck metric to the PDs allows to distinguish between different dissolution scenarios.

Let $X \epsilon \mathbb{R}^d$ be a *triangulable* topological space (i.e. one homomorphic with a finite simplicial complex), and $f, g : X \to \mathbb{R}$ two "*well-behaved*" functions. In this context the "well-behavedness" of a function $\varphi : X \to \mathbb{R}$ requires that $\varphi(x)$ has only a finite number of such *homological critical values* $\{r_i\}_{i=1}^N$ where the homology of the sub-level sets $X_{f \le r_i} = \{x \epsilon X | f(x) \le r_i\} = f^{-1}(-\infty, r_i]$ changes.

Denote by $PD(X, f)$ and $PD(X, g)$ the PDs constructed from X using the sublevel sets of f, and g, respectively. Then we have the following important theorem:

Stability Theorem for Persistence Diagrams (Cohen-Steiner et al. 2007, 2010; Edelsbrunner and Harer 2008): *The bottleneck distance of the two persistence diagrams satisfies*

$$d_B(PD(X, f), PD(X, g)) \le \|f - g\|_\infty, \tag{3.3.14}$$

where

$$\|f(x) - g(x)\|_\infty = sup_{x \epsilon X} |f(x) - g(x)|. \tag{3.3.15}$$

The essence of the theorem is that the persistence diagrams are stable under small perturbations.

The similarity of persistence diagrams can also be assessed using the *Structural Similarity Index* (SSIM) developed by Wang et al. (2004) for the comparison of images. Consider two images, $\mathcal{I}_1$ and $\mathcal{I}_2$, each consisting of $N \times N$ pixels, where each pixel $\{x, y\}$ has an *intensity value* $I(x, y) \epsilon \{0, 1, \ldots, 255\}$. Denote by μ_1, μ_2 the average intensities of pixels in the images, and by σ_1, σ_2 their standard deviations, let σ_{12} be the covariance between them.

The SSIM index is the product of three components: *luminance* $l(\mathcal{I}_1, \mathcal{I}_2)$, *contrast* $c(\mathcal{I}_1, \mathcal{I}_2)$, and *structure* $c(\mathcal{I}_1, \mathcal{I}_2)$ defined as:

$$SSIM(\mathcal{I}_1, \mathcal{I}_2) = l(\mathcal{I}_1, \mathcal{I}_2)^\alpha \cdot c(\mathcal{I}_1, \mathcal{I}_2)^\beta \cdot s(\mathcal{I}_1, \mathcal{I}_2)^\gamma \tag{3.3.16.a}$$

$$l(\mathcal{I}_1, \mathcal{I}_2) = \frac{2\mu_1\mu_2 + C_1}{\mu_1^2 + \mu_2^2 + C_1}; \; c(\mathcal{I}_1, \mathcal{I}_2) = \frac{2\sigma_1\sigma_2 + C_2}{\sigma_1^2 + \sigma_2^2 + C_2};$$
$$s(\mathcal{I}_1, \mathcal{I}_2) = \frac{\sigma_{12} + C_3}{\sigma_1\sigma_2 + C_3} \quad (3.3.16.b,c,d)$$

In Eq. (3.3.16.b) the constant C_1 is a small positive number to avoid numerical instability when $\mu_1^2 + \mu_2^2 \ll 1$; the constants C_2 and C_3 in Eqs. (3.3.16.c,d) are selected for the same reason. Wang et al. (2004) recommended the *default setting* $\alpha = \beta = \gamma = 1$ and $C_3 = C_2/2$, in this case the expression for $SSIM\,(\mathcal{I}_1, \mathcal{I}_2)$ simplifies to

$$SSIM\,(\mathcal{I}_1, \mathcal{I}_2) = \frac{(2\mu_1\mu_2 + C_1)(2\sigma_{12} + C_2)}{\left(\mu_1^2 + \mu_2^2 + C_1\right)\left(\sigma_1^2 + \sigma_2^2 + C_2\right)} \quad (3.3.16.e)$$

The range of the index is $SSIM\,(\mathcal{I}_1, \mathcal{I}_2)\epsilon[-1, 1]$, $SSIM\,(\mathcal{I}_1, \mathcal{I}_2) = 1$ indicates identical images.

Persistence diagrams and barcodes depend on sample size, in a rather unpredictable way. When estimating rock-physical properties from *pixelized* or *voxelized images*, the minimal volume for which the estimation becomes stable and smaller-scale fluctuations dampen out is called *Representative Elementary Volume* (REV), its size depends on the geology of the rock, on resolution, on the physical property, and the calculation technique. While there have been many studies on the REVs to compute porosity, permeability and electric conductance (Bear 1972; Zhang et al., 2000; Koestel et al. 2020; Marafini et al. 2020); apparently (Moon et al. 2019) was the first study where the REV for PDs had been determined. They used the SSIM index to measure the similarities between subsequent PDs as the size of the underlying voxel complex was increased, and accepted the size as the REV if the similarity index rose above 0.9 ("weak" similarity) or above 0.95 ("strict" similarity). For three different rock types they reported the REV sizes shown in Table 3.9. We note that Yoon and Dewers (2013), who determined five rock properties for chalk samples (porosity, permeability, tortuosity, anisotropy, and specific surface area) from voxelized images, defined *REV size in the weak sense* when *three of these properties* (porosity, tortuosity, and specific surface area) had a coefficient of variation, $CV = \sigma/\mu < 0.15$; and a *REV size in the strict sense* when *all five properties* had a coefficient of variation, $CV = \sigma/\mu < 0.15$. For the same chalk formation analyzed by Moon's group they found: $REV_{weak} \geq 400^3$ voxels, $REV_{strict} \geq 600 \times 520 \times 600$ voxels.

There are no strict guide-lines for how to interpret PDs, the method and its interpretation are still mysterious for most practicing petrophysicists. In some publications, the *color scale* used to display the points on the PD is not even explained in figure captions or the main text, let alone involved in the interpretation. Actually, the color scale represents *multiplicity*, i.e., how many points share the same (birth, death) values, it can be *linear*—as in (Moon et al. 2019)—or *logarithmic*, as in (Robins et al. 2016; Herring et al. 2019). A recent work (Suzuki et al. 2021) complains:

Table 3.9 Representative Elementary Volume to compute Persistence Diagrams from the voxalized image of the rock

Rock type	REV size (in voxels) needed for *SSIM* > 0.9	REV size (in voxels) needed for *SSIM* > 0.95
Bentheimer sandstone	$\geq 110^3$	$\geq 160^3$
Doddington sandstone	$\geq 150^3$	$\geq 230^3$
Selma group chalk	$\geq 330^3$	$\geq 550^3$

SSIM is defined in Eqs. 3.3.16.a–3.3.16.e; data are from Moon et al. (2019)

"Although persistent homology is useful for revealing the topological and geometric information, it is difficult to interpret the parameters of persistent homology themselves and difficult to directly relate the parameters to physical properties." We refer the Reader to a few carefully explained rules for how to interpret PDs, in case when the persistence parameter is the *signed Euclidean distance*, and its sign is negative in pore space, positive inside grains (Herring et al. 2019; Moon et al. 2019; Fig. 3.8; Table 3.10).

In these examples Persistence Diagrams are constructed for all Betti numbers, $D\left(\beta_0^{(3)}\right), \left(\beta_1^{(3)}\right), PD\left(\beta_2^{(3)}\right)$, and they are divided to *quadrants* according to the signs of the *birth* and *death* variables:

$$\{\{b, d\}|b_{min} \leq b \leq b_{max}, d_{min} \leq d \leq d_{max}, b \leq d, 0 \leq b, 0 \leq d\} \qquad (3.3.17a)$$

$$\{\{b, d\}|b_{min} \leq b \leq b_{max}, d_{min} \leq d \leq d_{max}, b \leq d, b \leq 0, 0 \leq d\} \qquad (3.3.17b)$$

Table 3.10. Examples for the interpretation of the $PD\left(\beta_0^{(3)}\right)$; $PD\left(\beta_1^{(3)}\right)$ and $PD\left(\beta_2^{(3)}\right)$ Persistence Diagrams*

Betti numberand quadrant (Fig. 3.8.)	Interpretation
β_0, Q2	Persistence of disconnected pore bodies
β_0, Q3	(a) Birth values give size distribution of pores, death values give size distribution of throats; (b) The *percolating radius*** of the pore space can be estimated from the *death value* of the most persistent features
β_1, Q2	(a) Persistence of redundant pore-space loops; (b) The *percolating radius*** of the pore space can be estimated from the birth value of the most highly persistent features
β_2, Q1	Useful to estimate (a) grain numbers, and (b) grain size (from *death values*)

*After Herring et al. 2019; Rock type: Bentheimer, Berea and Leopard sandstones; with the kind permission of Dr. Annalisa Herring, and Wiley Publishers
**Defined in (Robins et al. 2016) as the "radius of the grains in the vicinity of their contact"

$$\{\{b, d\}|b_{min} \leq b \leq b_{max}, d_{min} \leq d \leq d_{max}, b \leq d, b \leq 0, d \leq 0\} \tag{3.3.17c}$$

(see Fig. 3.8). Table 3.10 shows how to interpret the different quadrants, the interested Reader is referred to the cited papers for details.

Rather than using difficult-to-interpret PDs obtained from the PH analysis, it could be more convenient to derive from them *simple numerical measures* to establish correlation with physical properties (Takahashi et al. 2018; Herring et al. 2019; Suzuki et al. 2021), or to enter them as input to *Statistical Learning* or *Neural Networks* (Moon et al. 2019).

Herring and her group (2019) extracted from the PDs of the non-wetting fluid phase the measure

$$PH_{\beta_1/\beta_0} = \frac{\sum_i [b_i \times (d_i - b_i)]}{\sum_j \left[b_j \times \left(d_j - b_j\right)\right]} \times (voxel\ resolution)^3 \tag{3.3.18}$$

where index i runs over persistent topologic features in the PD for β_1, j runs over persistent features in the PD for β_0, the births and death values $\{b_i, d_i, b_j, d_j\}$ correspond to the radii of the inscribed circle for which features appear or disappear. Near-diagonal, noisy features, for which $d - b \leq 1$, were not considered. According to the interpretation of the authors, "larger, more numerous, and more persistent β_1 features (redundant pathways) result in larger PH_{β_1/β_0} values, indicating enhanced mobilization, whereas larger, more numerous, and more persistent β_0 features lead to decreasing PH_{β_1/β_0} values, indicating that structure facilitates capillary trapping" (*op. cit.*: 568).

Takahashi and co-workers (2018) used persistent homology to study the evolution of force networks in granular structures during simulated impact. They defined *Total Persistence* (TP) as the sum of all lifespans on the persistence diagram (neglecting "noise bursts" with life span of 1):

$$TP(PD) = \sum_{(b,d)\epsilon PD} (|b - d|) \tag{3.3.19}$$

They offered a picturesque justification for using TP: "If we think of the original image as a landscape (with the altitude corresponding to pixel brightness), then the TP essentially tells us how flat this landscape is: larger TP corresponds to the landscape that is more mountain-like, with lots of hills and valleys, while small TP corresponds to the landscape without too many features" (*op. cit.:* 012,906–3, 4).

As of to date, the rock-physical applications of Persistent Homology (PH) (Table 3.11), have been mostly *qualitative.* Suzuki's group (2021) has recently come within a hair's breadth of fulfilling the "boldest dream" (Korvin 2016) of petrophysics: to *quantitatively* estimate permeability "solely from image data". They studied fractured rock, subjected both the 3D voxelized pore space, Π, and its complement, the solid material of grains, Γ , to PH analysis. From the joint PH analysis of $\beta_1(\Pi)$ and $\beta_1(\Gamma)$ (details in Suzuki et al. 2020, 2021) they could eliminate stagnant ring-shaped holes through which there is no flow, and estimated the number of *active flow channels* $\mathcal{N}_{fc}$ leading from inlet to outlet, together with their narrowest apertures h_i,

$$h_i = 2d_i \cdot \delta \tag{3.3.20}$$

where d_i (in voxels units) is the time of death of the i-th flow channel, and δ is voxel size. If a fracture is parallel to flow direction, has smooth and parallel walls, an aperture h and width w, and there is a uniform pressure gradient dP/dx in one direction say from bottom to top, then the total volumetric flow rate in this fracture can be written as

$$Q_x = -\frac{wh^3}{12\mu}\frac{dP}{dx}, \tag{3.3.21}$$

where μ is the fluid's viscosity. (Eq. 3.3.21 is the *cubic law of hydraulics*, Renshaw 1995). By Darcy's law,

$$Q_x = -\frac{KA}{\mu}\frac{dP}{dx}, \tag{3.3.22}$$

Table 3.11. Applications of persistent homology

Application	References
Capillary fluid trapping	Herring et al. (2019)
Chemical dissolution of pore space	Khachkova et al. (2020); Lisitsa et al. (2020); Prokhorov et al. (2021)
CO_2 sequestration	Ushizima et al. (2012)
Estimating elastic modulus	Jiang et al. (2018)
Estimating permeability	Robins et al. (2016); Robinson et al. (2016); Moon et al. (2019); Suzuki et al. (2021); Thakur et al. (2021)
Force networks in granular media	Kramár et al. (2013, 2014a, b); Papadopoulos et al. (2018);Takahashi et al. (2018)
Fracture network	Suzuki et al. (2020)
Geospatial data	Feng and Porter (2021)
Granular crystallization	Saadatfar et al. (2017a, b)
Multi-phase fluid flow	Kim (2012); Herring et al. (2019)
Pore space geometry	Bazaikin et al. (2017); Jiang et al. (2018); Tsuji et al. (2018)
Rock modeling	Torquato (2002a, b)

and from Eqs. (3.3.21–3.3.22) the permeability of a single fracture is

$$wh^3/12A. \tag{3.3.23}$$

The cross-sectional area is $A = wh$, and the equations simplify to

$$K = h^2/12. \tag{3.3.24}$$

Assume that the $\mathcal{N}_{fc}$ active flow channels are *in parallel*, and neglect their *tortuosity*. We get

$$K = \sum_{i=1}^{\mathcal{N}_{fc}\gamma} \frac{w_i h_i^3}{12\mathcal{A}}, \tag{3.3.25}$$

where $\mathcal{A}$ is the cross sectional area of the material considered, w_i is the width of the i-th flow channel, h_i is its aperture. In Eq. (3.3.25) the apertures h_i are known from the death values d_i (through Eq. 3.3.20), their widths w_i however are unknown. We can substitute for each of them the same effective width $\overline{w}$ as follows. Assume the fractured material is included in a cube $\mathcal{C}$ of volume $Vol(\mathcal{C}) = V \times \delta^3$, where δ is voxel size. $\mathcal{C}$ consists of $N^3 = N_{pores} + N_{grains}$ voxels, from the number of voxels in Π we find porosity $\Phi = N_{pores}/N^3$. The total volume of pore-voxels in the cube is

$\Phi N^3\delta^3 = \sum_{i=1}^{\mathcal{N}_{fc}} N \cdot h_i \cdot w_i \cdot \delta^3$. The height of every flow channel is N because the channels penetrate the cube from bottom to top. There are N horizontal, 1-voxel-wide layers in the cube, in an average layer there are ΦN^2 voxels (according to the *De Lesse principle*: Delesse 1847; Stoyan et al. 1989; Oleschko 1998–1999), and the mean width of channels in such an average layer is $\overline{w}$. Thus we have for this average layer $\Phi N^2\delta^2 = \sum_{i=1}^{\mathcal{N}_{fc}}[(h_i\delta) \cdot (\overline{w}\delta)]$, from where

$$\overline{w}\delta = \frac{\Phi N^2\delta^2}{\sum_{i=1}^{\mathcal{N}_{fc}} h_i\delta}. \tag{3.3.26}$$

Noting that $\Phi N^2\delta^2$ is the *average cross sectional area* of pores, $\langle A_{pore}\rangle$, in a horizontal layer, Eq. (3.3.26) can be written as

$$\overline{w}\delta = \frac{\langle A_{pore}\rangle}{\sum_{i=1}^{\mathcal{N}_{fc}} h_i\delta}. \tag{3.3.27}$$

Suzuki et al. (*op. cit.*) recommend, instead, to use

$$\overline{w}\delta = \frac{min\left(A_{pore}\right)}{\sum_{i=1}^{\mathcal{N}_{fc}} h_i\delta} \tag{3.3.28}$$

where $min\left(A_{pore}\right)$ is *the minimum of total area of pore space for all layers.* This is in agreement with the view (of Nishiyama and Yokoyama 2017) adopted in Suzuki et al.'s paper that the narrowest width in flow channels, called *critical pore radius*, correlates with permeability better than other kind of pore radius.

Combining Eqs. (3.3.20 and 3.3.25) and using either (3.3.27 or 3.3.28) we obtain an expression for permeability:

$$K = \frac{\overline{w}}{12A}\sum_{i=1}^{\mathcal{N}_{fc}}(2d_i\delta)^3. \tag{3.3.29}$$

As the authors claim, all variables in Eq. (3.3.29) can be derived "solely from image data through persistent homology analysis". (Suzuki et al.: Fig. 8c) shows a comparison of the permeabilities predicted from PH with computed and measured results. Their consequent over-estimation of permeability might be due to *neglecting tortuosity* and *assuming parallel conduits.*

References

Aboufoul M, Garcia A (2017) Factors affecting hydraulic conductivity of asphalt mixture. Mater Struct 50:1–16

Aktas ME, Akbas E, Fatmaoui AE (2019) Persistence homology of networks: methods and applications. Appl Netw Sci 4(1):1–28

Alexandroff P, Hilbert D (1932) Einfachste Grundbegriffe der Topologie. Julius Springer, Berlin

Alqahtani N, Armstrong RT, Mostaghimi P (2018) Deep learning convolutional neural networks to predict porous media properties. SPE Asia Pacific Oil and Gas Conference and Exhibition.

Amorim E, Moreira RA, Santos FAN (2019) The Euler characteristic and topological phase transitions in complex systems. J Phys Complex 3:025003

Andreeva MV, Kalyuzhnyuk AV, Krutko VV, Russkikh NE, Taimanov IA (2021) Representative elementary volume via averaged scalar Minkowski functionals. arXiv: 2008.03727v2

Andrew M (2019) Comparing organic-hosted and intergranular pore networks: topography and topology in grains, gaps and bubbles. In: Dowey P, Osborne M, Volk H (eds) Application of analytical techniques to petroleum systems, vol 484. Geological Society, London, Special Publications, pp 241–253

Archie G (1942) The electrical resistivity log as an aid in determining some reservoir characteristics. Trans AIME 146(1):54–62

Armstrong RT, McClure JE, Robins V, Liu Z, Arns CH, Schlüter S, Berg S (2018) Porous media characterization using Minkowski functionals: theories, applications and future directions. Transp Porous Media 2018:1–31

Arns CH (2002) The influence of morphology on physical properties of reservoir rocks. Doctor of Philosophy Thesis in Petroleum Engineering. School of Petroleum Engineering. The University of New South Wales

Arns CH, Knackstedt MA, Pinczewski WV, Mecke KR (2001) Euler-Poincaré characteristics of classes of disordered media. Phys Rev E 63(031112):1–13
Arns CH, Knackstedt MA, Mecke KR (2002) Characterising the morphology of disordered materials. In: Mecke KR, Stoyan D (eds) Lecture notes in physics, vol. 600. Springer-Verlag, Berlin-Heidelberg, pp 37–74
Arns CH, Knackstedt MA, Pinczewski WV (2002a) V_P: V_s relationships for model sandstones. Geophys Res Lett 29(8):44-1–44-4
Bachmat Y, Bear J (1986) Macroscopic modelling of transport phenomena in porous media. 1: The continuum approach. Transp Porous Media 1:213–240
Bakke S, Øren P (1997) 3-d pore-scale modeling of sandstones and flow simulations in the pore networks. SPE J 2:136–149
Bazaikin Y, Gurevich B, Iglauer S, Khachkova T, Kolyukhin D, Lebedev M, Lisitsa V, Reshetova G (2017) Effect of ct image size and resolution on the accuracy of rock property estimates. J Geophys Res Solid Earth 122:3635–3647
Bear J (1972) Dynamics of fluids in porous media. Parts 1–2. American Elsevier Publishing Company, New York
Bengio Y (2009) Learning deep architectures in AI. Found Trends Mach Learn 2(1):1–127
Beran M (1965) Use of the variational approach to determine bounds for the effective permittivity in random media. Nuovo Cim 38:771–782
Berchtold MA (2007) Modelling of random porous media using Minkowski-Functionals. Doctoral Thesis. ETH, Zürich
Berg CF (2014) Permeability description by characteristic length, tortuosity, constriction and porosity. Transp Porous Media 103:381–400
Berryman JG, Blair SC (1986) Use of digital image analysis to estimate fluid permeability of porous materials. Application of the two-point correlation function. J Appl Phys 60(6):1930–1938
Blasquez I, Poiraudeau J-F (2003) Efficient processing of Minkowski functionals on a 3d binary image using binary decision diagrams. J WSCG 11(No1). ISSN 1213–6972. WSCG'2003, Feb 3–7, 2003, Plzen, Czech Republic
Bobrowski O, Skraba P (2020) Homological percolation and the euler characteristic. Phys Rev E 101:032304
Boelens MP, Tchelepi HA (2021) QuantImPy: Minkowski functionals and functions with Python. SoftwareX 16:100823
Borges J, Pires L, Cássaro F, Roque W, Heck R, Rosa J, Wolf F (2018) X-ray microtomography analysis of representative elementary volume (REV) of soil morphological and geometrical properties. Soil and Tillage Research https://doi.org/10.1016/j.still.2018.05.004
Bubenik P (2009) Persistent homology of functions. (Ppt presentation, Cleveland State University, August 4, 2009). https://academic.csuohio.edu/bubenik_p/talks/functions.pdf
Carlsson G (2009) Topology and data. Am Math Soc Bull New Series 46(2):255–308
Carlsson G, Zomorodian A, Collins A, Guibas LJ (2005) Persistence barcodes for shapes. Int J Shape Model 11(02):149–187
Carman PC (1936) Fluid flow through granular beds. Trans Inst Chem Eng 15:150–166
Chazal F, Michel B (2021) An Introduction to Topological Data Analysis: Fundamental and Practical Aspects for Data Scientists Front. Artif. Intell., Sec. Machine Learning and Artificial Intelligence vol. 4
Chen L, Rong Y (2010) Digital topological method for computing genus and the Betti numbers. Topol Appl 157:1931–1936
Cohen-Steiner D, Edelsbrunner H, Harer J (2007) Stability of persistence diagrams. Discret Comput Geom 37(1):103–120
Cohen-Steiner D, Edelsbrunner H, Harer J, Mileyko Y (2010) Lipschitz functions have Lp-stable persistence. Found Comput Math 10:127–139
Costa A (2006) Permeability-porosity relationship: a re-examination of the Kozeny-Carman equation based on a fractal pore-space geometry assumption. Geophys Res Lett 33:L02318
Coxeter HSM (1973) Regular polytopes, 3rd edn. Dover Publication Inc., New York

Dahrabou A, Viseur S, Gonzalez-Lorenzo A, Rohmer J, Bac A, Real P, Mari J-L, Audigane P (2016) Topological comparisons of fluvial reservoir rock volumes using betti numbers: application to CO_2 storage uncertainty analysis. In: Bac A, Mari JL (eds) Computational topology in image context. CTIC 2016. Lecture notes in computer science, vol 9667. Springer, Cham, pp 101–112
Daigle H (2016) Application of critical path analysis for permeability prediction in natural porous media. Adv Water Resour 96:43–54
Delesse MA (1847) Procédé mécanique pour déterminer la composition chimique des roches. CR Acad Sci (Paris) 25:545–565
Delgado-Friedrichs O, Robins V, Sheppard A (2014) Morse theory and persistent homology for topological analysis of 3D images of complex materials. In: Proceedings of 2014 IEEE international conference on image processing (ICIP), pp 4872–4876
Delgado-Friedrichs O, Robins V, Sheppard A (2015) Skeletonization and partitioning of digital images using discrete Morse theory. IEEE Trans Pattern Anal Mach Intell 37(3):654–666
Dohnalik M, Jarzyna J (2015) Determination of reservoir properties through the use of computed X-ray microtomography—eolian sandstone examples. Geol Geophys Environ 41(3):223–248
Edelsbrunner H, Harer J (2008) Persistent homology—a survey. Contemp Math 453:257–282
Edelsbrunner H, Harer JL (2010) Computational topology: an introduction. AMS Press, Providence
Edelsbrunner H, Letscher D, Zomorodian A (2002) Topological persistence and simplification. Discret Comput Geom 28(4):511–533
Epstein N (1989) On tortuosity and the tortuosity factor in flow and diffusion through porous media. Chem Eng Sci 44:777–779
Euler L (1758) Elementa doctrinae solidorum. Novi Commentarii Acad. Scientiarum Petropolitanae 109–140
Fabbri R, Costa LDF, Torelli JC, Bruno OM (2008) 2D Euclidean distance transform algorithms: a comparative survey. ACM Comput Surv 40(1):2-1–2-44
Feng M, Porter MA (2021) Persistent homology of geospatial data: a case study with voting. SIAM Rev 63(1):67–99
Flegg HG (1974) From geometry to topology. The English Universities Press Ltd., London
Garboczi EJ (1990) Permeability, diffusivity, and microstructural parameters: a critical review. Cem Concr Res 20:591–601
Ghanbarian B, Hunt AG, Ewing RP, Sahimi M (2013) Tortuosity in porous media: a critical review. Soil Sci Soc Am J 77:1461–1477
Ghanbarian B, Hunt AG, Ewing RP, Skinner TE (2014) Universal scaling of the formation factor in porous media derived by combining percolation and effective medium theories. Geophys Res Lett 41(11):3884–3890
Ghrist R (2008) Barcodes: the persistent topology of data. Bull Am Math Soc 45:61–75
Ghrist RW (2014) Elementary applied topology, vol 1, 10th edn. Createspace, Seattle
Gilmanov R, Kalyuzhnyuk A, Taimanov I, Yakovlev A (2018) Topological characteristics of digital models of geological core. In: 2nd international cross-domain conference for machine learning and knowledge extraction (CD-MAKE), Hamburg, Germany, pp 273–281
Gong L, Nie L, Xu Y (2020) Geometrical and topological analysis of pore space in sandstones based on X-ray computed tomography. Energies 13:3774
Hadwiger H (1957) Vorlesungen über Inhalt, Oberfläche und Isoperimetrie. Springer, Berlin
Haruzi P, Katsman R, Halisch M, Waldmann N, Spiro B (2021) Supplement of Benchmark study using a multi-scale, multi-methodological approach for the petrophysical characterization of reservoir sandstones. Suppl Solid Earth 12:665–689
Hashin Z, Shtrikman S (1962) A variational approach to the theory of the elastic behaviour of polycrystals. J Mech Phys Solids 10(4):343–352
Hatcher A (2002) Algebraic topology. Cambridge University Press, Cambridge, UK
Hausmann J-C (1995) On the Vietoris–Rips complexes and a cohomology theory for metric spaces. In: Prospects in topology: proceedings of a conference in honour of william browder, annals of mathematics studies, vol 138. Princeton University Press, Princeton, NJ, pp 175–188
Felix H (1914) Grundzüge der Mengenlehre. Veit & Co, Leipzig

Herring AL, Harper EJ, Andersson L, Sheppard A, Bay BK, Wildenschild D (2013) Effect of fluid topology on residual nonwetting phase trapping: implications for geologic CO_2 sequestration. Adv Water Resour 62:47–58
Herring AL, Andersson L, Schlüter S, Sheppard A, Wildenschild D (2015) Efficiently engineering pore-scale processes: the role of force dominance and topology during nonwetting phase trapping in porous media. Adv Water Resour 79:91–102
Herring AL, Robins V, Sheppard AP (2019) Topological persistence for relating microstructure and capillary fluid trapping in sandstones. Water Resour Res 55(1):555–573
Huber N (2018) Connections between topology and macroscopic mechanical properties of three-dimensional open-pore materials. Front Mater 5:69
Hyde ST, Barnes IS, Ninham BW (1990) Curvature energy of surfactant interfaces confined to the plaquettes of a cubic lattice. Langmuir 6:1055–1062
Ioannidis MA, Chatzis I (2000) On the geometry and topology of 3D stochastic porous media. J Colloid Interface Sci 229:323–334
Ioannidis MA, Lang E (1998) Microgeometry and topology of statistically homogeneous porous media. Trans Ecol Environ 17:223–230
Ivonin D, Kalnin T, Grachev E, Shein E (2020) Quantitative analysis of pore space structure in dry and wet soil by integral geometry methods. Geosciences 10(9):365
Jiang F, Tsuji T, Shirai T (2018) Pore geometry characterization by persistent homology theory. Water Resour Res 54(6):4150–4163
Jiang H, Arns CH (2020) A fast FFT method for 3D pore-scale rock-typing of heterogeneous rock samples via Minkowski functionals and hydraulic attributes. Web Conf 146: 04002 (2020) SCA 2019
Kac M (1966) Can one hear the shape of a drum? Am Math Monthly 73:1–24
Kaczynski T, Mischaikow K, Mrozek M (2006) Computational homology, vol 157. Springer Science Business Media
Katz A, Thompson A (1986) Quantitative prediction of permeability in porous rock. Phys Rev B34(11):8179
Kellerer HG (1984) Minkowski functionals of Poisson processes. Z Wahr Verw Gebiete 67:3–84
Khachkova TS, Bazaikin YV, Lisitsa VV (2020) Use of the computational topology to analyze the pore space changes during chemical dissolution. Numer Methods Program 21:41–55
Khanamiri HH, Torsæter O (2018) Fluid topology in pore scale two-phase flow imaged by synchrotron X-ray microtomography. Water Resour Res 54:1905–1917
Kim J (2012) Phase-field models for multi-component fluid flows. Commun Comput Phys 12:613–661
Kirkpatrick S (1973) Percolation and conduction. Rev Modern Phys 45(4):574–588
Koestel J, Larsbo M, Jarvis N (2020) Scale and REV analyses for porosity and pore connectivity measures in undisturbed soil. Geoderma 366:114206
Kondic L, Goullet A, O'Hern CS, Kramar M, Mischaikow K, Behringer RP (2012) Topology of force networks in compressed granular media. Europhys Lett 97(5):54001
Kong TY, Rosenfeld A (1989) Digital topology: introduction and survey. Comput Vision, Graph Image Process 48(3):357–393
Konkle SF, Moran PJ, Hamann B, Joy KI (2003) Fast methods for computing isosurface topology with Betti numbers. In: Post F, Nielsen GM, Bonneau GP (eds) Data visualization: the state of the art. proc. Dagstuhl seminar on scientific visualization. Kluwer Academic Publishers, Dordrecht, The Netherlands, pp 363–376
Korvin G (1982) Axiomatic characterization of the general mixture rule. Geoexploration 19(4):267–276
Korvin G (1992) Fractal models in the earth sciences. Elsevier, Amsterdam
Korvin G (2016) Permeability from microscopy: review of a dream. Arab J Sci Eng 41(6):2045–2065
Korvin G, Oleschko K, Abdulraheem A (2014) A simple geometric model of sedimentary rock to connect transfer and acoustic properties. Arab J Geosci 7:1127–1138

Korvin G, Sterligov B, Oleschko K, Cherkasov S (2013) Entropy of shortest distance (ESD) as pore detector and pore-shape classifier. Entropy 15(6):2384–2397

Kozeny J (1927) Über kapillare Leitung des Wassers im Boden. Sitzungsber Akad Wiss (Wien) 136:271–306

Kramár M, Goullet A, Kondic L, Mischaikow K (2013) Persistence of force networks in compressed granular media. Phys Rev E 87:042207

Kramár M, Goullet A, Kondic L, Mischaikow K (2014a) Evolution of force networks in dense particulate media. Phys Rev E 90:052203

Kramár M, Goullet A, Kondic L, Mischaikow K (2014b) Quantifying force networks in particulate systems. Phys D 283:37–55

Legland D, Kiêu K, Devaux M-F (2011) Computation of Minkowski measures on 2D and 3D binary images. Image Anal Stereol 26:83

Lehmann P, Berchtold M, Ahrenholz B, Tölke J, Kaestner A, Krafczyk M, Flühler H, Künsch H (2008) Impact of geometrical properties on permeability and fluid phase distribution in porous media. Adv Water Resour 31(9):1188–1204

Li M, Tang YB, Bernabé Y, Zhao JZ, Li XF, Bai XY, Zhang LH (2015) Pore connectivity, electrical conductivity, and partial water saturation: network simulations. J Geoph Res Solid Earth 120(6):4056–4068

Likos CN, Mecke KR, Wagner H (1995) Statistical morphology of random interfaces in microemulsions. J Chem Phys 102:9350–9361

Lisitsa V, Bazaikin Y, Khachkova T (2020) Computational topology-based characterization of pore space changes due to chemical dissolution of rocks. Appl Math Model 88:21–37

Liu Z, Herring AVR, Armstrong R (2017) Prediction of permeability from Euler characteristic of 3D images. The International Symposium of the Society of Core Analysts

Liu Z, Herring A, Arns C, Berg S, Armstrong RT (2017b) Pore-scale characterization of two-phase flow using integral geometry. Transp Porous Media 118(1):99–117

Lozano-Durán A, Borrell G (2016) Algorithm 964: an efficient algorithm to compute the genus of discrete surfaces and applications to turbulent flows. ACM Trans Math Softw 42(4):Article 34

Macdonald IG (1995) Symmetric functions and hall polynomials, 2nd edn. Clarendon Press, Oxford

Marafini E, La Rocca M, Fiori A, Battiato I, Prestininzi P (2020) Suitability of 2D modelling to evaluate flow properties in 3D porous media. Transp Porous Media 134:315–329

Martys NS, Torquato S, Bentz DP (1994) Universal scaling of fluid permeability for sphere packings. Phys Rev E 50:403

Matheron G (1975) Random sets and integral geometry. Wiley, New York

MATHWORKS (2022) 3-D volumetric image processing—MATLAB & Simulink. Available online: https://www.mathworks.com/help/images/3d-volumetric-image-processing.html?s_tid=CRUX_lftnav

McClure JE, Armstrong RT, Berrill MA, Schlüter S, Berg S, Gray WG, Miller CT (2018) A geometric state function for two-fluid flow in porous media. Phys Rev Fluids 3:084306

Mecke KR (1998) Integral geometry and statistical physics. Int J Mod Phys B 12:861–899

Mecke KR (2000) Additivity, convexity, and beyond: applications of Minkowski functionals in statistical physics. In: Mecke KR, Stoyan D (eds) Statistical physics—the art of analyzing and modeling spatial structures. Lecture notes in physics, vol 554. Springer, Berlin, pp 111–184

Mecke KR (2001) Exact moments of curvature measures in the Boolean Model. J Stat Phys 102(5/6):1343–1381

Mecke K, Arns C (2005) Fluids in porous media: a morphometric approach. J Phys Condens Matter 17(9):503–534

Mecke KR, Seyfried A (2002) Strong dependence of percolation thresholds on poly-dispersity. Europhys Lett 58:28–34

Mecke KR, Stoyan D (eds) (2000) Statistical physics—the art of analyzing and modeling spatial structures. Lecture notes in physics, vol 554. Springer, Berlin

Mecke K, Wagner H (1991) Euler characteristic and related measures for random geometric sets. J Stat Phys 64(3–4):843–850

Mecke KR, Buchert T, Wagner H (1994) Robust morphological measures for large-scale structure in the Universe. Astron Astrophys 288:697–704
Michielsen K, De Raedt H (2001) Integral-geometry morphological image analysis. Phys Rep General Maths 347(2001):461–538
Miles RE (1976) Estimating aggregate and overall characteristics from thick sections by transmission microscopy. J Microscopy 107(3):227–233
Milton GW (1982) Bounds on the elastic and transport properties of two-component composites. J Mech Phys Solids 30(3):177–191
Moon C, Mitchell SA, Heath JE, Andrew M (2019) Statistical inference over persistent homology predicts fluid flow in porous media. Water Resour Res 55:9592–9603
Munch E (2017) A user's guide to topological data analysis. J Learn Anal 4(2):47–61
Munkres JR (1993) Elements of algebraic topology. Addison Wesley, Menlo Park, CAL
Nair P, Mühlbauer S, Roy S, Pöschel T (2021) Can Minkowski tensors of a simply connected porous microstructure characterize its permeability? Phys Fluids 33: 042010
Nakahara M (2003) Geometry, topology and physics. Institute of Physics Publishing, Bristol-Philadelphia
Nishiyama N, Yokoyama T (2017) Permeability of porous media: role of the critical pore size. J Geophys Res Solid Earth 122:6955–6971
Noiriel C, Gouze P, Bernard D (2004) Investigation of porosity and permeability effects from microstructure changes during limestone dissolution. Geophys Res Lett 31:L24603
Ohser J, Mücklich F (2000) Statistical analysis of microstructure in materials science. John Wiley and Sons Ltd., New York
Ohser J, Nagel W, Schladitz K (2002) The Euler number of discretized sets—on the choice of adjacency in homogeneous lattices. In: Mecke KR, Stoyan D (eds) Morphology of condensed matter. Springer, Heidelberg
Joachim O, Nagel W, Schladitz K (2003) The Euler number of discretised sets—surprising results in three dimensions. Image Anal Stereol 22:11–19
Okuma G, Kadowaki D, Hondo T, Tanaka S, Wakai F (2017) Interface topology for distinguishing stages of sintering. Sci Rep 7:11106
Okuma G, Tanaka S, Wakai F (2022) Domain coarsening in viscous sintering as a result of topological pore evolution. J Eur Ceramic Soc 42:729–733
Oleschko K (1998–1999) Delesse principle and statistical fractal sets: Part 1. Dimensional equivalents. Soil Tillage Res 49:255–266; Part 2. Unified fractal model for soil porosity. Ibid 52:247–257
Ott H, Kharrat A, Borji M, Clemens T, Arnold P (2019) Screening of EOR potential on the pore scale by statistical and topological means. SCA 2019-011
Ott H, Kharrat A, Borji M, Arnold P (2020) Fluid-phase topology of complex displacements in porous media. Phys Rev Res 2:023240
Otter N, Porter MA, Tillmann U, Grindrod P, Harrington HA (2017) A roadmap for the computation of persistent homology. EPJ Data Sci 6:1–38
Øren PE, Bakke S, Arntzen OJ (1998) Extending predictive capabilities to network models. SPE J 3:324–336
Papadopoulos L, Porter MA, Daniels KE, Bassett DS (2018) Network analysis of particles and grains. J. Complex Netw 6:485–565
Patania A, Vaccarino F, Petri G (2017) Topological analysis of data. EPJ Data Sci 6(1):1–6
Pérez-Rosales C (1982) On the relationship between formation resistivity factor and porosity. SPE J 22(04):531–536
Pleijel A (1954) A study of certain Green's functions with applications in the theory of vibrating membranes. Ark Mat 2(1954):553–539
Porter M, Wildenschild D, Grant G, Gerhard J (2010) Measurement and prediction of the relationship between capillary pressure, saturation, and interfacial area in a NAPL-water-glass bead system. Water Resour Res 46:W08512

Pothuaud L, Rietbargen BV, Mosekilde L, Beuf O, Levitz P, Benhamou CL, Majumdar S (2002) Combination of topological parameters and bone volume fraction better predicts the mechanical properties of trabecular bone. J Biomech 35:1091–1099
Prokhorov D, Lisitsa V, Bazaikin Y (2021) Digital image reduction for analysis of topological changes in pore space during chemical dissolution. International Conference on Computational Science ICCS 2021, pp 382–393
Protter MH (1987) Can one hear the shape of a drum? Revisited. SIAM Rev 29(2):185–197
Renshaw CE (1995) On the relationship between mechanical and hydraulic apertures in rough-walled fractures. J Geophys Res 100:629–636
Robins V (1999) Towards computing homology from finite approximations. Topol Proc 24:503–532
Robins V, Wood PJ, Sheppard AP (2011) Theory and algorithms for constructing discrete Morse complexes from grayscale digital images. IEEE Trans Pattern Anal Mach Intell 33(8):1646–1658
Robins V, Saadatfar M, Delgado-Friedrichs O, Sheppard AP (2016) Percolating length scales from topological persistence analysis of micro-CT images of porous materials. Water Resour Res 52(1):315–329
Robinson J, Slater L, Johnson T, Shapiro A, Tiedeman C, Ntarlagiannis D, Johnson C, Day-Lewis F, Lacombe P, Imbrigiotta T, Lane J (2016) Imaging pathways in fractured rock using three-dimensional electrical resistivity tomography. Groundwater 54(2):186–201
Roy S, Tarafdar S (1997) Archie's law from a fractal model for porous rocks. Phys Rev 55:8038–8041
Saadatfar M, Takeuchi H, Robins V, Francois N, Hiraoka Y (2017a) Pore configuration landscape of granular crystallization. Nat Commun 8:15082
Saadatfar M, Takeuchi H, Hanifpour M, Robins V, Francois N, Hiraokam Y (2017b) Granular compaction and the topology of pore deformation. EPJ web of conferences, powders & grains, p 16009
Sahimi M (1985) Flow and transport in porous media and fractured rock: from classical methods to modern approaches. VCH, New York
Santaló LA (1976) Integral geometry and geometric probability. Addison-Wesley, Reading, MA
Schindelin J, Rueden CT, Hiner MC, Eliceiri KW (2015) The Image J ecosystem: an open platform for biomedical image analysis. Mol Reprod Dev 82:518–529
Schlüter S, Berg S, Rücker M, Armstrong R, Vogel HJ, Hilfer R, Wildenschild D (2016) Pore-scale displacement mechanisms as a source of hysteresis for two-phase flow in porous media. Water Resour Res 52(3):2194–2205
Schmalzing J, Kerscher M, Buchert T (1995) Minkowski functionals in cosmology. astro-ph/ 9508154(05 OCT 1995): 1–10
Schmidhuber J (2015) Deep learning in neural networks: ano. Neural Netw 61:85–117
Scholz C (2014) Fluss und Transport in mikrofluidischen porösen Medien . Von der Fakultät Mathematik und Physik der Universität Stuttgart zur Erlangung der Würde eines Doktors der Naturwissenschaften (Dr. rer. nat.) genehmigte Abhandlung
Scholz C, Wirner F, Götz J, Rüde U, Schröder-Turk GE, Mecke K, Bechinger C (2012) Permeability of porous materials determined from the Euler characteristic. Phys Rev Lett 109:264504
Scholz C, Wirner F, Klatt M, Hirneise D, Schröder-Turk G, Mecke K, Bechinger C (2015) Direct relations between morphology and transport in Boolean models. Phys Rev E 92:043023
Schröder-Turk GE, Mickel W, Kapfer SC, Klatt MA, Schaller FM, Hoffmann MJF, Kleppmann N, Armstrong P, Inayat A, Hug D, Reichelsdorfer M, Peukert W, Schwieger W, Mecke K (2011) Minkowski tensor shape analysis of cellular, granular and porous structures. Adv Mater 23:2535–2553
Schröder-Turk G, Mickel W, Kapfer S, Schaller F, Breidenbach B, Hug D, Mecke K (2013) Minkowski tensors of anisotropic spatial structure. New J Phys 15(8):083028
Schwartz LM, Martys N, Bentz DP, Garboczi EJ, Torquato S (1993) Cross-property relations and permeability estimation in model porous media. Phys Rev E 48:4584–4591
Sen PN, Straley C, Kenyon WE (1990) Surface-to-volume ratio, charge density, nuclear magnetic relaxation, and permeability in clay-bearing sandstones. Geophysics 55:61–69
Serra J (1982) Image analysis and mathematical morphology. Academic Press, London

Slotte PA, Berg CF, Khanamiri HH (2020) Predicting resistivity and permeability of porous media using Minkowski functionals. Transp Porous Media 131:705–722

Smalley JJ (1967) A simple model of a diagenetic system. Sedimentology 8:27–33

Soares JA, Rayane P, de Lopes A (2017) A model for permeability of carbonate rocks based on pore connectivity and pore size. In: 15th international congress of the brazilian geophysical society held in Rio de Janeiro, Brazil, 31 July to 3 Aug 2017

Sossa-Azuela JH, Cuevas-Jiménez EB, Zaldivar-Navarro D (2011) Alternative way to compute the Euler number of a binary image. J Appl Res Technol 9(3):335–341

Stauffer D, Aharony A (1992) Introduction to percolation theory, 2nd edn. Taylor and Francis, London

Stoyan D, Kendall WS, Mecke J (1989) Stochastic geometry and its applications. Akademie Verlag, Berlin

Sudakov O, Burnaev E, Koroteev D (2019) Driving digital rock towards machine learning: Predicting permeability with gradient boosting and deep neural networks. Comput Geosci 127:91–98

Sun WC, Andrade JE, Rudnicki JW (2011) Multiscale method for characterization of porous microstructures and their impact on macroscopic effective permeability. Int J Numer Methods Eng 88:1260–1279

Suzuki A, Miyazawa M, Okamoto A, Shimizu H, Obayashi I, Hiraoka Y, Tsuji T, Kang PK, Ito T (2020) Inferring fracture forming processes by characterizing fracture network patterns with persistent homology. Comput Geosci 143:104550

Suzuki A, Miyazawa M, Minto JM, Tsuji T, Obayashi I, Hiraoka Y, Ito T (2021) Flow estimation solely from image data through persistent homology analysis. Sci Rep 11(1):1–13

Takahashi T, Clark AH, Majmudar T, Kondic L (2018) Granular response to impact: topology of the force networks. Phys Rev E 97:012906

Thakur MM, Kim F, Penumadu D, Herring A (2021) Pore space and fluid phase characterization in round and angular partially saturated sands using radiation-based tomography and persistent homology. Transp Porous Media 137:131–155

Thovert J-F, Wary F, Adler PM (1990) Thermal conductivity of random media and regular fractals. J Appl Phys 68:3872–3883

Thovert J-F, Yousefian F, Spanne P, Jacquin CG, Adler PM (2001) Grain reconstruction of porous media: Application to a low-porosity Fontainebleau sandstone. Phys Rev E 63:061307

Tiab D, Donaldson EC (2004) Petrophysics: theory and practice of measuring reservoir rock and fluid transport properties. Elsevier, Oxford, UK

Torquato S (1984) Bulk properties of two-phase disordered media. I. Cluster expansion for the effective dielectric constant of dispersions of penetrable spheres. J Chem Phys 81(11):5079–5088

Torquato S (2002a) Statistical description of microcstructures. Ann Rev Mater Res 32:77–111

Torquato S (2002b) Random heterogeneous materials: microstructure and macroscopic properties. Interdisciplinary applied mathematics, vol 16. Springer, New York

Torquato S, Stell G (1983) Microstructure of two-phase random media. III. the n-point matrix probability functions for fully penetrable spheres. J Chem Phys 79:1505–1510

Tsuji T, Jiang F, Suzuki A, Shirai T (2018) Mathematical modeling of rock pore geometry and mineralization: applications of persistent homology and random walk. In: Anderssen R, Broadbridge P, Fukumoto Y, Kajiwara K, Simpson M, Turner I (eds) Agriculture as a metaphor for creativity in all human endeavors. Mathematics for Industry, vol 28. Springer, Singapore, pp 95–109

Ushizima DM, Morozov D, Weber GH, Bianchi AGC, Sethian JA, Wes Bethel E (2012) Augmented topological descriptors of pore networks for material science. IEEE Trans Visual Comput Graphics 18(12):2041–2050

Vejdemo-Johansson M, Skraba P (2016) Topology, big data and optimization. In: Emrouznejad A (ed) Big data optimization: recent developments and challenges. Springer, Cham, Switzerland, pp 147–176

Vietoris L (1927) Über den höheren Zusammenhang kompakter Räume und eine Klasse von zusammenhangstreuen Abbildungen. Math Ann 97:454–472

Vogel HJ (1997a) Digital unbiased estimation of the Euler-Poincaré characteristic in different dimensions. Acta Stereol 16(2):97–104

Vogel HJ (1997b) Morphological determination of pore connectivity as a function of pore size using serial sections. Eur J Soil Sci 48:365–377

Vogel HJ (2000) A numerical experiment on pore size, pore connectivity, water retention, permeability, and solute transport using network models. Eur J Soil Sci 51:99–105

Vogel H-J (2002) Morphology of condensed matter, physics and geometry of spatially complex systems. In: Mecke K, Stoyan D (eds) Topological characterization of porous media. Springer-Verlag, Berlin-Heidelberg, pp 75–92

Vogel H-J, Roth K (2001) Quantitative morphology and network representation of soil pore structure. Adv Water Resour 24:233–242

Vogel H-J, Cousin I, Roth K (2002) Quantification of pore structure and gas diffusion as a function of scale. Eur J Soil Sci 53(3):465–473

Vogel HJ, Weller U, Schlüter S (2010) Quantification of soil structure based on Minkowski functions. Comput Geosci 36(10):1236–1245

Vogel LE, Makowski D, Garnier P, Vieublé-Gonod L, Coquet Y, Raynaud X, Nunan N, Chenu C, Falconer R, Pot V (2015) Modeling the effect of soil meso- and macropores topology on the biodegradation of a soluble carbon substrate. Adv Water Resour 83:123–136

Volkhonskiy D, Muravleva E, Sudakov O, Orlov D, Burnaev E, Koroteev D (2021) Reconstruction of 3D porous media from 2D slices. arXiv: 1901.10233v4 [cs.CV] 6 Aug 2021

Waldir LR, de Souza ACA, Barbieri DX (2009) The Euler-Poincaré characteristic applied to identify low bone density from vertebral tomographic images. Rev Bras Reumatol 49(2):140–152

Wang Z, Bovik AC, Sheikh HR, Simoncelli EP (2004) Image quality assessment: from error visibility to structural similarity. IEEE Trans Image Processing 13:600–612

Wanner T, Fuller ER, Saylor DM (2010) Homology metrics for microstructure response fields in polycrystals. Acta Materialia 58(1):102–111

Weinberger S (2011) What is … persistent homology? Not Am Math Soc 58(36):36–39

Weissberg HL (1963) Effective diffusion coefficient in porous media. J Appl Phys 34:2636–2639

Weyl H (1911) Über die asymptotische Verteilung der Eigenwerte. Gött Nachr 1911:110–117

Wirner F (2015) Flow and transport of colloidal suspensions in porous media. Von der Fakultät Mathematik und Physik der Universität Stuttgart zur Erlangung der Würde eines Doktors der Naturwissenschaften (Dr. rer. nat.) genehmigte Abhandlung

Yang Y, Yang H, Tao L, Yao J, Wang W, Zhang K, Luquot L (2019) Microscopic determination of remaining oil distribution in sandstones with different permeability scales using computed tomography scan. J Energy Resour Technol 141(9):092903

Yonezawa F, Cohen MH (1983) Granular effective medium approximation. J Appl Phys 54(6):2895–2899

Yoon H, Dewers TA (2013) Nanopore structures, statistically representative elementary volumes, and transport properties of chalk. Geophys Res Lett 40:4294–4298

Zhang D, Zhang R, Chen S, Soll WE (2000) Pore scale study of flow in porous media: Scale dependency, REV, and statistical REV. Geophys Res Lett 27:1195–1198

Zhao Y (2019) Application of Euler-Poincaré characteristic in the prediction of permeability of porous media. Intell Autom Soft Comput 25(4):835–845

Zomorodian AJ (2005) Topology for computing. Cambridge monographs on applied and computational mathematics. Cambridge University Press, Cambridge

Zomorodian A, Carlsson G (2005) Computing persistence homology. Discrete Comput Geom 33(2):249–274

Chapter 4
Random Network Models

Chapter Highlights The common way to compute hydraulic and electric flow in porous media is to substitute the pore structure by a network consisting of *nodes* (for pores) and *bonds* (for the interconnecting throats). Different ways of constructing such models, and computation of their hydraulic and electric conductivity, are reviewed in Sect. 4.1. Section 4.2 introduces more than a dozen measures used to characterize the randomness and connectedness of such models. Section 4.3 treats three recently discovered phenomena arising in very large-scale networks: the *emergence of connectivity* and *appearance of a giant component* in evolving random networks; the *scale-free distribution* of the nodes' coordination numbers; and the networks' "*small-world*" property (Barabási 2018).

4.1 Pore Network Models and their Generation

Pore Network Models (PNMs) first appeared in Earth Sciences for modelling the flow of fluids through porous clastic rocks or soil. Since then their applications have become common in many other fields, including adsorption, biomass growth, capillary pressure, dual porosity (as in carbonates), dissolution/precipitation, drainage and imbibition, electric conduction, evaporation, force chains in granular packs, formation factor, fracture network, ganglion dynamics, geochemistry, hydraulic fracturing, hygric properties of building materials, liquid dissolution, rock failure, multiphase flow, evolution of crack networks, propagation of partial melts, percolation on pore networks, permeability, reactive transport, relative permeability, small-world features in rocks, soil hydraulics, solute transport, tortuosity, viscous forces, wettability (Table 4.1).

While PNMs idealize and simplify the complex *geometry* of the pore space, it is a basic requirement that these models should yield a true representation of its essential topological properties (average pore coordination number, connectivity: Celia et al. 1995; Claes and Janssen 2019; Joekar-Niasar and Hassanizadeh 2012; Thauvin and Mohanty 1998; Xiong et al. 2016).

G. Korvin, *Statistical Rock Physics*, Earth and Environmental Sciences Library,
https://doi.org/10.1007/978-3-031-46700-4_4

Table 4.1 Applications of random network models in rock physics

Application	Authors
Adsorption	Acharya et al. (2005), Raoof et al. (2013)
Biomass growth	Dupin et al. (2001a, b), Rosenzweig et al. (2013)
Capillary pressure	Fatt (1956a, b, c), Reeves and Celia (1996), Van der Marck et al. (1997)
Carbonates	Békri and Vizika (2006), Youssef et al. (2007)
Comparison of techniques	Al-Raoush et al. (2003)
Dissolution/precipitation	Kang et al. (2010), Knutson et al. (2001), Mehmani et al. (2012), Varloteaux et al. (2013); Zhou et al. (2000)
Drainage and imbibition	Hilpert and Miller (2001), Rostami et al. (2015)
Dual-porosity model	Wang et al. (2014)
Electric conduction	Edwards and Berg (2021), Fatt (1956a, b, c), Gackiewicz et al. (2021), Kirkpatrick (1973)
Evaporation	Nowicki et al. (1992)
Flow	Acharya et al. (2004), Al-Gharbi and Blunt (2005), Bakke and Øren (1997), Blunt (2001), Blunt et al. (2002), Gao et al. (2012), Heath et al. (2011), Hughes and Blunt (2000), Joekar-Niasar and Hassanizadeh (2012), Joekar-Niasar et al. (2010a, b), Koplik and Lasseter (1985), Lerdahl et al. (2000), Meyer (2021), Mogensen and Stenby (1998), Øren et al. (1996, 1997), Piri and Blunt (2005a, b), Ryazanov et al. (2009), Singh and Mohanty (2003), Song et al. (2016), Steven and Blunt (1992), Steven et al. (1993), Svirsky et al. (2004), Valvatne and Blunt (2004), Zareei et al. (2021)
Force chains in porous media	Laubie et al. (2017)
Formation factor	Bernabé et al. (2011), Yi et al. (2017)
Fracture network	Lin and Fairhurst (1991), Mo et al. (1998)
Ganglion dynamics	Dias and Payatakes (1986b), Li and Wardlaw (1986a, b)
Geochemistry	Peng et al. (2020)
Granular packings/matters	Arévalo et al. (2013), Papadopoulos et al. (2018)
Hydraulic conductivity	Bryant et al. (1993)
Hydraulic fracturing	Rabbel et al. (2020)
Hygric properties of building materials	Steven and Janssen (2019)
Liquid dissolution	Dillard and Blunt (2000)
Metal failure	Erickson et al. (2020)
Multiphase flow	Aghaei and Piri (2015), Jiang et al. (2012), Joekar-Niasar et al. (2009)
Network evolution	Levitz et al. (2012)
Partial melts	Skemer et al. (2017)

(continued)

Table 4.1 (continued)

Application	Authors
Percolation on pore networks	Vo et al. (2013)
Permeability	Al-Kharusi and Blunt (2007, 2008), Bernabé et al. (2010, 2011), Fatt (1956a, b, c), Jiang et al. (2017), Thompson and Fogler (1997), Thomson et al. (2018, 2020), van der Linden et al. (2016), Wang et al. (2019), Yi et al. (2017)
Reactive transport	de Vries et al. (2017), Mehmani et al. (2015), Raoof et al. (2012), Xiong et al. (2016)
Relative permeability	Arns et al. (2004), Fischer and Celia (1999), Mahmud et al. (2007), Øren and Bakke (2003)
Representative Elementary Volume (REV)	Gong et al. (2020)
Small-world fractures	Valentini et al. (2007), Ye et al. (2020), Zhu et al. (2021)
Small-world sandstone	Feng et al. (2020)
Soil hydraulics	Arns et al. (2004), Vogel and Roth (2001)
Solute transport	Köhne et al. (2011)
Tortuosity	Shanti et al. (2014)
Two-phase flow	Dias and Payatakes (1986a, b), Laroche and Vizika (2005)
Viscous forces	Van der Marck et al. (1997)
Wettability	Al-Futaisi and Patzek (2003a, b)

Network models are built up of two elements, *nodes* (also called *vertices* or *sites*) and *bonds* (or *edges*). In PNMs (*Pore Network Models*) the nodes substitute the actual pores, the bonds correspond to the throats between nearby pores. In the study of *granular packs* the nodes usually correspond to grains, and there is a bond between any two of them if and only if the two grains are in contact (CN, *Contact Network*"). In the network representation of a set of *fractures* (FN, *Fracture Network*"), the starting and terminating points of the traces of fractures on a selected plane as well as their points of intersections are the nodes, the traces themselves the bonds.

In geostatistical applications (Peng et al. 2020), let $\{X_i\}_{i=1}^{M}$ be observed random variables. If the observed values of x and y are $\{x_1, \cdots, x_N\}$ and $\{y_1, \cdots, y_N\}$, and $\overline{x}$ denotes the expected value of x, then the Pearson *correlation coefficient* between x,y is defined as (Jensen et al. 2000):

$$R_{xy} = \frac{\sum_{i=1}^{N}(x_i - \overline{x})(y_i - \overline{y})}{\sqrt{\sum_{i=1}^{N}(x_i - \overline{x})^2}\sqrt{\sum_{i=1}^{N}(y_i - \overline{y})^2}}. \qquad (4.1.1)$$

Let $0 \leq \lambda < 1$ be a *control parameter*, and define X_{ij}^{*} as

$$\left\{X_{ij}^*(\lambda)\right\}_{i,j=1}^{M} = \begin{cases} R_{X_iX_j} & when\ R_{X_iX_j} > \lambda \\ 0 & else \end{cases}. \tag{4.1.2}$$

To show the inter-dependencies between the M variables $\{X_i\}_{i=1}^M$ we can represent them with a *Cross Correlation Net* consisting of M nodes uniformly distributed in a circle or a square, such that for a given λ the nodes X_i and X_j are connected with a bond if and only if $\left\{X_{ij}^*(\lambda)\right\}_{i,j=1}^{M} > 0$.

Joekar-Niasar and Hassanizadeh (2012) classified network models according to (i) the spatial location of the nodes (whether they are points of a *structured* or *non-structured* lattice), and depending on (ii) whether their *coordination number* (number of bonds emerging from a node) is constant (*regular*) or variable (*irregular*). They introduced four classes: (a) *structured regular*, (b) *structured irregular*, (c) *unstructured regular*, (d) *unstructured irregular*.

If the network is needed for some petrophysical calculation, instead of using dimensionless points for pores, and lines of zero thickness for throats, we should represent, more realistically, the pores by spheres of given radii, and the throats by cylinders with given constant diameters and lengths.

Early applications of PNMs to study convection, diffusion, permeability, etc. simply represented pore-throats with straight cylindrical channels of constant circular cross section (Bakke and Øren 1997; Bryntesson 2002; De Josselin de Jong 1958; Jivkov and Xiong 2014; Laudone et al. 2008; Meyers and Liapis 1999; Meyers et al. 2001; Xiong et al. 2016) and the pores were substituted with nodes without volume (Vogel and Roth 2001) or were assumed to have cubic (Hughes and Blunt 2000; Joekar-Niasar et al. 2010a; Mogensen and Stenby 1998; Singh and Mohanty 2003) or spherical (Koplik and Lasseter 1985) shapes. In a few works (Jivkov et al. 2013; Joekar-Niasar and Hassanizadeh 2011; Xiong et al. 2014) the pore bodies were truncated octahedrons.

Joekar-Niasar and Hassanizadeh (2012) list the amazing variety of shapes used to approximate pore throats (*op. cit.*: Fig. 5). While in most models throats have *circular cross sections* (e.g., Dias and Payatakes 1986a, b; Koplik and Lasseter 1985), *triangular cross sections* which allow corner flow modeling and estimation of residual oil trapped in the corners are also common. (Al-Gharbi and Blunt 2005; Al-Kharusi and Blunt 2007; Bakke and Øren 1997; Békri and Vizika 2006; Laroche and Vizika 2005; Mason and Morrow 1991; Van der Marck et al. 1997; Xiong et al. 2016). Al-Kharusi and Blunt (2007) assigned the angles in the corners to match the measured *residual oil saturation*. Bakke and Øren (1997) used a *dimensionless shape factor G* to characterize the cross sections of both the pore bodies and throats. The *Pore Shape Factor* (PSF, introduced in Mason and Morrow 1991, see also Gong et al. 2020) is defined as

$$G = A/P^2, \tag{4.1.3}$$

where A is area, P is perimeter length in the cross section. (For cross sections shaped as a slit, an equilateral triangle, a square, or a circle, one has: $G_{-} \sim 0$; $G_{\blacktriangle} \sim 0.05$; $G_{\blacksquare}$

~ 0.06; G● ~ 0.08). When a wetting, and a non-wetting fluid are occupying the same pore space, the wetting fluid ensconces itself in the angular corners. Let r denote the inscribed radius of the throat, and r_w the radius of curvature of the interface between the phases. By the Young–Laplace equation

$$r_w = \frac{\gamma}{P_{nw} - P_w}, \tag{4.1.4}$$

where P_{nw} and P_w are pressures in the two phases, and γ is interfacial tension. By simple geometry, the cross-sectional areas available for the two fluids (A_{nw} and A_w) are given by

$$A_w = r_w^2\left(\frac{1}{4G} - \pi\right), \tag{4.1.5a}$$

$$A_{nw} = \frac{r^2}{4G} - A_w. \tag{4.1.5b}$$

The hydraulic conductance of fluid i, g_i, ($i = w$ or $i = nw$) is given by Poiseuille's law

$$g_i = \frac{R_i^2 A_i}{8\mu_i l}, \tag{4.1.6}$$

where μ_i is *fluid viscosity*, l the length of the throat, R_i are the equivalent hydraulic radii of the respective cross sections available for the two phases in the flow. In a similar example for corner-flow, Joekar-Niasar et al. (2010b) used cross sections bounded by *hyperbolic arcs* to derive relations between capillary pressure and saturation in glass beads.

Gao et al. (2012) carefully distinguished *bond* from *throat*: „a bond is defined as the passage connecting two neighboring pores, while a throat is defined as the narrowest part of a bond … the smallest constriction that will be encountered [by the fluid] when traversing a passage into the central pore". They introduced three different *levels of interconnectivity* between two neighboring spherical pores, P_1 and P_2. Let the radii of the two pores be r_1, r_2, and d_{12} the distance between their centers. The three levels are:

$$\begin{cases} Level\ 0 \quad d_{12} \geq r_1 + r_2 \\ Level\ 1 \quad |r_1 - r_2| \leq d_{ij} < r_1 + r_2 \\ Level\ 2 \quad d_{ij} < |r_1 - r_2| \end{cases} \tag{4.1.7}$$

For *Level 0* the two pores are separate from each other; for *Level 1* the two pores are interconnected but the center of one pore is not included in the other pore, for *Level 2* the two pores are interconnected, and the center of one pore is included in the other one.

According to Steven et al. (1993), the effective radius r_{eff} which is used to compute the flow through such an irregular cross section, lies between the radius of the largest inscribed circle r_c, and r_e which is radius of the '*equivalent circle*' having the same area as the irregular cross section. Approximately,

$$r_{eff} \approx \frac{1}{2}\{r_c + r_e\} \tag{4.1.8}$$

In addition to the *Pore Shape Factor G* (Eq. 4.1.3), a further important geometrical parameter of the PNM's is *the ratio of pore body diameter to pore throat diameter*, referred to as *aspect ratio* (Joekar-Niasar and Hassanizadeh 2012). The average *pore-throat aspect ratio* AR is calculated as:

$$AR = \frac{1}{N}\sum\nolimits_{i=1}^{N} \frac{r_{p,i}}{t_{t,i}}. \tag{4.1.9}$$

Joekar-Niasar et al. (2009) reported that they „did not find any trapped non-wetting phase at the end of imbibition in a high-porosity domain with small aspect ratio"; according to Gong et al. (2020) „the large value of this parameter may be an important factor leading to poor connectivity in sandstones".

Tortuosity τ (see Chap. 2) is also an important parameter, it describes how twisted are the bonds. *Geometric tortuosity* is the expected value of a ratio,

$$\tau = \left\langle \frac{L}{R} \right\rangle \tag{4.1.10}$$

where L is the length of a randomly selected flow channel and R is the length of the straight line between its two ends (Korvin 1992a, 2016; Shanti et al. 2014). Xiong et al. (2016) warns us that „the neglection of tortuosity of the connection … can lead to erroneous local conductivities". This can be one of the reasons that—as noted in (Gackiewicz et al. 2021)—the Kozeny-Carman formula sometimes overestimates (by a factor more than two) permeability.

If the nodes of the network are *ellipsoids* (as in Doyen's [1987] pore model discussed in Sect. 8.2) one would need the *degree of anisotropy* (DA) to describe the system (Harrigan and Mann 1984; Gong et al. 2020). The DA value is given by:

$$DA = \frac{L_l}{L_S} \tag{4.1.11}$$

where L_s and L_l are the shortest and longest axes, respectively, of an average ellipsoid.

What are those favorable properties what we expect of the PNMs extracted from 2D or 3D rock images? I list here a few expert opinions, in chronological order:

- Bakke and Øren (1997): "sandstone [pore network] models [should be] analogs of actual sandstones … realistic and fully characterized."

- Thauvin and Mohanty (1998): “Although it [a PNM] does not describe the exact morphology of a porous medium, it is able to take into account essential features of the pore space geometry and topology.”
- Vogel and Roth (2001): [PNMs should correspond] “to the metric and topological properties measured at the serial sections, i.e. the pore-size distribution and the connectivity function.”
- Oren and Bakke (2002): [PNMs should give] “a good representation of the connectivity of the rock and as a consequence accurately predict transport properties.”
- Békri and Vizika (2006): [The PNM should be] “a simple geometry 3D network in which flow calculations can be made. This network is not an exact representation of the pore space; however it contains significant pore structure information such as connectivity and pore size distribution.”
- Jiang et al. (2012): [The PNM should] “accurately predict multi-phase flow properties of rocks by network flow simulation. [It is] not very computationally intensive, and computations on representative volumes are feasible.”
- Joekar-Niasar and Hassanizadeh (2012): [The PNM should be] “computationally less demanding.”
- Blunt et al. (2013): “large-scale pore networks will serve as a bridging element between pore scale physics and effective field-scale models”.
- Xiong et al. (2016): “computationally not very demanding”.
- Yi et al. (2017): “A physically realistic pore network is not only a representation of the pore space in the sense of topology and morphology, but also a good tool for predicting transport properties accurately.”
- Claes and Janssen (2019): “The generated networks can be arbitrarily large and are not limited by the size of the original image.”
- Gong et al. (2020): [The PNM must be estimated from] “a suitable and accurate size of the REV [(*Representative Elementary Volume*)]. It should give] representative networks that make reasonable estimates of pore geometry and transport properties.”

The “*connectivity function*”, mentioned in Vogel and Roth’s classic (2001) paper is the 3D Euler number depending on pore size,

$$\chi_V(r) = \frac{N_N(r) - N_B(r)}{V} \tag{4.1.12}$$

(see Sect. 3.2) that is the *specific Euler number* calculated for the *reduced pore space* where all pores have a radius equal to or larger than a given value r. In Eq. (4.1.12) $N_N(r)$ is the number of nodes with radii larger than or equal to r, and $N_B(r)$ is the number of their interconnecting bonds whose radius is also larger than or equal to r.

Of course, we are restricted by our experimental techniques and computing power when we try to construct representative PNMs that satisfy, to a certain degree, the above-quoted requirements (Xiong et al. 2016). If a 3D voxelated image is available, say, from *X-ray micro-tomography*, we can construct the network directly from this image. In the lack of such data, we can create a synthetic 3D rock image by simulating the geologic processes (sedimentation, compaction, diagenesis) leading to the formation of sedimentary rocks—this would yield a PNM which is "analog" to the pore net of the actual rock, in the sense of Bakke and Øren (1997) mentioned above. Sometimes we have only one, or just a few, 2D sections and want to extract from them a 3-D PNM; or the network found in a millimeter-sized volume must be upscaled to a larger size. In such cases the PNM will be only *statistically representative*. If the pore data come from elsewhere than imaging, say from MIP (*Mercury Intrusion Porosimetry*), GA (*Gas Adsorption*) or NMR (*Nuclear Magnetic Relaxation*), we shall have reasonable knowledge about pore- and throat-radius distribution but no information at all on the pores' spatial distribution and their interconnectivity (coordination number distribution, see Chap. 6), so we have to build a PNM without knowing these parameters (Jivkov and Olele 2012).

(1) In case of *PNMs directly extracted from a 3D voxelated rock image*, there exists a one-to-one spatial correspondence between the network elements and the rock image. Usually, nodes correspond to centers of the pores, bonds correspond to the *medial axes* proceeding through the midpoints of the pores' and throats' cross sections. Two nodes are connected with bonds if and only if there exist throats between the two pores. The set of nodes and bonds is called *skeleton*, its extraction from the image starts with finding the medial axes, then defining nodes as points of intersection of two or more medial axes (Baldwin et al. 1996; Lindquist et al., 1996; Al-Raoush and Willson 2005; Jiang et al. 2017; Gackiewicz et al. 2021). In *Mathematical Morphology* (Serra 1982) the skeleton of the pore space is defined as a *central line*, the set of points (voxels) at equal distances from two or more points of the solid wall. The skeleton faithfully represents the topology of the original pore-space, but for the subsequent calculations the nodes and bonds still need to be assigned a non-zero size or thickness. This is achieved with the *maximal ball algorithm* (Al-Kharusi and Blunt 2007; Dong and Blunt 2009) where the 1-voxel-sized nodes and 1-voxel-wide medial axes are maximally extended with *morphological dilation* until they fill out the available space between neighboring grains.

A beautiful example for *skeletonization* is found in the works of (Zhu et al. 2011; Skemer et al. 2017) who studied the network topology of melt objects in partially molten rocks. They reduced the melt network to its skeleton consisting of *nodes* and *tubules* and proved the theoretical prediction (of Bulau et al. 1979) that the most frequent coordination number (CN) of the nodes is four. The result is important for *melt propagation*, because by the power-law relationship found by Bernabé et al. (2010) between permeability k and the average CN, $\overline{z}$, of pore networks,

$$k \sim (\overline{z} - z_c)^{\beta}, \tag{4.1.13a}$$

where the critical coordination number in 3D is $z_c = 1.5$ (4.1.13b). The exponent β depends on the *normalized standard deviation* $\sigma_r = \sigma / \langle r \rangle$ (4.1.13c) of the tubule diameter distribution as

$$\beta = 1.2343 + 0.93462\sigma_r + 1.4755\sigma_r^2. \qquad (4.1.13b)$$

Simulation of rock forming processes (sedimentation, compaction, diagenesis) is treated in (Bryant et al. 1993; Bakke and Øren 1997; Øren and Bakke 2003). There are four main modeling steps: (i) generation of the grain size distribution curve from the 2-D BSE (*Back Scattered Electronic*) image of the sandstone sample; (ii) *low energy-* or *high energy- deposition of the grains*; (iii) *compaction*; and (iv) *diagenesis*. In a *low-energy environment* (sand or coarse sand deposition) the grain, whose diameter r is randomly drawn from the grain-size distribution, is dropped vertically from a random position $(X.Y)$ onto the already accumulated grain-bed until it stops at a stable position. To help the grain falling through openings wider than its diameter, we reduce it to a point and increase the radii of all grains already in the stack by r. When the grain is deposited, we reset all radii to their original values and continue with the next grain, until a box of predefined size is filled with grains. In *high-energy deposition* (as in case of fine sand and silt) we allow for the grains a small horizontal freedom: they can roll down to the nearest local minimum of the bed's upper surface where they settle. (In some other works, deposition of non-spherical particles has been simulated: Coelho et al. 1997; Pilotti 2000).

Compaction which is the grain-bed's response to vertical stresses from the overburden is simply modeled by a linear transformation: The Z-coordinate of every sand-grain center (X_i, Y_i, Z_i) is shifted vertically downwards by an amount proportional to their original Z-position: $(X_i, Y_i, Z_i) \Rightarrow (X_i, Y_i, (1-\lambda)Z_i)$ (4.14), $\lambda \epsilon [0, 1]$ is the *compaction factor*. The grains remain spherical, their radii do not change, but the down-shifted grains might interpenetrate (as in *pressure dissolution*, Bjorkum 1996).

Out of the very complicated *diagenetic processes* the two simplest can be easily modeleled: *quartz cement overgrowth* and *clay coating* on the free quartz surfaces (Bryant et al. 1993). To model *quartz cement overgrowth* we uniformly increase the radii of all deposited and already compacted sand-grains with a prescribed amount, allowing overlaps; clay coating is simulated by random precipitation on the free quartz surfaces. Finally, since the locations of the grains are known, it is easy to locate the pore centers from a *Voronoi tessellation* (Delerue and Perrier 2002) around the grain centers, and the pore network can be constructed.

In the *statistical reconstruction* of a network, new pore networks are created which are supposed to have the same statistical characteristics, such as distributions of the geometrical properties (volume, radius, shape factor of the nodes, lengths and radii of the bonds, etc.) as the original network, and the same correlations between these properties for both nodes, bonds, and between nodes and bonds. The first such methods were based on two-point- and multiple-point correlations (see Sects. 1.2–1.3; Adler et al. 1992; Vogel and Roth 1997; Ioannidis and Chatzis 2000; Okabe and Blunt 2004). Øren, Bakke and co-workers (Bakke and Øren 1997; Lerdahl et al., 2000; Øren and Bakke, 2002; Øren et al. 1998), who pioneered physical-process-based reconstructions, compared the results of "geological" and statistical reconstructions and found that stochastic reconstructions strongly differed from the original rock as regards their connectivity, while "the geological reconstruction gave a good representation of the connectivity of the rock and as a consequence could accurately predict transport properties" (Øren and Bakke 2002). In a seminal paper, Vogel and Roth (2001) proposed a way to first generate a network, and then adapt it to the *connectivity function.* As already mentioned, the "*connectivity function*" is the 3D specific Euler number depending on pore size,

$$\chi_V(r) = \frac{N_N(r) - N_B(r)}{V} \tag{4.1.12}$$

(see Sect. 3.2), that is the *specific Euler number* calculated for such a *reduced pore space* where all pores have a radius equal to or larger than a given value r. Here, $N_N(r)$ is the number of nodes with radii larger than or equal to r, $N_B(r)$ is the number of their interconnecting bonds whose radius is also larger than or equal to r. For simplicity, we only discuss the case when in (4.1.12) $r = 0$, i.e. $\chi_V = \frac{N_n - N_b}{V}$ is computed considering all pores, and the constructed network also has this prescribed Euler number χ_V. The interested Reader is referred to (Vogel and Roth 2001; Jiang et al. 2012) for more details.

The Vogel-Roth model is generated using the nodes and selected edges of a *face-centered cubic* (*fcc*) *grid* with coordination number Z = 12, which is much higher than the usual range of pore coordination numbers in porous rock (2–7 in most rocks, according to Yanuka et al. 1986) so that we must remove in a later step the unnecessary bonds. First, we need a simple geometric relation between pore distribution, mean coordination number $\langle Z \rangle$, grid constant λ , and Euler characteristics χ_V. Let the total number of nodes and bonds be N_n and N_b, nodes are extension-less points, so all porosity is distributed between bonds. The volume of the network is

$$V = \frac{1}{\sqrt{2}} N_n \lambda^3 \tag{4.1.14}$$

and the total number of bonds is

$$N_b = \frac{1}{2} N_n \langle Z \rangle. \tag{4.1.15}$$

Suppose there are M distinct pore-size classes Π_i with corresponding median radii $\{r_1 < r_2 < \cdots < r_M\}$, and respective volume fractions $\Delta\Pi_i$. The probability that a bond belongs to class Π_i is

$$P_i = \frac{1}{r_i^2} \Delta\Pi_i / \sum_{j=1}^{M} \frac{1}{r_j^2} \Delta\Pi_j. \tag{4.1.16}$$

The total volume of the i-th pore-size class is

$$V_i = N_b P_i \pi r_i^2 \lambda, \tag{4.17}$$

it satisfies

$$V_i = V \cdot \Delta\Pi_i. \quad (4.18)$$

From Eqs. (4.1.14)–(4.1.18) we obtain

$$\lambda^2 = \frac{\pi}{\sqrt{2}}\langle Z\rangle\left[\sum\nolimits_j \frac{\Delta\Pi_j}{r_j^2}\right]^{-1}. \quad (4.1.19)$$

By the classical formula, the Euler number of the network model is

$$\chi_V = \frac{N_n - N_b}{V}, \quad (4.1.20)$$

that is, using Eqs. (4.1.14 and 4.1.15),

$$\chi_V = \sqrt{2}\frac{1-(1/2)\langle Z\rangle}{\lambda^3}. \quad (4.1.21)$$

In the knowledge of χ_V and the pore-size distribution $\left\{r_j;\ \Delta\Pi_j\right\}_{j=1}^{M}$, the system of Eqs. (4.1.19) and (4.1.21) can be solved for λ and $\langle Z\rangle$. Next, we find the number of nodes N_n in the desired volume V from Eq. (4.1.14), and randomly remove bonds from the regular *fcc* grid in a way that, independently of each other, every bond is removed with a probability $(12 - \langle Z\rangle)/12$ and kept with a probability $\langle Z\rangle/12$. Finally, we assign radii to the remaining N_b bonds according to the probability distribution $\left\{P_i = \frac{1}{r_i^2}\Delta\Pi_i / \sum_{j=1}^{M}\frac{1}{r_j^2}\Delta\Pi_j\right\}_{i=1}^{M}$.

More sophisticated algorithms to get a "diluted network" which contains only the required number of bonds are found in (Arns et al. 2004; Gao et al. 2012). In the algorithm of Arns et al. (2004) nodes are extension-less points. A more general "dilution algorithm" to match the CN distribution is provided in (Gao et al. 2012), where the nodes correspond to pores of different size and the rejection probability of any given bond also depends on the size of the two pores at its endpoints.

After the network had been generated, we can easily compute its hydraulic conductivity K. If there is a pressure gradient ΔP across opposite ends of the network, the flow q_{ij} through a bond b_{ij} connecting nodes n_i and n_j is, by Poiseuille's law:

$$q_{ij} = \frac{\pi}{8\mu}r_{ij}^4\frac{\Delta p_{ij}}{\lambda}, \quad (4.1.22)$$

where r_{ij} is the radius of the bond, μ is viscosity of the fluid and λ is the grid constant. The mass balance equation at each node i requires

$$\sum_j q_{ij} = 0. \quad (4.1.23)$$

This system of linear equations can be solved for the pressure distribution $\{\Delta p_{ij}\}$ using the *conjugate gradient method* (Press et al. 1992), and the hydraulic conductivity is obtained from the total flux Q through a plane perpendicular to the pressure gradient. The permeability K is determined from *Darcy's law*:

$$K = \frac{Q\mu L}{A \cdot \Delta P}, \tag{4.1.24}$$

where A is the cross-sectional area of the domain and L is length of the sample (Bakke and Øren 1997; Vogel and Roth 2001; Meyer 2021).

If the network is fully saturated with brine of conductivity σ_w, its electrical conductivity is similarly found (Bakke and Øren 1997), using current I instead of Q, a constant voltage gradient ΔV instead of ΔP, and Ohm's law

$$g_e = \frac{A\sigma_w}{L} \tag{4.1.25}$$

instead of Darcy's. The current conservation equations are relaxed using a *conjugate gradient method* to evaluate the voltages at each pore body, from the voltage distribution one gets the total current, and finally, the *formation factor*, $F = \sigma_w/\sigma_{rock}$.

In addition to these three main approaches (direct extraction of the network from 3D image; creating physical models by simulating rock formation; and stochastic reconstruction) there are several alternative ways of pore network generation, for example *nonlinear optimization* to minimize the differences between the CT image of the porous medium and the generated pore network. Hu et al. (2019) used *simulated annealing* (Sect. 6.1) for PN generation, Nejad Ebrahimi et al. (2013) applied a *genetic algorithm.*

There have been a few attempts (Laroche and Vizika 2005; Békri and Vizika 2006; Korvin et al. 2014) to substitute the rock's pore structure with an *equivalent* PNM. *Equivalent pore networks* are not exact representations of the pore space, they have very simple structure and their geometry and topology is constructed with an aim to fit the rock's measured properties (porosity, permeability, formation factor, etc.). In the Laroche-Vizika-Békri model (LVB) it is assumed that (i) the pores are situated at the nodes of a 3D cubic lattice; (ii) each pore has six outgoing throats of the same inscribed diameter *d,* (iii) pore-diameter D and throat diameter d has the same aspect ratio at all nodes: $D/d = \alpha$; (iv) the pore shape is irregular and obeys the scaling law

$$V_p(D) \sim D^{\lambda_p} \tag{4.1.26}$$

between pore volume and pore diameter found for real pores (Heiba et al. 1992; Fenwick and Blunt 1998). Writing out Eq. (4.1.26) in terms of d, and with a more precise prefactor,

$$V_p(D) \sim (\alpha \cdot d)^{\lambda_p} = C_p D^{(3-\lambda_p)} (AR)^{\lambda_p} d^{\lambda_p}, \tag{4.1.27a}$$

$$C_p \approx \frac{\Phi L^3}{\langle D\rangle^{(3-\lambda p)}\langle D^{\lambda p}\rangle}, \tag{4.1.27b}$$

where Φ is porosity, L is the node-to-node distance in the lattice and $\langle\cdot\rangle$ indicates average. The LVB model allows three different dependencies between throat-length l_t and throat diameter d, depending on whether a control parameter λ_t is 0, 1 or -1

$$l_{t_{\lambda_t}} = \begin{cases} C_t(L - \langle D\rangle)\cdot(d/\langle d\rangle) & for \quad \lambda_t = 1 \\ C_t(L - \langle D\rangle) & for \quad \lambda_t = 0 \\ C_t(L - D) & for \quad \lambda_t = -1 \end{cases} \tag{4.1.28a,b,c}$$

In these equations, if the scaling exponent λ_p and the pore throat distribution are known, the factors C_p in (4.1.27a, b), C_t in (4.1.28), and the parameter L are adjustable parameters used to reproduce the measured porosity, permeability, and electric formation factor.

The principal element of the equivalent network proposed in (Korvin et al. 2014) to fit macroscopic rock properties is a simplified version of Doyen's model (1987), and it further develops the model of Akbar et al. (1994). It consists of identical spherical pores connected with cylindrical tubes of the same radius. The model has three geometric parameters (average pore radius r, average distance between nearest pores d, and average throat radius δ), and one topological parameter, average coordination number Z of the pores (see Fig. 3.2.1 in Sect. 3.2). These parameters can be uniquely and explicitly determined in terms of four measured macroscopic rock properties: porosity Φ, permeability k, electric formation factor F, and cementation exponent m (Korvin et al. 2014: Eqs. 24–26).

4.2 How Random is a Pore Network Model?

In the mushrooming literature on networks one frequently comes across numerical measures that characterize the degree of order or disorder of an entire net, or the importance of a selected node or bond. Out of the plethora of such measures, I only list here (in alphabetic order) a selected few which have already been applied, or referred to, in rock-physics-related works.

(a) *Closeness centrality* (Papadopoulos et al. 2018). In *network analysis* (Freeman 1977; Newman 2010) this is one of the *centrality measures* which quantify the importance of particular nodes, edges, or other structures in a network. For a given node $\mathcal{N}_i$ *closeness centrality* is defined as the inverse of the sum over the *shortest-path lengths* from node $\mathcal{N}_i$ to all other nodes

$$\mathcal{N}_j : H_i = \frac{N-1}{\sum_{j\neq i} d_{ij}}, \tag{4.2.1}$$

where d_{ij} is the shortest-path distance between the two nodes measured in number of bonds along the path.

(b) *Clustering coefficient* (Dorogovtsev et al. 2008). The degree of *connectedness* between the nearest neighbors of a node $\mathcal{N}_i$ is described by its clustering coefficient

$$C(q_i) = t_i / \binom{q_i}{2}, \tag{4.2.2a}$$

where q_i is the coordination number of $\mathcal{N}_i$, t_i is the number of triangles (loops of length 3) attached to $\mathcal{N}_i$ and $\binom{q_i}{2} = q_i(q_i - 1)/2$ is the maximum possible number of such triangles.

Peng et al. (2020) define the *clustering coefficient* somewhat differently:

$$c_i = \frac{e_i}{\binom{k_i}{2}} = \frac{2e_i}{k_i(k_i - 1)}, \tag{4.2.2b}$$

where e_i is the actual number of edges between the neighboring nodes of node $\mathcal{N}_i$.

(c) *Clustering information entropy* (CIE) (Peng et al. 2020; Omar and Plapper 2020).

Denote by l_i the average length of the shortest paths connecting node $\mathcal{N}_i$ with other nodes. The CIE of node $\mathcal{N}_i$ is defined as

$$H_{l_i} = -\frac{l_i}{\sum_{j=1}^{n} l_j} log\left(\frac{l_i}{\sum_{j=1}^{n} l_j}\right). \tag{4.2.3}$$

(d) *Connection number*. Levitz et al. (2012) introduced this number to study *phase transitions* in the evolution of giant 3D *skeleton graphs*. It is the negative *Euler characteristics* of the net, normalized by the number of nodes:

$$C_t = -\frac{(N_n - N_b)}{N_n} = -\frac{\chi}{N_n}, \tag{4.2.4}$$

where N_n and N_b are the number of nodes and bonds, respectively. If k_i is the coordination number of node $\mathcal{N}_i$, and there are no isolated nodes or "hanging" nodes ($k = 0$, or $k = 1$), it is easy to check that $C_t = \frac{\langle k_i \rangle}{2} - 1 \geq 0$. On the other hand if there is a large number of isolated nodes and only relatively few bonds between the other ones, we shall have $N_n \gg N_b$ in Eq. (4.2.4) so that $C_t \in (-1, 0)$. The negativity of C_t thus indicates that the net consists of many dissociated parts.

(e) *Connectivity information entropy* (AIE) (Peng et al., 2020; Omar and Plapper 2020).

The AIE of node $\mathcal{N}_i$ is calculated as

$$H_{k_i} = -\frac{k_i}{\sum_{j=1}^{n} k_i} log\left(\frac{k_i}{\sum_{j=1}^{n} k_j}\right), \quad (4.2.5)$$

where k_i is the coordination number of node $\mathcal{N}_i$.

(f) *Efficiency* (Latora and Marchiori 2001; Valentini et al. 2007). For *weighted networks*, i.e., those networks in which each bond has a *Euclidean length*, the "*path length*" between two nodes $\mathcal{N}_i$ and $\mathcal{N}_j$ is the sum of lengths of the bonds forming the path connecting the nodes. The *efficiency of connection* between $\mathcal{N}_i$ and $\mathcal{N}_j$ is

$$\varepsilon_{ij} = \frac{1}{d_{ij}}, \quad (4.2.6)$$

where d_{ij} is the shortest path length between the two nodes.

(g) *Global and local clustering coefficient* (Watts and Strogatz 1998; Newman 2003; Papadopoulos et al. 2018). For an *unweighted, undirected network*, the *local clustering coefficient* C_i is defined as the *number of triangles involving node* $\mathcal{N}_i$, divided by the *number of triples centered at node* $\mathcal{N}_i$. (A triple is a set of three nodes that can include either three edges which form a 3-cycle, or just two of them). The *local clustering coefficient* is

$$C_i = \begin{cases} \frac{\sum_{hj} A_{hj} A_{ih} A_{ij}}{k_i(k_i-1)} & if\ k_i \geq 2 \\ 0 \quad if\ k_i = 0\ or\ k_i = 1 \end{cases}, \quad (4.2.7)$$

where $A_{ij} = \left\{\begin{matrix} 1\ if\ \mathcal{N}_i\ and\ \mathcal{N}_j\ are\ connected\ with\ a\ bond \\ 0 \qquad else \end{matrix}\right\}$ is the net's *adjacency matrix*, and k_i is coordination number of $\mathcal{N}_i$.

The *global clustering coefficient* is the mean value of C_i over all nodes:

$$C = \frac{1}{N} \sum_{i=1}^{N} C_i \quad (4.2.8)$$

(h) *Global and local efficiency* (Latora and Marchiori 2001; Valentini et al. 2007; Zhu et al. 2021). It is the average of the *efficiencies* $\varepsilon_{ij} = \frac{1}{d_{ij}}$:

$$E_{glob} = \frac{\sum_{i \neq j} \varepsilon_{ij}}{N(N-1)} = \frac{1}{N(N-1)} \sum_{i \neq j} \frac{1}{d_{ij}}. \quad (4.2.9a)$$

In the ideal case the network has all $N(N-1)/2$ possible bonds between its nodes and the length of shortest path between $\mathcal{N}_i$ and $\mathcal{N}_j$ is their Euclidean distance, l_{ij}. This is the largest possible global efficiency:

$$E_{glob(max)} = \frac{1}{N(N-1)} \sum\nolimits_{i \neq j} \frac{1}{l_{ij}}. \tag{4.2.9b}$$

We can characterize the *local properties* of a network G around each of its nodes $\mathcal{N}_i$ with the *efficiency* (Γ_i) computed as in (4.2.9a), which is the efficiency of the *subnet* Γ_i consisting of the nodes neighboring $\mathcal{N}_i$ and their interconnecting bonds. We define the *local efficiency* of the whole net $\mathcal{N}$ as the average efficiency of such local subnets:

$$E_{loc} = \frac{1}{N} \sum\nolimits_{\mathcal{N}_i \in \mathcal{N}} E(\Gamma_i). \tag{4.2.10}$$

It has been found (Latora and Marchiori 2001; Valentini et al. 2007) that for *regular, random and small-world* (see next Sect. 4.3) *networks* one typically has

$$\begin{cases} regular\ networks & E_{glob} < 0.1;\ E_{loc} > 0.1 \\ random\ networks & E_{glob} > 0.1;\ E_{loc} < 0.1\ . \\ small-world\ networks & E_{glob} > 0.1;\ E_{loc} > 0. \end{cases} \tag{4.2.11}$$

(i) *Mean intervertex distance* (Dorogovtsev et al. 2008). If the nodes $\mathcal{N}_i$ and $\mathcal{N}_j$ are connected by at least one path, the *intervertex distance* l_{ij} counts the number of edges in the shortest path. The *mean intervertex distance* $\langle l_{ij} \rangle = L(N)$ characterizes the *compactness* of a large network having $N \gg 1$ nodes.

(j) *Mean shortest-path distance* (Watts and Strogatz 1998; Papadopoulos et al. 2018). Denoting the unweighted shortest-path distance between nodes $\mathcal{N}_i$ and $\mathcal{N}_j$ as d_{ij}, the *mean shortest-path distance* L in a network is

$$L = \frac{1}{N(N-1)} \sum\nolimits_{i,j(i \neq j)} d_{ij}. \tag{4.2.12}$$

(k) *Neighborhood integrity information entropy* (NIE) (Peng et al. 2020; Omar and Plapper 2020). The NIE of node $\mathcal{N}_i$ is given by:

$$H_{c_i} = -\frac{c_i}{\sum_{i=1}^{n} c_i} log\left(\frac{c_i}{\sum_{i=1}^{n} c_i}\right), \tag{4.2.13}$$

where c_i is the *clustering coefficient* defined in Eq. (4.2.2b).

(l) *Order parameter* (Zareei et al. 2021). The authors considered low-Reynolds-number fluid flow through a porous network. The flow rate in the bond between $\mathcal{N}_i$ to $\mathcal{N}_j$ is $q_{ij} = C_{ij}(P_i - P_j)$ where P_i, P_j are pressures at the nodes, and by Poiseuille's law $C_{ij} = \pi r_{ij}^4/8\mu l_{ij}$, with r_{ij}, l_{ij}, μ being bond radius, bond length, and viscosity. To find the transition of the flow between *channeling instability* and *homogenization*, they defined the *order parameter*

$$\mathcal{O} = \frac{1}{N_b - 1}\left(N_b - \frac{\left(\sum_{ij} q_{ij}^2\right)^2}{\sum_{ij} q_{ij}^4}\right), \tag{4.2.14}$$

where N_b is the number of bonds. The *order parameter* $\mathcal{O} = 0$ when the flux is the same through every edge. When fluid flux becomes *highly localized* with only a few edges with non-zero flux, $\mathcal{O} \to 1$. Not surprisingly, (4.2.14) formally agrees with the formula for the *participation ratio* (PR) which quantifies the *localization of an eigenstate* in the theory of *Anderson localization* (Kramer and MacKinnon 1993).

(m) *Susceptibility* (Janson 2009; Manik et al. 2017). The susceptibility $\mathcal{S}$ of a network $\mathfrak{N}$ is the mean size of the largest connected component containing a randomly selected node:

$$\mathcal{S}(\mathcal{N}) := N_n^{-1} \sum\nolimits_{i=1}^{N_n} |\mathcal{C}(\mathcal{N}_i)|, \tag{4.2.15}$$

where N_n is the number of nodes, $\mathcal{N}_i$ are the nodes, $\mathcal{C}(\mathcal{N}_i)$ is the largest connected subnet containing $\mathcal{N}_i$, and $|\mathcal{C}(\mathcal{N}_i)|$ is the size (in number of nodes) of this component. If $\mathfrak{N}$ has $K = K(\mathfrak{N})$ distinct connected components $\mathcal{C}_1, \mathcal{C}_2, \cdots, \mathcal{C}_K$ then

$$\mathcal{S}(\mathcal{N}) = \sum\nolimits_{i=1}^{K} \frac{|\mathcal{C}_i|}{N_n}|\mathcal{C}_i| = \frac{1}{N_n}\sum\nolimits_{i=1}^{K} |\mathcal{C}_i|^2. \tag{4.2.16}$$

4.3 Giant Components and Small Worlds: Large-Scale Properties of Networks

Since the seminal results of Erdős and Rényi (1959, 1960; also found—independently—by Solomonoff and Rapoport 1951), the strange, unexpected properties of very large random graphs and networks have gained interdisciplinary attention among scholars of at least three closely related fields: *Graph Theory* (Bollobás 1998, 2001), *Percolation Theory* (Stauffer and Aharony 1994), and *Theory of Networks* (Watts 1999; Newman 2010, Barabási 2018). We briefly discuss three amazing phenomena observed in very large-scale social and natural networks, including pore- and fracture networks of rocks: (a) The *emergence of connectivity* and *appearance of a giant component*, in evolving random networks, (b) *"scale-free" distribution of the coordination numbers, (c)* the *"small-world"* property.

(A) The *emergence of connectivity*, and the *appearance of a giant component*, in evolving random networks.

Start out from $N_n = n(n \gg 1)$ distinct nodes, and add bonds randomly between them until the net becomes connected. Of course, we can achieve this using only $N_n - 1$ bonds, but this is a very special construction, and if the bonds are selected randomly, a very unlikely outcome. Erdős and Rényi (1959, 1960) found that one does not need "much more" bonds than that—$O(n \cdot \ln(n))$ random bonds will already ensure that the network is *almost surely* connected. More exactly, they introduced the so-called $\mathcal{N}(n, p)$ model, a network with $n \gg 1$ nodes which is constructed by connecting its nodes randomly, so that any possible bond is included with probability p, independently from the other bonds. $\mathcal{N}(n, p)$ has on the average $\binom{n}{2} p$ bonds, the contact number k_i of any node N_i follows binomial distribution:

$$P(k_i = k) = \binom{n-1}{k} p^k (1-p)^{n-1-k}; \ \langle k_i \rangle = (n-1)p. \qquad (4.3.1\text{a,b})$$

Erdős and Rényi (1960) proved that

- If $np < 1$, then $\mathcal{N}(n, p)$ will almost surely have no connected components of size larger than $O(\ln n)$ (4.3.2a);
- If $np = 1$, then a subnet in $\mathcal{N}(n, p)$ will almost surely have a largest component whose size is of $O(n^{2/3})$ (4.3.2b)
- If $np \to c > 1$, where c is a constant, then a part of $\mathcal{N}(n, p)$ will almost surely have a unique *giant component* containing a *positive fraction of the vertices.* No other component will contain more than $O(\ln n)$ vertices. (4.3.2c)
- If $p < \frac{(1-\varepsilon)\ln n}{n}$, then a part of $\mathcal{N}(n, p)$ will almost surely contain *isolated vertices,* and thus be disconnected. (4.3.2d)
- If $p > \frac{(1-\varepsilon)\ln n}{n}$, then a part of $\mathcal{N}(n, p)$ will almost surely be connected. (4.3.2e)
- By Eqs. (4.3.2d and e), $p = \ln n/n$ is a *sharp threshold* for the connectedness of $\mathcal{N}(n, p)$.

The exact mathematical proof (Bollobás 2001) of the full ER Theorem would be outside the scope of this book, I can only present a heuristic argument, "physicist style", showing that during the random addition of bonds there is a phase transition point at $\langle k \rangle = 1$ when a *giant component* (GC) appears. Suppose the GC is already there, its size is $|GC| = N_G$. The fraction of nodes *not* in the GC is $\lambda = 1 - N_G/n$. Select two nodes, $\mathcal{N}_i$ and $\mathcal{N}_j$ randomly. If $\mathcal{N}_i \notin GC$ there are two possibilities: (i) there is no link between $\mathcal{N}_i$ and $\mathcal{N}_j$ (this has the probability $1 - p$), or (ii) there is a link between them but $\mathcal{N}_j \notin GC$. By the construction rules of $\mathcal{N}(n, p)$, the total probability that $\mathcal{N}_i \notin GC$ is

$$Prob(\mathcal{N}_i \notin GC) = 1 - p + p\lambda. \qquad (4.3.3)$$

Carrying this out with all possible nodes $\mathcal{N}_j \neq \mathcal{N}_i$ the fraction of nodes not in the GC, λ , satisfies the equation

$$\lambda = (1 - p + p\lambda)^{n-1}. \tag{4.3.4}$$

Using $p = \langle k \rangle/(n-1)$ (4.3.1b) and taking the logarithm of both sides of (4.3.4),

$$\ln \lambda = (n-1)\ln\left[1 - \frac{\langle k \rangle}{n-1}\right](1-\lambda) \approx (n-1)\left[-\frac{\langle k \rangle}{N-1}(1-\lambda)\right] = -\langle k \rangle(1-\lambda) \tag{4.3.5}$$

From (4.3.5), $\lambda = exp[-\langle k \rangle(1-\lambda)]$. If $\mu = 1 - \lambda$ is the fraction of nodes in the GC, then

$$\mu = 1 - e^{-\langle k \rangle \mu}. \tag{4.3.6}$$

This nonlinear equation has no closed-form solution, its graphical solution presented in (Barabási 2018) shows that the phase transition when the GC appears occurs for $\langle k \rangle = 1$. For a similar derivation, see (Dorogovtsev et al. 2008) whose proof is based on the "*tree ansatz*", which assumes the absence of finite loops in the network in the thermodynamic limit.

We mentioned before the *connection number* $C_t = -\frac{(N_n - N_b)}{N_n}$ (Eq. 4.2.4). For large N_n and N_b we have $N_b/N_n = \langle k \rangle$, i.e.

$$C_t = -(1 - \langle k \rangle). \tag{4.3.7}$$

By the ER Theorem, $C_t = 0$ indicates the appearance of the GC (Levitz et al. 2012).

(B) "Scale-free" distribution of the pore coordination numbers

The ER random graph $\mathcal{G}(n, p)$, or random nets $\mathcal{N}(n, p)$, have on the average $\binom{n}{2}p$ edges (bonds), their degree (bond) distribution is *binomial*:

$$P(k_i = k) = \binom{n-1}{k} p^k (1-p)^{n-1-k}, \tag{4.3.1a}$$

if $n \to \infty$ such that $np = const$, the distribution becomes *Poisson*,

$$P(k_i = k) \to \frac{(np)^k}{k!} e^{-np}. \tag{4.3.8}$$

In many real rocks however the contact number distribution of the extracted PN (pore network) follows a *power-function probability distribution*,

$$P(k) \propto k^{-\gamma}. \tag{4.3.9}$$

In Eq. (4.3.9) the exponent γ is independent of system size. This distribution, reported by Guannan et al. (2020) and Ye et al. (2020) for the pore nets of sandstone and limestone, is characteristic of *self-organized systems and critical phenomena* (Korvin 1992b; Bunde and Havlin 1994).

In a landmark paper Barabási and Albert ("BA", 1999) called attention to the ubiquity of power-function distribution of the node coordination numbers in a wide range of social and information networks (including the *World Wide Web*) and pointed out that the (then) existing network models fail to reproduce this scale-free distribution.

They proposed a new "BA" model that had two basic properties: (i) the network *continuously grows* by the addition of new nodes; (ii) the new nodes are *preferentially connected* to such vertices that are already well connected. These two properties will lead to power-like coordination number distribution in the evolving network. Barabási and Albert (1999) also demonstrated with simple counterexamples, that the two properties, (i) *growth*, and (ii) *selective attachment* are not only sufficient but also necessary conditions for a scale-free distribution.

The two conditions are formalized as follows:

(i) *Growth*: Initially (at time step $K = 0$), there are a finite number $N_n^{(0)}$ of nodes, at every time step K we add a new node which has $m(\leq m_0)$ bonds. In step K, we shall get the net $\mathcal{N}^{(K)}$, it has $N_n^{(K)} = m_0 + K$ nodes, and $N_b^{(K)} = Km$ bonds.

(2) *Preferential attachment*: In time step K we assume that the probability P that the new node will be connected to node $\mathcal{N}_i$ depends on the connectivity $k_i^{(K-1)}$ of $\mathcal{N}_i$, such that

$$P(\mathcal{N}_{new}\ and\ \mathcal{N}_i\ connected) = k_i^{(K-1)} / \sum\nolimits_{\mathcal{N}_j \in \mathcal{N}^{(K)}} k_j^{(K-1)} \tag{4.3.10}$$

(for a more precise definition of this construction, see Bollobás and Riordan 2004).

Assume that the contact number is a *continuous variable of time*, then its *rate of change* satisfies

$$\frac{\partial k_i}{\partial t} = \frac{mk_i}{\sum_{j=1}^{m_0+t-1} k_j}. \tag{4.3.11}$$

As $\sum_j k_j = 2mt$ and during a time-step Δt the increase in N_b is m, $k_i(t)$ obeys the equation

$$\frac{\partial k_i}{\partial t} = \frac{k_i}{2t}, \tag{4.3.12a}$$

with the initial condition that at time t_i when $\mathcal{N}_i$ was included in the system, its contact number was m, that is

$$k_i(t_i) = m. \tag{4.3.12b}$$

The solution of (4.3.12a, b) is

$$k_i(t) = m \cdot \left(\frac{t}{t_i}\right)^{0.5}. \tag{4.3.13}$$

From Eq. (4.3.13) we find the distribution of $k_i(t)$,

$$P(k_i(t) < k) = P\left(t_i > \frac{m^2 t}{k^2}\right). \tag{4.3.14}$$

If the vertices are added one by one in successive intervals of unit time, the times $\{t_i\}$ are uniformly distributed with density

$$P_i(t_i) = \frac{1}{m_0 + t}. \tag{4.3.15}$$

Thus,

$$P(k_i(t) < k) = P\left(t_i > \frac{m^2 t}{k^2}\right) = 1 - P\left(t_i \le \frac{m^2 t}{k^2}\right) = 1 - \frac{m^2 t}{k^2(t + m_0)}. \tag{4.3.16}$$

The probability density function (pdf) for *P(k)* is obtained as

$$P(k) = \frac{\partial P(k_i(t) < k)}{\partial k} = \frac{2m^2 t}{m_0 + t} \cdot \frac{1}{k^3} \tag{4.3.17}$$

which is of the form$P(k) \propto k^{-\gamma}$. Note that the exponent γ (in this case $\gamma = 3$) is independent of the size of the network, that is the law is *scale free*.

The BA model has been generalized in 2000 (independently by Dorogovtsev et al., and Krapivsky et al.) so that it could describe power law distributions of the contact number with arbitrary $\gamma \epsilon [2, \infty]$. In the BA model, when a new node $\mathcal{N}_{new}$ is added, it is linked with m "old" nodes in a way that each node is chosen with a probability which is *proportional* to its contact number: $P(\mathcal{N}_{new}\ and\ \mathcal{N}_i\ connected) \sim k_i$ (see Eq. 4.3.10). In the Dorogovtsev et al. (2000) variant of the model, the probability of connection is proportional to a *linear function* of k_i:

$$P(N_{new}\ and\ N_i\ connected) \sim (k_i + A) \quad \text{where} \quad A > -m. \tag{4.3.18}$$

In this case, it can be proved, that the degree distribution is again a power law, with exponent

$$\gamma = 3 + \frac{A}{m}. \tag{4.3.19}$$

For the BA model $A = 0$, and (4.3.19) reduces to $P(k) \sim k^{-3}$ as in Eq. (4.3.17). Note that except for the BA model, the power law is not scale-free, because γ depends on a system-parameter, m.

(C) The small-world property of large random networks.

If $\mathcal{N}_i$ and $\mathcal{N}_j$ are two nodes of a network $\mathcal{N}$ which are connected by at least one path along nodes, we can define the *shortest path length* l_{ij} (also called *inter-node distance*) between them. The distribution of l_{ij}, in particular its mean value $\bar{l}(\mathcal{N})$, is an important network property, it characterizes the *compactness* of $\mathcal{N}$ (see Sect. 4.2). If the nodes are regular lattice points in $\mathbb{R}^d$, by dimensional arguments we expect that both $\bar{l}$, and its maximal value, the network's *diameter* $\Delta(\mathcal{N}) = \max_{\mathcal{N}_i \in \mathcal{N}, \mathcal{N}_j \in \mathcal{N}} l_{ij}$ scale with system size as

$$\bar{l} \propto N_n^{1/d};\ \Delta \propto N_n^{1/d}. \tag{4.3.20}$$

It has been found, however, that in many large networks describing social or physical systems, the average inter-node distance and the diameter grow much slower than any power of N_n, as $\propto \ln(N_n)$. This unexpected property of networks is called the *small-world phenomenon* (Watts and Strogatz 1998; Watts 1999; Albert et al. 2000), in rock networks it has been observed by (Valentini et al. 2007; Feng et al. 2020; Ye et al. 2020).

The essence of the "small-world" effect is easily understood from a simple example (Barabási 2018). Take a random ER (Erdős - Rényi) network $\mathcal{N}(N, p)$ with N nodes and with connection probability $= \langle k \rangle / n$, where the mean coordination number $\langle k \rangle > 1$. Suppose the diameter of $\mathcal{N}(N, p)$ is $d_{max} \equiv \Delta(\mathcal{N})$. Starting out from an arbitrary node $\mathcal{P}$, there are $\approx \langle k \rangle$ nodes at distance 1, $\approx \langle k \rangle^2$ nodes at distance 2, ..., $\approx \langle k \rangle^{d_{max}}$ nodes at distance d_{max}. (We assume the "*tree ansatz*"—see Dorogovtsev et al. 2008—that is with high probability there are no loops, so the distance is well defined between any two points.). As in at most d_{max} steps we can reach all points, and almost all points that can be reached are counted only once, we have (for $N \gg d_{max} \gg 1$)

$$1 + \langle k \rangle + \langle k \rangle^2 + \cdots + \langle k \rangle^{d_{max}} = \frac{\langle k \rangle^{d_{max}+1} - 1}{\langle k \rangle - 1} \approx \langle k \rangle^{d_{max}} = N_n, \tag{4.3.21}$$

that is

$$d_{max} \approx \frac{\ln N_n}{\ln \langle k \rangle} \tag{4.3.22}$$

and the constructed net is "*small-world*".

A more precise expression for the diameter of $\mathcal{N}(N, p = \langle k \rangle / N)$ is due to Fernholz and Ramachandran (2007):

$$\Delta\left(\mathcal{N}\left(N, \frac{\langle k\rangle}{N}\right)\right) = ln\, N\left\{\frac{1}{ln\,\langle k\rangle} + \frac{2}{ln[-W(\langle k\rangle exp(-\langle k\rangle))]}\right\} + o(ln\, N) = c \cdot ln\, N + o(ln\, N), \tag{4.3.23}$$

where $W(z)$ is the *Lambert W-function*, defined as the principal inverse of $f(z) = z \cdot \exp(z)$ (Roy and Olver 2010). The coefficient c can be expressed for two important cases:

$$c = \begin{cases} \frac{3-o(1)}{\ln\langle k\rangle} & as \quad \langle k\rangle \to 1 \\ \frac{1}{ln\langle k\rangle} + \frac{2}{\langle k\rangle} + O\left(\frac{ln\langle k\rangle}{\langle k\rangle^2}\right) & as \quad \langle k\rangle \to \infty, N \to \infty, \frac{\langle k\rangle}{N} \ll 1 \end{cases} \tag{4.3.24a, b}$$

Equation (4.3.24a) shows, that for $\langle k\rangle \approx 1$, when the GC (*giant component*) emerges,

$$\Delta = d_{max} \approx 3\frac{ln\, N}{ln\langle k\rangle}, \tag{4.3.25}$$

that is the predicted value $d_{max} \approx \frac{\ln N}{\ln\langle k\rangle}$ (in Eq. 4.3.22) is in error by a factor of 3. On the other hand, when $\langle k\rangle \to \infty$, $N \to \infty$, $\frac{\langle k\rangle}{N} \ll 1$, the $\frac{1}{ln\langle k\rangle}$ term dominates in the coefficient c, that is indeed $d_{max} \approx \frac{\ln N}{\ln\langle k\rangle}$.

Denote by $\mathcal{N}_{BA}^{(N,m)}$ the Barabási and Albert (BA, 1999) net on N nodes, constructed by randomly adding $m > 2$ bonds in each step, as described above. Bollobás and Riordan (2004: Theorem 1) proved that, for every $m > 2$ and $\varepsilon > 0$, almost surely $\mathcal{N}_{BA}^{(N,m)}$ is connected, and

$$(1-\varepsilon)\frac{log\, N}{log\, log\, N} \le \Delta\left(\mathcal{N}_{BA}^{(N,m)}\right) \le (1-\varepsilon)\frac{log\, N}{log\, log\, N}, \tag{4.3.26}$$

that is the diameter is asymptotically $log\, N/log\, log\, N$.

One of the first *small-world* network models described in literature was due to Watts and Strogatz (1998), constructed as follows:

(1) Start with a regular network on N nodes uniformly distributed along a circle, where each node is connected to its K closest neighbors.
(2) Make random rewiring in a way that each bond is rewired with probability p. One end of a rewired bond remains in its original position, the other end is taken to a randomly selected node of the network. Double bonds between two points and self-loops are not permitted.

The choice of p allows us to proceed from regularity ($p = 0$) to full randomness ($p = 1$). Denote the constructed net by $\mathcal{N}_{WS}^{(N,K,p)}$, and assume that

$$N \gg K \gg \ln(N \gg 1. \tag{4.3.27}$$

The condition $K \gg \ln(N)$ guarantees that the WS net is almost certainly connected (Bollobás 2001), so we can introduce the *average inter-node distance* (measured in number of bonds in the shortest connecting path) between its nodes, $\bar{l} = \langle l_{PQ} \rangle_{\mathcal{P},\mathcal{Q} \in \mathcal{N}_{WS}}$. Watts and Strogatz (1998) found in their *Monte Carlo study* of such random rewirings (using 12 realizations, each with $N = 1,000$ nodes and on the average 10 bonds per node) that if (4.3.27) is satisfied,

$$\bar{l} \sim \begin{cases} N/2K & as \quad p \to 0 \\ \ln(N)/\ln(K) & as \quad p \to 1 \end{cases}. \tag{4.3.28}$$

Thus, the regular lattice at $p \approx 0$ is *large-world*, close to the fully random case ($p \to 1$) the net is *small-world.*

In the eye of the petrophysicist, the two steps of the WS construction are similar to the development of *primary and secondary porosities* in dual-porosity sedimentary rocks, because of three reasons: (i) New connections are randomly added between hitherto unconnected pores; (ii) the first step, when only closest neighbors are connected, is similar to the primary inter-grain pore structure, and (iii) the second step, rewiring, is similar to the appearance of secondary connections (leached-out channels and/or fractures) which might connect even far-away pores. The contact numbers in the WS model are not power-distributed, actually, this distribution is not too frequently found in the PNMs of sedimentary rock. I foresee that a suitable modification of the WS model will have an important role in studies of secondary porosity, limestone porosity, *hydro-fracturing*, *fracking* and "*super k*".

References

Acharya RC, Sjoerd E, Anton L (2004) Porosity-permeability properties generated with a new 2-parameter 3D hydraulic pore-network model for consolidated and unconsolidated porous media. Adv Water Resour 27:707–723

Acharya R, Van der Zee S, Leijnse A (2005) Transport modeling of nonlinearly adsorbing solutes in physically heterogeneous pore networks. Water Resour Res 41(2)

Adler PM, Jacquin CG, Thovert JF (1992) The formation factor of reconstructed porous media. Water Resour Res 28(6):1571–1576

Aghaei A, Piri M (2015) Direct pore-to-core up-scaling of displacement processes: Dynamic pore network modeling and experimentation. J Hydrol 522:488–509

Akbar N, Mavko G, Nur A, Dvorkin J (1994) Seismic signatures of reservoir properties and pore fluid distribution. Geophysics 59(8):1222–1236

Albert R, Jeong H, Barabasi A (2000) Error and attack tolerance of complex networks. Nature 406:378–382

Al-Futaisi A, Patzek TW (2003) Extension of Hoshen-Kopelman algorithm to non-lattice environments. Physica A 321:665–678

Al-Futaisi A, Patzek T (2003b) Impact of wettability alteration on two-phase flow characteristics of sandstones: a quasi-static description. Water Resour Res 39:1042

Al-Gharbi MS, Blunt MJ (2005) Dynamic network modeling of two-phase drainage in porous media. Phys Rev E 71:016308–016308

Al-Kharusi AS, Blunt MJ (2007) Network extraction from sandstone and carbonate pore space images. J Petrol Sci Eng 56:219–231
Al-Kharusi AS, Blunt MJ (2008) Multiphase flow predictions from carbonate pore space images using extracted network models. Water Resour Res 44:W06S01
Al-Raoush RI, Willson CS (2005) Extraction of physically realistic pore network properties from three-dimensional synchrotron X-ray microtomography images of unconsolidated porous media systems. J Hydrol 300:44–64
Al-Raoush R, Thompson K, Willson CS (2003) Comparison of network generation techniques for unconsolidated porous media. Soil Sci. Soc. Amer. J. 67:1687–1700
Arévalo R, Pugnaloni LA, Maza D, Zuriguel I (2013) Tapped granular packings described as complex networks. Phil Mag 93(31–33):4078–4089
Arns J-Y, Robins V, Sheppard AP, Sok RM, Pinczewski WV, Knackstedt MA (2004) Effect of network topology on relative permeability. Transp Porous Media 55:21–46
Bakke S, Øren P-E(1997) 3-D Pore-Scale Modelling of Sandstones and Flow Simulations in the Pore Networks. SPE J 2:136–149
Baldwin CA, Sederman AJ, Mantle MD, Alexander P, Gladden LF (1996) Determination and characterization of the structure of a pore space from 3D volume images. J Colloid Interface Sci 181(1):79–92
Barabási A-L (2018) Network science. Cambridge University Press, Cambridge
Barabasi A-L, Albert R (1999) Emergence of Scaling in Random Networks. Science 286:509–512
Békri S, Vizika O (2006) Pore network modeling of rock transport properties: application to a carbonate. SCA 22:2006
Bernabé Y, Li M, Maineult A (2010) Permeability and pore connectivity: a new model based on network simulations. J Geophys Res 115:B10203
Bernabé Y, Zamora M, Li M, Maineult A, Tang Y (2011) Pore connectivity, permeability, and electrical formation factor: a new model and comparison to experimental data. J Geophys Res 116:B11204
Bjorkum, Per Arne (1996) How important is pressure in causing dissolution of quartz in sandstones? J Sediment Res 66(1):147–154
Blunt MJ (2001) Flow in porous media—pore-network models and multiphase flow. Curr Opin Colloid Interface Sci 6:197–207
Blunt MJ, Jackson M, Piri M, Valvatne PH (2002) Detailed physics, predictive capabilities and macroscopic consequences for pore-network models of multiphase flow. Adv Water Resour 25:1069–1089
Blunt MJ, Bijeljic B, Dong H, Gharbi O, Iglauer S, Mostaghimi P, Paluszny A, Pentland C (2013) Pore-scale imaging and modelling. Adv Water Resour 51:197–216
Bollobás B (1998) Modern graph theory. Springer-Verlag, Berlin
Bollobás B (2001) Random graphs, 2nd edn. New York Academic Press, Washington, DC, USA
Béla B, Riordan O (2004) The diameter of a scale-free random graph. Combinatorica 24(1):5–34
Bryant SP, King R, Mellor DW (1993) Network model evaluation of permeability and spatial correlation in a real random sphere packing. Transp Porous Media 11:53–70
Bryntesson LM (2002) Pore network modelling of the behaviour of a solute in chromatography media: transient and steady-state diffusion properties. J Chromatogr A 945(1):103–115
Bulau JR, Waff HS, Tyburczy JA (1979) Mechanical and thermodynamic constraints on fluid distribution in partial melts. J Geophys Res 84:6102–6108
Bunde A, Havlin S (1994) Fractals in Science. Springer, Berlin
Celia MA, Reeves PC, Ferrand LA (1995) Recent advances in pore scale models for multiphase flow in porous media. Rev Geophys 33:1049–1058
Steven C, Janssen H (2019) Towards stochastic generation of 3D pore network models of building materials. In: MATEC web of conferences, vol 282, p 0222
Coelho D, Thovert J-F, Adler P (1997) Geometrical and transport properties of random packings of spheres and aspherical particles. Phys Rev E 55:1959–1978

De Josselin de Jong G (1958) Longitudinal and transverse diffusion in granular deposits. Trans Am Geophys Union 39:67–74
Delerue J, Perrier E (2002) DXSoil, a library for 3D image analysis in soil science. Comput Geosci 28:1041–1050
de Vries ET, Raoof A, van Genuchten MT (2017) Multiscale modelling of dual-porosity porous media; a computational pore-scale study for flow and solute transport. Adv Water Resour 105:82–95
Dias MM, Payatakes AC (1986) Network models for two-phase flow in porous media part 1. immiscible microdisplacement of non-wetting fluids. J Fluid Mech 164:305–336
Dias MM, Payatakes AC (1986) Network models for two-phase flow in porous media part 2. motion of oil ganglia. J Fluid Mech 164:337–358
Dillard LA, Blunt M (2000) Development of a pore network simulation model to study nonaqueous phase liquid dissolution. Water Resour Res 36(2):439–454
Dong H, Blunt M (2009) Pore-network extraction from micro-computerized-tomography images. Phys Rev E 80:036307
Dorogovtsev SN, Mendes JFF, Samukhin AN (2000) Exact solution of the Barabasi-Albert model. Phys Rev Lett 85:4633
Dorogovtsev SN, Goltsev AV, Mendes JF (2008) Critical phenomena in complex networks. Rev Mod Phys 80(4):1275
Doyen PE (1987) Crack geometry of igneous rocks: a maximum entropy inversion of elastic and transport properties. J Geophys Res 92(B8):8169–8181
Dupin HJ, Kitanidis PK, McCarty PL (2001) Pore-scale modeling of biological clogging due to aggregate expansion: a material mechanics approach. Water Resour Res 37(12):2965–2979
Dupin HJ, Kitanidis PK, McCarty PL (2001) Simulations of two-dimensional modeling of biomass aggregate growth in network models. Water Resour Res 37(12):2981–2994
Edwards J, Berg P (2021) Pore-network models and effective medium theory: a convergence analysis. arXiv: 2109.07599
Erdős P, Rényi A (1959) On random graphs, I. Publ Math (debrecen) 6:290–297
Erdős P, Rényi A (1960) On the evolution of random graphs. Publ Math Inst Hung Acad Sci 5:17–61
Erickson JM, Rahman A, Spear AD (2020) A void descriptor function to uniquely characterize pore networks and predict ductile-metal failure properties. Int J Fract 225:47–67
Fatt I (1956) The network model of porous media I. Capillary pressure characteristics. Trans AIME 207(7):144–159
Fatt I (1956) The network model of porous media II. Dynamic properties of a single size tube network. Trans AIME 207:160–163
Fatt I (1956) The network model of porous media III. Dynamic properties of networks with tube radius distribution. Trans AIME 207:164–181
Feng G, Guannan L, Dayu Y, Xin L, Xiaoqian Z (2020) Study on microstructure mechanism of sandstone based on complex network theory. J. Meas. Eng. 8:27–33
Fenwick DH, Blunt MJ (1998) Three-dimensional modeling of three-phase imbibition and drainage. Adv Water Resour 21(2):121–143
Fernholz D, Ramachandran V (2007) The diameter of sparse random graphs. Random Struct Algorithms 31:482–516
Fischer U, Celia M (1999) Prediction of relative and absolute permeabilities for gas and water retention curves using a pore-scale network model. Water Resour Res 35(4):1089–1100
Freeman LC (1977) A set of measures of centrality based on betweenness. Sociometry 40:35–41
Gackiewicz B, Lamorski K, Sławiński C, Hsu S-Y, Chang L-C (2021) An intercomparison of the pore network to the Navier-Stokes modeling approach applied for saturated conductivity estimation from X-ray CT images. Sci Rep 11:5859
Gao S, Meegoda JN, Hu L (2012) Two methods for pore network of porous media. Int J Numer Anal Meth Geomech 36(18):1954–1970
Gong L, Nie L, Xu Y (2020) Geometrical and topological analysis of pore space in sandstones based on X-ray computed tomography. Energies 13:3774

Guannan L, Dayu Y, Xin L, Xiaoqian Z, Feng G (2020) Study on microstructure mechanism of sandstone based on complex network theory. J Meas Eng 8(1):27–33
Harrigan T, Mann R (1984) Characterization of microstructural anisotropy in orthotropic materials using a second rank tensor. J Mater Sci 19:761–767
Heath JE, Dewers TA, McPherson BJ, Petrusak R, Chidsey TC Jr, Rinehart AJ, Mozley PS (2011) Pore networks in continental and marine mudstones: Characteristics and controls on sealing behavior. Geosphere 7(2):429–454
Heiba AA, Sahimi M, Scriven LE, Davis HT (1992) Percolation theory of two-phase relative permeability. SPE Reserv Eng 7:123–130
Hilpert M, Miller CT (2001) Pore-morphology-based simulation of drainage in totally wetting porous media. Adv Water Resour 24:243–255
Hu MC, Shen CH, Hsu SY, Yu HL, Lamorski K, Sławiński C (2019) Development of Kriging-approximation simulated annealing optimization algorithm for parameters calibration of porous media flow model. Stoch Env Res Risk Assess 33(2):395–406
Hughes RG, Blunt MJ (2000) Pore scale modeling of rate effect in imbibition. Transp Porous Media 40:295–322
Ioannidis M, Chatzis I (2000) On the geometry and topology of 3D stochastic porous media. J Colloid Interface Sci 229(2):323–334
Janson S (2009) Susceptibility of random graphs with given vertex degrees. J Comb 1:357–387
Jensen JL, Lake LW, Corbett PWM, Goggin DJ (2000) Statistics for Petroleum Engineers and Geoscientists, 2nd edn. Elsevier, Amsterdam
Jiang Z, van Dijke MIJ, Wu K, Couples GD, Sorbie KS, Ma J (2012) Stochastic pore network generation from 3D rock images. Transp Porous Med 94:571–593
Jiang Z, Van Dijke MIJ, Geiger S, Ma J, Couples GD, Li X (2017) Pore network extraction for fractured porous media. Adv Water Resour 107:280–289
Jivkov AP, Olele JE (2012) Novel lattice models for porous media. MRS Online Proceedings Library (OPL):1475
Jivkov AP, Xiong Q (2014) A network model for diffusion in media with partially resolvable pore space characteristics. Transp Porous Media 105(1):83–104
Jivkov AP, Hollis C, Etiese F, McDonald SA, Withers PJ (2013) A novel architecture for pore network modelling with applications to permeability of porous media. J Hydrol 486:246–258
Joekar-Niasar V, Hassanizadeh SM (2011) Effect of fluids properties on non-equilibrium capillarity effects; dynamic pore-network modeling. Int J Multiph Flow 37:198–214
Joekar-Niasar V, Hassanizadeh SM (2012) Analysis of fundamentals of two-phase flow in porous media using dynamic pore-network models: a review. Crit Rev Environ Sci Technol 42(18):1895–1976
Joekar-Niasar V, Hassanizadeh SM, Pyrak-Nolte LJ, Berentsen C (2009) Simulating drainage and imbibition experiments in a high-porosity micromodel using an unstructured pore network model. Water Resour Res 45:W02430
Joekar-Niasar V, Hassanizadeh SM, Dahle HK (2010) Non-equilibrium effects in capillarity and interfacial area in two-phase flow: dynamic pore-network modeling. J Fluid Mech 655:38–71
Joekar-Niasar V, Prodanović M, Wildenschild D, Hassanizadeh SM (2010) Network model investigation of interfacial area, capillary pressure and saturation relationships in granular porous media. Water Resour Res 46:W06526
Kang Q, Lichtner PC, Viswanathan HS, Abdel-Fattah AI (2010) Pore scale modeling of reactive transport involved in geologic CO_2 sequestration. Transp Porous Media 82(1):197–213
Kirkpatrick S (1973) Percolation and conduction. Rev Mod Phys 45(4):574–588
Knutson CE, Werth CJ, Valocchi AJ (2001) Pore-scale modeling of dissolution from variably distributed nonaqueous phase liquid blobs. Water Resour Res 37(12):2951–2963
Köhne JM, Schlüter S, Vogel H-J (2011) Predicting solute transport in structured soil using pore network models. Vadose Zo J 10:1082
Koplik J, Lasseter TJ (1985) Two-phase flow in random network models of porous media. Soc Petrol Eng J 25:89–110

Korvin G (1992) A percolation model for the permeability of kaolinite-bearing sandstones. Geophys Trans 37(2–3):177–209

Korvin G (1992) Fractal models in the earth sciences. Elsevier, Amsterdam

Korvin G (2016) Permeability from microscopy: review of a dream. Arabian J Sci Eng 41(6):2045–2065

Korvin G, Oleschko K, Abdulraheem A (2014) A simple geometric model of sedimentary rock to connect transfer and acoustic properties. Arab J Geosci 7(3):1127–1138

Kramar M, Goullet A, Kondic L, Mischaikow K (2013) Persistence of force networks in compressed granular media. Phys Rev E 87:042207

Kramer B, MacKinnon A (1993) Localization: theory and experiment. Rep Prog Phys 56:1469

Krapivsky PL, Redner S, Leyvraz F (2000) Connectivity of growing random networks. Phys Rev Lett 85:4629

Laroche C, Vizika O (2005) Two-phase flow properties prediction from small-scale data using pore-network modeling. Transp Porous Media 61(1):77–91

Latora V, Marchiori M (2001) Efficient behavior of small-world networks. Phys Rev Lett 87(1987011–1987014):440–442

Laubie H, Radjai F, Pellenq R, Ulm FJ (2017) Stress transmission and failure in disordered porous media. Phys Rev Lett 119:075501

Laudone GM, Matthews GP, Gane PAC (2008) Modelling diffusion from simulated porous structures. Chem Eng Sci 63(7):1987–1996

Lerdahl TR, Øren PE, Bakke S (2000) A predictive network model for three-phase flow in porous media. SPE59311. In: Proceedings of the SPE/DOE symposium on improved oil recovery, Tulsa, OK (April 2nd–5th)

Levitz P, Tariel V, Stampanoni M, Gallucci E (2012) Topology of evolving pore networks. Eur Phys J Appl Phys 60:24202

Li Y, Wardlaw NC (1986) Mechanisms of nonwetting phase trapping during imbibition at slow rates. J Colloid Interface Sci 109:461–472

Li Y, Wardlaw NC (1986) The influence of wettability and critical pore-throat size ratio on snap-off. J Colloid Interface Sci 109:473–486

Lin D, Fairhurst C (1991) The topological structure of fracture systems in rock, rock mechanics as a multidisciplinary science. In: Roegiers J-C (ed) Proceedings of 32nd US symposium on rock mechanics. Norman, Oklahoma, pp 1155–1163

Lindquist WB, Lee SM, Coker DA, Jones KW, Spanne P (1996) Medial axis analysis of void structure in three-dimensional tomographic images of porous media. J Geophys Res 101(B4):8297–8310

Mohamed MW, Arns JY, Sheppard A, Knackstedt MA, Val Pinczewski W (2007) Effect of network topology on two-phase imbibition relative permeability. Transp Porous Med 66:481–493

Manik D, Rohden M, Ronellenfitsch H, Zhang X, Hallerberg S, Witthaut D, Timme M (2017) Network susceptibilities: theory and applications. Phys Rev E 95:012319

Mason G, Morrow NR (1991) Capillary behaviour of a perfectly wetting liquid in irregular triangular tubes. J. of Colloid and Interface Science 141:262–274

Mehmani Y, Sun T, Balhoff M, Eichhubl P, Bryant S (2012) Multiblock pore-scale modeling and upscaling of reactive transport: application to carbon sequestration. Transp Porous Media 95(2):305–326

Mehmani A, Mehmani Y, Prodanović M, Balhoff M (2015) A forward analysis on the applicability of tracer breakthrough profiles in revealing the pore structure of tight gas sandstone and carbonate rocks. Water Resour Res 51:4751–4767

Meyer DW (2021) Random generation of irregular natural flow or pore networks. Adv Water Resour 152:103936

Meyers J, Liapis A (1999) Network modeling of the convective flow and diffusion of molecules adsorbing in monoliths and in porous particles packed in a chromatographic column. J Chromatogr A 852(1):3–23

Meyers J, Nahar S, Ludlow D, Liapis AI (2001) Determination of the pore connectivity and pore size distribution and pore spatial distribution of porous chromatographic particles from nitrogen sorption measurements and pore network modelling theory. J Chromatogr A 907(1):57–71
Mo H, Bai M, Lin D, Roegiers J-C (1998) Study of flow and transport in fracture network using percolation theory. Appl Math Model 22:277–291
Mogensen K, Stenby EH (1998) A dynamic two-phase pore-scale model for imbibition. Transp Porous Media 32:299–327
Nejad Ebrahimi A, Jamshidi S, Iglauer S, Boozarjomehry RB (2013) Genetic algorithm-based pore network extraction from microcomputed tomography images. Chem Eng Sci 92:157–166
Newman MEJ (2003) The structure and function of complex networks. SIAM Rev 45:167–256
Newman MEJ (2010) Networks: an introduction. Oxford University Press, Oxford, UK
Nowicki SC, Davis HT, Scriven LE (1992) Microscopic determination of transport parameters in drying porous media. Drying Technol 10(4):925–946
Okabe H, Blunt MJ (2004) Prediction of permeability for porous media reconstructed using multiple-point statistics. Phys Rev E 70(6):066135
Omar YM, Plapper P (2020) A survey of information entropy metrics for complex networks. Entropy 22(12):1417
Øren P-E, Bakke S (2002) Process based reconstruction of sandstones and prediction of transport properties. Transp Porous Media 46(2–3):311–343
Øren P, Bakke S (2003) Reconstruction of Berea sandstone and pore-scale modelling of wettability effects. J Pet Sci Eng 39:177–199
Øren PE, Bakke S, Nilsen LS, Henriquez A (1996). Prediction of relative permeability and capillary pressure from pore-scale modelling. In: Proceedings of the 5th European conference on the mathematics of oil recovery, Leoben, Austria
Øren P, Bakke S, Artzen O (1997) Extending predictive capabilities to network models. SPE. 38880
Øren P-E, Bakke S, Arntzen OJ (1998) Extending predictive capabilities to network models. SPE J 3(04):324–336
Papadopoulos L, Porter MA, Daniels KE, Bassett DS (2018) Network analysis of particles and grains. J Complex Netw 6:485–565
Peng N, Zhu X, Liu Y, Nie B, Cui Y, Geng Q, Yu C (2020) Complex network dynamics of the topological structure in a geochemical field from the Nanling area in South China. Sci Rep 10:19826
Pilotti M (2000) Reconstruction of clastic porous media. Transp Porous Media 41(3):359–364
Piri M, Blunt MJ (2005a) Three-dimensional mixed-wet random pore-scale network modeling of two- and three-phase flow in porous media. I. Model description. Phys Rev E Stat Phys Plasmas Fluids Relat Interdiscip Topics 71:026301
Piri M, Blunt MJ (2005b) Three-dimensional mixed-wet random pore-scale network modeling of two- and three-phase flow in porous media. II. Results. Phys Rev, E Stat Phys Plasmas Fluids Relat Interdiscip Topics 71:026331
Press WH, Teukolsky SA, Vetterling WT, Flannery BP (1992) Numerical Recipes in C. The Art of Scientific Computing, 2nd edn. Cambridge University Press, Cambridge
Rabbel O, Mair K, Galland O, Grühser C, Meier T (2020) Numerical modeling of fracture network evolution in organic-rich shale with rapid internal fluid generation. JGR Solid Earth 125(7):e2020JB019445
Raoof A, Nick HM, Wolterbeek TKT, Spiers CJ (2012) Pore-scale modeling of reactive transport in wellbore cement under CO_2 storage conditions. Int J Greenh Gas Control 11:S67–S77
Raoof A, Nick H, Hassanizadeh S, Spiers C (2013) PoreFlow: a complex pore-network model for simulation of reactive transport in variably saturated porous media. Comput Geosci 61:160–174
Reeves PC, Celia MA (1996) A functional relationship between capillary pressure, saturation, and interfacial areas as revealed by a pore-scale network model. Water Resour Res 32(8):2345–2358
Rosenzweig R, Furman A, Shavit U (2013) A channel network model as a framework for characterizing variably saturated flow in biofilm-affected soils. Vadose Zone J 12(2):1–15

Rostami A, Habibagahi G, Ajdari M, Nikooee E (2015) Pore network investigation on hysteresis phenomena and influence of stress state on the SWRC. Int J Geomech 15:04014072
Roy R, Olver FWJ (2010) Lambert W function. In: Olver FWJ, Lozier DM, Boisvert RF, Clark CW (eds) NIST handbook of mathematical functions. Cambridge University Press, Cambridge, Cambridge, Sec. 4.13
Ryazanov AV, van Dijke MIJ, Sorbie KS (2009) Two-phase pore-network modelling: existence of oil layers during water invasion. Transp Porous Med 80:79–99
Serra J (1982) Image analysis and mathematical morphology. Academic Press, New York
Shanti NO, Chan VW, Stock SR, de Carlo F, Thornton K, Faber KT (2014) X-ray micro-computed tomography and tortuosity calculations of percolating pore networks. Acta Mater 71:126–135
Singh M, Mohanty KK (2003) Dynamic modeling of drainage through three-dimensional porous materials. Chem Eng Sci 58:1–18
Philip S, Chaney MM, Emmerich AL, Miller KJ, Zhu W-l (2017) Network topology of olivine–basalt partial melts. Geophys J Int 210:284–290
Solomonoff R, Rapoport A (1951) Connectivity of random nets. Bull Math Biol 13:107–117
Song R, Liu J, Cui M (2016) Single- and two-phase flow simulation based on equivalent pore network extracted from micro-CT images of sandstone core. Springerplus 5:817
Stauffer D, Aharony A (1994) Introduction to percolation theory. CRC Press, Boca Raton, FL, USA
Steven B, Blunt MJ (1992) Prediction of relative permeability in simple porous media. Phys Rev A 46(4):2004–2011
Claes S, Janssen H (2019) Towards stochastic generation of 3D pore network models of building materials. MATEC Web of Conferences 282(02):222
Steven LB, King PR, Mellor DW (1993) Network model evaluation of permeability and spatial correlation in a real random sphere packing. Transp Porous Media 11:53–70
Svirsky D, van Dijke MIJ, Sorbie KS (2004) Prediction of three phase relative permeabilities using a pore scale network model anchored to two phase data. SPE 89992
Thauvin F, Mohanty KK (1998) Network modeling of non-Darcy flow through porous media. Transp Porous Media 31:19–37
Thompson KE, Fogler HS (1997) Modeling flow in disordered packed beds from pore-scale fluid mechanics. AIChE J 43:1377–1389
Thomson P-R, Aituar-Zhakupova A, Hier-Majumder S (2018) Image segmentation and analysis of pore network geometry in two natural sandstones. Front Earth Sci 6:Article 58
Thomson P-R, Jefferd M, Clark BL, Chiarella D, Mitchell TM, Hier-Majumder S (2020) Pore network analysis of Brae Formation sandstone, North Sea. Mar Pet Geol 122:104614
Valentini L, Perugini D, Poli G (2007) The "'small-world'" topology of rock fracture networks. Physica A 377:323–328
Valvatne PH, Blunt MJ (2004) Predictive pore-scale modelling of two-phase flow in mixed wet media. Water Resour Res 40:W07406
van der Linden JH, Narsilio GA, Tordesillas A (2016) Machine learning framework for analysis of transport through complex networks in porous, granular media: a focus on permeability. Phys Rev E 94:022904
Van der Marck SC, Matsuura T, Glas J (1997) Viscous and capillary pressures during drainage: Network simulations and experiments. Phys Rev E 56:5675–5687
Varloteaux C, Békri S, Adler PM (2013) Pore network modelling to determine the transport properties in presence of a reactive fluid: From pore to reservoir scale. Adv Water Resour 53:87–100
Vo K, Walker DM, Tordesillas A (2013) Transport pathways within percolating pore space networks of granular materials. AIP Conf Proc 1542:551–554
Vogel HJ, Roth K (1997) A new approach for determining effective soil hydraulic functions. Eur J Soil Sci 49(4):547–556
Vogel H-J, Roth K (2001) Quantitative morphology and network representation of soil pore structure. Adv Water Resour 24:233–242

Wang W, Tang M, Zhang HF, Gao H, Do Y, Liu ZH (2014) Epidemic spreading on complex networks with general degree and weight distributions. Phys Rev E 90:042803

Wang YD, Chung T, Armstrong RT, McClure JE, Mostaghimi P (2019) Computations of permeability of large rock images by dual grid domain decomposition. Adv Water Resour 126:1–14

Watts DJ (1999) Small worlds: the dynamics of networks between order and randomness. Princeton University Press, Princeton, NJ

Watts DJ, Strogatz SH (1998) Collective dynamics of 'small-world' networks. Nature 393:440–442

Xiong Q, Jivkov AP, Yates JR (2014) Discrete modelling of contaminant diffusion in porous media with sorption. Microporous Mesoporous Mater 185:51–60

Xiong Q, Baychev TG, Jivkov AP (2016) Review of pore network modelling of porous media: experimental characterisations, network constructions and applications to reactive transport. J Contam Hydrol 192:101–117

Yanuka M, Dullien FA, Elrick DE (1986) Percolation processes and porous media. I. Geometrical and topological model of porous media using a three-dimensional joint pore size distribution. J Colloid Interface Sci 1986(112):24–41

Ye D, Liu G, Gao F, Zhu X, Hu Y (2020) A study on the structure of rock engineering coatings based on complex network theory. Coatings 10(12):1152

Yi Z, Lin M, Jiang W, Zhang Z, Li H, Gao J (2017) Pore network extraction from pore space images of various porous media systems. Water Resour Res 53:3424–3445

Youssef S, Rosenberg E, Gland N, Kenter J, Skalinski M, Vizika O (2007) High resolution CT and pore-network models to assess petrophysical properties of homogeneous and heterogeneous carbonates. In: LPE/EAGE reservoir characterization and simulation conference (society of petroleum engineers)

Zareei A, Pan D, Amir A (2021) Temporal evolution of flow in pore-networks: from homogenization to instability. *arXiv*: 2106.09745

Zhou D, Dillard LA, Blunt MJ (2000) A physically based model of dissolution of nonaqueous phase liquids in the saturated zone. Transp Porous Media 39(2):227–255

Zhu W, Gaetani GA, Fusseis F, Montesi LGJ, De Carlo F (2011) Microtomography of partially molten rocks: three-dimensional melt distribution in mantle peridotite. Science 332:88–91

Zhu W, Khirevich S, Patzek TW (2021) Impact of fracture geometry and topology on the connectivity and flow properties of stochastic fracture networks. Water Resour Res 57:e2020WR028652

Chapter 5
Menger Sponge Models

Chapter Highlights *Menger Sponge* (in 3D) and *Sierpiński Carpet* (in 2D) are the most useful fractal models of the rocks' pore space. Sections 5.1 and 5.2 treat these models, their combinatorial and topological properties. Random and multifractal generalizations are discussed in Sect. 5.3, where the permeability of these models is also studied. A mathematical Appendix 5.4 introduces a useful measure, *lacunarity*, a tool to distinguish between textures of the same *fractal dimension* but different geometry. Petrophysical applications are summarized in Table 5.1.

5.1 Menger Sponge and Its Cousin, The Sierpiński Carpet

Menger's Sponge, a *fractal object* embedded in the 3-dimensional Euclidean space $\mathbb{R}^3$ with mass-fractal dimension $2 < D_m < 3$, and a similar *fractal construction* embedded in $\mathbb{R}^2$, the *Sierpiński carpet* with $1 < D_m < 2$ are versatile and popular *fractal models* (Mandelbrot 1977; Korvin 1992; Turcotte 1993; Gouyet 1996; Meakin 1998) which have found an amazingly wide range of applications in rock physics (Table 5.1).

In the traditional Sierpiński carpet (Fig. 5.1), we start with a solid *initiator square* of size $l \times l$. The *initiator square* is divided into b^2 squares of size l/b, where $b = 3$ is the *scaling factor*. The center square of the original square is removed, this will be considered a *pore* of size l/b. For the next iteration, each of the remaining $b^2 - 1$ squares of size $(l/b)^2$ are divided into b^2 squares of size $\left(l/b^2\right)^2$. Altogether $b^2 - 1$ "regions" are thus created, in each there are b^2 smaller squares yielding a total of $b^2\left(b^2 - 1\right)$ squares. Again, the center square is removed from each, creating $b^2 - 1$ pores of area $\left(l/b^2\right)^2$ and the process is repeated.

To calculate the fractal dimension of this model, we use the power-law relation of Mandelbrot (1977) between the scale factor b and the amount of solid phase of the medium:

$$\left(b^i\right)^{D_m} = \left(b^i\right)^2 - n \tag{5.1.1}$$

G. Korvin, *Statistical Rock Physics*, Earth and Environmental Sciences Library,
https://doi.org/10.1007/978-3-031-46700-4_5

Table 5.1. Rock physics applications of the Menger Sponge and Sierpiński Carpet Models

Application	Authors
Capillary absorption	Pia et al. (2014)
Capillary pressure–saturation relation	Tyler and Wheatcraft (1990), Bird et al. (1996), Perfect (2005)
CAT scan of soil	Gibson et al. (2006)
Collapsed zone in coal mine	Xing-shang and Jia-lin (2009)
Drainage	Bird and Dexter (1997), Sukop et al. (2001)
Fractal dimension of porous media	Xia et al. (2018)
Fractured rock	Doughty and Karasaki (2002)
Fragmentation	Perrier and Bird (2002)
Gas adsorption isotherms	Paz-Ferreiro et al. (2013), Liu et al. (2018)
Gravity anomaly of fractal body	Mostafa (2008)
Heat conduction	Pia and Sanna (2013a, b), 2014b; Du et al. (2019)
Hydraulic conductivity	Perfect et al. (2006)
Hydraulic dispersivity	Wheatcraft and Tyler (1988)
Mercury Intrusion Porosimetry of Coal	Friesen and Mikula (1987)
NMR	Zhao et al. (2017)
Particle size distribution	Rieu and Sposito (1991a, b, c)
Permeability	Cihan et al. (2009), Pia and Sanna (2014a), Jin et al. (2015), Cousins (2016), Li et al. (2018), Hu et al. (2020); Zhou et al. (2021)
Porosity	Vita et al. (2012)
Reservoir rock pore space	Garrison et al. (1992, 1993)
Shockwave dissipation	Dattelbaum et al. (2020)
Transport properties	Pfeifer et al. (1984), Toledo et al. (1990), Wheatcraft et al. (1991), Garrison et al. (1992), Garza-López et al. (2000), Cihan et al. (2009)
Water retention curve	Tyler and Wheatcraft (1990), Rieu and Sposito (1991a, b), Perrier et al. (1996, 1999), Bird et al. (2000), Huang et al. (2006), Cihan et al. (2007), Fallico et al. (2010), Ghanbarian et al. (2011)

where D_m is the *mass fractal dimension* and n is the minimal number of squares of size $\left(l/b^i\right)^2$ that must be used to cover all pores of size $\left(l/b^i\right)^2$. For the Sierpiński carpet D_m becomes

$$D_m = \frac{log\left(\left(b^i\right)^2 - n\right)}{log\left(b^i\right)} \tag{5.1.2}$$

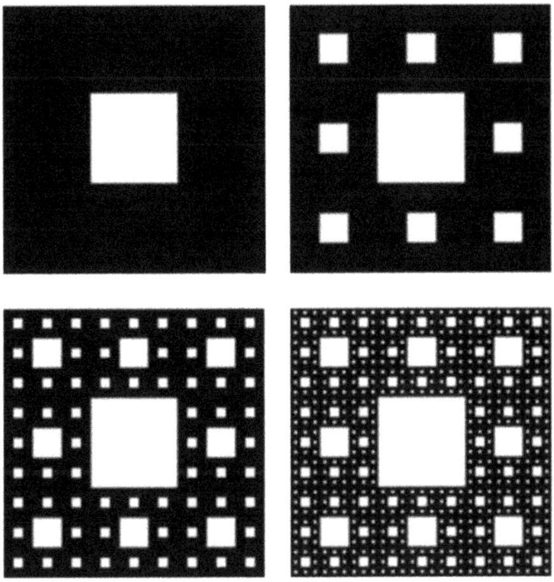

Fig. 5.1 First four iterations of a traditional Sierpiński's Carpet. Pore space is represented by white

In the traditional case (Fig. 5.1), $b = 3$. For the first iteration ($i = 1$) a single square ($n = 1$) of size $(l/b)^2$ covers the single pore, and Eq. (5.1.2) with $b = 3, i = 1, n = 1$ yields $D_m = 1.893$. But Sierpiński carpets are *self similar*, the mass fractal dimension is scale invariant and remains constant across all iterations. This allows Eq. (5.1.2) to be simplified (Cousins 2016: Eq. 2.8) to

$$D_m = \frac{log\left(b^2 - n\right)}{\log(b)} \tag{5.1.3}$$

where D_m is expressed in terms of b and the number of pores n removed from each "region" in every iteration. The porosity for the carpet can be found at any iteration i by counting the pores of size $\left(l/b^i\right)^2$ per total area (Tyler and Wheatcraft 1990; Ghanbarian et al. 2011)

$$\Phi_i = \frac{\left(l/b^i\right)^2 \mathrm{n}}{l^2} = \frac{n}{\left(b^i\right)^2} \tag{5.1.4}$$

where n is the minimal covering number defined above. Combining Eqs. (5.1.2) and (5.1.4) results in

$$\Phi_i = 1 - \left(b^i\right)^{(D_m - 2)} \tag{5.1.5}$$

The construction of a *traditional Menger sponge* starts with a solid cube of size l^3. It is divided into b^3 cubes of size $(l/b)^3$, the center cube from each face of the *initiator cube*, as well as the innermost center cube are removed, that is altogether 7 pores are created. In the next iteration, each of the remaining, $b^3 - 7$ cubes are divided into b^3 "regions" of size $(l/b^2)^3$, and—exactly as before—the center cubes are removed. As in case of the Sierpiński carpet, the process is repeated until the *desired number of iterations or porosity is reached*. In any iteration step *i*, $7(b^3 - 7)^{i-1}$ pores of size $(l/b^i)^3$ are removed. Equations (5.1.1) through (5.1.5) can be expressed in a more general form as:

$$\begin{cases} (b^i)^D = (b^i)^E - n \\ D_m = \dfrac{log\left((b^i)^E - n\right)}{log(b^i)} \\ D_m = \dfrac{log(b^E - n)}{log(b)} \\ \Phi_i = \dfrac{(l/b^i)^E n}{l^2} = \dfrac{n}{(b^i)^E} \\ \Phi_i = 1 - (b^i)^{D_m - E} \end{cases} \quad (5.1.6, 7, 8, 9, 10)$$

where E is the embedding dimension, equal to 2 for the Sierpiński carpet and 3 for the Menger sponge. The fractal dimension of the traditional *Menger Sponge* is defined through the following limit process:

$$D_f = -\lim_{n\to\infty} \frac{lnN_n}{\ln 3} = \frac{\ln 20}{ln3} = 2.7268\cdots \quad (5.1.11)$$

where

$$N_n = 20^n \quad (5.1.12)$$

is the number of filled boxes (Sergeyev 2009; Vita et al. 2012: Eq. 11). The *n*th stage of the Menger sponge, M_n, is made up of 20^n smaller cubes, each with a side length of $(1/3)^n$. The total volume of M_n is, for a *unit Menger cube* (i.e. if $l = 1$)

$$Vol(M_n) = (20/27)^n \quad (5.1.13)$$

The *total surface area* of M_n is given by the expression (for the *unit Menger cube*)

$$S(M_n) = 2 \cdot (20/9)^n + 4 \cdot (8/9)^n \quad (5.1.14)$$

therefore the sponge's volume approaches zero while its surface area increases without bound. However, this huge internal "surface" will be so many times punctured as the iterative construction continues, that the limit is *neither a solid nor a surface and no collection of distinct scattered points*; it has a topological dimension of 1 and is accordingly identified as a *curve*. It is the strangest curve in the world, Menger called it *universal curve*, and constructed it as an example of a curve containing the *topological image of all other possible curves* (Menger 1932; Anderson 1958).

Formally, the Menger sponge can be defined as follows:

$$M := \bigcap_{n\in\mathbb{N}} M_n \tag{5.1.15a}$$

where

$$M_{n+1} := \left\{ (x, y, z) \in \mathbb{R}^3 \;\middle|\; \begin{array}{l} \exists i, j, k \in \{0, 1, 2\} | (3x - i, 3y - j, 3z - k) \in M_n) \\ \text{and at most one of } i, j, k \text{ is equal to } 1 \end{array} \right\} \tag{5.1.15b}$$

Recall (from Sect. 3.1) the definition of Minkowski functionals:

$$W_0 := \int_K d^3\tau;\; W_1 := \frac{1}{3}\int_{\partial K} d^2\tau;\; W_2 := \int_{\partial K} G_2 d^2\tau;\; W_3 := \int_{\partial K} G_3 d^2\tau \tag{5.1.16a–d}$$

Here K is a compact body with a smooth bounding surface ∂K in 3D Euclidean space $\mathbb{R}^3$, $G_2 = \frac{\kappa_1+\kappa_2}{2}$ is the mean curvature on ∂K, and $G_3 = \kappa_1 \cdot \kappa_2$ is the pointwise Gaussian curvature (Schröder-Turk et al. 2013). For the traditional *unit Menger sponge* Schönhöfer (2014: App. B) analytically determined the dependence of these functionals on iteration number *n*:

$$\begin{cases} W_0(n) = 20^n \\ W_1(n) = \frac{2}{3}20^n + \frac{4}{3}8^n \\ W_2(n) = -\frac{1}{51}\pi 20^n + \frac{4}{15}\pi 8^n + \frac{64}{85}\pi 3^n \\ W_3(n) = -\frac{4}{19}\pi 20^n - \frac{8}{21}\pi 8^n + \frac{256}{133}\pi \end{cases} \tag{5.1.17a–d}$$

The *Euler characteristic* $\chi(n) \equiv \chi(M_n)$ and *genus*

$$g(n) = 1 - \frac{\chi(n)}{2} \tag{5.1.18}$$

(see Sect. 3.2) can be also determined using combinatorial arguments (Li and Tauraso 2006) or by a generating-function approach (Mackeprang and Myers 2007). First, find the number of vertices *v(n)*, sides *s(n)*, and faces $f(n)$ of the surface M_n. Since M_0 is the unit cube, $v(0) = 8$, $s(0) = 12$, $f(0) = 6$. As the construction proceeds, for n $\geq$ 1, we have the recursion

$$\begin{cases} v(n) = v(n-1) + 2s(n-1) + 4f(n-1) + 8 \cdot 20^{n-1} \\ s(n) = 3s(n-1) + 12f(n-1) + 36 \cdot 20^{n-1} \\ f(n) = 8f(n-1) + 24 \cdot 20^{n-1} \end{cases} \tag{5.1.19a–c}$$

From the last equation, $f(n) = 4 \cdot 8^n + 2 \cdot 20^n$, that is by its definition the Euler characteristic is $\chi(n) = v(n) - s(n) + f(n) = \chi(n-1) - f(n-1) - 4 \cdot 20^{n-1} = \chi(n-1) - 4 \cdot 8^n - 6 \cdot 20^n$. Using (5.1.18):

$$g(n) = g(n-1) + 2 \cdot 8^n + 3 \cdot 20^n = 2 \cdot \frac{8^n - 1}{7} + 3 \cdot \frac{20^n - 1}{19} \quad (5.1.20)$$

The first terms of the sequence $g(n)$ are:

$$g(n) = \left\{ \begin{matrix} 0, 5, 81, 1409, 26433, 514625, 10180161, 202704449, \\ 4046898753, 80880453185, \cdots \end{matrix} \right\} \quad (5.1.21)$$

(Tee 2015; OEIS 2022).

In rock-physical modeling we only use the first few steps of the Menger construction. What is the reasonable number n of iterations when modeling the fractal pore space of a rock? If the pore space is fractal in the range $L_1 \leq diameter_{pore} \leq L_2$, the reasonable number of steps is

$$n = \frac{\log\left(\frac{L_2}{L_1}\right)}{\log b} \quad (5.1.22)$$

(*Gupta's formula*, quoted in Cowan 1987: Eq. 5.15). The *range of fractality* of the porosity in rocks is summarized in Korvin (1992: 285–287). Katz and Thompson (1985) estimated the value of L_1 to be approximately 20 Ångströms for sandstones $\left(1\ \text{Å} = 10^{-10}\ m = 0.1nm\right)$; according to Krohn (1988a, b) $100 < L_2/L_1 < 35,000$ for most reservoir rocks. Thus, by Eq. (5.1.22), the reasonable range for n is Eq. (5.1.23). If we want to match a given porosity Φ, n can be found from Eq. (5.1.6 and 5.1.10), solving $\Phi = 1 - (b^n)^{D_m - E} =$, that is

$$4 \leq n \leq 9 \quad (5.1.23)$$

$$n = \frac{1}{D_m - E} \cdot \frac{log(1 - \Phi)}{\log b} \quad (5.1.24)$$

If we know the *topological complexity* of the pore space, expressed by the *genus* $g(n)$ or Euler characteristics $\chi(n)$, the sequence (5.1.21) will help to select a sponge with sufficiently complex internal structure.

5.2 A More General Menger Sponge

The traditional Menger Sponge construction took a cube as *initiator*, divided it into 27 equal parts, and then removed 7 small cubes which lay in the center of the initial cube and its surface, so only 20 small cubes remained. Repeated operations until infinity led to the fractal Menger Sponge with fractal dimension D = 2.7768.

Jin et al. (2015) introduced the useful notation for the generalization of this type of fractal media *DdVmSqLnRl*. In this nomenclature, for an object embedded in Euclidean space of dimension $D(= d)$, the initiator (Fig. 5.2a) is of linear size $R(= l)$, it is divided to m^d equal parts $(V = m)$. At the 1st iteration step, the generator (Fig. 5.2b) divides the m^d parts of the initiator into two sets of q^d *phase-1* $(S = q)$ and $m^d - q^d$ *phase-2 sub-regions*. (The two phases are usually colored by black and white, respectively). At the next step(s) (when the iteration number is $L = n$) each of the *phase 2* sub-regions is replaced by a reduced replica of the generator. In this terminology, the object in Fig. 5.2c is called *D2V5S3L2Rl* because the object (*Sierpiński carpet*) is embedded to $\mathbb{R}^d = \mathbb{R}^2$, the initiator is of linear size l, it is divided to $m^d = 5^2$ equal parts, out of these $q^d = 3^2$ are selected as being *phase-1*, and $m^d - q^d = 5^2 - 3^2$ as *phase-2*. Finally, as Fig. 5.2c represents the 2nd iteration, we have $L = n = 2$. Note that in the symbol *D2V5S3L2Rl* "q" (=5) does not specify the locations of the left-out subregions, only their number. The first four iterations of the *classic Sierpiński carpet* (Fig. 5.1) are thus denoted as $2V3S1L(1, 2, 3, 4)R1$; the first four iterations of the traditional *Menger Sponge* are $D3V3S1L\{1, 2, 3, 4\}R1$.

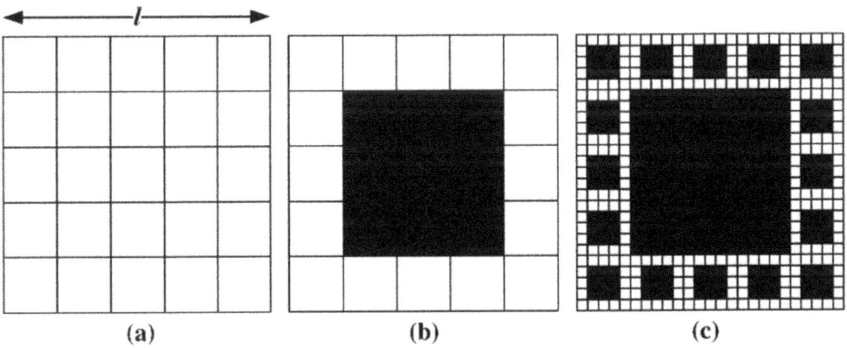

Fig. 5.2 Construction of the *DdVmSqLnRl* type fractal porous media. The *two phases* are denoted by *black* and *white*, respectively. For an object embedded in Euclidean dimension $D = d$ the initiator (**a**) is of linear size $R(= l)$, it is divided to m^d equal parts $(V = m)$. At the 1st iteration step, the generator (**b**) divides the m^d parts of the initiator into two sets of q^d *phase-1* $(S = q)$ and $m^d - q^d$ *phase-2 subregions*. At the next steps (iteration number $L = n$) each of the *phase-2* subregions is replaced by a reduced replica of the generator. This notation is due to Jin et al. (2015), the object in (**c**) is called *D2V5S3L2Rl*. (After Jin et al. 2015: Fig. 1; reproduced with kind permission of Dr. Y. Jin and *Springer Nature*)

We shall discuss the scaling and other properties of the fractal object *DdVmSqLnRl* through a study of its *hydraulic permeability*. The estimation of permeability K of porous media is based on the classic *Kozeny equation*, and its later modifications by Carman (Kozeny 1927; Carman 1937, 1939, 1956; Jin et al. 2015; Korvin 2016):

$$K = \frac{\Phi^3}{k(1-\Phi)^2 S^2} \tag{5.2.1}$$

where Φ is the *porosity*, S is the *specific surface* and $k = C_f \tau^2$ is the *KC constant*, in which τ is the *hydraulic tortuosity*, and C_f a shape factor. Assuming the porous media consists of solid particles having the same shape, and size, yields

$$K = \frac{\Phi^3}{C_0 \tau^2 (1-\Phi)^2} \delta^2 \tag{5.2.2}$$

where C_0 is a constant. Similarly to the flow of electric current, flow rate Q in a porous media can be written in terms of *flow resistance* (Bear and Verruijt 1987),

$$Q = \frac{\Delta P}{R} \tag{5.2.3a}$$

where ΔP is pressure difference between inlet and outlet. Using *Darcy's law*, from Eq. (5.2.3a) we have

$$R = \mu L / AK \tag{5.2.3b}$$

where μ is *dynamic viscosity*, A is cross sectional area, L is the inlet—outlet distance. We define *fluid flow resistivity* R_f by assuming unit cross sectional area and unit sample length:

$$R_f = \mu / K, \tag{5.2.3c}$$

i.e. from (5.2.2):

$$R_f = \frac{C_0 (1-\Phi)^2 \mu}{\Phi^3} \cdot \frac{\tau^2}{\delta^2} \tag{5.2.4}$$

In a *fractal pore space*, the *cumulative distribution of pore size* λ follows a fractal scaling law. The medium is composed of *representative units* of linear size L in which the physical properties, such as porosity, permeability etc. are statistically the same as in the whole porous medium, but they contain pores only of the largest size. The fractal dimension of the pore size distribution, D_f, is a function of the porosity and the ratio of the lower limit to upper limit of such self-similar regions (Yu and Li 2001):

$$\Phi(\lambda \geq r) = \left(\frac{r}{L}\right)^{D-D_f} \tag{5.2.5}$$

where r is the scale, D is the Euclidean space dimension. In Wang et al. (2014) a new method has been proposed to determine D_f from two quantities, F_λ and P_λ, where F_λ is the ratio of the number of pores of size λ_i to that of size λ_{i-1}; and P_λ is the ratio of λ_{i-1}/λ_i, where λ_i and λ_{i-1} are two successive pore sizes in a fractal porous media, $\lambda_{i-1}/\lambda_i > 1$. The fractal dimension can be calculated by

$$D_f = \frac{\log F_\lambda}{\log P_\lambda} \tag{5.2.6}$$

In the Sierpiński carpet or Menger sponge models if the smallest *pore size* is r, the smallest *particle size* will be ξr in a *pore fractal* (Ghanbarian et al. 2013a), where $\xi \geq 1$ is a constant (in the standard Sierpiński carpet, $\xi = 1$), thus, the number vs. size distributions of the pores and particles are power laws with identical exponent (Perrier and Bird 2002). After n iterations, the cumulative pore/particle surface area in a representative unit of a fractal porous medium is:

$$A_s = \sum_{i=1}^{n} \beta_s \delta_i^{d-1} N(\delta_i); \text{ where } N(\delta_i) = F_\lambda^{i-1}, \delta_i = \delta_{\max} P_\lambda^{-(i-1)} \tag{5.2.7}$$

where $N(\delta_i) = F_\lambda^{i-1}$, $\delta_i = \delta_{max} P_\lambda^{-(i-1)}$, δ_i and $N(\delta_i)$ are the size and number of the minimum solid particles after i iterations, δ_{max} is the largest size of a solid particle in a representative unit, β_s is the shape factor, for square solid particles $\beta_s = 2D$ (Mortensen et al. 2005).

In (5.2.7), A_s is the sum of a geometric series. By the definitions of F_λ and P_λ, and using (5.2.6).

$$\delta_{max} = \delta_{min} P_\lambda^{n-1} \tag{5.2.8}$$

Equation (5.2.7) is rearranged into:

$$A_s(\delta) = \beta \delta_{\max}^{d-1} \frac{\left(1 - \left(\frac{\delta}{\delta_{\max}}\right)^{D-1-D_f}\right)}{1 - P_\lambda^{D_f+1-D}} \approx \zeta \beta_s \delta_{\max}^{d-1} r \tag{5.2.9}$$

where ζ is function of P_λ, that is it depends on pore structure (Wang et al. 2014). Substituting the relationships $\delta = \xi r$ and $P_\lambda \delta_{max}/\xi = L$ into Eq. (5.2.9), the pore/particle surface area is

$$A_s(\delta) = \beta_s \delta_{max}^{d-1} \frac{\left(1 - \left(\frac{r}{L}\right)^{D-1-D_f}\right)}{1 - P_\lambda^{D_f+1-D}} \tag{5.2.10}$$

Consequently, the specific area of fractal porous media is expressed by:

$$S(\delta) = \frac{A_s(\delta)}{L^\delta} = \zeta \beta_s \frac{1}{L} \left(\frac{\delta}{\delta_{max}}\right)^{D-1} \left(\frac{\delta}{\delta_{max}}\right)^{D-1-D_r} \tag{5.2.11}$$

At a measuring scale of *pore size* r, the minimum measurable size of solid particles δ is ξr. Thus, together with Eq. (5.2.5) and written out with respect to the pore size, Eq. (5.2.11) becomes:

$$S(\lambda \geq r) = S(\delta) = \zeta \beta_s \xi_1^{D_f} \xi^{D-1-D_f} \frac{\Phi}{r} = C_s \frac{\Phi}{r} \tag{5.2.12}$$

where $\xi_1 = \frac{\delta_{max}}{L}$ is a constant in a fractal porous media, for example, $\xi_1 = 1/3$ in a standard Sierpiński carpet.

Rewrite now these formulae for a general *DdVmSqLnRl*-type fractal porosity model. If the size of the *representative region* is l, at the beginning we have $\lambda_{max} = l, \lambda_{min} = l$. In the 1st iteration $\lambda_{min} = l/m; r = \frac{l}{m}; \delta_{min} = \delta_{max} = \frac{lq}{m}$; after n iterations $\lambda_{min} = l \cdot m^{-n}; \delta_{min} = l \cdot q \cdot m^{-n}; \delta_{max} = \frac{lq}{m}; \delta_{min} = q \cdot \lambda_{min}$. The factors ξ and ξ_1 become

$$\xi = q; \xi_1 = q/m \tag{5.2.13}$$

$$F_\lambda \text{ and } P_\lambda \text{ become } P_\lambda = m^d;\ F_\lambda = m^d - q^d \tag{5.2.14}$$

Thus, by (5.2.6), the fractal dimension of pore area and the porosity, respectively, are given by

$$D_f = \frac{ln F_\lambda}{ln P_\lambda} = \frac{ln\left(m^d - q^d\right)}{ln(m)} \tag{5.2.15}$$

and

$$\Phi = \left(m^{D_f-d}\right)^n = \left(\frac{\lambda_{max}^d - \delta_{max}^d}{\lambda_{max}^d}\right)^n \tag{5.2.16}$$

As shown in Perrier et al. (1999), $F_\lambda \in \left(0, m^d - q^d\right)$,

when,

$$F_\lambda \to 0^+, D_f \to -\infty \tag{5.2.17}$$

(see Ghanbarian-Alavijeh and Hunt 2012a, b). In this limiting case the specific area of the *DdVmSqLnRl*-type porous media is distributed as

$$S(\lambda \geq \lambda_{min}) = \zeta \beta_s \frac{q^{d-1}}{m^{D_f}} \cdot \frac{\Phi}{\lambda_{min}} \tag{5.2.18}$$

To compute *permeability*, we first estimate *flow resistance*. In a *DdVmSqLnRl*-type fractal medium, the space is filled with *VmSqL1R*(l^{m-n}) voids and solids, as shown in Fig. 5.2. To explicitly refer to the scaling of porosity and tortuosity with size l we shall write Φ_l and τ_l. Equation (5.2.4) can be rewritten to the form of

$$R_f = C_l \frac{1}{\delta^2} \propto \frac{1}{\delta^2} \tag{5.2.19}$$

where $\delta = \xi_1 l$ (5.2.20a); $C_1 = \frac{C_0(1-\Phi_l)^2 \mu \tau_l^2}{\Phi_l^3}$(5.2.20b). Denote by $R_f^{\left(\frac{d}{i}\right)}$ he flow resistivity of the Dd*VmSqLiRl*-type porous medium. When a Dd*VmSqL1Rl*-type porous medium becomes Dd*VmSqL2Rl*-type in the 1st iteration, its flow resistivity changes from $R_f^{\left(\frac{d}{1}\right)}$ to $R_f^{\left(\frac{d}{2}\right)}$. According to Eq. (5.2.19), $R_f^{\left(\frac{d}{1}\right)}$ takes the value of $C_1(qd/m)^{-2}$, while $R_f^{\left(\frac{d}{2}\right)}$ can be calculated by the parallel–series method:

$$R_f^{\left(\frac{d}{2}\right)} = R_f^{((dm^{-1})/1)} \left(\frac{m-q}{m} + \frac{q}{m-q} \right) \tag{5.2.20}$$

Replacing the expression $\frac{m-q}{m} + \frac{q}{m-q}$ by $f(m, q)$ for short, the average flow resistivity of *DdVmSqLnRl*-type porous media is

$$R_f^{\left(\frac{d}{n}\right)} = \left(m^2 f(m, q)\right)^{n-1} \cdot R_f^{\left(\frac{d}{1}\right)} = C_1 \left(m^2 f(m, q)\right)^{n-1} \frac{1}{\delta_{\max}^2} \tag{5.2.21}$$

Substituting $\delta_{max} = qL/m$ into Eq. (5.2.21) yields

$$R_f^{\left(\frac{d}{n}\right)} = C_1 \frac{\left(m^2 f(m, q)\right)^n}{q^2 f(m, q)} \cdot \frac{1}{L^2} \tag{5.2.22}$$

where L is the linear size of the representative unit of a fractal porous medium, also the size of the largest pore.

Recalling the relationship between the permeability (K) and flow resistivity $R_f^{\left(\frac{d}{n}\right)}$ ($R_f = \mu/K$, Eq. 5.2.3c), and using Eqs. 5.2.12, 5.2.16 and 5.2.22) the permeability estimation for the

$$Dn\ VmSqLnRl \text{ model becomes } K = \frac{\zeta^2 \beta_s^2}{C_0} \frac{\Phi_1 \Phi^2}{S(\lambda)^2 T^2} \tag{5.2.23}$$

where

$$T^2 = \left(\tau_l^2 f(m,q)^{n-1}\right) \tag{5.2.24}$$

is a parameter accounting for hydraulic tortuosity. Let us check Eqs. (5.2.23 and 5.2.24) for the *non-fractal* case that is for $n = 1$, $\Phi_l = \Phi$, $S = S_0$, when Eq. (5.2.23) reduces to

$$K = \frac{\Phi_l^3}{C_0\tau_l^2} \frac{\delta_{max}^2}{\left(\left(\frac{\delta_{max}}{L}\right)^d\right)^2} = \frac{\Phi_l^3}{C_0\tau_l^2} \frac{\delta_{max}^2}{(1-\Phi_l)^2} \tag{5.2.25}$$

This is identical with the KC equation in Eq. (5.2.25) for a *non-fractal porous medium* consisting of *monosized particles.* To make Eqs. (5.2.23 and 5.2.23) applicable for the *DdVmSqLnRl* model, we must express the tortuosity τ_l in terms of the system parameters m, q, n, l. First we note that hydraulic tortuosity depends only on the arrangement of the solid particles, not on a general magnification factor, that is, it is independent of l and can only depend on m, q and n, but not on the linear size of the representative unit L. Also, while τ correlates with porosity Φ (see Chap. 2), it is not a *model-independent function of porosity* (Valdés-Parada et al. 2011). For the standard *Sierpiński carpet* (*D2V3S1*), Li and Yu (2011) proposed a tortuosity–porosity formula

$$\tau = (19/18)^{\ln\Phi/\ln(8/9)} \tag{5.2.26}$$

Note that, for a D2V3S1-type porous medium, in Li and Yu's equation the argument "8/9" is equal to Φ_l so that

$$\tau = (19/18)^{ln\Phi/ln\Phi_l} \tag{5.2.27}$$

Assuming that T is a function of (Φ_l/Φ) of the form,

$$T = (\Phi_l/\Phi)^{\varepsilon} \tag{5.2.28}$$

and using the size-distribution rule

$$\frac{\Phi_l}{\Phi} = \left(\frac{\delta_{max}}{\delta_{min}}\right)^{d-D_f} \tag{5.2.29}$$

the hydraulic tortuosity for the fractal pore model can be expressed as

$$\tau = C_\tau T = C_\tau\left(\frac{\delta_{max}}{\delta_{min}}\right)^{\varepsilon(d-D_f)} \tag{5.2.30}$$

where ε and C_τ ($C_\tau > 1$) are fitting parameters.

Returning to *permeability* (Eqs. 5.2.23 and 5.2.24), some further algebra gives

$$K = \frac{1}{C_0 \tau_l^2} \cdot \frac{\Phi_l^{3-2\varepsilon} \Phi^{2\varepsilon}}{(1-\Phi_l)^2} \cdot \left(\frac{\Phi_l}{\Phi}\right)^{2/(D_f - d)} \cdot \delta_{\max}^2 \tag{5.2.31}$$

(Equation 38 in Jin et al. 2015a).

In this equation C_0 is not a constant, but—as reported by Jin et al. (2015: Eq. 39)—it approximately depends on porosity as

$$C_0 \approx 32\Phi^{1/(d-D_f)} \tag{5.2.32}$$

As another application of the regular Menger sponge (and to practice the *DdVmSqLnRl* formalism), let us review *void distributions of fragmented rock* (Xing-shang and Jia-lin 2009), in the collapsed zone of a coal mine. We use a *DdVmSqLnRl*-type Menger Sponge embedded to $\mathbb{R}^3$ (that is $d = 3$), the initial cube has side-length x ($l = x$), divide the side to k equal parts ($m = k$), thus the initial cube will consist of k^3 small cubes, each of linear size x/k. Remove ν small cubes selected according to some rules ($q = \nu^{1/d}$), consider the removed cubes as *broken rock*, and the rest the *voids* in the *collapsed zone*. This is repeated for the remaining $k^3 - \nu$ pieces of rock, for $n = 1, 2, \cdots$. With standard notation, the n-th *prefractal* will be

$$D3VkS\upsilon^{1/d}LnRx \tag{5.2.33}$$

The void volume in the collapsed zone scales with the fractal dimension

$$D = \ln\left(k^3 - \nu\right)/lnk \tag{5.2.34}$$

To find the fractal dimension of the broken rock, suppose the number of rock particles whose diameter is greater than r is

$$N(r) = N(\geq r) = \int_r^\infty f(r)dr \propto r^{-D} \tag{5.2.35}$$

Denote by *M(r)* the cumulative mass of rocks smaller than r, and by M the total mass of rock in the collapsed zone, so

$$M(r)/M \propto r^b \tag{5.2.36}$$

where b is the regression coefficient in the relation

$$ln(M(r)/M) = b \cdot lnr + c \tag{5.2.37}$$

From Eq. (5.2.35)

$$dN \propto r^{-D-1}dr \tag{5.2.38}$$

Since

$$dM \propto r^3 dN, r^{b-1}dr \propto r^3 r^{-D-1}dr \tag{5.2.39}$$

and the fractal dimension satisfies

$$D = 3 - b \tag{5.2.40}$$

From Eqs. (5.2.36) and (5.2.40) the fractal dimension of fragments in the collapsed zone is

$$ln(M(r)/M) = (3 - D)lnr + c \tag{5.2.41}$$

where c is constant, and

$$\frac{M(r)}{M} = cr^{3-D} \tag{5.2.42}$$

When $r = r_{\max}$, $M(r) = M$, so that $\frac{M(r)}{M} = \left(\frac{r}{r_{\max}}\right)^{3-D}$ (5.2.43)

5.3 Random and Multifractal Versions

There are two approaches to construct *random Sierpiński carpets* in $\mathbb{R}^2$ (or *random Menger sponges* in $\mathbb{R}^3$), one leads to a *monofractal object with homogeneous mass distribution*, the other results in a *multifractal object with heterogeneously distributed mass* (Mandelbrot 1977, 1982, 1989; Gouyet 1996; Perfect et al. 2006). Consider the Sierpiński carpet's *mass distribution*. At the *i*-th iteration level the *fraction of mass* in the *j*-th cell is

$$f_{j,i} = \begin{cases} \frac{1}{N\left(\frac{1}{b^i}\right)} = \left(b^i\right)^{-D_f} & if 1 \le j \le \left(b^i\right)^{D_f} \\ 0 & if \left(b^i\right)^{D_f} < j \le \left(b^i\right)^2 \end{cases} \tag{5.3.1}$$

where $N(r)$ is the number of r-sized particles, D_f is mass fractal dimension. Note that, at any iteration level i, the mass fractions sum to one: for

$$\forall i \quad \sum_{j=1}^{(b^i)^{D_f}} f_{j,i} = 1, \tag{5.3.2}$$

and—if $f_{j,1} > 0$—then

$$D_f = -\frac{log(f_{j,1})}{\log(b)}. \tag{5.3.3}$$

To obtain a *random Sierpiński carpet* we use the following *heterogeneous algorithm* (Mandelbrot 1977, 1982, 1989; Gouyet 1996; modified by Perfect et al. 2006) based on *binary decisions* (*Bernoulli trials*: Freund 1971):

(5.3.4a) Divide the initiator into b^2 parts.

(5.3.4b) Choose a probability $p \quad \left(p = 1 - \frac{n}{b^2}\right)$.

(5.3.4c) For each part independently, generate a uniformly distributed random number in the interval [0,1].

(5.3.4d) If the random number is greater than p, remove that part.

The probability $B(N, b^2, p)$ of the event that exactly N parts will be retained out of b^2 is given by the *binomial distribution*:

$$B(N, b^2, p) = \frac{(b^2)!}{N!(b^2 - N)!} p^N (1 - p)^{b^2 - N} \tag{5.3.5}$$

Any given realization of the algorithm (5.3.4a–d) could lead with non-zero probability to the result that $N = 0$, that is the whole Sierpiński carpet would disappear. For example, for $p = 1/9$ we have $B(0, 9, p) = 0.346$ that is, this problem would occur with 0.346 probability. To avoid this, Perfect et al. (2006) constrained the set of possible outcomes of the random trials (5.3.4a–d) in a way which excludes $N = 0$. They used the *truncated binomial distribution* B_T (Patil 1962), defined as

$$B_T(N, b^2, p) = \frac{\frac{(b^2)!}{N!(b^2-N)!} p^N (1 - p)^{b^2 - N}}{\sum_{N=1}^{b^2} \frac{(b^2)!}{N!(b^2-N)!} p^N (1 - p)^{b^2 - N}} \tag{5.3.6}$$

(It can be checked that $B(0, 9, 0) = 1$; $B_T(0, 9, 0) = 0$) Averaging the mass fractions from several realizations of the heterogeneous algorithm at iteration step $i = 1$ gives:

$$f_{j,1} = \sum_{k=0}^{j-1} B_T(b^2 - k, b^2, p) \frac{1}{b^2 - k} \tag{5.3.7}$$

where $f_{j,1}$ is the average mass fraction associated with the *j*-th part of the generator, $1 \leq j \leq b^2$. We always have that

$$\sum_{j=1}^{(b^i)^{D_f}} \langle f_{j,1} \rangle = 1 \tag{5.3.8}$$

Recent analyses have revealed that the pore space of sedimentary rocks and soils, and the local permeability distribution, are *multifractal* (Saucier 1992a,b; Liu and Molz 1997; Giménez et al. 1999; Boufadel et al. 2000; Mukhopadhyay and Sahimi 2000; Tennekoon et al. 2003; Veneziano and Essiam 2003, 2004). *Multifractals* cannot be characterized by a single fractal dimension but require a spectrum of *generalized dimensions* for their description (Mandelbrot 1989; Evertsz and Mandelbrot 1992; Korvin 1992: 308–317; Gouyet 1996; Turcotte 1997). We first calculate the *generalized moments* for the average mass fractions $f_{j,1}$ in the generator:

$$M_1(q) = \sum_{j=1}^{b^2} f_{j,1}^q \tag{5.3.9}$$

where $M_1(q)$ is the *generalized moment of order q* for $i = 1$, q being any integer between $-\infty$ and ∞. The *q*-th order *generalized dimension*, D_q is defined as the exponent satisfying

$$M_i(q) = \left(\frac{1}{b^i}\right)^{(q-1)D_q} \tag{5.3.10}$$

For $i = 1$ this gives

$$D_q = \frac{1}{q-1} \cdot \frac{\log\left(M_1(q)^{-1}\right)}{\log(b)}; q \neq 1 \tag{5.3.11a}$$

For the case $q = 1$ (Gouyet 1996),

$$D_1 = -\sum_{j=1}^{b^2} \langle f_{j,1} \rangle \cdot \frac{\log(\langle f_{j,1} \rangle)}{\log(b)} \tag{5.3.11b}$$

D_1 is called the *entropy dimenson*, or *information* dimension. If all the values of D_q are equal, the mass distribution is *monofractal,* if the D_q are different it is *multifractal*. Perfect et al. (2006) derived the $f_{j,1}$ values using the Sierpiński parameters b and $p = 1 - \frac{n}{b^2}$. In some other works (Saucier 1992a, b; 1996) these values are chosen arbitrarily.

Randomized Menger sponges have been also applied to simulate flow and transport in complex, *self similar* pore geometries (for *reservoir rocks* by Garrison et al. 1992;

tortuosity modeling: Garza-López et al. 2000; *water retention and permeability in soil*: Cihan et al. 2007, 2009).

The surface fractal dimensions D_2 inside slices of a randomized Menger sponge change between slices, and between realizations of the sponge. The average is almost constant, indicating that $D_2 \approx D_3 - 1$. Heuristically, this result easily follows from *Mandelbrot's rule* (Mandelbrot 1982: 365; Miyazima and Stanley 1987; Korvin 1992: 125), stating that *for two statistically independent fractal sets, A and B*, $\dim(A \cap B) = dim(A) + \dim(B) - D$, where D is the Euclidean dimension. Select A as the Menger sponge $\mathbb{M}$, and B as a random Euclidean plane $\mathbb{P}$ cutting through the sponge, then

$$D_2 = \dim(\mathbb{M} \cap \mathbb{P}) = \dim(\mathbb{M}) + \dim(\mathbb{P}) - E = D_3 + 2 - 3 = D_3 - 1.$$

Cihan et al. (2009) formally proved the assertion $D_2 = D_3 - 1$, using a simple randomized Menger sponge. In the classical (*deterministic*) Menger sponge, six of the removed cubes are in the middle of the six faces of the initiator, while the last removed cube is taken from the center of the initiator. In a random version of this sponge, the seven removed cubes are *randomly chosen* from the 27 solid cubes of side-length b^{-1}. At the first iteration level ($i = 1$) each slice is divided to $n_{boxes}^{(1)} = b^2$ *boxes* (solid or pore), the entire Menger sponge generator contains b^{D_3} *solids*. The probability that a randomly selected box in the generator is solid is $p_1 = b^{D_3}/b^3$. The probability of finding x_1 pieces of solids with sides b^{-1} among the b^2 boxes of a random slice using *sampling without replacement* is provided by the *hypergeometric distribution*. Using this distribution, the expected value of x_1 is $x_1 = n_{boxes}^{(1)} \cdot p_1 = b^{(D_3-1)}$, that is the *mean surface fractal dimension* is

$$D_2 = \frac{\log x_1}{\log b} = D_3 - 1 \qquad (5.3.12)$$

With similar dimensional arguments one can prove that the *mean areal porosity is equal to the volumetric porosity*, $\Phi_2 = \Phi_3$, which is the famous "*De Lesse principle*" of stereology (Delesse 1847; Oleschko 1998 – 1999).

It has been frequently reported (Paz-Ferreiro et al. 2013; Pia and Sanna 2013a,b, 2014a,b; Zhao et al. 2017; Li et al. 2018; Liu et al. 2018; Zhou et al. 2021; Ren et al. 2022) that the pore structure has a *multi-scale fractal* character, where there are different fractal dimensions in increasing ranges of radii. An early treatment of this phenomenon, in the context of *coastal erosion*, was (Korvin1992: 47 –55). Multi-scale fractal models have found wide applications: capillary absorption (Pia et al. 2014); gas adsorption isotherms of shale (Liu et al. 2018); MIP and NMR of coal (Ren et al. 2022); nitrogen adsorption and desorption isotherms in soil (Paz-Ferreiro et al. 2013); permeability prediction (Pia and Sanna 2014a; Li et al. 2018; Zhou et al. 2021); thermal conductivity (Pia and Sanna 2013a, b). Most of these models are versions of the IFU (*Intermingled Fractal Units*) model introduced by Pia and co-workers (Pia and Sanna 2013b, 2014a; Pia 2016; Pia and Casnedi 2017; Pia et al., 2016a, b, 2015).

The original IFU models consisted of *randomly superimposed Sierpiński* or Menger fractals, selected from three or four independently running iterations based on different generators but having the same b value. Pia and Sanna (2014b) simulated real porous media with this IFU model, and derived from it a formula for permeability. In this model, the factor b was always three. This is a very simple way of iteration, because the side lengths of cells generated in successive steps are in a rigid geometric progression (Atzeni et al. 2008), so it cannot describe more general pore-size spectra. In a recent study Zhou et al. (2021) have tried to circumvent this problem by altering the iteration rule of the *Sierpiński carpet* and allowing the values of b to change during iterations according to a prescribed program.

Cousins (2016) studied permeability for single-phase fluid flow through randomly generated Menger sponges with $2.579 \leq D_m \leq 2.893$ He followed the pioneering idea of Chen et al. (2009) that after generating the sponge, one should subsequently *roughen the pore/solid boundary* by substituting the planar faces of the small cubes with a *Weierstrass-Mandelbrot surface* of surface fractal dimension D_s and root-mean-square roughness amplitude σ. Permeability was calculated for these media using the *Lattice-Boltzmann Method* (Sect. 11.2). In the generation of the *random Menger sponge* the n cubes removed from each region were selected randomly. In the i-th iteration $n\left(b^3-n\right)^{i-1}$ cubes of size $\left(l/b^i\right)^3$ were removed. Thus, in this step the porosity becomes $\Phi_i = 1-\left(b^i\right)^{(D_m-3)}$ where $D_m = \frac{log\left(\left(b^i\right)^3-n\right)}{log\left(b^i\right)}$ (recall Eqs. 5.1.6–5.1.10).

To construct a random surface $\sigma(x, y)$ over a rectangular element $\mathcal{S} = (0 \leq \mathrm{x} \leq \mathrm{L}) \times (0 \leq \mathrm{y} \leq \mathrm{L})\mathbb{R}^2$, we use the *Weierstrass-Mandelbrot function* (*WM function*, Mandelbrot 1982; Korvin 1992: 154–155). For some given x_0, $0 \leq x_0 \leq L$ let the roughness profile $R(y) = \sigma(x_0, y)$ $(0 \leq y \leq L)$ be represented by

$$R(y) = G^{D_s-1}\sum_{n=n_1}^{\infty}\frac{cos(2\pi\gamma^n y)}{\gamma^{(2-D_s)n}};\ 1 < D_s\langle 2;\gamma\rangle 1 \tag{5.3.13}$$

where D_s is the *self-affine fractal dimension*, G is a scaling constant, and γ is a frequency-scaling parameter, in this case it is set equal to 1.5 (as recommended by Majumdar and Tien 1990). Note that the summation starts (instead of $-\infty$) only at $n = n_1$; n_1 is related to the *low-cut frequency* and it depends on the parameter γ and domain size L: as

$$\gamma^{n_1} = 1/L. \tag{5.3.14}$$

Similarly, for computational reasons there is also a high-cut limit $n_2 < \infty$, $n_2 \sim 1/\Delta\sigma(x, y)$, where $\Delta\sigma(x, y)$ is the measurable accuracy of surface roughness. *Surface roughness* is characterized by the RMS value of $R(y)$, it is given by

$$\sigma = \left[\frac{G^{2(D_s-1)}}{2ln\gamma} \cdot \frac{1}{4-2D_s}\left(\frac{1}{w_l^{4-2D_s}} - \frac{1}{w_h^{4-2D_s}} \right) \right], \tag{5.3.15}$$

where σ is the RMS *roughness height*, w_l is related to the *low frequency cut-off*, w_h to the *high frequency cut-off* as

$$w_l = \frac{1}{L}; \, w_h = \frac{1}{\Delta\sigma(x, y)} \tag{5.3.16}$$

, here $\Delta\sigma(x, y)$ is the maximum resolution at which the surface roughness can be measured (Chen et al. 2009). Equation (5.3.13) describes a *profile* along one of the coordinate directions. It is a *fractal curve,* its dimension satisfies $1 \le D_s \le 2$. To create a *fractal surface* $\sigma(x, y)$ over the square $\mathcal{S} = (0 \le \mathrm{x} \le \mathrm{L}) \times (0 \le \mathrm{y} \le \mathrm{L})\mathbb{R}^2$, we proceed as follows. Denote pixel-size by δ, and let $L = l \cdot \delta$, that is $\mathcal{S}$ contains l^2 pixels. We create a fractal curve $R(y)$ of length $l^2 \cdot \delta$, and separate it to l segments, which are then placed side by side in the l rows of $\mathcal{S}$. If the *fractal curve* $R(y)$ has fractal dimension D_s, the dimension of the constructed *fractal surface* $\sigma(x, y)$ will be, by *Mandelbrot's rule*, $D_{surface} = D_s + 1$.

With the two parameters σ and $\mathrm{D_s}$ one can create a large range of reasonable surface roughness profiles for the pore-solid interfaces inside the Menger sponges. It is natural to expect that the RMS roughness of the wall of a small pore (of size L_1) is smaller than in case of a large pore (of size L_2). Based on the findings of Power et al. (1988), who reported a linear increase of fracture surface roughness with fracture length, we can hypothesize a linear relation

$$\frac{\sigma_1}{\sigma_2} = \frac{L_1}{L_2}, \tag{5.3.17a}$$

or a more general *allometric power-law* (Korvin 1992: 1–11; Korvin 2021)

$$\frac{\sigma_1}{\sigma_2} = \left(\frac{L_1}{L_2}\right)^{\lambda}. \tag{5.3.17b}$$

A regular and a random Sierpiński carpet might be both characterized by the same D_m value, but still could have very different arrangements of the pores so that their simulation would yield quite different mechanical and transfer properties. The randomness of the geometrical arrangement of pores can be characterized by the *lacunarity* (see next Sect. 5.4), which "measures the extent at which a set is not translationally invariant" (Allain and Cloitre 1991). Lacunarity is calculated with a *gliding box algorithm* (of Allain and Cloitre 1991), where a box of linear size l "glides" over the lattice, and the mass M contained in the box is measured at each location. Here, "mass" M is defined as the number of *solid pixels* (pixels of value 1). Denote by $N(M, l)$ the total number of l-sized boxes containing a mass M. By dividing $N(M, l)$ by the total number of boxes, we get a probability distribution

(M, l). Denote the *moments* of this distribution by

$$Z_Q^{(q)}(l) = \sum_M M^q Q(M, l), \tag{5.3.18}$$

and define *lacunarity* as

$$\Lambda(l) = \frac{Z_Q^2(l)}{\left[Z_Q^1(l)\right]^2}. \tag{5.3.19}$$

For mass fractal models, lacunarity is inversely proportional to the gliding box size l (Turcotte 1997, Kim et al. 2011). Lacunarity is maximal when l is the pixel length, $l = \delta$, in this case $\Lambda(\delta) = 1/\Phi$; it is minimal when l is equal to the lattice size L, when $\Lambda(L) = 1$ (Plotnick et al. 1996).

Lacunarity analysis is a powerful tool to analyze the random geometry of spatial patterns, with an amazing range of applications in Earth Sciences, from *Structural Geology* (Roy et al. 2010) to *Well Logging* (Arizabalo et al. 2006). As it is not very well known among Petrophysicists, the next section offers a brief introduction.

5.4 Appendix: Introduction to Lacunarity

We shall need some concepts from *Probability Theory*. The n-th *moment* $M^{(n)}$ of a probability distribution $P(x)$ is defined as $M^{(n)} = \langle x^n \rangle$ where $\langle \cdots \rangle$ is *expected value*, $\langle f(x) \rangle = \int f(x) P(x) dx$. That is:

$$M^{(n)} = \langle x^n \rangle = \int x^n P(x) dx \tag{5.4.1}$$

In particular, $M^{(1)} = \mu$ is the *mean*. The so-called *central moments* $m^{(n)}$ are defined as

$$m^{(n)} = (x - \mu)^n = \int (x - \mu)^n P(x) dx \tag{5.4.2}$$

The second central moment is the *variance* or *standard variation*, $m^{(2)} = \sigma^2$, where σ is called the *scatter*. *Central moments* can be expressed in terms of the *moments*:

$$\begin{aligned}
m^{(2)} &= -M^{(1)^2} + M^{(2)} \\
m^{(3)} &= 2M^{(1)^3} - 3M^{(1)} M^{(2)} + M^{(3)} \\
m^{(4)} &= -3M^{(1)^4} + 6M^{(1)^2} M^{(2)} - 4M^{(1)} M^{(3)} + M^{(4)}; \ldots
\end{aligned} \tag{5.4.3}$$

The *characteristic function* is defined as the *Fourier Transform* of the *probability density function* (*pdf*):

$$\Phi(t) = \mathcal{F}[P(x)] = \int_{-\infty}^{\infty} e^{itx} P(x)dx = \sum_{k=0}^{\infty} \frac{(it)^k}{k!} \int_{-\infty}^{\infty} x^k P(x)dx = 1 + itM^{(1)} - \frac{1}{2}t^2 M^{(2)} - \frac{i}{3!}t^3 M^{(3)} + \cdots \tag{5.4.4}$$

If we know *all* moments, we can compute the characteristic function $\Phi(t)$, and then by inverse Fourier transform we get the *pdf* $P(x)$ as:

$$P(x) = \mathcal{F}^{-1}[\Phi(t)] = \frac{1}{2\pi} \int_{-\infty}^{\infty} e^{-itx} \Phi(t)dt \tag{5.4.5}$$

But in practice, only the first four moments are needed to characterize the most important properties, namely the *mean*, SD (*Standard Deviation*), *skew*, *kurtosis*, and *width* of an observed histogram:

$$mean = M^{(1)};\ SD = m^{(2)};\ skew = \frac{m^{(3)}}{\left[m^{(2)}\right]^{3/2,}};$$
$$width\left(if\, M^{(1)} \neq 0\right) = \frac{M^{(2)}}{\left[M^{(1)}\right]^2};\ kurtosis = \frac{m^{(4)}}{\left[M^{(2)}\right]^2} \tag{5.4.6a–c}$$

Note that for any four normally distributed variables A, B, C, D we have the identity $\langle ABCD\rangle = \langle AB\rangle\langle CD\rangle + \langle AC\rangle\langle BD\rangle + \langle AD\rangle\langle BC\rangle - 2\langle AB\rangle\langle C\rangle\langle D\rangle$, (5.4.7) which implies that for the normal distribution the *kurtosis* is 3 (Bendat and Piersol 1966; Korvin 1993).

After these preparations, consider now a point set $\mathfrak{F}$ which is distributed as a D_m—dimensional *mass fractal* on the 2-D plane. Consider an $(R \times R)$—sized square $\mathcal{S}_{\mathrm{R}}$, it contains $M_0 \propto R^{D_m}$ points of $\mathfrak{F}$. Let $r \ll R$. The part of $\mathfrak{F}$ which is inside the $(R \times R)$-square can be covered by $K(r) \propto \left(\frac{R}{r}\right)^{D_m}$ small $(r \times r)$-squares $\mathcal{S}_{\mathrm{r}}$. Denote by $M_k(k = 1, 2. \cdots ; K(r))$ the number of $\mathfrak{F}$ points in the k-th small square. Then

$$\sum_{i=1}^{K(r)} \left(\frac{M_k}{M_0}\right)^q \propto \left(\frac{r}{R}\right)^{(q-1)D_m} \tag{5.4.7}$$

This scaling law was found by Hentschel and Procaccia (1983). (In case of *multifractal point distributions* they had a q-dependent $D_m(q)$ exponent, what they called *generalized fractal dimension*). To prove (5.4.7), note that the average number of points in a small square is $\propto r^{D_m}$, so we have

$$\sum_{i=1}^{K(r)}\left(\frac{M_k}{M_0}\right)^q \approx \sum_{i=1}^{K(r)}\left(\frac{M_k}{M_0}\right)^q \propto \sum_{i=1}^{K(r)}\left(\frac{r^{D_m}}{R^{D_m}}\right)^q =$$
$$K(r)\cdot\left(\frac{r}{R}\right)^{D_m q} \propto \left(\frac{r}{R}\right)^{-D_m}\cdot\left(\frac{r}{R}\right)^{D_m q}.$$

Consider again a D_m-dimensional mass fractal $\mathfrak{F}$ inside a large $(R \times R)$ square $\mathcal{S}_\mathrm{R}$. Place a small $(r \times r)$-square $\mathcal{S}_\mathrm{r}$ randomly inside S_R, denote its center-point by $(x, y) \in \mathcal{S}_\mathrm{r} \subset \mathcal{S}_\mathrm{R}$. As long as $\mathcal{S}_\mathrm{r}$ is fully inside $\mathcal{S}_\mathrm{R}$, the number of points of $\mathfrak{F}$ in $\mathcal{S}_\mathrm{r}$ grows with increasing r values as $N\{(x_i, y_i)|(x_i, y_i) \in \mathcal{S}_r \cap \mathfrak{F}\} \propto r^{D_m}$, or, written out with the proper *prefactor*,

$$N\{(x_i, y_i)|(x_i, y_i) \in \mathcal{S}_r \cap \mathfrak{F}\} = P \cdot r^{D^m} = P(x, y, r)\cdot r^{D^m} \tag{5.4.8}$$

where the *prefactor* $P = P(x, y, r)$ depends on the spatial position of $\mathcal{S}_r$ and on its size r. The meaning of Eq. (5.4.8) is that while all parts of the fractal $\mathfrak{F}$ are D_m-dimensional according to the scaling exponent of their mass, the actual point density can change from point-to-point. There might even occur large gaps (*lacunae*) where $P(x,y,r) = 0$ that is there are no $\mathfrak{F}$ points in them at all. Mandelbrot (1977) identified the changing prefactor with "*lacunarity*", he predicted that it *may scale* with R and r, but the first workable *measure of lacunarity* was proposed only much later by Allain and Cloitre (1991), who could also deduce its scaling law. Allain and Cloitre interpreted Eq. (5.4.8) in a probabilistic sense: Take a small r-sized square $\mathcal{S}_\mathrm{r}$, drop it randomly so that its center is $(x, y)\mathcal{S}_\mathrm{R}$ and count the number N of points of $\mathfrak{F}$ in it. By repeating this experiment a large number of times, we shall find that N is a *random variable* whose values range from $N = 0$ to $N = r^2$;

$$\mathrm{Prob}(N = i) = p_i;\ \left(i = 0, 1, \cdots, r^{2)}\right);\ \sum_{i=0}^{r^2} p_i = 1 \tag{5.4.9}$$

If the *histogram* of N is very narrow, the fractal $\mathfrak{F}$ is *not lacunary*, if it is very wide *it is lacunary*. Allain and Cloitre proposed to use the *width* of the distribution of N to measure the *extent of lacunarity*:

$$\Lambda(r) = \frac{M^{(2)}}{\left[M^{(1)}\right]^2} \tag{5.4.10}$$

Suppose we make $n \gg 1$ random trials, using independently dropped boxes, to estimate $\Lambda(r)$. Then

$$\Lambda(r) = \frac{M^{(2)}}{\left[M^{(1)}\right]^2} \approx \frac{\sum_{i=1}^{n} N_i^2}{\left(\sum_{i=1}^{n} N_i\right)^2} \tag{5.4.11}$$

where N_i is number of points in the i-th box. In practice the boxes S_r are not "randomly thrown" onto $\mathcal{S}_R$ but are regularly "gliding" across its whole area with overlap (*gliding box method*). To find the scaling law of the lacunarity with r/R, consider again the fractal $\mathfrak{F}$ inside the big square $\mathcal{S}_R$. Denote the total number of points of $\mathfrak{F} \cap \mathcal{S}_R$ by M_0, of course it scales as $M_0 \propto R^{D_m}$. Select $r \ll R$ in such a way that $\frac{R}{r} = M$ is an integer. Divide the square $\mathcal{S}_R$ to M^2 small $(r \times r)$-squares. Denote by n_{ij}, $(i,\ j = 1, 2, \cdots, M)$ the number of $\mathfrak{F}$—points in the small square at position (i, j). The q-th moment can be estimated as

$$M^{(q)} \approx \frac{1}{M^2} \cdot \sum_i \sum_j n_{ij}^{(q)} = \frac{M_0^{(q)}}{M^2} \cdot \sum_i \sum_j \left(\frac{n_{ij}}{M_0} \right)^q .$$

Using the scaling law (5.4.7) of Hentschel and Procaccia (1983), $\sum_i \sum_j \left(\frac{n_{ij}}{M_0}\right)^q \propto \left(\frac{r}{R}\right)^{(q-1)D_m}$, we get

$$M^{(q)} \propto \frac{R^{D_m}}{\left(\frac{R}{r}\right)^2} \cdot \left(\frac{r}{R} \right)^{(q-1)D_m} = R^{qD_m} \left(\frac{r}{R} \right)^{(q-1)D_m+2} \qquad (5.4.12)$$

In the special cases $q = 1, q = 2$ Eq. (5.4.12) reads as $M^{(1)} = R^{D_m}\left(\frac{r}{R}\right)^2; M^{(2)} = R^{2D_m}\left(\frac{r}{R}\right)^{D_m+2}$, and we get the *scaling law of lacunarity*:

$$\Lambda = \frac{M^{(2)}}{\left[M^{(1)}\right]^2} \propto \left(\frac{r}{R} \right)^{D_m-2} \qquad (5.4.13)$$

References

Allain C, Cloitre M (1991) Characterizing the lacunarity of random and deterministic fractal sets. Phys Rev A 44(6):3552–2558

Anderson R (1958) One-dimensional continuous curves and a homogeneity theorem. Ann Math 68:1–16

Arizabalo RD, Oleschko K, Korvin G, Lozada M, Castrejón R, Ronquillo G (2006) Lacunarity of geophysical well logs in the Cantarell oil field, Gulf of Mexico. Geofísica Int 45(2):99–113

Atzeni C, Pia G, Sanna U, Spanu N (2008) A fractal model of the porous microstructure of earth-based materials. Constr Build Mater 22(8):1607–1613

Bear J, Verruijt A (1987) Modeling groundwater flow and pollution. Springer, New York

Beck C, Schlögl F (1993) Thermodynamics of chaotic systems: an introduction. Cambridge University Press, New York

Bendat JS, Piersol AG (1966) Measurement and analysis of random data. Wiley, New York

Bird NRA, Dexter AR (1997) Simulation of soil water retention using random fractal networks. Eur J Soil Sci 48:633–641

Bird NRA, Bartoli F, Dexter AR (1996) Water retention models for fractal soil structures. Eur J Soil Sci 47:1–6

Bird N, Perrier E, Rieu M (2000) The water retention function for a model of soil structure with pore and solid fractal distributions. Eur J Soil Sci 55:55–63

Boufadel MC, Lu S, Molz FJ, Lavallee D (2000) Multifractal scaling of the intrinsic permeability. Water Resour Res 36:3211–3222

Carman PC (1937) Fluid flow through granular beds. Trans Inst Chem Eng 15:150–166

Carman PC (1939) Permeability of saturated sands, soils and clays. J Agric Sci 29:262–273

Carman PC (1956) Flow of gases through porous media. Butterworths Scientific Publications, London

Chen Y, Zhang C, Shi M, Peterson GP (2009) Role of surface roughness characterized by fractal geometry on laminar flow in microchannels. Phys Rev E 80(2):026301

Cihan A, Perfect E, Tyner JS (2007) Water retention models for scale-variant and scale-invariant drainage of mass prefractal porous media. Vadose Zone J. 6:786–792

Cihan A, Sukop MC, Tyner JS, Perfect E, Huang H (2009) Analytical predictions and lattice Boltzmann simulations of intrinsic permeability for mass fractal porous media. Vadose Zone Journal 8(1):187–196

Cousins TA (2016) Effect of rough fractal pore-solid interface on single-phase permeability in random fractal porous media. Master of Science in Engineering Thesis. The University of Texas at Austin

Cowan Kenneth Lee, 1987. The Fractal Dimension as a Petrophysical Parameter. MSc in Engineering Thesis, The University of Texas at Austin.

Danielson RE, Sutherland PL (1986) Porosity. In: Klute A (ed) Methods of soil analysis: Part 1 – Physical and mineralogical methods. American Society of Agronomy, Madison, WI, pp 443–461

Dattelbaum DM, Ionita A, Patterson BM, Branch BA, Kuettner L (2020) Shockwave dissipation by interface-dominated porous structures. AIP Adv 10(7):075016

Delesse A (1847) Procédé mécanique pour déterminer la composition des roches. (Extrait). C.r. hebd. séanc. Acad. Sci., Paris 25:544

Doughty C, Karasaki K (2002) Flow and transport in hierarchically fractured rock. J Hydrol 263:1–22

Du ZY, Yang ML, Liu QG, Yu M (2019) Numerical investigation on thermal conduction in fractal-like porous media. Arch Environ Sci Environ Toxicol: AESET-109. https://doi.org/10.29011/AESET-109 100009

Evertsz CJG, Mandelbrot BB (1992) Multifractal measures. In: Peitgen H-O, Jurgens H, Saupe D (eds) Chaos and fractals: new frontiers of science. Springer-Verlag, New York, pp 921–969

Fallico C, Tarquis AM, De Bartolo S, Veltri M (2010) Scaling analysis of water retention curves for unsaturated sandy loam soils by using fractal geometry. Eur J Soil Sci 61:425–436

Freund JE (1971) Mathematical statistics. Prentice-Hall Inc., Englewood Cliffs, NJ

Friesen I, Mikula RJ (1987) Fractal dimensions of coal particles. J Colloid Interface Sci 120(1):263–271

Garrison Jr JR, Pearn WC, von Rosenberg DW (1992) The fractal Menger sponge and Sierpinski carpet as models for reservoir rock/pore systems: I. Theory and image analysis of Sierpinski carpets. In Situ 16(4):351–406

Garrison JR Jr, Pearn WC, Rosenberg DU (1993) The fractal Menger sponge and Sierpinski carpet as models for reservoir rock pore systems: II. Image analysis of natural fractal reservoir rocks. In Situ 17:1–53

Garza-López RA, Naya L, Kozak JJ (2000) Tortuosity factor for permeant flow through a fractal solid. J Chem Phys 112(22):9956–9960

Ghanbarian-Alavijeh B, Hunt AG (2012) Comments on "More general capillary pressure and relative permeability models from fractal geometry" by Kewen Li. J Contam Hydrol 140–141:21–23

Ghanbarian-Alavijeh B, Hunt A (2012b) Unsaturated hydraulic conductivity in porous media: percolation theory. Geoderma 187–188:77–84

Ghanbarian-Alavijeh B, Millán H, Huang G (2011) A review of fractal, prefractal and pore-solid-fractal models for parameterizing the soil water retention curve. Can J Soil Sci 91(1):1–14

Ghanbarian B, Hunt AG, Ewing RP, Sahimi M (2013a) Tortuosity in porous media: a critical review. Soil Sci Soc Am J 77:1461–1477
Ghanbarian B, Hunt AG, Sahimi M, Ewing RP, Skinner TE (2013b) Percolation theory generates a physically based description of tortuosity in saturated and unsaturated porous media. Soil Sci Soc Am J 77:1920–1929
Gibson JR, Lin H, Bruns MA (2006) A comparison of fractal analytical methods on 2- and 3-dimensional computed tomographic scans of soil aggregates. Geoderma 134:335–348
Giménez D, Rawls WJ, Lauren JG (1999) Scaling properties of saturated hydraulic conductivity in soil. Geoderma 88:205–220
Gouyet J-F (1996) Physics and fractal structures. Springer, New York
Hentschel HGE, Procaccia I (1983) The infinite number of generalized dimensions of fractals and strange attractors. Physica 8:435–444
Hu Y, Wang Q, Zhao J, Xie S, Jiang H (2020) A novel porous media permeability model based on fractal theory and ideal particle pore-space geometry assumption. Energies 13(3):510
Jin Y, Zhu YB, Li X, Zheng JL, Dong JB (2015) Scaling invariant effects on the permeability of fractal porous media. Transp Porous Media 109(2):433–453
Katz AJ, Thompson AH (1985) Fractal sandstone pores: implications for conductivity and pore formation. Phys Rev Lett 54(12):1325–1328
Kim JW, Sukop MC, Perfect E, Pachepsky YA, Choi H (2011) Geometric and hydrodynamic characteristics of three-dimensional saturated prefractal porous media determined with Lattice Boltzmann modeling. Transp Porous Media 90(3):831–846
Korvin G (1992) Fractal models in the earth sciences. Elsevier, Amsterdam
Korvin G (1993) The kurtosis of reflection coefficients in a fractal sequence of sedimentary layers. Fractals 1:263–268
Korvin G (2016) Permeability from microscopy: review of a dream. Arab J Sci Eng 41(6):2045–2065
Korvin G (2021) Allometric power laws. In: Daya Sagar B, Cheng Q, McKinley J, Agterberg F (eds) Encyclopedia of mathematical geosciences. Encyclopedia of Earth Sciences Series. Springer, Cham
Kozeny J (1927) Über Kapillare Leitung des Wassers in Boden Stizungsber. Akad. Wiss. Wien 136:271–306
Krohn CE (1988a) Sandstone fractal and Euclidean pore volume distributions. J Geophys Res: Solid Earth 93(B4):3286–3296
Krohn CE (1988b) Fractal measurements of sandstones, shales, and carbonates. J Geophys Res: Solid Earth 93(B4):3297–3305
Li J, Yu B (2011) Tortuosity of flow paths through a Sierpinski carpet. Chin Phys Lett 28:34701
Li ZL, Tauraso R (2006) Problem 11208. American Mathematical Monthly 113
Li LC, Lin M, Ji L, Jiang W, Ca G (2018) Rapid evaluation of the permeability of organic-rich shale using the 3D intermingled-fractal model. SPE J 23(6):2175–2187
Liu HH, Molz FJ (1997) Multifractal analyses of hydraulic conductivity distributions. Water Resour Res 33:2483–2488
Liu K, Ostadhassan M, Zou J, Gentzis T, Rezaee R, Bubach B, Carvajal-Ortiz H (2018) Multifractal analysis of gas adsorption isotherms for pore structure characterization of the Bakken Shale. Fuel 219:296–311
Mackeprang C, Myers K (2007) Problem 11208 from Volume 133 No. 3 p. 267 Menger Sponge Map-Coloring. American Mathematical Monthly
Majumdar A, Tien CL (1990) Fractal characterization and simulation of rough surfaces. Wear 136(2):313–327
Mandelbrot BB (1974) Intermittent turbulence in self-similar cascades: divergence of high moments and dimension of the carrier. J Fluid Mech 62:331–358
Mandelbrot BB (1977) Fractals: form, chance, and dimensions. W. H. Freeman and Co., San Francisco, California
Mandelbrot BB (1982) The fractal geometry of nature. W.H. Freeman, New York
Mandelbrot BB (1989) Multifractal measures, especially for the geophysicist. PAGEOPH 131:5–42

Meakin P (1998) Fractals, scaling and growth far from equilibrium. Cambridge University Press, Cambridge, UK

Menger K (1932) Kurventheorie. Teubner, Leipzig

Miyazima S, Eugene Stanley H (1987) Intersection of two fractal objects: Useful method of estimating the fractal dimension. Phys Rev B Condens Matter 35(16):8898–8900

Mortensen NA, Okkels F, Bruus H (2005) Reexamination of Hagen-Poiseuille flow: shape dependence of the hydraulic resistance in microchannels. Phys Rev E 71:057301

Mostafa ME (2008) Finite cube elements method for calculating gravity anomaly and structural index of solid and fractal bodies with defined boundaries. Geophys J Int 172:887–902

Mukhopadhyay S, Sahimi M (2000) Calculation of effective permeabilities of field-scale porous media. Chem Eng Sci 55:4495–4513

OEIS®, 2022. A135918 - Genus of stage-n Menger sponge. The On-Line Encyclopedia of Integer Sequences

Oleschko K (1998–1999) Delesse principle and statistical fractal sets: Part 1. Dimensional equivalents. Soil & Tillage Research 49:255–266. Part 2. Unified fractal model for soil porosity. Ibid. 52:247–257

Patil GP (1962) Maximum likelihood estimation for generalized power series distributions and its application to the truncated binomial distribution. Biometrika 49:227–237

Paz-Ferreiro J, da Luz L, Lado M, Vázquez EV (2013) Specific surface area and multifractal parameters of associated nitrogen adsorption and desorption isotherms in soils from Santa Catarina Brazil. Vadose Zone J 12:185–192

Perfect E (2005) Modeling the primary drainage curve of prefractal porous media. Vadose Zone J 4:959–966

Perfect E, Gentry W, Sukop MC, Lawson JE (2006) Multifractal Sierpinski carpets: Theory and application to upscaling effective saturated hydraulic conductivity. Geoderma 134:240–252

Perrier EMA, Bird NRA (2002) Modelling soil fragmentation: the pore solid fractal approach. Soil Tillage Res. 64:91–99

Perrier E, Rieu M, Sposito G, Marsily G (1996) Models of the water retention curve for soils with a fractal pore size distribution. Water Resour Res 32:3025–3031

Perrier E, Bird N, Rieu M (1999) Generalizing the fractal model of soil structure: the PSF approach. Geoderma 88:137–164

Pfeifer P, Avnir D, Farin D (1984) Scaling behaviour of surface irregularity in the molecular domain: from adsorption studies to fractal catalysis. J Stat Phys 36:699–716

Pia G (2016) High porous Yttria-stabilized Zirconia with aligned pore channels: morphology directionality influence on heat transfer. Ceram Int 42(10):11674–11681

Pia G, Casnedi L (2017) Heat transfer in high porous alumina: experimental data interpretation by different modelling approaches. Ceram Int 43(12):9184–9190

Pia G, Sanna U (2013a) Intermingled fractal units model and electrical equivalence fractal approach for prediction of thermal conductivity of porous materials. Appl Therm Eng 61:186–192

Pia G, Sanna U (2013b) A geometrical fractal model for the porosity and thermal conductivity of insulating concrete. Constr Build Mater 44:551–556

Pia G, Sanna U (2014a) An Intermingled Fractal Units Model and method to predict permeability in porous rock. Int J Eng Sci 75:31–39

Pia G, Sanna U (2014b) An Intermingled Fractal Units Model to evaluate pore size distribution influence on thermal conductivity values in porous materials. Appl Therm Eng 65(1/2):330–336

Pia G, Sassoni E, Franzoni E, Sanna U (2014) Predicting capillary absorption of porous stones by a procedure based on an intermingled fractal units model. Int J Eng Sci 82:196–204

Pia G, Casnedi L, Ionta M et al (2015) On the elastic deformation properties of porous ceramic materials obtained by pore-forming agent method. Ceram Int 41(9):11097–11105

Pia G, Casnedi L, Sanna U (2016a) Porosity and pore size distribution influence on thermal conductivity of Yttria-stabilized Zirconia: experimental findings and model predictions. Ceram Int 42(5):5802–5809

Pia G, Siligardi C, Casnedi L, Sanna U (2016b) Pore size distribution and porosity influence on sorptivity of ceramic tiles: from experimental data to fractal modelling. Ceram Int 42(8):9583–9590
Plotnick RE, Gardner RH, Hargrove WW, Prestegaard K, Perlmutter M (1996) Lacunarity analysis: a general technique for the analysis of spatial patterns. Phys Rev E 53(5):5461–5468
Power WL, Tullis TE, Weeks JD (1988) Roughness and wear during brittle faulting. J Geophys Res: Solid Earth 93(B12):15268–15278
Ren W, Zhou H, Zhong J, Xue D, Wang C, Liu Z (2022) A multi-scale fractal approach for coal permeability estimation via MIP and NMR methods. Energies 15(8):2807
Rieu M, Sposito G (1991a) Fractal fragmentation, soil porosity and soil water properties: I Theory. Soil Sci Soc Am J 55:1231–1238
Rieu M, Sposito G (1991b) Fractal fragmentation, soil porosity and soil water properties: II Applications. Soil Sci Soc Am J 55:1239–1244
Rieu M, Sposito G (1991c) Relation pression capillaire-teneur en eau dans les milieux poreux fragmentés et identification du caractère fractal de la structure des sols. Metals Technology 1:462–467
Roy A, Perfect E, Dunne WM, Odling N, Kim J-W (2010) Lacunarity analysis of fracture networks: evidence for scale-dependent clustering. J Struct Geol 32(10):1444–1449
Saucier A (1992a) Effective permeability of multifractal porous media. Physica A 183:381–397
Saucier A (1992b) Scaling of the effective permeability in multifractal porous media. Physica, A 191:289–294
Saucier A (1996) Scaling properties of disordered multifractals. Physica A 226:34–63
Schönhöfer P (2014) Fractal Geometries: Scaling of Intrinsic Volumes. Masterarbeit aus der Physik, Institut für Theoretische Physik I, Friedrich-Alexander-Universität Erlangen-Nürnberg
Schröder-Turk GE, Mickel W, Kapfer SC, Schaller FM, Breidenbach B, Hug D, Mecke K (2013) Minkowski tensors of anisotropic spatial structure. New J Phys 15(8):083028
Sergeyev and D., , 2009 Sergeyev DY (2009) Valuating the exact infinitesimal values of area of Sierpinski's carpet and volume of Menger's sponge. Chaos Soliton. Fract 2:3042–3046
Sukop MC, Perfect E, Bird NRA (2001) Water retention of prefractal porous media generated with the homogeneous and heterogeneous algorithms. Water Resour Res 37:2631–2636
Tee GJ (2015) Tunnel numbers for fractal polyhedra, 1–3. Appendix A in the online version of H. Molina-Abral, P. Real, A. Nakamura and R. Klette, Connectivity calculus of fractal polyhedrons, Pattern Recognition 48 No. 4 (April 2015):1146–1156
Tennekoon L, Boufadel MC, Lavalle D, Weaver J (2003) Multifractal anisotropic scaling of hydraulic conductivity. Water Resour Res 39(7):1193
Toledo PG, Novy RA, Davis HT, Scriven LE (1990) Hydraulic conductivity of porous media at low water content. Soil Sci Soc Am J 54:673–679
Turcotte DL (1993) Fractals and chaos in geology and geophysics. Cambridge University Press, Cambridge UK
Tyler SW, Wheatcraft SW (1990) Fractal processes in soil water retention. Water Resour Res 26:1047–1054
Valdés-Parada FJ, Porter ML, Wood BD (2011) The role of tortuosity in upscaling. Transp Porous Med 88:1–30
Veneziano D, Essiam AK (2003) Flow through porous media with multifractal hydraulic conductivity. Water Resour Res 39(6):1166
Veneziano D, Essiam AK (2004) Nonlinear spectral analysis of flow through multifractal porous media. Chaos, Solitons FrActals 19:293–307
Vita MC, De Bartolo S, Fallico C, Veltri M (2012) Usage of infinitesimals in the Menger's Sponge model of porosity. Appl Math Comput 218(16):8187–8195
Wang BY, Jin Y, Chen Q, Zheng JL, Zhu YB, Zhang XB (2014) Derivation of permeability–pore relationship for fractal porous reservoirs using series-parallel flow resistance model and lattice Boltzmann method. Fractals 22:1440005

Wheatcraft SW, Tyler SW (1988) An explanation of scale-dependent dispersivity in heterogeneous aquifers using concepts of fractal geometry. Water Resour Res 24:566–578

Wheatcraft SW, Sharp GA, Tyler SW (1991) Fluid flow and solute transport in fractal heterogeneous porous media. In: Bear J, Corapcioglu MY (eds) Transport processes in porous media., Kluwer Academic, Dordrecht, The Netherlands, pp 695–722

Xia YX, Cai JC, Wei W, Hu XY, Wang X, Ge XM (2018) A new method for calculating fractal dimensions of porous media based on pore size distribution. Fractals 26:1850006

Li and Jia-lin, 2009 Xing-Shang L, Xu J-l (2009) A model of void distribution in collapsed zone based on fractal theory. Procedia Earth Planet Sci 1:203–210

Yu BM, Li JH (2001) Some fractal characters of porous media. Fractals 9:365–372

Zhao et al., 2017 Zhao Y, Zhu G, Dong Y, Danesh NN, Chen Z, Zhang T (2017) Comparison of low-field NMR and microfocus X-ray computed tomography in fractal characterization of pores in artificial cores. Fuel 210:217–226

Zhou Y, Wu S, Li Z, Zhu R, Xie S, Zhai X, Lei L (2021) Investigation of microscopic pore structure and permeability prediction in sand-conglomerate reservoirs. J Earth Sci 32(4):818–827

Chapter 6
Coordination Number of Grains

Chapter Highlights This chapter reviews the combinatorial geometry of *granular packings* in 2- and 3-D, and the average value of the coordination number, $\langle Z \rangle$, for different cases. The mechanical stability of random packs requires the satisfaction of a set of linear equations, and the consistency of these equations determines the possible values of $\langle Z \rangle$. The coordination number is also determined in the presence of *friction* and *adhesive forces*. Recent attempts to find an *Equation of State* (EOS) between packing fraction and mean coordination number are also discussed.

Before discussing *grain coordination number* and its different kinds, we start with the definition of *packings* in *n*-dimensional Euclidean space (Rogers 1964; Fejes Tóth 1972; Groemer 1986).

Let E^n denote the *n*-dimensional Euclidean space, and let $\boldsymbol{K} = \{K_\nu\}_{\nu \in \mathcal{N}}$ be a collection of convex bodies in E^n, where each K_ν is a nonempty compact convex subset of E^n, and $K_\nu = K_\mu$ if and only if $\nu = \mu$. Let I denote the group of all isometries and T the group of all translations of E^n, and let G be a subgroup of I that contains T (for example *translations* and *rotations*). In the traditional theory of packings one permits *repeated use* of the elements of the set $\boldsymbol{K} = \{K_\nu\}$.

A class $\boldsymbol{C} = \{C_i\}_{i=1,2,\cdots}$ of convex bodies will be called a $\boldsymbol{K}^r$—*packing* in E^n (the superscript "*r*" indicates that the sets from **K** may be used *repeatedly*) if the following conditions are satisfied:

(a) For each $C_i \epsilon \boldsymbol{C}$ there exist a convex body $K_\nu \epsilon \boldsymbol{K}$ and a transformation $g_{\nu \to i} \epsilon G$ such that $g_{\nu \to i}(K_\nu) = C_i$.
(b) Every ball in E^n has nonempty intersection with only finitely many sets of $\boldsymbol{C}$.
(c) The interiors of any two sets from $\boldsymbol{C}$ are disjoint. (Conditions 6.1a–c)

Note that this mathematical definition does not contain the requirements that some or all of the objects touch each other, or that the packing should be *gravitationally stable* (Berryman 1983; Nolan and Kavanagh 1992). The collection of convex bodies $\boldsymbol{K} = \{K_\nu\}_{\nu \in \mathcal{N}}$ is the *supply set* where the packing elements are taken from, and then put to their place with a rigid movement. In Rock Physical applications (Table 6.1) we use packings made up of different kinds of convex bodies (Table 6.2).

G. Korvin, *Statistical Rock Physics*, Earth and Environmental Sciences Library,
https://doi.org/10.1007/978-3-031-46700-4_6

Table. 6.1 Rock physics applications of the grain coordination number

Application	Author
Avalanche risks	Oger et al. (2007)
Boundary migration in crystalline rocks	Flinn (1969)
Compaction	Gaither (1953); Taylor (1949, 1950); Wilson and McBride (1988)
Dispersivity	Vasilyev et al. (2012)
Elastic moduli	Madadi and Christy (2012)
Granule mass fraction	Roy et al. (2009)
Gravitational stability of sphere packings	Nolan and Kavanagh (1992)
Heat Transfer	Wakao and Kaguei (1982); Zhou et al. (2009)
Hysteresis	Darling et al. (2004)
Permeability	Pinson et al. (1997)
Phase formation	German (1989)
Porosity	Suzuki and Oshima (1989)
Shear strength	Oda (1977)
Solid–solid reactions	Hao and Tanaka (1990)
Stress transmission	Wiącek et al. (2020)
Tensile strength	Johnson et al. (1971); Cundall and Strack (1979)

Table. 6.2 Convex shapes used in random packing studies

Shape	References
Deformable spheres	Mukhopadhyay and Peixinho (2011)
Mono-sized disks	Meyer et al. (2010); Atkinson et al. (2014)
Mono-sized spheres	Bernal and Mason (1960); Suzuki and Oshima (1989); Nolan and Kavanagh (1992); Aste et al. (2005); Nguyen et al. (2011); Yi et al. (2011); Hanifpour et al. (2013); Song et al. (2016); Liu et al. (2017); Blétry et al. (2018); Migal et al. (2020); Wiącek et al. (2020)
Multi-sized disks	Ueda et al. (2012)
Multi-sized spheres	Bezrukov et al. (2001); Brouwers (2013); Burtseva et al. (2015); Taiebat et al. (2017); Wang et al. (2008); Roy et al. (2009); Migal et al. (2019)
Non-spherical particles	Baule et al. (2013); Stroeven and He (2013)
Rods	Wouterse et al. (2009)
Superellipsoids	Yuan et al. (2019)

The most important parameters characterising the packings of convex particles are their *packing density* (or *packing fraction*) η and mean *coordination number* $\langle Z\rangle$ (the coordination number is the number of particles that contact a given particle). If there are a finite number of particles $\{C_i\}_{i=1,2,\cdots,N}$ packed into a bounded, closed, measurable set $X \subset E^n$, the *packing density* is the fraction of the packing volume X occupied by the particles:

$$\eta = \frac{\sum_{i=1}^{N} vol(C_i)}{vol(X)} \tag{6.1}$$

If $\boldsymbol{C} = \{C_i\}_{i=1}^{\infty}$ then, because of Condition 6.1b, X cannot be bounded, and the packing density is defined as

$$\eta(\boldsymbol{C}) = \lim_{t\to\infty} \frac{1}{vol(B_t)} \sum_{C_i \epsilon C} vol(B_t \cap C_i) \tag{6.2}$$

If the limit in Eq. (6.2) does not exist, we use the *upper packing density* and the *lower packing density* of $\boldsymbol{C}$. These are defined, respectively, by

$$\overline{\eta}(\boldsymbol{C}) = \frac{\overline{lim}}{t \to \infty} \frac{1}{vol(B_t)} \cdot \sum_{C_i \epsilon C} vol(B_t \cap C_i), \tag{6.3a}$$

and

$$\underline{\eta}(\boldsymbol{C}) = \frac{\underline{lim}}{t \to \infty} \frac{1}{vol(B_t)} \cdot \sum_{C_i \epsilon C} vol(B_t \cap C_i), \tag{6.3b}$$

see Groemer (1986).

Packing porosity, $\Phi(\boldsymbol{C}) = 1 - \eta(\boldsymbol{C})$ is the volume fraction of that part of X *not occupied* by the particles. The basic characteristics of *regular sphere packings* have been compiled by (von Seckendorff and Hinrichsen 2021: Table 1). The *coordination number* is $N_c = 8$ for both the *orthorhombic* and *body-centered cubic* arrangements, while their packing densities η are different ($\eta_{orthorh.} = 0.395$; $\eta_{bcc} = 0.395$), that is we cannot expect a unique functional relation $N_c = f(\eta)$ between the two characteristics.

A simple concept of coordination number is (Migal et al. 2020) when we "count the number of neighboring, or "nearest" particles in a system of particles located around a selected particle". In this case there is no necessary physical contact between neighboring spheres, the system can be subjected to tessellation and triangulation and we can define the coordination number of each sphere as "the number of its triangulation edges that belong to the contact network" (Picka 2012).

But in practice, we almost always require a *direct contact* between neighboring particles, and the *coordination number* thus becomes a "*contact number*". In regular packings, this coordination number is the same for all particles, in irregular, random packings it fluctuates locally and we are interested in its *distribution* and the *average coordination number*, usually denoted by $\overline{Z}$ or $\langle Z \rangle$.

In the voxelized (pixelized) image of a packing of convex bodies we say that "two particles are in contact if they are *not separated by any background voxel/pixel*" (Nguyen et al. 2011). Consider two nearby spheres $\mathcal{S}_1(O_1, r_1)$; $\mathcal{S}_2(O_2, r_2)$ centred at O_1 and O_2, and with respective radii r_1, r_2. We say (Aste et al. 2005; Isola 2008) that $\mathcal{S}_1$ and $\mathcal{S}_2$ are in contact if

$$r_1 + r_2 \leq \overline{O_1 O_2} \leq r_1 + r_2 + v \tag{6.4}$$

where $\overline{AB}$ means Euclidean distance, v is voxel-size.

Aste et al. (2005: Table 1) analyzed 82,626 mono-sized spheres in the central part of a container to eliminate wall effects, the packing density was 0.617 ± 0.005, and they got $N_c^{(a)} = 3.23$; $N_c^{(b)} = 6.11$; $N_c^{(c)} = 7.05$ where $N_c^{(a)}, N_c^{(b)}$ and $N_c^{(c)}$ were, respectively, the average numbers of neighbors $\mathcal{S}_i(O_i, r)$; $\mathcal{S}_j(O_j, r)$ satisfying

$$\overline{O_i O_j} = 2r;\ 2r \leq \overline{O_i O_j} \leq 2r + v/2;\ 2r \leq \overline{O_i O_j} \leq 2r + v. \tag{6.5a,b,c}$$

These authors also studied the change in the average number of neighbors $n_t(x)$, where $x = \overline{O_i O_j} \epsilon (0.98 \times 2r, 1.1 \times 2r)$ and found a more precise estimate for the number of spheres in contact (*ibid.*: Fig. 3): above $x = 0.98 \times 2r$ the number of neighbors grows very steeply up to a *knee* at about$1.02 \times 2r$, then a slower growth follows (*ibid.*: Fig. 3).

Wang et al. (2008) introduced a *Modified Radial Distribution Function* (MRDF) $g_z(r)$ which, as they claimed, distinguishes between *physically touching grains* for which a "mechanical coordination number" is determined, and those which are only considered neighbors because of their geometric proximity. It is defined as

$$g_z(r) = \frac{1}{N}\left(\frac{R}{r}\right)^2 \sum \sum_{i,j=1,\ldots,N;i \neq j} \Theta\left(\frac{r_{ij}}{r-R}\right)\Theta\left(\frac{r+R}{r_{ij}}\right) \tag{6.6}$$

where R is the (identical) radius of the N spheres, r_{ij} the distance between the centers of two of them, and Θ the Heaviside step function.

Hanifpour et al. (2013) numerically determined the MRDF $g_z(\Delta(r))$, where $\Delta(r) = \frac{r-R}{2R}$, in different sub-volumes of an experimentally generated sphere pack. In the limit $\Delta(r) \to 0$ the MRDF tends to the *mechanical coordination number* g_{mech}, while its maximum is attained for the *geometrical coordination* number g_{geom}. Hanifpour et al. (2013) have found however (*ibid.*: Fig. 6) that this method fails even for the slightest poly-dispersivity in sphere size distribution.

Any mechanical effect in a stable pack of spheres must propagate through neighboring grains, from grain to grain. To understand, e.g., *sound propagation* in such a pack, we must reach out "beyond first neighbors", and use the concept of *topological distance*, and *topological density* (Aste et al. 2005: 061,302–6; Picka 2012). *Topological distance* from a given sphere is defined recursively, by identifying the increasing shells radiating out from it. The first shell contains only those spheres which touch the central sphere, the second contains all spheres in contact with the first shell but not already contained within the first shell, and so on. Consider a pack of spheres, each of diameter d, the spheres in the j-th shell ($j = 1, 2, 3, \cdots$) have a topological distance $dist_{top} = j \cdot d$ from the central sphere. (Mathematics enthusiasts will note the similarity with the "Erdős number": an author's Erdős number is 1 if he has co-authored a paper with Erdős, 2 if he has co-authored a paper with someone who has co-authored a paper with Erdős, etc.; see Hoffman 1998: 13). In general, spheres at shell ($j + 1$) are those which are in direct contact to spheres in shell j but have not been counted previously. In *Crystallography*, it has been observed (Brunner, 1979; Brunner and Laves 1971; O'Keeffe 1991a, 1995) in infinite, periodic, crystalline structures with no boundaries, that the number of atoms N_j per such shells increases with topological distance, and in many 3-D crystalline structures one has

$$N_j = a_j \cdot j^2 + b_j \cdot j + c_j \tag{6.7}$$

where the coefficients a_j, b_j, c_j are bounded, and they only slightly change with j. For instance, for spheres on a *cubic lattice* $N_j = 4j^2 + 2$; the number of spheres in a *body centered cubic arrangement* is

$$N_j = 6j^2 + 2. \tag{6.8a,b}$$

Following the definition of O'Keeffe (1991b), both for crystalline systems and sphere packs, the *asymptotic behavior* of N_j is characterized by the *topological density*,

$$TD = \langle a_j \rangle / 3, \tag{6.9}$$

where a_j is the leading coefficient of the quadratic expression (6.7). For several experimental packs containing ~150,000 spheres the values of TD are compiled in (Aste et al. 2005: Table II.)

Nolan and Kavanagh (1992) developed a model for *random, gravitationally stable packings of monosized spheres*. Their model predicted that the two governing parameters, *packing fraction* η and *mean coordination number* $\langle Z \rangle$, are inside a quadrilateral in the (η, $\langle Z \rangle$) plane with packing densities ranging from 0.509 to 0.638 and mean coordination numbers from 4.4 to 5.9. These packings occur between two extreme states, *random loose packing and random close packing* (Scott 1960; Berryman 1983). Their different structures try to satisfy two very different sets of constraints: the *close-packed structure* optimally fills out the available space in a lattice while the *loose-packed structure* only obeys the less stingent requirements of gravitational stability. (A sphere is *gravitationally stable* if its centre is above the plane passing through the centres of any three of its supporting spheres.). In experimental sphere packs, we frequently encounter *bridged structures*, especially in the presence of inter-particle adhesive forces (Liu et al. 2017). A bridge is formed when two unstable spheres contact and form a stable structure, before either could roll down to settle on a deeper-lying site (Tory et al. 1968; Nolan and Kavanagh 1992). Bridging causes a simultaneous decrease in packing density and in average coordination number.

In the theory of random packings of *mono-disperse frictionless spheres* it is generally assumed that a mean coordination number $\langle Z \rangle = 6$ is *necessary requirement* for the stability of a random pack (Bennett's hypothesis 1972). This can be made plausible using *Maxwell's constraint counting arrguments* (Maxwell 1864; Alexander 1998; Aste 2005: S2372; Isola 2008): At mechanical equilibrium, Newton's equations for the balance of the force and torque acting on each sphere must be satisfied. To achieve stability, the number of degrees of freedom must agree with the number of constraints. In a packing of N *ideal, smooth* spheres, there are $3N$ degrees of freedom (*translational modes*) and $\langle Z \rangle N/2$ equations (one for the normal force at each contact). Thus, $\langle Z \rangle$ must be equal to 6.

On the other hand, in a *real pack* the spheres unavoidably have *rotational modes* and therefore the total number of degrees of freedom is $6N$ (3 *translational modes* and 3 *rotational modes* for each sphere). Also, some *friction* is always present, giving rise to *tangential forces*. Newton's law yields $3\langle Z \rangle N/2$ equations, and the equilibrium requires that $\frac{3\langle Z \rangle N}{2} = 6N$, that is

$$\langle Z \rangle = 4. \tag{6.10}$$

Generalizing the argument to d dimensions (Silbert et al. 2002), it can be shown that *the minimal average coordination number required to obtain static packings of d-dimensional spheres* is

$$\left\langle Z^{(d)}_{smooth} \right\rangle = 2d; \left\langle Z^{(d)}_{frictional} \right\rangle = d + 1; \tag{6.11a,b}$$

in particular

$$\left\langle Z^{(3)}_{smooth} \right\rangle = 6; \left\langle Z^{(3)}_{frictional} \right\rangle = 4. \tag{6.12a,b}$$

Sphere packings satisfying this condition are called *isostatic*, those with smaller or larger average coordination number are called, respectively, *ipostatic* and *hyperstatic*. Silbert et al. (*ibid*.: p. 1) claimed that the "isostaticity hypothesis is important in theories focusing on the *macroscopic response* of packings".

Based on a delightful thought-experiment (Song et al. 2016: Fig. 2), we can speak of two types of coordination numbers: the *geometrical coordination number*, Z_{geom}, counting all contacts between spheres, and the *mechanical coordination number*, Z, counting *only those contacts that carry forces*.

Song et al. (*op. cit.*) recognized that the *coordination number* appearing in Eqs. (6.10 and 6.11a) is the *geometrical coordination number* Z_{geom}, which might differ from the *mechanical coordination number* Z_{mech} that counts the number of contacts per sphere with non-zero force in the isostatic state.

Imagine a finite, *frictionless* stable packing of uniform spheres with volume fraction $\eta \approx 0.64$, where all contacts carry forces, here the mechanical stability requires $\langle Z \rangle = 6$ contacts.

Then we introduce *tangential forces* between the spheres and increase the *coefficient of friction* μ to infinity. For $\mu = \infty$ the consistency of the force and torque balance equations would yield again the *isostatic condition*: at $Z = 4$ the number of equations equals the number of force variables. Thus, while there are still $Z_{geom} = 6$ actual contacts, only $Z_{mech} = 4$ out of them carry non-zero forces. This thought experiment suggests that, in 3-D, for any $\mu \epsilon (0, \infty)$ we have:

$$4 \leq \langle Z_{mech}(\mu) \rangle \leq \left\langle Z_{geom} \right\rangle \leq 6. \tag{6.13}$$

(Song et al. 2016: Eq. 12). Liu et al. (2017: Table 3) extended constraint counting to the more general case when there are *normal*, *tangential*, and *adhesive* forces. They introduced a *dimensionless adhesion number*

$$Ad = \gamma / \left(\rho U_0^2 R\right), \tag{6.14a}$$

where γ is *adhesive surface energy*, ρ particle density, U_0 velocity, R radius, and distinguished three cases:

$$\left.\begin{array}{l} \mu_f = 0, ad = any \ then \ \langle Z \rangle^{iso} = 6 \\ \mu_f = \infty, ad = 1 \ \ then \ \langle Z \rangle^{iso} = 4 \\ \mu_f = \infty, ad = \infty \ then \ \langle Z \rangle^{iso} = 2 \end{array}\right\}, \tag{6.14b,c,d}$$

where μ_f is coefficient of friction.

From among the many applications of constraint counting to packings of *nonspherical bodies*, we only discuss the mechanically stable random packings of *hard rods* (Blouwolff and Fraden 2006; Wouterse et al. 2009; Baule et al. 2013). In the *frictionless* case, at (almost) every contact we have the trivial constraint that the two axes of the rods are separated by a distance of one rod diameter. For N rods we have to satisfy $N \langle Z \rangle /2$ such constraint equations. In 3D each rod has five degrees of freedom (three spatial cooordinates, two angles). The number of constraints must be less than or equal to the number of degrees of freedom, that is $N \langle Z \rangle /2 \leq 5N$ which yields an upper bound for

$$\langle Z \rangle : \langle Z \rangle \leq 10. \tag{6.15}$$

To get a lower bound we use the mechanical stability condition that the sum of the forces and of the torques on each rod must be zero. For forces this gives three equations per rod, for torques only two equations per rod because without friction there is no torque around the symmetry axis. The number of mechanical stability equations cannot exceed the number of force and torque variables, thus $5N \leq N \langle Z \rangle /2$ from where

$$\langle Z \rangle \geq 10 \tag{6.16}$$

which, together with (6.15) proves that

$$\langle Z \rangle = 10. \tag{6.17}$$

In case of friction, the constraints are the same, that is once more $\langle Z \rangle \leq 10$ (6.15), but there are three force equations, and three torque equations that is altogether 6 mechanical equations per rod, and simple algebra yields

$$4 \leq \langle Z \rangle \leq 10. \tag{6.18}$$

A general treatment of the mechanical stability of packings with N *nonspherical particles* (*superellipsoids*) in d spatial dimensions, based on calculating the *eigenvalue spectrum of the dynamical matrix for the packing* is found in (Yuan et al. 2019).

The structure of monodispersive spherical packing is characterized by two main parameters: the *packing density* η (or *porosity* $\Phi = 1 - \eta$, also called *voidage ratio*), and *mean coordination number*, $\langle Z \rangle$. There have been many attempts to relate these two parameters through physical experiments and *sphere packing algorithms*. There are several algorithms for sphere packing. The *force-biased algorithm* (e.g. Mościński et al. 1989; Mościński and Bargieł 1991) starts with a dense configuration of spheres which can even overlap. It moves the spheres stepwise in order to reduce overlaps, and it can produce very dense, nearly isotropic packings. Another frequently used algorithm (e.g. Jodrey and Tory 1979) simulates the packing of a container with equal spheres falling into it in the gravitational field. A sphere with three contacts with already packed spheres is considered as being in its final position, otherwise it rolls lower down. This algorithm results in loose packings of volume fraction $\eta = 0.58$ in the case of equal spheres. A third popular way to simulate packings is by means of the *Discrete Element Method* (Stroeven and Stroeven 2000; Yi et al. 2011; Hanifpour et al. 2013; Stroeven and He 2013; Wiącek 2016).

As the coordination number cannot be uniquely estimated from packing density, these efforts only resulted in empirical formulae, available in tabulated and graphical forms in (Suzuki and Oshima 1989: Table 1, Fig. 5; Migal et al. 2020: Table 1; von Seckendorff and Hinrichsen 2021: Table 2, Figs. 1 and 3). Most equations contain empirically determined fitting constants, the only exceptions being Kolmogorov's Eq. (1937),

$$\langle Z \rangle = -8 \ln(1 - \eta), \tag{6.19}$$

and a formula by Bondarev et al. (2015),

$$\langle Z \rangle = 6(6\eta/\pi)^2, \tag{6.20}$$

see (Migal et al. 2020: Table 2, rows 2 and 11).

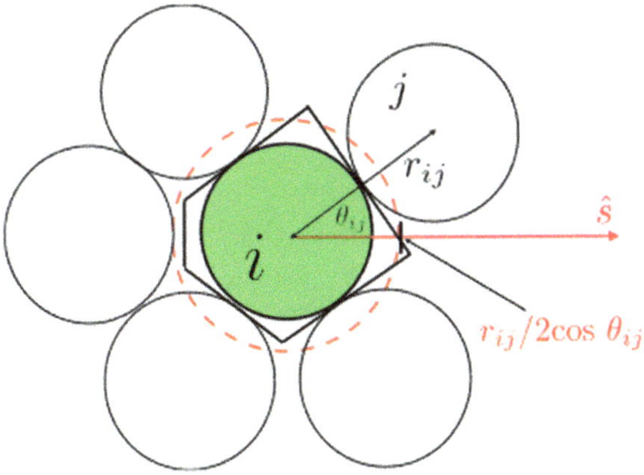

Fig. 6.1 Two-dimensional sketch of the *Voronoi volume* and the *orientational volume* associated with particle *i*. The boundary of the Voronoi cell is the irregular pentagon in black which defines the *Voronoi volume* $\mathcal{W}_i^{vor}$. The limit of the Voronoi cell of particle *i* in the direction $\hat{s}$ is the minimum of $r_{ij}/(2cos\theta_{ij})$ over *all the particles in the packing*. $\mathcal{W}_i^s$ is the volume of the sphere of radius $r_{ij}/(2cos\theta_{ij})$ defined by the dashed red circle. (After Song et al. 2016: Fig. 1a; reproduced with the kind permission of Prof. Hernán Makse and *AIP Publishing*)

The most spectacular results in this field have been achieved by co-workers and students of Hernán Makse who, in a series of ground-breaking papers (Song et al. 2008, 2016; Wang et al. 2008, 2011; Meyer et al. 2010; Liu et al. 2017) succeeded to derive in some ideal limiting cases EOS (*Equation of State*)-like general relations between packing fraction and average coordination number. They recognized that at grain level *the free volume* $\mathcal{W}(\vec{r}_i)$ ("*free volume function*" in the sense of Edwards and Oakeshott 1989) associated with each particle $\vec{r}_i$ is provided by the Voronoi volume of the particle, analytically given in 3D as

$$\mathcal{W}_i^{vor} = \frac{1}{3}\oint \left(\min_{\hat{s}\cdot\hat{r}_{ij}>0} \left(\frac{r_{ij}}{2R\hat{s}\cdot\hat{r}_{ij}} \right) \right)^3 ds, \tag{6.21}$$

where $\hat{r}_{ij}$ is the vector from the center of particle *i* to that of particle *j*, the integration is taken over all the directions $\hat{s}$ forming an angle θ_{ij} with$\hat{r}_{ij}$, and *R* is the radius. In Fig. 6.1 the Voronoi cell bounded by the pentagon in black defines $\mathcal{W}_i^{vor}$. The limit of the Voronoi cell of particle *i in the direction* $\hat{s}$ is the minimum of $r_{ij}/(2cos\theta_{ij})$ *taken over all particles in the packing*. This defines the *orientational volume* $\mathcal{W}_i^s$ which is the volume of the sphere of radius $r_{ij}/(2cos\theta_{ij})$ indicated by the dashed red circle in the figure.

In *isotropic systems* there is no need to average over $\hat{s}$. We define the *reduced free orientational volume function* as

$$w^s \equiv \frac{\mathcal{W}_i^s - V_g}{V_g} \tag{6.22}$$

with

$$\mathcal{W}_i^s \equiv V_g\left(\frac{1}{2R}\cdot\min_{\hat{s}\cdot\hat{r}_{ij}>0}\left(\frac{r_{ij}}{2\hat{s}\cdot\hat{r}_{ij}}\right)\right)^3, \tag{6.23}$$

where V_g is the particle volume.

The next steps are summarized in (Song et al. 2008: 629) as follows: "We first derive the Voronoi volume in terms of the particle positions and then use statistical analysis to coarse-grain the Voronoi volume over a mesoscopic length scale, obtaining a mesoscopic free volume function that is analytically tractable."

For *monodisperse hard frictionless spheres* of volume V_g, they found that the *free volume per particle* satisfies the *Equation of State*

$$W\langle\langle Z\rangle\rangle = \frac{2\sqrt{3}}{\langle Z\rangle}\cdot V_g \tag{6.24}$$

In case of friction, in a *random loose pack* (RLP) the packing density and mean coordination number are related as

$$\eta_{RLP}\langle\langle Z\rangle\rangle \approx \frac{\langle Z\rangle}{\langle Z\rangle + 2\sqrt{3}} \tag{6.25}$$

Liu et al. (2017) derived an *Equation of State* for *uniform sphere packings with adhesion and friction*:

$$\frac{1}{\eta} - 1 = \frac{C}{\langle Z\rangle^{\gamma}}, \tag{6.26}$$

where C and the exponent γ depend on the *dimensionless adhesion number* $Ad = \gamma/(\rho U_0^2 R)$ (defined in Eq. 6.14). In the limit of *no adhesion*,

$$\lim_{Ad\to 1} C = 2\sqrt{3};\ \lim_{Ad\to 1}\gamma = 1, \tag{6.27}$$

and (6.25) reduces to the adhesionless EOS, $\eta = \langle Z\rangle/(\langle Z\rangle + 2\sqrt{3}$.

The left-hand-side term $\left(\frac{1}{\eta} - 1\right)$ in Eq. (6.26) is the *reduced free volume*, defined as

$$\omega = \left(\langle W\rangle - V_g\right)/\langle W\rangle, \tag{6.28}$$

where $\langle W\rangle$ is the average volume of the Voronoi cells, and V_g is the volume of a particle. Equation (6.26) yields a power-law relation between ω and

$$\langle Z\rangle : \omega = \frac{C}{\langle Z\rangle^{\gamma}}, \tag{6.29}$$

Very few exact relations are known about the relation between *packing fraction* and *coordination number*, or the *distribution of the coordination numbers* in stable packings of *polydisdersive spheres*. The following treatment is based on the general comments of (Burtseva et al. 2015; Migal et al. 2019); on results of (Taiebat et al. 2017) on uniformly distributed spheres; (Bezrukov et al. 2001; Brouwers 2013; Wiącek 2016) on bimodal distribution of spheres; (Bezrukov et al. 2001; Ueda et al. 2012) on lognormal sphere packs; (Yi et al. 2011; Wiącek et al. 2020) on packings of ternary mixtures of spheres.

The only analytic treatment to date of the relation between packing fraction and coordination number has been given by Brouwers (2013) who considered *random close packings* (RCP) of *bimodal spheres*. He started out from a stack of equal spheres in a lattice of unit cell volume V_{cell}. If there are N spheres of volume Ω in the cell, the packing fraction is

$$f_1^{cp} = \frac{N\Omega}{V_{cell}} = \frac{N \cdot \frac{\pi}{6} \cdot d^3}{l^3} \tag{6.30}$$

where l is *the lattice constant*. For the *fcc* structure $N = 4$ and $l = 2^{1/2}d$, for the *bcc* structure $N = 2$ and $l = 2d/3^{1/2}$, and for the *sc* structure $N = 1$ and $l = d$, that is the *monosized* packing fractions are

$$f_1^{fcc} = 2^{1/2} \cdot \pi/6,\ f_1^{bcc} = 3^{1/2} \cdot \pi/8 \text{ and} f_1^{sc} = \pi/6. \tag{6.31}$$

For a *bimodal distribution* of a fraction X_L of large spheres and a fraction X_S of small spheres, the *mean sphere volume* is

$$\langle\Omega\rangle = X_L\Omega_L + X_S\Omega_S = X_L\Omega_L + (1 - X_L)\Omega_S = \frac{\pi\left[X_L d_L^3 + (1 - X_L)d_L^3\right]}{6} \tag{6.32}$$

The *cell volume*, accomodating the most probable combinations of small and large spheres is

$$V_{cell} = \sum_{i=0}^{n}\left\{\binom{n}{i}X_L^{n-1}(1 - X_L)^i \times \left[\frac{n-1}{n}l_L^3 + \frac{i}{n}l_S^3 + \lambda_i\left(l_L^3 - l_S^3\right)\right]\right\} \tag{6.33}$$

where n is the number of spheres in the cell. The factor λ_i takes care of the *lattice distortions* for those cells in which both kinds of spheres appear, the distortion is proportional to the volume mismatch $\left(l_L^3 - l_S^3\right)$. Brouwers assumed that the distortion parameter λ_i is proportional to the fraction of non-matching contacts (b_{LS}) among the total number of contacts (b_t) between spheres in the unit cell, i.e. $\lambda_i \sim b_{LS}(i)/b_t(i)$. A combinatorial counting of the possible mismatching configurations for the *fcc*, *bcc*, and *sc* lattices (Brouwers 2013: 032,202–2) gave

$$\lambda_i = C \cdot \frac{\binom{n-2}{i-1}}{\binom{n}{i}} = C \cdot \frac{i(n-i)}{n(n-1)}, \; (1 \le i \le n-1) \tag{6.34}$$

where C is a constant. Using this value in Eq. (6.33) yields

$$V_{cell} = X_L l_L^3 + (1 - X_L) l_S^3 + C X_L (1 - X_L)\left(l_L^3 - l_S^3\right) \tag{6.35}$$

The first two terms on the right-hand side of Eq. (6.35) correspond to the so-called *Retgers' law* (1889).

Introducing a variable

$$u = l_L / l_S = d_L / d_S \tag{6.36}$$

measuring size ratio, from Eqs. 6.33 and 6.35 we can express the ratio of *bimodal packing fraction* η and *unimodal packing fraction*

$$f_1 : \frac{\eta^{cp}}{f_1^{cp}} = \frac{X_L\left(u^3 - 1\right) + 1}{X_L\left(u^3 - 1\right) + 1 + C(1 - X_L)(u^3 - 1)}, \tag{6.37}$$

cp refers to "crystalline packing". To find the value of the constant C, observe that when we add small spheres to a crystalline structure of large spheres only, this will not change the cell volume, because the small spheres will rattle in the cages formed by the large ones:

$$\left.\frac{dV_{cell}}{dX_L}\right|_{X_L=1} = 0. \tag{6.38}$$

Applying this condition to Eq. (6.35) yields $C = 1$. The *contraction coefficient* β is defined as

$$\beta = \frac{1}{f_1(1 - f_1)} \cdot \left.\frac{d\eta}{du}\right|_{u=1, X_L=0.5}, \tag{6.39}$$

that is Eq. (6.37) yields

$$\beta^{cp} = -3/4\left(1 - f_1^{cp}\right) \tag{6.40}$$

For *monodispersive spheres with friction* (Song et al. 2008) we have already encountered the relation (6.25) between the mean number of mechanical contacts $\overline{Z}$ and random packing fraction: $f_1^{rp} = \frac{\overline{Z}}{\overline{Z}+2\sqrt{3}}$, with $\overline{Z}$ ranging from 4 (RLP) to 6 (RCP). Combining this rule with (6.40), Brouwers (2013) found the following relation between β and $\overline{Z}$ for bimodal random sphere packings:

$$\beta^{rp} = \beta^{rcp}\left(1 - f_1^{rcp}\right)\left(1 + \frac{\overline{Z}}{2\sqrt{3}}\right) = 0.072\left(1 + \frac{\overline{Z}}{2\sqrt{3}}\right) \tag{6.41}$$

According to this equation *there is a linear relation between the mean number of mechanical contacts and the contraction coefficient* β^{rp}.

Yi et al. (2011) found that the *coordination number distribution* in *packings of ternary mixtures of spheres* can be described by the so-called *Johnson SB function.* ("SB" stands for "System Bounded".) It is a modified Gaussian distribution, its *cdf* (*cumulative distribution function*) can be written as (Johnson et al. 1970; Yu and Johnson 1994):

$$F_{SB}(C) = -\frac{1}{\sqrt{2\pi}} \int_{-\infty}^{\xi} exp\left(-t^2/2\right)dt, \tag{6.42a}$$

with

$$\xi = \sigma\left[ln\left(\frac{C - C_{min}}{C_{max} - C}\right) - ln\left(\frac{\lambda - \lambda_{min}}{1 - \lambda}\right)\right] \tag{6.42b}$$

where C_{min} and C_{max} represent the minimum and maximum contact numbers, λ is ratio of the median M to the maximum contact number C_{max}, λ_{min} is the ratio of C_{min} to C_{max}, and σ is the distribution parameter.

Wiącek et al. (2020) studied coordination number distribution in ternary granular packings. The samples were characterized by g, defined as the ratio between the diameter of the largest and the smallest spheres, and they distinguished between geometrical coordination number CN, and mechanical coordination number (what they termed "*corrected coordination number*", CN^*). The *corrected coordination number increased* with the g value, similarly to the case of binary sphere packings, studied in (Wiącek 2016).

Ueda et al. (2012) studied the coordination number distribution in *poorly graded* and *well-graded* soils. In the *poorly graded samples* most of the particles had approximately the same diameter, $D_{50} \pm \sigma$, the grain-size distribution was *lognormal* with pdf

$$f(D) = \frac{1}{\sqrt{2\pi\sigma^2} \cdot D} exp\left(-\frac{(log_{10}D - log_{10}D_{50})^2}{2\sigma^2}\right). \tag{6.43}$$

In this case the mean coordination number is governed by, and decreasing with, both for loose and tight packings.

Consider now a *well-graded soil* where particles of all size are present, and take randomly a typical particle $\mathcal{P}$ (for simplicity we consider 2D disks) inside the pack with radius R_f which is in contact with M other particles of respective radii $R_1, R_2, \cdots, R_M$. Substitute each of these surrounding particles with the same particle of radius $R_c \equiv \frac{1}{M}(R_1 + R_2 + \cdots + R_M) \approx \frac{1}{N}\sum_{i=1}^{N} R_i = R_m$ where N is the number of particles in the system. The expected number of contacts between particle

$\mathcal{P}$ and particles of average radius R_m is a linear function of the perimeter of $\mathcal{P}$ (i.e. of R_f) in the 2D case, and of its surface (i.e. of R_f^2) in the 3D case, so that it is plausible to anticipate a linear relation $\langle Z \rangle = A \cdot R_f + B$ for disks,

$$\langle Z \rangle = C \cdot R_f^2 + D \tag{6.44a,b}$$

for spherical particles. Indeed, Ueda et al. (2012: Fig. 12) found experimentally $\langle Z \rangle = 2.24\left(\frac{R_f}{R_m}\right) + 2.0$ for loose packings,

$$\langle Z \rangle = 2.72\left(\frac{R_f}{R_m}\right) + 2.0 \text{ for dense packing} \tag{6.45a,b}$$

and established the *upper limit*

$$\langle Z \rangle \le \frac{\pi}{sin^{-1}\left(\frac{1}{1+\frac{R_f}{R_m}}\right)}. \tag{6.46}$$

The group Taiebat et al. (2017) used *Molecular Dynamics Simulation* of frictionless spherical particles to study packings where the particle size distribution was *uniform* in particle volumes: in a range $[d_{min}, d_{max}]$ of particle diameters

$$P(d) = \frac{2}{d_{min}^{-2} - d_{max}^{-2}} d^{-3}. \tag{6.47}$$

Normalizing all particle diameters by d_{min}: $d_r = d/d_{min}$, the reduced diameters are distributed as

$$P(d_r) = \frac{2}{1 - \alpha^{-2}} d_r^{-3}, \tag{6.48}$$

where

$$\alpha = d_{max}/d_{min}. \tag{6.49}$$

In the range $1 \le \alpha \le 5$ the packing factor approximately linearly increased from $\eta|_{\alpha=1} = 0.643$ to $\eta|_{\alpha=5} = 0.71$, the shape of the contact number distribution curve showed no systematic change with α.

A Russian group at the Belgorod State University (Bondarev et al. 2008; Migal et al. 2019, 2020) studied the relation between η and $\langle Z \rangle$ for packings of mono-dispersive and poly-dispersive spheres. In the mono-dispersive case they found two empirical equations, the first expressed $\langle Z \rangle$ of a central particle in terms of the interparticle distances; (Migal et al. 2019: Eq. 1)

$$\langle Z \rangle = \sum_{i=1}^{m} exp\{-\gamma[(r_i/\delta)^6 - 1]\} \tag{6.50}$$

the second related $\langle Z \rangle$ to packing density; Migal et al. 2019: Eq. 2).

$$\langle Z(n) \rangle = n_{max}\left[\left(\frac{2n}{\pi}\eta(n)\right)^2 - 1\right] + 2n \tag{6.51}$$

In (6.50) γ is a constant; m is the total number of particles in the first and second coordination spheres; r_i is the *interparticle distance* between a selected central sphere and the i-th sphere; δ is the particle diameter. In (6.51) n_{max} (taken as $n_{max} = 6$) is a constant, η is the packing density, n is the dimension of space. The last equation expresses $\langle Z \rangle$ in terms of η, however the authors state "the causality inherent in this equation shows that the *dependent quantity* here is the packing density whose values can be determined from the coordination number and the dimension of the space" (*op.cit.*: 3188).

To treat the case of poly-dispersive sphere packings the authors consider an elementary, closely packed, connected cluster inside the pack, consisting of four spheres. "Closely packed" here means that the cluster contains no inter-particle "holes" larger than equal to the smallest sphere among the four.

We are looking for a differential equation describing the change in the coordination number of the central sphere if we slightly perturb the positions of the spheres around the central sphere. The perturbation will also change the total volume of the cluster, and the total volume of empty space inside it. Instead of the integer values of the number of contacts, the authors introduced a continuous variable, called *coordination index*, that describes the contribution of the outer spheres to the coordination number of the *central sphere* S_1. The maximum value of the coordination index, f_c is unity in the case of direct contact with the central particle, otherwise $0 \leq f_c < 1$. The coordination number Z of the central particle is the sum of the coordination indexes of all neighbors located nearby:

$$Z = \sum_{i=1}^{m} f_{ci} \tag{6.52}$$

if there are m neighboring particles in the first and second coordination spheres.

Suppose we have a packing in which all particles are spherical and have different size. Let us choose some particle S_1 inside the packing and call it the *central particle*, select the *coordinated particle* S_2 and the two *neighbors* S_3, S_4. If we infinitesimally change the distance r_{12} between the central and the coordinated particle, this will cause a change: (a) in the total volume W of the cluster; and (b) in the volume of the void space Φ. There will also be a displacement of the neighboring particles but because we want to keep the cluster close-packed, they must keep their contacts with the central and coordinated particles. We also assume that the magnitude of these changes depend on the dimension of space, n. What will be change in the coordination index $f_c \equiv f_c(O_2)$ of the coordinated particle?

After the small perturbation of r_{12} with an amount Δr, the original *cluster volume* W_0 will change to W, *porosity* will change from Φ_0 to Φ and the coordination index can be represented as a function:

$$f_c = F(W, \Phi, W_0, \Phi_0, n, r_{12}). \tag{6.53}$$

To write up a differential equation for f_c we assume that (i) the infinitesimal changes in the coordination index Δf_c and the relative volume increments ΔW are directly proportional, as well as (ii) that small changes in the coordination index Δf_c are proportional to the coordination index f_c itself and to the porosity, that is:

$$df_c = -A\frac{f_c \cdot \Phi \cdot \omega(n)}{\Phi_0 W_0} dW, \tag{6.54}$$

where A is a coefficient of proportionality, $\omega(n)$ is a function describing the effect of dimension n of the considered space on the coordination index. Upon integration we get

$$f_c = exp\left[-A\frac{\Phi \cdot \omega(n)}{\Phi_0 W_0}(W - W_0)\right] \tag{6.55}$$

If we can express the volumes W, W_0, Φ, Φ_0 in terms of the geometrical parameters $(r_1, r_2, r_3, r_4, r_{12})$ and know the function $\omega(n)$, f_c can be computed for all the m spheres in the proximity of S_1 and Eqs. (6.52 and 6.55) would yield the coordination number as

$$Z = Z(r_1, r_2, r_3, r_4, r_{12}). \tag{6.56}$$

To get the *expected coordination number* $\langle Z \rangle$, we have to find the expected value of $Z(r_1, r_2, r_3, r_4, r_{12})$ with respect to the size distribution function $P(x) = Prob(r \leq x)$. In a special case this would yield Eq. (6.50), that is Eq. 1 of (Migal et al. 2019).

In 2011 Mukhopadhyay and Peixinho studied disordered packings of deformable spheres (hydrogel spheres immersed in water). At first glance, the evolution of the coordination number with increasing packing fraction ϕ seems linear, but a close scrutiny, together with studying the coordination number distributions, reveal the significant jump in coordination number around $\phi = 0.74$ and its saturation above $\phi = 0.80$ where $Z \approx 10$ (Fig. 6.2). The jump at $\phi \approx 0.74$ signals a *phase transition* from a packing with point-like contacts to a packing where the particles are deformed under load, they are faceting, and touch each other along the newly developed facets.

This close pack of *deformable spheres* touching each other along small facets is closer to geological reality than any ideal packing with point contacts which, in compacted rocks is the exception rather than the rule. Fonseca et al. (2010)

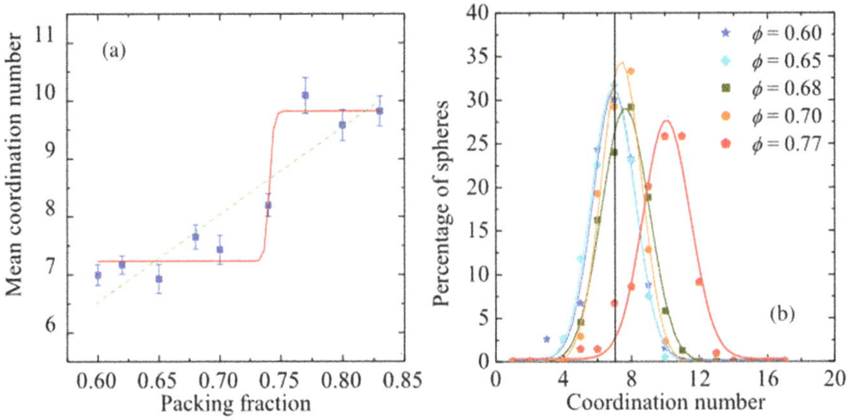

Fig. 6.2 **a** Mean coordination number *vs.* packing fraction ϕ for deformable spheres (hydrogel spheres immersed in water) under compression. **b** Probability distributions of $\langle Z \rangle$ for selected ϕ values. (After Mukhopadhyay and Peixinho 2011: Fig. 4; with the kind permission of the *American Physical Society*)

observed this in their study of the grain contacts in the "locked sand" from the Folkestone Bed formation, SE England: "there is a predominance of *straight contacts* and *concavo-convex* contacts giving high grain-grain contact areas, ….. *Tangential* or *point contacts* are less frequent and *floating particles* (i.e. particles with no contacts) are very rare". Their terminology stems from the classic compaction studies of Jane Taylor (1949, 1950), who was the first to describe and classify non-punctual grain contacts in Wyoming Mesozoic sandstones as (1) *tangential*, (2) *long*, and (3) *concavo-convex*. By definition (Taylor 1949, 1950; Gaither 1953): "*tangential contacts* are those that appear as a point in the plane of the thin section; *long contacts* appear as a relatively straight line; and *concavo-convex* (or "*embayed*") *contacts* as a curved line." *Sutured contacts*, the fourth type defined by Taylor, and believed by her to result from *pressure solution*, are relatively rare, and have not been found by (Fonseca et al. 2010) in the Folkestone Bed samples. Taylor's study showed the usefulness of identifying the *average number of contacts per grain*, later named the *contact index* (CI) by Pettijohn et al. (1972). More recent studies have introduced new indices for the grain contacts, emphasizing the relative importance of the different contact types. The *Tangential Index* (TI) that gives the proportion of tangential (or "point") contacts amongst all grain contacts was proposed by (Barton 1993). Wilson and McBride (1988) used both the *contact index* CI and the *tight packing index* (TPI), defined as the average number of *long*, *sutured*, and *concavo-convex contacts* per grain. For the Pliocene sandstones, Ventura Basin, California, they documented for samples with less than 10% cement that CI and TPI have distinctly different depth gradients:

$$CI = 2.05 \times 10^{-4} \cdot Depth(in feet) + 2.87 \tag{6.57}$$

$$TPI = 2.66 \times 10^{-4} \cdot Depth(infeet) + 0.22 \tag{6.58}$$

A further useful measure (Fonseca et al. 2010) is the *Contact Area Ratio* (CRA), the ratio between the contact surface area (S_C) and the particle surface area (S_P):

$$CRA_{ave} = 2\sum_{N_C} S_C / \sum S_P. \tag{6.59}$$

In some *crystalline rocks* the description of grain contacts is even more complicated. Flinn (1969) studied the distribution of contacts between mineral grains in gneiss samples from Shetland and Siberia. He found that instead of being randomly distributed, grains of the same phase tended not to get in contact with each other. This results from *boundary migration* (Drury and Urai 1990), because interfacial energies of contacts between like phases are greater than those between unlike phases.

Madadi and Christy (2012) recognized that the *non-point-wise grain contacts* must be properly modeled when computing the elastic moduli of granular composite materials. They modified the traditional *Coherent Potential Approximation* (CPA) method to compute elastic moduli by taking into consideration the grain coordination numbers and the *contact areas* at the grain-contact zones. The modified CPA was applied to 3D digital images of sintered quartz bead samples and of Fontainebleau sandstones. It has been proved that the contact zones around grains decrease the *local elastic moduli* and cause an overall *elastic weakening* of the bulk sample.

References

Alexander S (1998) Amorphous solids: their structure, lattice dynamics and elasticity. Phys Rep 296:65–236

Aste T (2005) Variations around disordered close packings. J Phys: Condens Matter 17:2361–2390

Aste T, Saadatfar M, Senden TJ (2005) Geometrical structure of disordered sphere packings. Phys Rev E 71:061302

Atkinson S, Stillinger FH, Torquato S (2014) Existence of isostatic, maximally random jammed monodisperse hard-disk packings. Proc Natl Acad Sci 111:18436–18441

Barton M (1993) Cohesive sands; the natural transition from sands to sandstones. In: Anagnostopoulos A, Scholosser F, Kalteziotis N, Frank R (eds) Geotechnical engineering of hard soils—soft rocks, vol 1: geological features, investigation and classification; mechanical properties and bchaviour. A.A. Balkcma, Rottcrdam, pp 367–374

Baule A, Mari R, Lin B, Portal L, Makse HA (2013) Mean-field theory of random close packings of axisymmetric particles. Nat Commun 4:2194

Bennett CH (1972) Serially deposited amorphous aggregates of hard spheres. J Appl Phys 43(6):2727–2734

Bernal DJ, Mason J (1960) Coordination of randomly packed spheres. Nature 188:910–911

Berryman JG (1983) Random close packing of hard spheres and disks. Phys Rev A 27:1053–1061

Bezrukov A, Stoyan D, Bargieł M (2001) Spatial statistics for simulated packings of spheres. Image Anal Stereol 20:203–206

Blétry M, Russier V, Barbé E, Blétry J (2018) Structure of sticky-hard-sphere random aggregates: the viewpoint of contact coordination and tetrahedral. Phys Rev E 98(1–1):012101

Blouwolff J, Fraden S (2006) The coordination number of granular cylinders. Europhys Lett 76(6):1095–1101

Brouwers HJH (2013) Random packing fraction of bimodal spheres: an analytical expression. Phys Rev E Stat Nonlinear Soft Matter Phys 87(3):032202-1/8

Brunner GO (1979) The properties of coordination sequence and conclusions regarding the lowest possible density of zeolites. J Solid State Chem 29:41–45

Brunner GO, Laves F (1971) Zum Problem der Koordinationszahl. Wiss z Tech Univ Dresden 20:387–390

Burtseva L, Salas BV, Romero R, Werner F (2015) Multi-sized sphere packings: models and recent approaches. Otto-von-Guericke-Universität, Fakultät für Mathematik

Cundall PA, Strack ODL (1979) Discrete numerical model for granular assemblies. Geotechnique 29:47

Darling TW, TenCate JA, Brown DW, Clausen B, Vogel SC (2004) Neutron diffraction study of the contribution of grain contacts to nonlinear stress-strain behavior. Geophys Res Lett 31(16):L16604

Drury MR, Urai JL (1990) Deformation-related recrystallization processes. Tectonophysics 172(3–4):235–253

Fejes Tóth L (1972) Lagerungen in der Ebene, auf der Kugel und im Raum, 2nd edn. Springer-Verlag, Berlin-Heidelberg-New York

Flinn D (1969) Grain contacts in crystalline rocks. Lithos 2(4):361–370

Fonseca J, O'Sullivan C, Coop MR (2010) Quantitative description of grain contacts in a locked sand. In: Reed AH (ed) Advances in computed tomography for geomaterials: GeoX 2010. Wiley, Hoboken NJ, pp 17–25

Gaither A (1953) A study of porosity and grain relationships in experimental sands. J Sediment Petrol 23(3):180–195

German RM (1989) Particle packing characteristics. Metal Powder Industries Federations, Princeton, NJ

Groemer H (1986) Some basic properties of packing and covering constants. Discret Comput Geom 1(2):183–193

Hanifpour M, Francois N, Allaei MV, Saadatfar M (2013) DEM simulation of experimental dense granular packing. AIP Conf Proc 1542:337–340

Hao YJ, Tanaka T (1990) A new experimental method to specify the diffusing component in a reacting particulate packing. Can J Chem Eng 68:81

Hoffman P (1998) The man who loved only numbers: the story of paul erdős and the search for mathematical truth. Hyperion, New York

Isola R (2008) Packing of granular materials. Ph.D. Thesis, University of Nottingham

Jodrey WS, Tory EM (1979) Simulation of random packing of spheres. J Simul 32:1–12

Johnson NL, Kotz S, Balakrishnan N (1970) Continuous univariate distributions, 2nd edn, vol 1. Wiley, New York

Johnson KL, Kendall K, Roberts AD (1971) Surface energy and contact of elastic solids. Proc R Soc London Ser A: Math Phys Sci 324:301

Liu W, Jin Y, Chen S, Makse HA, Li S (2017) Equation of state for random sphere packings with arbitrary adhesion and friction. Soft Matter 13:421–427

Madadi M, Christy AG (2012) A modified coherent potential approximation: Grain-contact moduli and coordination-number effect. Geophysics 77(3):WA141–WA148

Maxwell JC (1864) On the calculation of the equilibrium and stiffness of frames. Lond Edinburgh Dublin Philos Mag J Sci 27:182, 294–299

Meyer S, Song C, Jin Y, Wang K, Makse HA (2010) Jamming in two-dimensional packings. Physica A: Stat Mech Appl 389(22):5137–5144

Migal LV, Bondarev VG, Chekanov NA, Bondareva TP (2020) Simulation of the coordination number of random sphere packing. J Phys: Conf Ser 1479:012097

Migal LV, Bondarev VG, Bondareva TP, Belyaeva IN (2019) Mathematical model of coordination number of spherical packing. Compusoft Int J Adv Comput Technol 8(6):3187–3191

Mościński J, Bargieł M (1991) C-language program for the irregular close packing of hard spheres. Comput Phys Comm 64:183–192
Mościński J, Bargieł M, Rycerz ZA, Jacobs PWM (1989) The force-biased algorithm for the irregular close packing of equal hard spheres. Mol Simul 3(4):201–212
Mukhopadhyay S, Peixinho J (2011) Packings of deformable spheres. Phys Rev E: Stat Nonlinear Soft Matter Phys 84:011302
Nguyen TT, Tran TN, Willemsz TA, Frijlink HW (2011) A density based segmentation method to determine the coordination number of a particulate system. Chem Eng Sci 66:6385–6392
Nolan GT, Kavanagh PE (1992) Computer simulation of random packing of hard spheres. Powder Technol 72:149
O'Keeffe M (1991a) N-dimensional diamond, sodalite and rare sphere packings. Acta Cryst A 47:748–753
O'Keeffe M (1995) Coordination sequences for lattices. Zeit F Krist 210:905–908
Oda M (1977) Co-ordination number and its relation to shear strength of granular material. Soils Found 17(2):29–42
Oger L, Ippolito I, Vidales AM (2007) How disorder can diminish avalanche risks: effect of size distribution. Granular Matter 9:267–278
O'Keeffe M (1991b) Dense and rare four-connected nets. Z Kristallogr 196:21–37
Papiya R, Vashishtha M, Khanna R, Subbarao D (2009) Variation of granule mass fraction with coordination number in wet granulation process. Particuology 7:408–413
Pettijohn FJ, Potter PE, Siever R (1972) Sand and sandstone. Springer-Verlag, New York
Picka J (2012) Statistical inference for disordered sphere packings. Stat Surv 6:74–112
Pinson D, Yu AB, Zulli P, McCarthy MJ (1997) Powder entrapment in a multi-phase flow packed bed. In: Battle TP, Henein H (eds) Processing and handling of powders and dusts powder. TMS Publications, Warrendale, PA
Retgers JW (1889) Das spezifische Gewicht isomorpher Mischungen. Zeitschrift Für Physikalische Chemi 3(1):497–561
Rogers CA (1964) Packing and covering. Cambridge University Press, Cambridge
Scott GD (1960) Packing of spheres: packing of equal spheres. Nature 188(4754):908–909
Silbert LE, Ertaş D, Grest GS, Halsey TC, Levine D (2002) Geometry of frictionless and frictional sphere packings. Phys Rev E 65(031304):1–6
Song C, Wang P, Makse HA (2008) A phase diagram for jammed matter. Nature 453(7195):629–632
Song C, Wang P, Makse HA (2016) Theory of random packings. arXiv: 1001.5468v2[cond-mat.soft]
Stroeven P, He H (2013) Packing of non-spherical aggregate particles by DEM. In: Proceedings of the ACCTA 2013: international conference on advances in cement and concrete technology in Africa. Johannesburg, South Africa, pp 809–816
Stroeven P, Stroeven M (2000) Assessment of particle packing characteristics at interfaces by SPACE system. Image Anal Stereol 19:85–90
Suzuki M, Oshima T (1989) Relationship between average coordination number and void fraction in randomly packed systems of uniform-sized spheres developed by four kinds of computer simulation. KONA 7:22–28
Taiebat M, Mutabaruka P, Pellenq R, Radjai F (2017) Effect of particle size distribution on 3D packings of spherical particles. EPJ Web Conf 140:02030
Taylor JM (1950) Pore-space reduction in sandstones. AAPG Bull 34(4):701–716
Taylor JM (1949) Pore space reduction in sandstones. Masters Thesis, University of Cincinnati
Tory E, Cochrane N, Waddell S (1968) Anisotropy in simulated random packing of equal spheres. Nature 220:1023–1024
Ueda T, Matsushima T, Yamada Y (2012) Micro structures of granular materials with various grain size distributions. Powder Technol 217:533–539
Vasilyev L, Raoof A, Nordbotten JM (2012) Effect of mean network coordination number on dispersivity characteristics. Transp Porous Media 95(2):447–463
von Seckendorff J, Hinrichsen O (2021) Review on the structure of random packed beds. Can J Chem Eng 99:S703–S733

Wakao N, Kaguei S (1982) Heat and mass transfer in packed beds. Gordon and Breach Science Publishers, New York

Wang P, Song C, Briscoe C, Wang K, Makse HA (2008) From force distribution to average coordination number in frictional granular matter. Physica A-Stat Mech Appl 389:3972–3977

Wang, Song C, Jin Y, Makse HA (2011) Jamming II: Edwards' statistical mechanics of random packings of hard spheres. Physica A: Stat Mech Appl 390(3):427–455

Wiącek J (2016) Geometrical parameters of binary granular mixtures with size ratio and volume fraction: experiments and DEM simulations. Granul Matter 18:1–10

Wiącek J, Stasiak M, Kafashan J (2020) Structural and micromechanical properties of ternary granular packings: effect of particle size ratio and number fraction of particle size classes. Materials 13:339

Wilson JC, McBride EF (1988) Compaction and porosity evolution of Pliocene sandstones, Ventura Basin, California. AAPG Bull 72(6):664–681

Wouterse A, Luding S, Philipse AP (2009) On contact numbers in random rod packings. Granular Matter 11(3):169–177

Ye Y, Van der Werf K, Shattuck MD, O'Hern CS (2019) Jammed packings of 3D superellipsoids with tunable packing fraction, coordination number, and ordering. Soft Matter 15(47):9751–9761

Yi LY, Dong KJ, Zou RP, Yu AB (2011) Coordination number of the packing of ternary mixtures of spheres: DEM simulations versus measurements. Ind Eng Chem Res 50:8773–8785

Yu AB, Johnson SB (1994) Distribution function as applied in the mathematical representation of particle-size distributions. Part 1: theoretical background and numerical simulation. Part Part Syst Char 11:291–367; Part 2: Application of numerical results. Part Part Syst Char 11:367

Zhou ZY, Yu AB, Zulli P (2009) Particle scale study of heat transfer in packed and bubbling fluidized beds. AIChE J 55:868

Chapter 7
The Shape of Pebbles, Grains and Pores

Chapter Highlights This short chapter offers a wealth of material, many of them recent developments. We discuss and tabulate (Table 7.2) the different classification schemes and numerical formulae used to describe the *shape of grains, pebbles and rock fragments*; acquaint the Reader with the efforts of a Hungarian research group (of Professor Gábor Domokos) to find the *evolution equation* that explains the characteristic ovoid shape of beach pebbles. Further topics include the *number of stable and unstable points of equilibrium* of a pebble or rock fragment, dependence of the *settling velocity* of sediments in water on particle shape, and the physical–mathematical fundamentals of the techniques measuring *surface roughness* and *surface fractal dimension of irregular particles.*

"The esthetic shapes of beach pebbles", as Dobkins and Folk remarked in their landmark 1970 paper, "have an irresistible fascination for sensitive mankind" (Krynine 1960; Ball 2006; Durian et al. 2007; Roussillon et al. 2009; Szabó and Domokos 2010; Hill 2022). This fascination dates back to Newton (1642–1727) who is said to have been uttered a " little before he died" (Brewster 1965 vol. 2: 407)

> I do not know what I may appear to the world, but to myself I seem to have been only like a boy playing on the seashore, and diverting myself in now and then finding a smoother pebble or a prettier shell than ordinary, whilst the great ocean of truth lay all undiscovered before me.

Pascal (1623–1662) wrote «*Le moindre mouvement importe à toute la nature, la mer entière change pour une pierre*» ("The least movement is of importance to all nature. The entire ocean is affected by a pebble"), and even Aristotle (d. 322 BC) devoted *Question 15* of his *Mechanical Problems* to the roundness of pebbles:

> Why are the stones on the seashore which are called pebbles round, when they are originally made from long stones and shells? Surely it is because in movement what is further from the middle moves more rapidly. For the middle is the centre, and the distance from this is the radius. And from an equal movement the greater radius describes a greater circle. But that which travels a greater distance in an equal time describes a greater circle. Things travelling with a greater velocity over a greater distance strike harder; and things which strike harder are themselves struck harder. So that the parts further from the middle must always get worn down. As this happens to them they become round.

G. Korvin, *Statistical Rock Physics*, Earth and Environmental Sciences Library,
https://doi.org/10.1007/978-3-031-46700-4_7

> In the case of pebbles, owing to the movement of the sea and the fact that they are moving with the sea, they are perpetually in motion and are liable to friction as they roll. But this must occur most of all at their extremities (Aristotle 1936. Transl. by W.S. Hett.)

The shape of sedimentary particles is an important *geometrical property*, which provides information about their geologic history and the depositional environment (Graham and Midgley 2000), and determines the physical properties of the whole rock consisting of myriads of such particles (see Table 7.1 about the diverse uses of particle shape indexes).

In *Nature News* (published online on July 14, 2006) Philip Ball wrote "A seaside conundrum has been solved: what shape is a pebble?", reporting on the important result of Doug Durian of the University of Pennsylvania and his colleagues (Durian et al. 2006) who introduced the *probability distribution of local suface curvature* as a new descriptor of pebble shape (see Table 7.2, Eq. 7.7). Ball paraphrased Durian's message as "a trained geologist might be able to tell how a pebble was made", actually, in (Durian et al. 2006) we read

> The characterization of such eroded forms is of obvious importance in the subfield of geology known as sedimentology where pebbles are considered as witness to the geological conditions under which they were formed. It would therefore be of great practical consequence if there existed a method to decipher the message imprinted in a pebble shape, a method that would distinguish beach vs glacier vs river erosion, that would tell how far has a pebble traveled down a stream and perhaps even for how long it has been subjected to erosion forces.

These words bring back to me the bittersweet memories of 1972, when I spent a Semester at the University of Houston as an IREX Research Fellow, working with geophysicist Milton Dobrin (1915–1980) on seismic wave propagation, and sedimentologist Dewitt Van Siclen (1918–2001) on the physics of sedimentary rocks. "Van" took me to a field trip to the Galvestone beaches (Babe's Beach, Stewart Beach, Crystal Beach, Seawolf Park, ...) but prior to that he showed me, in his office, his collection of beach pebbles, and explained how their form and surface texture mirror the oceanological conditions controling a particular beach.

To scientifically deal with Aristotle's question "Why are the stones on the seashore which are called pebbles *round*, when they are originally made from *long* stones and shells?", we must have well-defined measures of the degree of *roundedness* of an irregular object (or the *circularity* of its silhouette), of its *elongation*, *width*, and *flatness*. In Table 7.2 we give a (far from exhaustive) list of the most common shape measures. We honor the authors' notations, and variably use $a \geq b \geq c$ or $L \geq I \geq S$ for the lengths of the *long*, *intermediate*, and *short* axes.

The determination of these three basic parameters (*length, width, thickness*) is not a trivial task for irregular shapes (such as "free gold" particles, McGrath et al. 2015, or pores, Schmitt et al. 2016). They must be perpendicular to each other but do not need to intersect at a common point (Blott and Pye 2008). There are two methods to measure L, I, and S of 3-D irregular objects: by means of the *bounding-box* (BB) or the *Feret caliper* (FC) geometries (Schmitt et al. 2016). In *bounding-box analysis*, the particle's system of coordinates is parallel to that of the whole 3-D volume of interest, that is L, I, and S are measured in the directions of the X, Y, and Z axes. The *Feret*

Table 7.1 The use of particle shape

Topic	Author
Adsorption	Ghanbarian (2019), Gu et al. (2019), Jaroniec and Kruk (1997), Liu and Nie 2016, Neimark (Neimark 1990a, b, c), Neimark and Unger (1993), Pfeifer and Avnir (1983a; b), Wang et al. (2007, 2019), Yang et al. (2014)
Asteroids	Domokos et al. (2009a, b)
Bifractal particles	Orford and Whalley (1983), Korvin (1992)
Bituminous coals	Zhu et al. (2019)
Capillary pressure	Deinert et al. (2008)
Carbon electrodes	Pyun and Rhee (2004)
Compressibility of coal	Guo et al. (2014), Zhang et al.(2018)
Drag coefficient	Chitra et al. (2010)
Fluvial transport	Attal and Lavé (2009)
Fragment shape	Cheng and Redner (1988), Domokos et al. (2015); Grady (2010), Inaoka et al. (1997), Kun and Herrmann (1999), Le Pen et al. (2013), Moss et al. (1973), Perfect (1997), Wittel et al. (2006)
Gold particles	McGrath et al. (2015)
Grain shape	Graham and Midgley (2000), Oakey et al. (2005), Willetts and Rice (1983)
Granular flow	Arabnia and Sklar (2016)
Image based fractal analysis	Peng et al. (2011), Sarkar and Chaudhuri (1992, 1994), Shi et al. (2018), Zhou et al. (2020)
Ion conductivity	Wiedenmann et al. (2013)
Kaolin surface fractal dimension	Sokołowska and Sokołowski (1999)
Loess	Howarth (2010), Rogers and Smalley (1993), Smalley (1990)
Martian pebbles	Szabó et al. (2015)
MIP (Mercury Intrusion Porosimetry)	Rootare and Prenzlow(1967), Zhang and Li (1995)
Modeling fractal porosity	Sun and Koch (1998)
Multifractal analysis of coal	Li (2015), Li et al. (2015), Zhu et al. (2015)
Multifractal analysis of soil	Bird et al. (2006), Caniego et al. (2001), Caniego (2003), Ferreiro and Vázquez (2010), Kravchenko et al. (1999), Posadas et al. (2003) San (2010) Ferreiro and Vázquez (2010)
Nitrogen isotherms	El Shafei et al. (2004)
NMR (Nuclear Magnetic Resonance)	Stallmach et al. (2002)
Ooids	Sipos et al. (2018)
Particulate solids	Ghadiri and Zhang (2002)

(continued)

Table 7.1 (continued)

Topic	Author
Pebble shape	Andrews (1999), Ball (2006), Black (1877), Bloore (1977), Bluck (1967), Cailleux (1945), Carr (1969), Cottet (2006), Dobkins and Folk (1970), Domokos and Gibbons (2012, 2018), Domokos et al. (2008, 2010, 2014), Durian et al. (2006, 2007), Firey (1974), Hill (2022), Illenberger (1991), Kaitanen and Strom (1978), Krapivsky and Redner (2006), Krumbein (1941a; b), Krumbein and Griffith (1938), Kuenen (1956, 1959), Landon (1930), Lewin and Brewer (2002), Litty and Schlunegger (2017), Litwin et al. (2014), Novák-Szabó et al. (2018), Palmer (1834), Pyôkàri (1980), Rice and Church (1998), Roussillon et al. (2009), Sipos et al. (2011, 2021), Sneed and Folk (1958), Strutt (Lord Rayleigh) (1942, 1944a; b), Szabó (2013), Szabó and Domokos (2010), Várkonyi and Domokos (2006, 2011), Várkonyi et al. (2008), Wadell (1935), Wald (1990), Wentworth (1922, 1923), Winzer (2013, 2017), Winzer and Hegerfeldt (2021)
Permeability	Pape et al. (1999), Yu and Liu (2004), Zhu et al. (2019)
Pore fractal dimension	Shen and Li (1994), Shen et al. (1995)
Pore morphology	Schmitt et al. (2016)
Pumice clasts	Manga et al. (2011)
Quartz sand	Scott and Smalley (1991), Smalley (1996)
Settling velocity	Baba and Komar (1981), Camenen (2007), Cheng (1997), Corey (1949), Dietrich (1982), Fok-Pun and Komar (1983), Fredsøe and Deigaard (1992), Gabitto and Tsouris (2008), Gerritsen (1978), Gibbs et al. (1971), Goldstein (1929), Graf (1971), Hallermeier (1981), Janke (1966), Komar and Reimers (1978), Krumbein (1942), Le Roux (1992, 1996, 1997), Nielsen (1992), Rubey (1933), Smith and Cheung (2002, 2003)
Shale gas	Cai et al. (2018), Wang et al. (2021), Yin et al. (2017)
Shape of rock particles	Wadell (1932)
SIP (Spectral Induced Polarization)	Öner et al. (2016), Weller et al. (2016)
Static equilibria of stones	Domokos (2015), Domokos and Lángi (2014), Domokos et al. (1994)
Surface fractal dimension	Avnir et al. (1984), Lai et al. (2018a; b), Shao et al. (2017), Yang et al. (2014), Zhang and Weller (2014), Zhao et al. (2017)
Surface fractal dimension of adsorbent	Neimark (1990b)
Surface fractal dimension of coal	Friesen and Mikula (1987), Mahamud and Novo (2008), Ouyang et al. (2016)
Surface fractal dimension of sandstone	Dou et al. (2021)
Surface fractal dimension of soil	Hajnos et al. (2000); Jozefaciuk et al. (2001)

(continued)

Table 7.1 (continued)

Topic	Author
Ventifacts	Laity and Bridges (2009), Várkonyi and Laity (2012)
Water retention	de Gennes (1985a; b), Ghanbarian and Millán (2017), Ghanbarian-Alavijeh and Millán (2009), Huang et al. (2006), Perrier et al. (1996), Tyler and Wheatcraft (1990)

caliper (Feret 1930) of a 3-D particle *in a given direction* is defined as the distance between two parallel tangent planes, normal to that direction, that touch the particle's surface. As this depends on the particle's orientation, measurements are made in all possible directions (most image analysis software use 31 or more directions) to find the particle's *maximal, medium,* and *minimal* Feret calipers. Schmitt et al. (2016) found that for pore bodies the largest Feret caliper is about 10% higher than the "L" value obtained from the bounding-box.

Most formulae in Table 7.2 can be computed in terms of a, b, c (or L, I, S): *Aschenbrenner shape factors* (7.1) and (7.2); *Corey shape factor* (7.6); *Disc-rod index* (7.8); *Ellipticity* (7.9); *Elongation* (7.10); *Equancy* (7.12); *Flatness* (7.13); *Intercept sphericity* (7.14); *modified NASA shape factor* (7.21); *Oblate-Prolate Index* (7.22); *Rod index* (7.25); *Shape entropy* (7.28); different *Shape factors* (7.29–31); *Triangular shape diagram* (7.35; see Fig. 7.1); *Wentworth flatness index* (7.36); and the *Zingg categories* (7.39–40; see Fig. 7.2). Some shape factors are computed from *Area* and *Perimeter* in 2-D, or *Surface* and *Volume* in 3-D: *Circularity* (7.3–4), *NASA shape factors* (7.19–20); *Sphericity* (7.33). In the shape descriptors *Convexity* (7.5); *Equivalent diameter* (7.11); *Morphological roundness* (7.18); *Perimeter ratio* (7.23); *Surface convexity index* (7.34) the object is compared to some idealized shape (convex hull, fitting ellipse, etc.). The measures *Cumulative curvature* (7.7); *Modified Wentworth roundness* (7.17); *Roundness* (7.26); *Wentworth flatness ratio* (7.37); *Wentworth roundness ratio* (7.38) are based on curvature. *The Simplified E-classification* (7.32) is based on L, I, S and on the numbers of stable resp. unstable *points of equilibrium* (Szabó and Domokos 2010). The *Roundness by Fourier transform* (7.27) was proposed by (Diepenbroek et al. 1992; Drevin and Vincent 2002). In this method one Fourier transforms the polar coordinates of 64 points on the particle's silhouette, spaced at equal angular intervals. Distances are measured from the centroid. The weighted sum of the amplitudes of the first 24 coefficients of the Fourier transform is the roundness estimate. To remove dependence on size, the coefficients are normalized by the zeroth coefficient, that is, by the *mean radius* of the rock fragment.

Finally, Table 7.2. mentions two frequently used *visual charts*, that of Krumbein (Eq. 7.15), and of Powers (Eq. 7.24).

Aristotle claimed that "the parts [of stones] further from the middle must always get worn down. As this happens to them they become round". Many mathematical models for the evolving shapes of 2- and 3-dimensional eroding "stones", both purely mathematical models ("*curve-shortening flows*": Gage 1984; Huisken 1984; Andrews 1999; Andrews et al. 2013) and physical models of *frictional abrasion*

Table 7.2 Mathematical and Visual Description of Grain and Pore Size ($a \geq b \geq c$ or $L \geq I \geq S$ are the lengths of the long, intermediate, or short axes)

Name	Eq#	Equation	Author	Notes
Aschenbrenner shape factor	7.1	$\frac{13.392\sqrt[3]{\frac{c^2}{ab}}}{1+(\frac{c}{b})\left(1+\frac{b}{a}\right)+6\sqrt{1+(\frac{c}{b})^2\left(1+\left(\frac{b}{a}\right)^2\right)}}$	Aschenbrenner (1956)	$a \geq b \geq c$
Aschenbrenner shape factor	7.2	LS/I^2	Aschenbrenner (1956)	$L \geq I \geq S$
Circularity	7.3	$\frac{P_s}{2\sqrt{A_S\pi}}$	Roussillon et al. (2009)	P_S perimeter, A_S *area*
Circularity	7.4	$4\pi A/P^2$	Blott and Pye (2008), Novák-Szabó et al. (2018)	A area, P perimeter
Convexity	7.5	A_s/A_{ch}	Roussillon et al. (2009)	A_s area, A_{ch} area of convex hull
Corey shape factor	7.6	$\frac{c}{\sqrt{ab}}$	Corey (1949)	$a \geq b \geq c$
Cumulative curvature	7.7	$f(K) = \int_0^K \rho(K')dK'$	Durian et al. (2006)	Fraction of the perimeter with curvature less than K
Discrod index	7.8	$(L-I)/(L-S)$	Sneed and Folk (1958)	$L \geq I \geq S$
Ellipticity	7.9	$\varepsilon = \left(1 - \frac{b^2}{a^2}\right)^{1/2}$	Winzer and Heger-feldt (2021)	$a \geq b$
Elongation	7.10	I/L	Zingg (1935)	$L \geq I \geq S$
Equivalent diameter	7.11	Diameter of a sphere with equal volume	Cnudde et al. (2011), Van Dalen and Koster2012)	
Equancy	7.12	S/L	Sneed and Folk (1958), Szabó and Domokos (2010)	$L \geq I \geq S$
Flatness	7.13	S/I	Zingg (1935), Szabó and Domokos (2010)	$L \geq I \geq S$
Intercept sphericity	7.14	$\sqrt[3]{bc/a^2}$	Krumbein (1942)	$a \geq b \geq c$

(continued)

Table 7.2 (continued)

Name	Eq#	Equation	Author	Notes
Krumbein's visual chart	7.15		Krumbein (1941a; b)	
Maximum Projection Sphericity	7.16	$\Psi_p = \sqrt[3]{S^2/LI}$	Pyôkàri (1980), Dobkins and Folk (1970)	$L \geq I \geq S$
Modified Wentworth roundness	7.17	$R_{wt} = D_k/D_i$	Pyôkàri (1980), Dobkins and Folk (1970)	D_k diameter of curvature of the sharpest corner, D_i diameter of largest inscribed circle
Morphological roundness	7.18	$r_D = \frac{A_{s \oplus C(0,r_c)\Theta(0,r_c)}}{A_s}$	Drevin and Vincent (2002), Roussillon et al. (2009)	A_S area, $A_{S \oplus C(0,r_c)\Theta C(0,r_c)}$ area after *opening* with a circular *structuring element*
NASA shape factor	7.19	A/V	Liou and Johnson (2006), Johnson et al. (2001)	Surface to volume ratio
NASA shape factor (modified)	7.20	$S_f = AR_g/V$	Wittel et al. (2006), Domokos et al. (2015)	R_g radius of gyration
NASA shape factor (modified for rectangular shapes)	7.21	$S_f = \frac{2\left(\frac{1}{S}+\frac{1}{I}+\frac{1}{L}\right)\sqrt{S^2+I^2+L^2}}{\left(2\sqrt{3}\right)}$	Domokos et al. (2015)	$L \geq I \geq S$
Oblate-Prolate Index	7.22	$\overline{OP} = \frac{10\left(\frac{L-I}{L-S}-0.50\right)}{S/L}$	Pyôkàri (1980), Sneed and Folk (1958)	$L \geq I \geq S$
Perimeter ratio	7.23	$r_p = P_S/P_e$	Roussillon et al. (2009), Cottet (2006)	P_S perimeter, P_e perimeter of best approximating ellipse
Powers' visual scale	7.24	Very Angular, Angular, Sub-Angular, Sub-Rounded, Rounded, Well-Rounded	Powers (1953)	
Rod index	7.25	$(S+I)/L$	Illenberger (1991)	$L \geq I \geq S$

(continued)

Table 7.2 (continued)

Name	Eq#	Equation	Author	Notes
Roundness	7.26	Ratio between mean radius of curvature of the corners and the radius of the largest inscribed circle: $r_W = \frac{1}{kR}\sum_{i=1}^{k} r_i$	Wadell (1932)	
Roundness by Fourier transform	7.27		Diepenbroek et al. (1992)	
Shape entropy	7.28	$-[P_a \ln P_a + P_b \ln P_b + P_c \ln P_c]/1.0986$	Hofmann (1994)	$a \geq b \geq c$; $P_a = \frac{a}{a+b+c}$, etc.; $1.0986 = \ln 3$
Shape factor	7.29	$\sqrt[3]{\frac{c^2}{ab}}$	Sneed and Folk (1958)	$a \geq b \geq c$
Shape factor	7.30	$\frac{c}{\sqrt{\frac{a^2+b^2+c^2}{3}}}$	Janke (1966)	$a \geq b \geq c$
Shape index	7.31	$(L+S)/I$	Illenberger (1991)	$L \geq I \geq S$
Simplified E-classification	7.32	E(I): E-classes {2,U}, U > 2 E(II): E-classes {S,U}, S, U > 2 E(III): E-class {2,2} E(IV): E-classes {S,2}, S > 2	Szabó and Domokos (2010)	S number of stable points of equilibrium; U number of unstable points of equilibrium
Sphericity	7.33	$\gamma = \frac{2}{3}\frac{\sqrt{A_S \pi}}{\sqrt[3]{\frac{16}{9}\pi^2 V}}$	Archilha et al. (2016)	Specific Surface Area normalized to a sphere. A_S surface area, V pore volume
Surface convexity index	7.34	$\beta = A_C/A_H,\ 0 \leq \beta \leq 1$	Domokos et al. (2014)	A_C area of the convex regions, AH total area of the convex hull

(continued)

Table 7.2 (continued)

Name	Eq#	Equation	Author	Notes
Triangular shape classification diagram	7.35	Compact (C), Compact-platy (CP), Compact-bladed (CB), Compact-elongated (CE), Platy (P), Bladed (B), Elongated (E), Very platy (VP), Very bladed (VB), Very elongated (VE)	Sneed and Folk (1958)	See Fig. 7.1
Wentworth flatness index	7.36	$(L+I)/2S$	Wentworth (1922)	$L \geq I \geq S$
Wentworth flatness ratio	7.37	r_2/R	Wentworth (1923)	r_2 largest radius of curvature, R mean radius
Wentworth roundness ratio	7.38	r_1/R	Wentworth (1923)	r_1 smallest radius of curvature, R mean radius
Zingg categories	7.39	class Im $a \leq b-r, b > c-r$ *disc* class IIm $a > b-r, b > c-r$ *sphere* class IIIm $a \leq b-r, b \leq c-r$ *blade* class IVm $a > b-r, b \leq c-r$ *rod*	Howarth (2010)	$a \geq b \geq c$
Zingg diagram with 4 classes	7.40	Disc (I), Sphere (II), Blade (III), Rod (IV)	Zingg (1935)	See Fig. 7.2

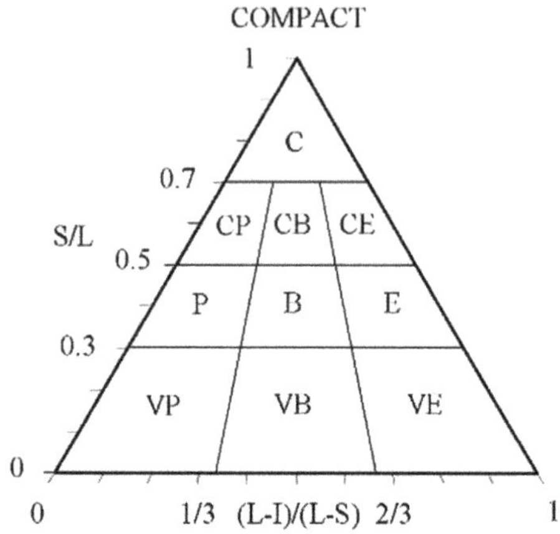

Fig. 7.1 Triangular shape classification diagram (of Sneed and Folk 1958; Table 7.2, Eqs. 7.8; 7.12; 7.35). Compact (C), Compact-platy (CP), Compact-bladed (CB), Compact-elongated (CE), Platy (P), Bladed (B), Elongated (E), Very platy (VP), Very bladed (VB), Very elongated (VE). (After Szabó and Domokos 2010: Fig. 1b; with the kind permission of Prof. Gábor Domokos)

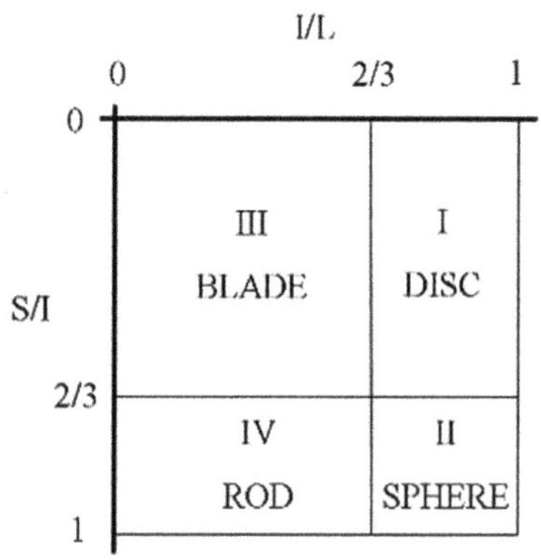

Fig. 7.2 Zingg diagram (Zingg 1935; Table 7.2, Eqs. 7.10; 7.13; 7.40). Disc (I), Sphere (II), Blade (III), Rod (IV). (After Szabó and Domokos 2010: Fig. 1a; with the kind permission of Prof. Gábor Domokos)

(Firey 1974; Bloore 1977) also predicted that the shapes will become spherical in the limit. Regarding *real beach stones*, however, researchers have reported that "pebbles never approach the spherical" (Wald 1990: 211), "one will never find stones in spherical form" (Winzer 2013: 1), and "there is little or no tendency for a pebble of nearly spherical form to get nearer to the sphere" (Strutt 1944b: 169). In fact, Landon (1930: 437) documented that "round pebbles become flat" and Rayleigh (Strutt 1942: 114) observed "a tendency to change away from a sphere" i.e., in the parlance of dynamic systems, that spheres are in unstable (repelling) equilibrium. Real beach stones typically have elongated oval shapes. Concerning these shapes Black (1877) wrote that this "ovoid shape seems to be taken by all sorts of stones, from the soft sandstone to the hard quartzite, and may therefore be independent of mineral composition, or relative hardness of the stone", while, according to Drake (1970), Litty and Schlunegger (2017) the pebbles derived from isotropic sedimentary rocks are "usually more spherical than the ones from anisotropic metamorphic rocks". Winzer and Hegerfeldt (2021) mathematically proved that "an elliptical pebble on an ocean beach will never become disk-like or spherical. This contradicts Aristotle's statement that the abrasion process on beaches ultimately leads to spheres."

Translating to modern terms, Aristotle stated that *erosion is more rapid at regions farther from the center of the stone, where greater impulses can be more readily delivered since the moment arm is longer* (Ball 2006; Durian et al. 2006; Winzer and Hegerfeldt 2021; Hill 2022). Mathematically (Domokos and Gibbons 2012: Eq. 1.1, 2018: Eq. 1), the inward rate of abrasion at time t and in unit direction u is an increasing function $f(h)$ of the distance $h(t, u)$ from the center of mass of the stone to the tangent plane in that direction:

$$\frac{\partial h}{\partial t} = -f(h) \tag{7.41}$$

This is not a conventional *partial differential equation*, because the center of mass is continuously changing and is determined by a time-variant integral, but a *non-local partial integro-differential equation* (Kavallaris and Suzuki 2018). In addition to this difficulty, we do not know the functional form of $f(h)$. It is more reasonable to assume, (i) as Aristotle did, that the ablation is normal to the surface of the stone, but unlike in Aristotle's model, (ii) its rate is *proportional to the curvature at the point of contact*, rather than being a function of the moment arm. (iii) Let us also assume that the shape of the stone, $\mathcal{S}$, is convex and (iv) the abrasion is *isotropic*, i.e. at all times each point on the surface of the stone is equally likely to get in contact with some abrasive plane. Taking the constant of proportionality in (ii) to be 1, we get Firey's (1974) PDE (Partial Differential Equation)

$$\frac{\partial h_0}{\partial t} = -\kappa(t, u) \tag{7.42}$$

where $h_0 = h_0(u, t)$ is the distance from a fixed origin in unit direction u and time t, and $\kappa(t, u)$ is the Gaussian curvature at time and unit direction (t, u). The functions h in Eq. (7.41) and h_0 in (7.42) are the *support functions* of the convex shape $\mathcal{S}$ taken from the *barycentric origin* (in case of h), and from a *fixed origin* (in case of h_0), respectively. The limiting support function h_0 under the *curve-shortening dynamic* flow (7.42) is constant for all reasonably smooth convex initial shapes $\mathcal{S}$ (Firey 1974; Andrews 1998, 2002; Fehér et al. 2020). Support functions uniquely determine convex bodies (Ghosh and Kumar 1998), and since spheres are the only convex bodies with constant support functions (with the origin at the center), the shape $\mathcal{S}$ of a convex stone eroding according to Eq. (7.42) becomes spherical in the limit.

Lord Rayleigh, son of the Noble Prize winner physicist, observed (Strutt 1942, 1944a; b) that the ultimate shape of pebbles is usually *not spherical*, sometimes not even *ellipsoidal*, but rather *discoid*. He warned that the Gauss curvature in itself is not sufficient to explain abrasion.

Firey's (1974) Partial Differential Eq. (7.42) modeled the shape evolution of a convex particle abraded by repeated collisions with an *infinitely large abrader* (i.e.

a flat plane). In his model, abrasion occurs in the direction normal to the surface at a rate $v = \frac{\partial h_0}{\partial t}$ which is proportional to the *Gaussian curvature*, $v \propto \kappa(t, u) = \boldsymbol{K}$. Bloore (1977) generalized this model for *finite-size abraders*; in 3-D the equation is:

$$v = 1 + 2fH + g\mathbf{K} \tag{7.43}$$

where H is the *mean curvature*, f and g are the *integrated mean curvature* and *surface area* of the abrader, respectively (Várkonyi and Domokos 2011; Szabó 2013; Novák-Szabó et al. 2018). *Mean curvature* is the average of the two principal curvatures κ_1 and κ_2, while the *Gaussian curvature* is their product:; $H = (\kappa_1 + \kappa_2)/2$; $K = \kappa_1\kappa_2$. In the limit of an infinitely large abrader, Eq. (7.43) reduces to Eq. (7.42) and the limiting shape is a sphere.

Bloore (1977) showed that a spherical shape of radius R abraded by an abrader of radius r is stable under the action of (7.43) if

$$\frac{R}{r} < 3 \tag{7.44}$$

In the general case, as Szabó (2013) noted, "[in 3-D] very little is known about analytical solutions of Bloore's PDE [(7.43)], in particular, it is not known whether it contains non-spherical, nontrivial shapes as stable attractors. Due to the abundance of initial shapes, [its] global, numerical investigation is still out of reach."

In *two dimensions* (Novák-Szabó et al. 2018) Bloore's equation reads

$$v = 1 + c\kappa \tag{7.45}$$

where κ is the local curvature of the eroding pebble and c is a typical perimeter of the impactor particles. In the limit of large impactors,

$$v \approx c\kappa \tag{7.46}$$

which is Firey's Eq. (7.42). Partial Differential Equations similar to (7.42) and (7.46) have been very thoroughly studied (Firey 1974; Gage 1983; Gage and Hamilton 1986; Grayson 1987) because attempts at the proof of Poincaré's conjecture (Perelman 2003) were based on such *curvature-driven evolution equations*. Using this arsenal of mathematical tools, Novák-Szabó et al. (*op. cit.*: "Methods for details") could sketch how *circularity* $R = 4\pi A/P^2$ (Table 7.2, Eq. 7.3) would generally increase with *relative mass loss* $\mu = \delta m/m_0$ (where δm is the eroded mass, m_0 the initial mass of a particle).

Further generalizations of Bloore's equation can be found in (Domokos and Gibbons 2012, 2018; Domokos et al. 2014; Hill 2022; Novák-Szabó et al. 2018; Sipos et al. 2021; Várkonyi and Domokos 2011; Várkonyi et al. 2008; Winzer 2013, 2017; Winzer and Hegerfeldt 2021).

Bloore's (1977) equation (7.43) described the abrasion of a single particle abraded during random collisions with finite abraders. (The case, when the abrader is an *infinite plane*, was dealt with by Firey 1974). There are two ways to generalize the equation: to describe the mutual abrasion by two interacting particles, and the collective abrasion of stones in a many-particle system. Further generalizations are when the collisions, and rolling of the stones are controlled by a *stochastic driving force*, namely the *random sea level changes* (Hill 2022). These developments are outside the scope of the chapter, and we must content ourselves with just sketching the *deterministic box equation* for binary abrasion (Domokos and Gibbons 2012; Szabó 2013), beause of its interesting use of the shape indices *c/a* (=*S/L*, Table 7.2, Eq. 7.12); and *b/a* (=*I/L*, Table 7.2, Eq. 7.10) to characterize particles, where $a \geq b \geq c(L \geq I \geq S)$ are sides of the *sedimentological bounding box*. Introduce the notations.

$$\frac{c}{a} = \frac{S}{L} = y_1; \frac{b}{a} = \frac{I}{L} = y_2; \frac{a}{2} = \frac{L}{2} = y_3 \tag{7.47}$$

Denote the first particle by $\boldsymbol{y}$, the second by $\boldsymbol{z}$, their mutual interaction is described by the coupled system of ODEs:

$$\dot{\boldsymbol{y}} = \boldsymbol{F}(\boldsymbol{y}, \boldsymbol{z}); \dot{\boldsymbol{z}} = \boldsymbol{F}(\boldsymbol{z}, \boldsymbol{y}), \boldsymbol{y} = (y_1, y_2, y_3), \boldsymbol{z} = (z_1, z_2, z_3) \tag{7.48.a}$$

where the coordinates are the box ratios and the semi-major axes, as in Eq. (7.47). The function $\boldsymbol{F} = \left(F_1\ F_2\ F_3\right)^T$ is defined as (Domokos and Gibbons 2012; Szabó 2013):

$$\dot{y}_i = F_i(y_1, y_2, y_3, z_1, z_2, z_3) = \frac{F_i^E}{y_3} + 2A\frac{F_i^M}{y_3^2} + B\frac{F_i^G}{y_3^3} \tag{7.48.b}$$

$$\dot{y}_3 = F_3(y_1, y_2, y_3, z_1, z_2, z_3) = -1 - \frac{A}{y_3}\frac{y_1^2 + y_2^2}{y_1^2 \cdot y_2^2} - \frac{B}{y_3^2}\frac{1}{y_1^2 \cdot y_2^2} \tag{7.48.c}$$

where

$$F_i^E = y_i - 1, F_i^M = \frac{1 - y_i^2}{2y_i}, F_i^G = \frac{1 - y_i^3}{y_i y_j^2}\ \mathit{for}\ i, j = 1, 2;\ i \neq j \tag{7.48.d}$$

and

$$A = \frac{z_3(z_1 + z_2 + 1)}{3}, B = \frac{z_3^2(z_1 + z_2 + z_1 z_2)}{3} \tag{7.48.e}$$

In the expression of the *shape entropy*

$$H = -[P_a \ln P_a + P_b \ln P_b + P_c \ln P_c]/1.0986$$

(Table 7.2, Eq. 7.28), $P_a = \frac{a}{a+b+c}$, $P_b = \frac{b}{a+b+c}$, $P_c = \frac{c}{a+b+c}$, and the normalizing factor $1.0986 = \ln 3$ has been selected to make $H = 1$ for a sphere. This shape index has been proposed to estimate the *settling velocity* of an ellipsoidal particle in fluid (Hofmann 1994; Le Roux 1996, 1997; Chitra et al. 2010).

The established relationship for the *settling velocity* of *spherical particles* in a viscous fluid is *Stokes Law*,

$$v_s\left(\frac{m}{s}\right) = \frac{2}{9} \cdot \frac{(\rho_p - \rho_f)gR^3}{\mu} \tag{7.49}$$

where $g\left(\frac{m}{s^2}\right)$ is gravitational acceleration; ρ_p and ρ_f are densities of the particle and the fluid (kg/m^3), respectively, R (m) radius, μ dynamic viscosity $(kg/m\,s)$ (Lamb 1994: 599). For *non-spherical particles*, the *settling velocity* v_s is a complicated function of both particle shape and *Reynolds number* (Le Roux 1992, 1996, 1997; Chitra et al. 2010). Le Roux (1996, 1997) compared the different published sphericity indices and concluded the *shape entropy* of Hofmann (1994) is best suited for predicting settling velocity of ellipsoidal grains through a linear regression

$$v_{s,ellipsoid} \approx v_{s,sphere} \times H + const \tag{7.50}$$

where H is the ellipsoid's *Hoffman shape-entropy* (Table 7.2, Eq. 7.28). According to Le Roux (1996: 15) "velocities predicted from this equation are more accurate than those based on the *Corey* or *Janke shape factors*, with the main advantage being that it can be employed at Reynold numbers of up to 50,000 or more. The method is applicable to ellipsoidal particles of any density settling in various fluids". (The other two referred shape factors are Eqs. 7.6 and 7.30 in Table 7.2).

Table 7.2 contains an interesting, and innovative shape descriptor for irregular convex bodies: the *simplified E-classification* (Eq. 7.32) based on the number S of *stable points of equilibrium*, and the number U of *unstable points of equilibrium* of the body. A typical convex body has three types of equilibria: of the *stable, unstable* and *saddle* types, and the *simplified E-classification* is based on the number of these equilibrium points (Várkonyi and Domokos 2006; Domokos et al. 1994, 2010; Szabó 2013; Domokos and Lángi 2014).

Stable equilibria are points on the surface where the body stays at rest on a horizontal, frictionless plane, and *will return there after a small perturbation.* A cube, for example, has 6 stable equilibria, on its 6 faces. At *unstable points of equilibria* the body appears as balanced when placed on a horizontal surface, but it does not return there after small perturbations. There are two kinds of unstable equilibria: *unstable points* and *saddles*. For example, the cube has 8 unstable points (its 8 vertices) and 12 saddle-type equilibria (the 12 edges).

Denote the number of stable equilibrium points by S, the number of unstable equilibria by U and the number of saddle-type equilibria by H. According to the *Poincaré-Hopf Theorem* (Arnold 1998) we have

Table 7.3 Distribution of convex pebbles collected on the coast of Rhodes, Greece, among the $\{S, U\}$ classes

S	U				
	1	2	3	4	5
1	–	–	–	–	–
2	$0.1 \pm 0.2\%$	$74.5 \pm 2.2\%$	$17.0 \pm 1.9\%$	$1.0 \pm 0.8\%$	$0.1 \pm 0.2\%$
3	–	$4.5 \pm 2.2\%$	$0.4 \pm 0.4\%$	$0.1 \pm 0.2\%$	–
4	–	$0.2 \pm 0.3\%$	–	$0.7 \pm 0.5\%$	$0.1 \pm 0.2\%$

After Várkonyi and Domokos (2006: Table 1; their. Figure 14b shows some of the pebbles involved in the analysis.) With kind permission of Prof. G. Domokos

$$S + U - H = 2 \tag{7.51}$$

that is S and U uniquely determine H. The class $\{S, U\}$ (Várkonyi and Domokos 2006) contains all convex bodies with S stable, and U unstable points of equilibria. If the shape of a crystal is a *minimal polyhedron* with F faces, V vertices and E edges then $S = F, U = V, H = E$ and the *Poincaré-Hopf relationship* is the same as Euler's formula for polyhedra

$$F + V - E = 2 \tag{7.52}$$

(see Sect. 3.2, Table 3.1, Eq. 6.b). For example, the cubical *halite* (*rock salt*, NaCl) crystals (Aquilano et al. 2016: Fig. 1) belongs to the class {6,8} with $S = 6$ stable and $U = 8$ unstable equilibria.

Várkonyi and Domokos (2006) have shown that all classes $\{S, U\}$ exist and can be physically realized. A fascinating spin-off of their research has been the discovery of the celebrated *Gömböc* (Hungarian, ~"spherule, roly-poly") shape which belongs to class {1, 1} and has 1 stable, and 1 unstable points of equilibrium. The tri-axial ellipsoid belongs to class {2, 2}.

The *simplified E-classification* system of Szabó and Domokos (2010) is based on Zingg's categories and the $\{S, U\}$ classes. Flat pebbless typically belong to simplified E-class E(I); elongated ones to the class E(IV); flat and elongated ones belong to E(III); pebbles neither flat nor elongated usually belong to E(II) (see Table 7.3).

In sediment transport, particles are subjected to three different attrition mechanisms, in order of increasing energy: *frictional abrasion*, *chippping*, and *fragmentation* (Novák-Szabó et al. 2018). The *size distributions* provided by the various *fragmentation models* were summarized in (Turcotte 1986, 1997; Korvin 1989, 1992). For the present treatment we need to recall the *exponential fragment size distribution* of (Gilvarry 1961), and the formula for fractal fragmentation in D-dimensions (Åström et al. 2004). Gilvarry assumed that in D dimensions, under sufficiently large external load the pre-existing flaws within a fragile D-dimensional body get activated and they initiate cracks of dimension $D - 1$, which propagate and merge to form *fragments*. The probability of finding a fragment of size v is proportional to the probability of *not finding* an activated flaw within the same volume. If flaws are uncorrelated and obey *Poisson distribution*, we obtain exponential size distribution

$$dn_1(v) = V_0 \gamma_v \exp(-\gamma_v) dv \tag{7.53}$$

for fragments of linear size l and volume $v \propto l^D$.

Åström et al. (2004) realized that nearby cracks *attract each other* in the sense that stress intensity in a thin strip of width $\in$ between two cracks diverges as $1/\in$. Adjacent branches would eventually *merge* and a single branch continues to propagate forward from each merging point. There arises a "second generation" of branches that continue to propagate and attract each other. A decreasing number of branches are created in each new "generation", and ever larger fragments are created by the mergings. If the main crack produces n_b "first-generation" branches with an average mutual distance of l_b, these will merge to form $n_b/2^{D-1}$ fragments of size $(l_b)^D$. Mergings of the farther propagating "second-generation" branches will produce $n_b/4^{D-1}$ fragments of size $(2l_b)^D$, and this process continues as a geometric series. This leads to the "*fractal*" or "*power function*" fragment-size distribution

$$dn_2(v) = n_b v^{-(2D-1)/D} dv \tag{7.54}$$

Because of *elastic relaxation* and *energy dissipation*, the branching/merging process is limited to a finite range, so we might asume the fragment size distribution will be power-like for small fragments, exponential for large ones.

Domokos et al. (2015) turned their attention to a much less studied question, *what is the characteristic fragment shape, and how are its shape descriptors distributed*? As they wrote, "Although the statistics of the mass of fragments has been found to show a universal scaling behavior, the comprehensive characterization of fragment shapes still remained a fundamental challenge." As they experimentally observed that the freshly created fragments are polyhedra with sharp edges and corners, they used the same *bounding-box measures* $L \geq I \geq S$ as for pebbles. They used the value of L to measure the fragments' linear extension, and the dimensionless ratios S/L and I/L to characterize the shape of the pieces. They experimentally established a universal *shape vs. size relation* shown in Eqs. (7.55a, b).

For small values of L the shape parameters S/L and I/L approach the value one that is small-sized fragments are nearly isotropic with $L \approx I \approx S$. As fragment size L increases, both S/L and I/L decrease and exponentially tend to finite constant values B_S and B_I, respectively:

$$\left.\begin{aligned} S/L &= A_S \exp[-(L - L_c)/L_0] + B_S \\ I/L &= A_I \exp[-(L - L_c)/L_0] + B_I \end{aligned}\right\} \tag{7.55a,b}$$

The best fit was obtained with the parameters $A_S = 0.45$; $A_I = 0.4$; $B_S = 0.43$; $B_I = 0.67$; $L_0 = 7.2$ mm; $L_c = 17.0$ mm. The results imply that for large fragments with $L \gg L_c + L_0 \approx 25$ mm the average fragment has an elongated shape with $S/L \approx B_S$; $I/L \approx B_I$, that is its bounding-box dimensions satisfy.

$$S : I : L \approx 1 : 1.56 : 2.33 \tag{7.56}$$

Domokos et al. (2015) also studied the mass distribution of the fragments. They found a *power law*,

$$p(m) \propto m^{-\tau} \tag{7.57}$$

with $\tau = 1.7 \pm 0.06$. The value of the exponent τ is close to the theoretical $\tau = \frac{2D-1}{D} = \frac{5}{3}$ of (Åström et al. 2004), see Eq. (7.54).

Readers familiar with *fractal geometry* will recall Mandelbrot's *area-perimeter* rule for fractal particle silhouettes embedded in 2-D, and his similar *volume-surface* rule for fractal particles embedded in 3-D (Mandelbrot 1982; Lovejoy 1982; Korvin 1992; Turcotte 1997). By these rules A and P (in 2-D), as well as V and S in 3-D, satisfy the scaling relations

$$P \propto \left(\sqrt{A}\right)^{D_f} \tag{7.58}$$

and

$$S \propto \left(\sqrt[3]{V}\right)^{D_f} \tag{7.59}$$

where in (7.58) $D_f \in [1, 2)$ is the *fractal dimension* of the particles' silhouette, in (7.59) $D_f \in [2, 3)$ is the *fractal surface dimension* of the particles. Because of Mandelbrot's rules some of the roundedness indices in Table 7.2 are not *scale-free*, but they scale with size and resolution. Such scale factors are: *circularity* (7.3) $\frac{P_s}{2\sqrt{A_S\pi}}$; *circularity* (7.4) $4\pi A/P^2$; *NASA shape factor* A/V; *modified NASA shape factor* $S_f = AR_g/V$; and *sphericity* $\gamma = \frac{2}{3}\frac{\sqrt{A_S\pi}}{\sqrt[3]{\frac{16}{9}\pi^2 V}}$.

For irregular particles with rough surface, their *surface fractal dimension* $D_s > 2$ is also an important shape descriptor. The *specific internal surface* of a porous rock with fractally rough pore walls, $S_{\text{spec}} :=$ (total pore surface area)/(total rock volume) will depend on resolution λ and linear rock size L. Fixing L, two specific surface values determined at different resolutions λ_1 and λ_2, respectively, are related as

$$S_{por}(\lambda_1) = S_{por}(\lambda_2)\left(\frac{\lambda_1}{\lambda_2}\right)^{E-D_s} \tag{7.60}$$

(Pape et al. 1999), where $E = 2$ *or* 3 is the Euclidean dimension. (As well known, see Chap. 1, the *specific internal surface* is an important input parameter for the *Kozeny-Carman equation* to compute *permeability*.)

There are many experimental techniques to determine the surface fractal dimension of porous materials, such as *nitrogen adsorption* (NA); *small-angle scattering* of X-rays (SAXS) or *neutrons* (SANS); *image processing* of scanning electronic microscopy (SEM) or *Micro X-ray computed tomography* data; *mercury intrusion porosimetry* (MIP), and some others. SAXS, SANS and NA cover the range of 1–100 nm; SEM images capture pore geometry in range of >10 μm; MIP can cover a large range of 1nm–1mm. (For example, in coal, we distinguish <10 nm *micropores*, *transition pores* between 10 and 100 nm, *mesopores* between 100 and 1000 nm, and >1000 nm *macropores*, see Zhang et al. 2018). The classic SAXS and SANS results were already discussed in (Korvin 1992). For recent developments in the other techniques see: in *Mercury Intrusion Porosimetry* (Li 2010); in *Micro X-ray computed tomography* (Zhao et al. 2017); in N_2 *adsorption and desorption* (Neimark 1990a, b, c); in *NMR* (Zhang and Weller 2014; Zhao et al. 2017; Lai et al. 2018a); in *Spectral Induced Polarization* Weller et al. 2016).

This is not the place to discuss the details of these techniques, their relative merits or rules of interpretation. We shall only briefly consider, how the fractal dimension of the pore surface appears in the different kinds of measurements.

First, following (Dou et al. 2021), consider some recent ideas how *Mercury Intrusion Porosimetry* (MIP) experts have attempted to find the surface fractal dimension of the pore space.

1. The *Brooks–Corey capillary pressure model* (Li 2010). The number of capillary tubes of radius r and length l filled with mercury is

$$N(r) = \frac{V_{Hg}}{\pi r^2 l} \tag{7.61}$$

where V_{Hg} is cumulative volume of mercury at a given pressure, tortuosity being neglected. The radius scales as

$$N(r) \propto r^{-D} \tag{7.62}$$

capillary pressure and tube radius are connected by

$$P_c = \frac{2\sigma cos\theta}{r} \tag{7.63}$$

where P_c is capillary pressure, σ is surface tension, and θ the contact angle. From Eqs. (7.61–7.63) follows

$$N(r) \propto P_c^{-(2-D)} \tag{7.64}$$

The *mercury saturation* is

$$V_S = V_{Hg}/V_P \tag{7.65}$$

where V_P is the pore volume of the core. Easy algebra gives

$$V_S = aP_c^{-(2-D)} \tag{7.66}$$

with a a constant, and D is obtained from the slope of the double logarithmic plot of the Hg saturation vs. pressure curve. Note that D is the exponent figuring in Eq. (7.62), *not* the surface fractal dimension.

2. The *Friesen and Mikula model* (1987, also discussed in Korvin 1992) simulated the pore space with a *Menger sponge model* (see Chap. 5) and derived a relation for calculating the surface fractal dimension

$$D : \log\left[-\frac{dV}{dr}\right] \propto (2-D)\log r \tag{7.67}$$

The relation between r and P is given by the *Laplace equation*

$$r = 2\sigma\cos\theta/P \tag{7.68}$$

so that Eq. (7.67) can be rewritten as:

$$\log\left[\frac{dV}{dr}\right] \propto (D-4)\log P \tag{7.69}$$

which provides the surface fractal dimension. Similar consideration is found in the *Pfeifer and Avnir* (1983a; b) *model.*

3. *Neimark's model* (Neimark 1990a, b, c; Dou et al. 2021) is based on thermodynamic considerations combined with the equations of Rootare and Prenzlow (1967) for the average curvature of *meniscus* at the mercury-sample interface. The surface area of the solid medium, $S(P)$, measured with spheres of radius r is given as:

$$S(P) = (1/\sigma\cos\theta)\int_0^V P\,dV \tag{7.70}$$

where P is pressure, σ the surface tension of the liquid, θ wetting angle, V is the accumulative mercury volume intruded into the sample, and r and P are connected through the Laplace equation (7.68). From Eqs. (7.70) and (7.68) the surface fractal dimension is calculated as

$$D = 2 + \frac{d[\log S(P)]}{d\log P} \tag{7.71}$$

4. *Zhang and Li's model* (1995) also used thermodynamic principles. They computed the following cumulative integrals through subsequent stages of mercury injection:

$$\sum_{i=1}^{n} \overline{P}_i \Delta V_i = -\gamma_L \cos\theta (S_E)_n \tag{7.72}$$

where $\overline{P}_i$ is average pressure, ΔV_i the intruded Hg volume, γ_L the surface free energy of liquid mercury *in vacuo*, and $(S_E)_n$ is the pore surface area. Aplying Mandelbrot's rule (7.59) to these surface areas we get the scaling relation.

$$\sum_{i=1}^{n} \overline{P}_i \Delta V_i = C' r_n^{2-D} V_n^{D/3} \tag{7.73}$$

Thus, D can be found from computing the two sequences $Q_n = r_n^{2-D} V_n^{D/3}$ and $W_n = \sum_{i=1}^{n} \overline{P}_i \Delta V_i$ for $n = 1, 2, \ldots,$ for, and plotting $\log W_n$ against $\log Q_n$.

5. In the *Shen and Li model* (Shen and Li 1994; Shen et al. 1995) pores are capillary tubes with radius r and length l. At a certain pressure P the number of such units filled by an amount $V_{HG}(P)$ of mercury is

$$N(r) = V_{Hg}(P)/\pi r^2 l \tag{7.74}$$

If the pore network is fractal,

$$N(r) \propto r^{-D} \tag{7.75}$$

and the fractal dimension is found by plotting $\log[N(r)]$ against $\log[V_{Hg}(P)]$.

6. *Li's Model* (2004, 2010)

From Eqs. (7.74) and (7.75)

$$V_{Hg} \propto r^{2-D} \tag{7.76}$$

In MIP, mercury saturation is defined as

$$S_{Hg} = V_{Hg}/V_p \tag{7.77}$$

where V_p is total pore volume in the core. Combining Eqs. (7.76 and 7.77) with the Laplace equation (7.68) yields

$$S_{Hg} \propto P^{-(2-D)} \tag{7.78}$$

7. *Zhang and Weller's model* (Zhang and Weller 2014)

The pore volume $V(r)$ within a fixed radius r contains

$$N(r) = V(r)/\pi r^2 l \tag{7.79}$$

capillary tubes of radius r, length l. If, for simplicity,

$$l = r, V(r) = N\pi r^3 = C\pi r^{3-D} \tag{7.80}$$

that is the differential pore volume distribution is

$$\frac{dV(r)}{dr} = C(3 - D)\pi r^{2-D} \tag{7.81}$$

The cumulative volume of pores with radii less than r is obtained by integrating:

$$V(< r) = C\pi \left(r^{3-D} - r_{\min}^{3-D}\right) \tag{7.82}$$

The *cumulative volume fraction* of pores with radii less than r is

$$V_c = \frac{V(c < r)}{V(c < r_{\max})} = \left(r^{3-D} - r_{\min^{3-D}}\right) / \left(r_{\max^{3-D}} - r_{\min^{3-D}}\right) \tag{7.83}$$

Assuming $r_{\min} \ll r$ and making use of the Laplace equation (7.68),

$$V_c = (P/P_{\min})^{D-3} \tag{7.84}$$

Zhang and Weller (2014) investigated the fractal behavior of the pore space of an Eocene sandstone formation using a combination of *capillary pressure curves* and *nuclear magnetic resonance* (NMR, see Coates et al. 1999; Dunn et al. 2002). The fractal dimension of the pore surface was determined based on the specific surface area S_{spec} (surface area per unit pore volume). They considered the *transverse relaxation time* T_2 describing the decrease of magnetization perpendicular to the static magnetic field,

$$B(t) = B_0 \cdot \sum_{j=1}^{n} b_j \exp\left(\frac{-t}{T_{2j}}\right) \tag{7.85}$$

where b_j is the signal amplitude corresponding to the j-th relaxation time T_{2j}. Relaxation times are related to surface-to-volume ratios as

$$\frac{1}{T_2} = \rho \cdot \left(\frac{A}{V}\right) \tag{7.86}$$

where ρ is surface relaxivity, assumed constant. For *total surface area* and *total volume of the pore space, A/V* is called *specific surface area per unit pore volume.*

$$S_{spec} = A/V \tag{7.87}$$

For a simple *capillary tube model* with straight cylindrical pores of radius r, we get

$$S_{por} = 2/r \tag{7.88}$$

From Eqs. (7.86–7.88)

$$T_2 \propto r \tag{7.89}$$

so that the cumulative volume fraction can be expressed in terms of T_2 as

$$V_c = \frac{V(c < r)}{V(c < r_{\max})} = \frac{\left(r^{3-D} - r_{\min^{3-D}}\right)}{\left(r_{\max^{3-D}} - r_{\min^{3-D}}\right)} = \frac{T_2^{3-D} - T_{2\min}^{3-D}}{T_{2\max}^{3-D} - T_{2\min}^{3-D}} \tag{7.90}$$

Neglecting $T_{2\min}$,

$$V_c = \frac{V(c < r)}{V(c < r_{\max})} = \frac{V(c < 2\rho T_2)}{V(c < 2\rho T_{2\max})} \propto \frac{T_2^{3-D}}{T_{2\max}^{3-D}} \tag{7.91}$$

wherefrom $D = D_{NMR}$ is found.

NMR and MIP are frequently combined in coal studies (Ouyang et al. 2016). The pore-fracture system of coal is modeled as a collection of spherical tubes of size r, their cumulative porosity is defined as

$$V_p = \frac{V(< r)}{V} == \frac{r^{3-D} - r_{\min}^{3-D}}{r_{\max}^{3-D} - r_{\min}^{3-D}} \approx \left(\frac{r}{r_{\max}}\right)^{3-D} \tag{7.92}$$

By Washburn's (1921) equation capillary pressure P_c and pore size r are related as

$$P_c = 2\sigma \frac{\cos\theta}{r} \tag{7.93}$$

that is

$$V_p = \left(\frac{P_c}{P_{\max}}\right)^{D-3} \tag{7.94}$$

By Eq. (7.86) the NMR transverse relaxation time T_2 satisfies

$$\frac{1}{T_2} = \rho\frac{A}{V} = F_S\frac{\rho}{r} \tag{7.95}$$

where ρ is surface relaxivity and F_S a geometric factor. Equations (7.94 and 7.95) establish a relation between T_2 and P_c, so that, with a proper constant factor,

$$P_{c\min} = C \cdot T_{2\max}^{-1} \tag{7.96}$$

Combining this with relation (7.94), we get

$$V_p = \left(\frac{T_{2\max}}{T_2}\right)^{D-3} \tag{7.97}$$

that yields the fractal dimension D.

There exist a wide range of *adsorption techniques* to determine pore surface fractal dimension D_s. Their common theoretical basis can be understood by studying the relation between D_s and the adsorbed water content, θ_{ads} (Ghanbarian 2019). The thickness of liquid films adsorbed on the pore/solid interface of a water-wet porous rock depends on (i) *van der Waals forces*, responsible for molecular interactions; (ii) *electrostatic forces*; and (iii) *hydration forces* (Tuller et al. 1999; Saramago 2010). For simplicity, condider only the van der Waals and electrostatic forces. In the first case the *disjoining tension* h_v (Pa) and *film thickness* e_v (m) satisfy

$$h_v = \frac{-A}{6\pi e_v^3} \tag{7.98}$$

where A (J) is the *Hamaker constant* (Hamaker 1937). In case of *electrostatic* films by Langmuir's (1938) *double-layer model*

$$h_e = \frac{\varepsilon_0 \varepsilon_r}{2}\left(\frac{\pi k_B T}{E_c z}\right)\frac{1}{e_e^2} \tag{7.99}$$

where h_e is the *disjoining tension* (Pa) and and e_e the *film thickness* (m) due to electrostatic forces, respectively, ε_0 is vacuum permittivity (8.85×10^{-12} C^2 J^{-1} m^{-1}), ε_r the *dielectric constant* of water (78.54 at 25 °C), E_c *electric charge* (1.602×10^{-19} C), k_B Boltzmann constant (1.381×10^{-23} J K^{-1}), T is the Kelvin temperature, and z *ion valence.*

The *water retention curve model* of de Gennes (1985b) relates *tension head*, *h*, to water content, θ, by a power-law

$$h = h_d\left(\frac{\theta}{\theta_s}\right)^{D_s-3} \tag{7.100}$$

where $D_s \in [2, 3)$ is the surface fractal dimension, h_d is the *displacement tension head* (Pa), and θ_s is the saturated water content. (For Eq. 7.100 see also Tyler and Wheatcraft 1990, Perrier et al. 1996; Ghambarian and Millán 2017). If we assume that the adsorbed water content θ_{ads} is mainly controlled by van der Waals and electrostatics forces, we can write $h_{ads} \approx h_v + h_e$ and, using Eqs. (7.98 and 7.99) and the logarithmized form of (7.100) we find

$$D_s = \frac{1}{C}\ln\theta_{ads} + \left(3 - \frac{\ln\theta_s}{C}\right) \tag{7.101}$$

where

$$C = \ln\left[\frac{-A}{6\pi h_d e_v^3} + \frac{\varepsilon_0 \varepsilon_r}{2h_d e_\theta^2}\left(\frac{\pi k_B T}{E_c z}\right)\right] \tag{7.102}$$

Here, e_ν(m) and e_e(m) are the adsorbed film thicknesses, respectively, at $h = h_{ads}$. For the use of *nitrogen adsorption* to find pore surface fractal dimension, see (Pfeifer and Avnir 1983b; El Shafei et al. 2004; Wang et al. 2007; Liu and Nie 2016; Zhu et al. 2019; Gu et al. 2019).

Korvin et al. (2001) examined the changes of pore surface fractal dimension under pressure. They determined D_s from the measured pressure-dependence of permeabilities.

Weller et al. (2016) determined on eight Upper Cretaceous carbonate rock samples from the Tushka area in Egypt the value of fractal dimension from *pore radius distribution* provided by different methods, such as *Mercury Porosimetry* (MP), *Nuclear Magnetic Resonance* (NMR), and *Spectral Induced Polarization* (SIP). With the SIP, the complex conductivity spectra were acquired in the frequency range between 3 MHz and 100 Hz using an impedance spectrometer with high phase accuracy. They transformed the *relaxation time distribution* (found by *Debye decomposition*, Nordsiek and Weller 2008) of the *complex conductivity spectra* into a distribution of pore radii r using the equation of Schwarz (1962), Revil et al. (2012):

$$r = \sqrt{2\tau D_{(+)}} \tag{7.103}$$

τ being relaxation time and $D_{(+)}$ the diffusion coefficient of the counter-ions in the *Stern layer* (the Stern Layer is the internal layer of the electric double layer, which forms at a charged surface in an ionic solution). They assumed a constant *diffusion coefficient* D(+) = 3.8 × 10 $^{-12}$ m^2/s proposed for clayey materials in (Revil 2013). *Chargeability* is attributed to pore volume. The volume fraction $V_c = V(< r)/V$ is the cumulative volume fraction of pores with radii less than r, it corresponds to the ratio of *cumulative intensity* to *total intensity*. In case of fractal pore volume distribution routine scaling argument gives

$$D_{SIP} = 3 - s \tag{7.104}$$

where s is the slope of the straight line fitting the $\log(V_c)$ versus $\log(r)$ plot.For reservoir characterization, as Dou et al. (2021) noted, "the value of fractal dimension could quantitatively characterize the reservoir heterogeneity. Increasing fractal dimensions often reflect a more complex pore structure (Li 2010; Lai et al. 2018a, b) and consequently reduce the permeability (Yang et al. 2014; Zhao et al. 2017; Shao et al. 2017; Lai et al. 2018b)".

In all cases of fractal analyses discussed above, the core idea of derivation was the inherent self-similarity of the pore structure which has led to an *allometric power relation* (Korvin 2021) between two rock properties of the form

$$y = \alpha \cdot x^{\beta D} \tag{7.105}$$

where D is the fractal dimension sought for. D could be found from the slope of the straight line fitting the $\log(y) vs. \log(x)$ plot. Howevever, in many practical cases the plot cannot be fitted with reasonable accuracy using a single line, but *we need at least two lines, resulting in two different D values in different ranges of x*. (Examples are: Ouyang et al. 2016: Fig. 3; Wang et al. 2019: Fig. 15.) Orford and Whalley (1983: Fig. 6), who were the first to propose "the use of fractal dimension to quantify the morphology of irregular-shaped particles" have already observed this, and spoke of "*texture fractal*", "*structural fractal*", and Euclidean regimes. Pyun and Rhee (2004) also noted this, when analysing the structure of *carbon electrodes*. They defined the two dimensions as *pore surface fractal dimension* and *pore space structural fractal dimension*, respectively. Korvin (1992) named such structures *bifractals*.

It has been realized later, for the two most complicated materials on Earth, coal and *soil*, that even bifractals cannot describe their internal structure, and one needs *multifractal spectra* or *generalized fractal dimensions* to give justice to their complexity (coal: Toda and Toyoda 1972; Li 2015; Li et al. 2015; Zhu et al. 2019; soil: Kravchenko et al. 1999; Caniego et al. 2001; Caniego 2003; Posadas et al. 2003; Bird et al. 2006; San 2010; Ferreiro and Vázquez 2010). For these advanced concepts see e.g. (Korvin 1992), here I can just give a very brief summary to help the Reader to appreciate the efforts behind these studies, in particular the results of Zhu et al. (2019: Figs. 5, 6) who computed the *multifractal singular spectra* and the *generalized fractal dimension spectra* of bituminous coal samples.

According to Zhu et al. (2019) "*multifractals* are an extension and a superposition of *monofractal*s, which means, geometrically, [that] multifractal structures can be decomposed into a set of intertwined fractal subsets that characterize the local variability and heterogeneity of the studied variables". They can be described, equivalently, as *multifractal singularity spectra* α and $f(\alpha)$ (Halsey et al. 1986) or *generalized fractal dimension spectra* q and $D(q)$ (e.g. Toda and Toyoda 1972). Let us follow (Zhu et al. 2019), and introduce these concepts in terms of MIP (Mercury Intrusion Porosimetry) data. There are three parameters to be calculated, the *probability quality distribution* $P_i(\varepsilon)$, the *singular exponent* α and the *dimension function* $f(\alpha)$. We proceed in the following steps: To find the *multifractal singularity spectra*: (i) we sort the measured values, say MIP values, into $N \gg 1$ bins of equal width ε; (ii) we construct for each bin the mass probability

$$P_i(\varepsilon) = \frac{N_i(\varepsilon)}{N_t} \tag{7.106}$$

where N_i is the content of mercury in the i-th box, N_t is the final, total content of mercury; (iii) we find numbers α_i satisfying

$$P_i(\varepsilon) = \varepsilon^{\alpha_i} \tag{7.107}$$

(iv) we construct a probability distribution

$$\left\{\omega_i(q,\varepsilon) = \frac{P_i^q(\varepsilon)}{\sum_{i=1}^{N(\varepsilon)} P_i^q(\varepsilon)}\right\}_{i=1}^{N(\varepsilon)} \tag{7.108}$$

(v) finally we compute, for a sufficient number of $q \in (-\infty, \infty)$ values,

$$\alpha(q) \propto \frac{\sum_{i=1}^{N(\varepsilon)} \omega_i(q,\varepsilon)\log[P_i(\varepsilon)]}{\log(\varepsilon)} \tag{7.109}$$

and

$$f\left[\alpha(q)\right] \propto \frac{\sum_{i=1}^{N(\varepsilon)} \omega_i(q,\varepsilon)\log\left[\omega_i(q,\varepsilon)\right]}{\log(\varepsilon)} \tag{7.110}$$

To find the *generalized fractal dimension spectra* $q \sim D(q)$: (v) we must solve the equations

$$\sum_{i=1}^{N(\varepsilon)} P_i(\varepsilon)^q = \varepsilon^{(q-1)D_q} \tag{7.111}$$

for D_q; (vi) if $q \neq 1$, the solution is

$$D_q = \frac{1}{q-1}\lim_{\varepsilon\to 0}\frac{\log\sum_{i=1}^{N(\varepsilon)} P_i(\varepsilon)^q}{\log\varepsilon} \tag{7.112}$$

(vii) if $q = 1$, L'Hopital's law gives $D_1 = f(\alpha_1)$; (viii) the value of D_q can be calculated by using

$$F\left(\alpha_{(q)}\right) = \lim_{\varepsilon\to 0}\frac{\sum_{i=1}^{N(\varepsilon)} \omega_i(q,\varepsilon)\log\left[\omega_i(q,\varepsilon)\right]}{\log\varepsilon} \tag{7.113}$$

Particularly important generalized dimensions are: D_0 called the *capacity dimension*, D_1 called the *information dimension* and D_2, the *correlation dimension* (Peitgen et al. 1992).

References

Andrews B (1998) Evolving convex curves. Calc Var Part Differ Equ 7(48):315–371
Andrews B (1999) Gauss curvature flow: the fate of the rolling stones. Invent Math 138(1):151–161

Andrews B (2002) Classification of limiting shapes for isotropic curve flows. J Am Math Soc 16(2):443–459
Andrews B, McCoy J, Zheng Y (2013) Contracting convex hypersurfaces by curvature. Calc Var Part Differ Equ 47(3–4):611–665
Aquilano D, Otálora F, Pastero L, García-Ruiz JM (2016) Three study cases of growth morphology in minerals: halite, calcite and gypsum. Prog Cryst Growth Charact Mater 62(2):227–251
Arabnia O, Sklar LS (2016) Experimental study of particle size reduction in geophysical granular flows. Int J Eros Control Eng 9:122–129
Archilha Nathaly L, Missagia RM, Hollis C, de Ceia MAR, McDonald SA, Lima Neto IA, Eastwood DS, Lee P (2016) Permeability and acoustic velocity controlling factors determined from x-ray tomography images of carbonate rocks. AAPG Bull 100(8):1289–1309
Aristotle (1936) Minor works, mechanical problems, question 15. (Transl. by W.S. Hett). Harvard University, Cambridge, MA
Arnold VI (1998) Ordinary differential equations (10th printing). MIT Press, Cambridge
Aschenbrenner BC (1956) A new method of expressing particle sphericity. J Sediment Petrol 26:15–31
Åström JA, Ouchterlony F, Linna RP, Timonen J (2004) Universal dynamic fragmentation in D dimensions. Phys Rev Lett 92:245506
Attal M, Lavé J (2009) Pebble abrasion during fluvial transport: experimental results and implications for the evolution of the sediment load along rivers. J Geophys Res 114:F04023
Avnir D, Farin D, Pfeifer P (1984) Molecular fractal surfaces. Nature 308:261–263
Baba J, Komar PD (1981) Measurements and analysis of settling velocities of natural quartz sand grains. J Sediment Petrol 51(2):631–640
Ball P (2006) What shape is a pebble? Nature News (published online) July 14, 2006
Bird N, Díaz MC, Saa A, Tarquis AM (2006) Fractal and multifractal analysis of pore-scale images of soil. J Hydrol 322(1–4):211–219
Black WT (1877) On rolled pebbles from the beach at Dunbar. Trans Edinb Geol Soc 3(1):122–123
Bloore FJ (1977) The shape of pebbles. J Int Assoc Math Geol 9(2):113–122
Blott SJ, Pye K (2008) Particle shape: a review and new methods of characterization and classification. Sedimentology 55:31–63
Bluck BJ (1967) Sedimentation of beach gravels; examples from South Wales. J Sediment Res 37:128–156
Brewster SD (1965) Memoirs of the life, writings, and discoveries of Sir Isaac Newton. Johnson Reprint Corporation, New York
Cai J, Lin D, Singh H, Wei W, Zhou S (2018) Shale gas transport model in 3D fractal porous media with variable pore sizes. Mar Pet Geol 98:437–447
Cailleux A (1945) Distinction des galets marins et fluviátiles. Soc Géol France Bull 5(15):375–404
Camenen B (2007) A simple and general formula for the settling velocity of particles in suspension. J Hydraul Eng 133(2):229–233
Caniego FJ (2003) Rényi dimensions of soil pore size distribution. Geoderma 112:205–216
Caniego FJ, Martín MA, San JF (2001) Singularity features of pore-size soil distribution: singularity strength analysis and entropy spectrum. Fractals 9:305–316
Carr AP (1969) Size grading along a pebble beach; Chesil beach, England. J Sediment Res 39:297–311
Cheng NS (1997) Simplified settling velocity formula for sediment particle. J Hydraul Eng 123(2):149–152
Cheng Z, Redner S (1988) Scaling theory of fragmentation. Phys Rev Lett 60:2450–2453
Chitra A, Prasad Kumar B, Narayana AC (2010) Influence of particle shape on drag coefficient for commonly occurring sandy particles in coastal areas. Int J Ocean Clim Syst 1(2):99–112
Cnudde V, Boone M, Dewanckele J, Dierick M, Van Hoorebeke L, Jacobs P (2011) 3D characterization of sandstones by means of X-ray computed tomography. Geosphere 7:1–8
Coates GR, Xiao L, Prammer G (1999) NMR logging principles and applications. Haliburton Energy Services, Houston

Corey AT (1949) Influence of shape on the fall velocity of sand grains. M.Sc. Thesis, Colorado A & M College
Cottet ML (2006) Mesure et structures spatiales et temporelles de l'émoussé des galets dans le réseau hydrographique du Bez (On the measurement, the spatial and temporal distribution of pebble roundness in the hydrographic network of the Bez). M.Sc. Thesis, Université Jean Moulin Lyon 3, Lyon, France
de Gennes PG (1985a) Wetting: statics and dynamics. Rev Mod Phys 57(3):827–863
de Gennes PG (1985b) Partial filling of fractal structure by a wetting fluid. In: Adler D, Fritzsche H, Ovshinsky SR (eds) Physics of disordered materials. Plenum, New York, pp 227–241
Deinert MR, Dathe A, Parlange J-Y, Cady KB (2008) Capillary pressure in a porous medium with distinct pore surface and pore volume fractal dimensions. Phys Rev E 77:021203
Diepenbroek M, Bartholomä A, Ibbeken H (1992) How round is round? A new approach to the topic 'roundness' by Fourier grain shape analysis. Sedimentology 39:411–422
Dietrich W (1982) Settling velocities of natural particles. Water Resour Res 18(6):1615–1626
Dobkins JE Jr, Folk RL (1970) Shape development on Tahiti-Nui. J Sediment Res 40(4):1167–1203
Domokos G (2015) Monotonicity of spatial critical points evolving under curvature-driven flows. J Nonlinear Sci 25:247–275
Domokos G, Gibbons GW (2012) The evolution of pebble size and shape in space and time. Proc Math Phys Eng Sci 468(2146):3059–3079
Domokos G, Gibbons GW (2018) The geometry of abrasion. In: Ambrus G, Bárány I, Böröczky KJ, Tóth GF, Pach J (eds) New trends in intuitive geometry, vol 27 of bolyai society mathematical studies. János Bolyai Mathematical Society, Budapest; and Springer-Verlag GmbH, Berlin, pp 125–153
Domokos G, Lángi Z (2014) The robustness of equilibria on convex solids. Mathematika 60:237–256
Domokos G, Papadopoulos J, Ruina A (1994) Static equilibria of rigid bodies: is there anything new? J Elast 36:59–66
Domokos G, Sipos AÁ, Várkonyi PL (2009a) Countinuous and discrete models for abrasion processes. Period Polytech Arch 40(1):3–8
Domokos G, Sipos AA, Gy S, Várkonyi P (2009b) Formation of sharp edges and planar areas of asteroids by polyhedral abrasion. Astrophys J 699(1):L13–L16
Domokos G, Sipos A, Szabó T, Várkonyi P (2010) Pebbles, shapes, and equilibria. Math Geosci 42:29–47
Domokos G, Jerolmack DJ, Sipos AÁ, Török Á (2014) How river rocks round: resolving the shape-size paradox. PLOS One 9(2):e88657
Domokos G, Kun F, Sipos AÁ, Szabó T (2015) Universality of fragment shapes. Sci Rep 5:9147
Dou W, Liu L, Jia L, Xu Z, Wang M, Du C (2021) Pore structure, fractal characteristics and permeability prediction of tight sandstones: a case study from Yanchang Formation, Ordos Basin, China. Mar Pet Geol 123:104737
Drake LD (1970) Rock texture: an important factor for clast shape studies. J Sediment Petrol 40:1356–1361
Drevin GR, Vincent L (2002) Granulometric determination of sedimentary rock particle roundness. In: International symposium on mathematical morphology, 315–325
Dunn KJ, Bergman DJ, LaTorraca GA (2002) Nuclear magnetic resonance—petrophysical and logging application. In: Handbook of geophysical exploration, seismic exploration, vol 32. Pergamon Press, Amsterdam
Durian DJ, Bideaud H, Duringer P, Schröder A, Thalmann F, Marques CM (2006) What is in a pebble shape? Phys Rev Lett 97(2):028001
Durian DJ, Bideaud H, Duringer P, Schröder AP, Marques CM (2007) Shape and erosion of pebbles. Phys Rev E 75:021301
El Shafei Gamal MS, Christine P, Moussa Nabawaya A (2004) Fractal analysis of hydroxyapatite from nitrogen isotherms. J Colloid Interface Sci 277:410–416
Fehér E, Domokos G, Krasukopf B (2020) Computing critical point evolution under planar curvature flows. arXiv:2010.11169 [math.DS]

Feret LR (1930) La grosseur des grains des matières pulvérulentes. Premières Communications de la Nouvelle Association Internationale pour l'Essai des Matériaux, Groupe D, pp 428–436

Ferreiro JP, Vázquez EV (2010) Multifractal analysis of Hg pore size distributions in soils with contrasting structural stability. Geoderma 160:64–73

Firey WJ (1974) Shapes of worn stones. Mathematika 21(41):1–11

Fok-Pun L, Komar PD (1983) Settling velocities of planktonic foraminifera; density variations and shape effects. J Foramin Res 13(1):60–68

Fredsøe J, Deigaard R (1992) Mechanics of coastal sediment transport. In: Advanced series on ocean engineering, vol 3. World Scientific, Singapore

Friesen WI, Mikula RJ (1987) Fractal dimensions of coal particles. J Colloid Interface Sci 120(1):263–271

Gabitto J, Tsouris C (2008) Drag coefficient and settling velocity for particles of cylindrical shape. Powder Technol 183(2):314–322

Gage ME (1983) An isoperimetric inequality with applications to curve shortening. Duke Math J 50:1225–1229

Gage ME (1984) Curve shortening makes convex curves circular. Invent Math 76(2):357–364

Gage M, Hamilton RS (1986) The heat equation shrinking convex plane curves. J Differ Geom 23:69–96

Gerritsen F (1978) Beach and surf parameters in Hawaii. Sea Grant Technical Report TR-78–02, p 178

Ghadiri M, Zhang Z (2002) Impact attrition of particulate solids. Part 1: a theoretical model of chipping. Chem Eng Sci 57:3659–3669

Ghanbarian, B., 2019. Surface fractal dimension and its theoretical relationship with adsorbed water content. arXiv:1909.09286.

Ghanbarian B, Millán H (2017) Fractal capillary pressure curve models. In: Ghanbarian B, Hunt AG (eds) Fractals: concepts and applications in geosciences. CRC Press, Boca Raton, FL

Ghanbarian-Alavijeh B, Millán H (2009) The relationship between surface fractal dimension and water content at permanent wilting point. Geoderma 151:224–232

Ghosh PK, Kumar KV (1998) Support function representation of convex bodies, its application in geometric computing, and some related representations. Comput vis Image Underst 72(3):379–403

Gibbs RJ, Matthews MD, Link DA (1971) The relationship between sphere size and settling velocity. J Sediment Petrol 41(1):7–18

Gilvarry JJ (1961) Fracture of brittle solids. I. Distribution function for fragment size in single fracture (theoretical). J Appl Phys 32(3):391–399

Goldstein S (1929) The steady flow of viscous fluid past a fixed spherical obstacle at small Reynolds numbers. Proc R Soc Lond, 123A

Grady DE (2010) Length scales and size distributions in dynamic fragmentation. Int J Fract 163:85–99

Graf W (1971) Hydraulics of sediment transport. McGraw-Hill, New York

Graham DJ, Midgley NG (2000) Graphical representation of particle shape using triangular diagrams: an Excel spreadsheat method. Earth Surf Proc Land 25:1473–1477

Grayson MA (1987) The heat equation shrinks embedded plane curves to round points. J Differ Geom 26:285–314

Gu Y, Ding W, Yin M, Jiao B, Shi S, Li A, Xiao Z, Wang Z (2019) Nanoscale pore characteristics and fractal characteristics of organic-rich shale: an example from the lower Cambrian Niutitang Formation in the Fenggang block in northern Guizhou Province, South China. Energy Explor Exploit 7(1):273–295

Guo XQ, Yao YB, Liu DM (2014) Characteristics of coal matrix compressibility: an investigation by mercury intrusion porosimetry. Energy Fuel 28:3673–3678

Hajnos M, Korsunskaia L, Pachepsky Y (2000) Soil pore surface properties in managed grasslands. Soil Tillage Res. 55:63–70

Hallermeier R (1981) Terminal settling velocity of commonly occurring sand grain. Sedimentology 28(6):859–865

Halsey TC, Jensen MH, Kadanoff LP, Procaccia I, Shraiman BI (1986) Fractal measures and their singularities—the characterization of strange sets. Phys Rev Gen Phys 33:1141–1151

Hamaker HC (1937) The London—van der Waals attraction between spherical particles. Physica 4(10):1058–1072

Hill TP (2022) On the oval shapes of beach stones. Appl Math 2(1):16–38

Hofmann HJ (1994) Grain-shape indices and isometric graphs. J Sediment Res A64:916–920

Howarth JJ (2010) The shape of loess particles reviewed. Cent Eur J Geosci 2(1):41–44

Huang G, Zhang R, Huang Q (2006) Modeling soil water retention curve with a fractal method. Pedosphere 16:137–146

Huisken G (1984) Flow by mean curvature of convex sources into spheres. J Differ Geom 20:237–266

Illenberger WK (1991) Pebble shape (and size!). J Sediment Petrol 61:756–767

Inaoka H, Toyosawa E, Takayasu H (1997) Aspect ratio dependence of impact fragmentation. Phys Rev Lett 78:3455–3458

Janke NC (1966) Effect of shape upon the settling velocity of regular convex geometric particles. J Sediment Petrol 36:370–376

Jaroniec M, Kruk M (1997) Fractal analysis of composite adsorption isotherms by using density functional theory data for argon in slitlike pores. Langmuir 13:1031–1035

Johnson NL, Krisko PH, Liou JC, Anz-Meador PD (2001) NASA's new breakup model of evolve 4.0. Adv Space Res 28:1377–1384

Jozefaciuk G, Muranyi A, Szatanik-Kloc A, Farkas C, Gyuricza C (2001) Changes of surface, fine pore and variable charge properties of a Brown forest soil under various tillage practices. Soil Tillage Res 59:127–135

Kaitanen V, Strom O (1978) Shape development of sandstone cobbles associated with the Sâkylâ-Mellilà esker, southwest Finland. Fennia 155:23–67

Kavallaris NI, Suzuki T (2018) Non-local partial differential equations for engineering and biology. Springer International Publishing, Cham, Switzerland

Komar P, Reimers C (1978) Grain shape effects on settling rates. J Geol 86(2):193–209

Korvin G (1989) Fractured but not fractal: fragmentation of the Gulf of Suez Basement. Pure and Applied Geophysics 131(1–2):289–305

Korvin G (1992) Fractal models in the earth sciences. Elsevier, Amsterdam

Korvin G (2021) Allometric power laws. In: Daya Sagar B, Cheng Q, McKinley J, Agterberg F (eds) Encyclopedia of mathematical geosciences. Encyclopedia of earth sciences series. Springer, Cham.

Korvin G, Mohiuddin MA, Abdulraheem A (2001) Experimental investigation of the fractal dimension of the pore surface of sedimentary rocks under pressure. Geophys Trans 44(1):3–19

Krapivsky PL, Redner S (2006) Smoothing rock by chipping. Phys Rev E 75(3 Pt 1):031119

Kravchenko AN, Boast CW, Bullock DG (1999) Multifractal analysis of soil spatial variability. Agron J 91:1033–1041

Krumbein WC (1941a) The effects of abrasion on the size, shape and roundness of rock fragments. J Geol 49(5):482–520

Krumbein WC (1941b) Measurement and geologic significance of shape and roundness of sedimentary particles. J Sediment Petrol 11:64–72

Krumbein WC (1942) Settling velocities and flume behavior of non-spherical particles. Trans Am Geophys Union 23(l):621–633

Krumbein WC, Griffith JS (1938) Beach environment in Little Sister Bay. Wisconsin Geol Soc Am Bull 49:629–652

Krynine PD (1960) On the antiquity of "sedimentation" and hydrology (with Some Moral Conclusions). Geol Soc Am Bull 71(11):1721–1726

Kuenen PhH (1956) Experimental abrasion of pebbles: 2 rolling by current. J Geol 64:336–368

Kuenen PH (1959) Experimental abrasion; 3, fluviatile action on sand. Am J Sci 257:172–190

Kun F, Herrmann HJ (1999) Transition from damage to fragmentation in collision of solids. Phys Rev E 59:2623

Lai J, Wang GW, Fan ZY, Zhou ZL, Chen J, Wang SC (2018a) Fractal analysis of tight shaly sandstones using nuclear magnetic resonance measurements. Am Assoc Pet Geol Bull 102(2):175–193

Lai J, Wang G, Wang Z, Chen J, Pang X, Wang S, Zhou Z, He Z, Qin Z, Fan X (2018b) A review on pore structure characterization in tight sandstones. Earth Sci Rev 177:436–457

Laity JE, Bridges NT (2009) Ventifacts on Earth and Mars: Analytical, field, and laboratory tudies supporting sand abrasion and windward feature development. Geomorphology 105:202–217

Lamb H (1994) Hydrodynamics, 6th ed. Cambridge University Press, Cambridge, p 599

Landon RE (1930) An analysis of beach pebble abrasion and transportation. J Geol 38(5):437–446

Langmuir I (1938) Repulsive forces between charged surfaces in water, and the cause of the Jones-Ray effect. Science 88:430–432

Le Pen LM, Powrie W, Zervos A, Ahmed S, Aingaran S (2013) Dependence of shape on particle size for a crushed rock railway ballast. Gran Matt 15:849861

Le Roux JP (1992) Settling velocity of spheres: a new approach. Sed Geol 81:11–16

Le Roux JP (1996) Settling velocity of ellipsoid grains as related to shape entropy. Sed Geol 101:15–20

Le Roux JP (1997) Comparison of sphericity indices as related to the hydraulic equivalence of settling grains. J Sediment Res 67(3):527–530

Lewin J, Brewer P (2002) Laboratory simulation of clast abrasion. Earth Surf Proc Land 27(2):145–164

Li K (2004) Characterization of rock heterogeneity using fractal geometry. In: Society of petroleum engineers international thermal operations and heavy oil symposium and western regional meeting, Bakersfield, California, March 16–18, 2004. SPE-86975-MS

Li K (2010) Analytical derivation of Brooks-Corey type capillary pressure models using fractal geometry and evaluation of rock heterogeneity. J Petrol Sci Eng 73(1–2):20–26

Li W (2015) Multifractal analysis of Hg pore size distributions of tectonically deformed coals. Int J Coal Geol 144:138–152

Li W, Liu H, Song X (2015) Multifractal analysis of Hg pore size distributions of tectonically deformed coals. Int J Coal Geol 144:138–152

Litty C, Schlunegger F (2017) Controls on pebbles' size and shape in streams of the Swiss Alps. J Geol 125:101–112

Litwin MK, Szabó T, Jerolmack DJ, Domokos G (2014) Quantifying the significance of abrasion and selective transport for downstream fluvial grain size evolution. J Geophys Res Earth Surf 119:2412–2429

Liou JC, Johnson NL (2006) Risks in space from orbiting debris. Science 311:340–341

Liu X, Nie B (2016) Fractal characteristics of coal samples utilizing image analysis and gas adsorption. Fuel 182:314–322

Lovejoy S (1982) Area-perimeter relation for rain and cloud areas. Science 4542:185–187

Mahamud MM, Novo MF (2008) The use of fractal analysis in the textural characterization of coals. Fuel 87(2):222–231

Mandelbrot BB (1982) The fractal geometry of nature. WH Freeman, New York

Manga M, Patel A, Dufek J (2011) Rounding of pumice clasts during transport: field measurements and laboratory studies. Bull Volcanol 73:321–333

McGrath TDH, O'Connor L, Eksteen JJ (2015) A comparison of 2D and 3D shape characterisations of free gold particles in gravity and flash flotation concentrates. Miner Eng 82:45–53

Moss AJ, Walker PH, Hutka J (1973) Fragmentation of granitic quartz in water. Sedimentology 20:489–511

Neimark A (1990a) Thermodynamic method for calculating surface fractal. JETP Lett 51(10)

Neimark AV (1990b) Calculating surface fractal dimensions of adsorbents. Adsorpt Sci Technol 7(4):210–219

Neimark AV (1990c) Determination of surface fractal dimension from the results of an adsorption experiment. Russ J Phys Chem 64:1398–1403
Neimark AV, Unger KK (1993) Method of discrimination of surface fractality. J Colloid Interface Sci 158:412–419
Nielsen P (1992) Coastal bottom boundary layers and sediment transport. In: Advanced series on ocean engineering. World Scientific, Singapore.
Nordsiek S, Weller A (2008) A new approach to fitting induced-polarization spectra. Geophysics 73(6):F235–F245
Novák-Szabó T, Sipos AÁ, Shaw S, Bertoni D, Pozzebon A, Grottoli E, Sarti G, Ciavola P, Domokos G, Jerolmack DJ (2018) Universal characteristics of particle shape evolution by bed-load chipping. Sci Adv 4(3):eaao4946
Oakey RJ, Green M, Carling PA, Lee MWE, Sear DA, Warburton J (2005) Grain-shape analysis—a new method for determining representative particle shapes for populations of natural grains. J Sediment Res 75:1065–1073
Öner Ü, Weller A, Sattler C-D, Kassab MA (2016) Petrographic and petrophysical investigation on carbonate samples (Upper Cretaceous) from the Tushka Area (Egypt) with special focus on the effective pore radius. Arab J Geosci 9:229
Orford JD, Whalley WB (1983) The use of the fractal dimension to quantify the morphology of irregular-shaped particles. Sedimentology 30:655–668
Ouyang Z, Liu D, Cai Y, Yao Y (2016) Fractal analysis on heterogeneity of pore−fractures in middle−high rank coals with NMR. Energy Fuels 30:5449−5458
Palmer HR (1834) Observations on the motions of shingle beaches. Phil Trans R Soc Lond 124:567–576
Pape H, Clauser C, Iffland J (1999) Permeability prediction based on fractal pore-space geometry. Geophysics 64:1447–1460
Peitgen H, Jürgens H, Saupe D (1992) Chaos and fractals. New frontiers of science. Springer, New York
Peng RD, Yang YC, Ju Y, Mao LT, Yang YM (2011) Computation of fractal dimension of rock pores based on gray CT images. Sci Bull 56(31):3346–3357
Perelman G (2003) Ricci flow with surgery on three-manifolds. arXiv preprint math/0303109
Perfect E (1997) Fractal models for the fragmentation of rocks and soils: a review. Eng Geol 48:185–198
Perrier E, Rieu M, Sposito G, de Marsily G (1996) Models of the water retention curve for soils with a fractal pore size distribution. Water Resour Res 32(10):3025–3031
Pfeifer P, Avnir D (1983a) Chemistry in noninteger dimensions between two and three. I. Fractal theory of heterogeneous surfaces. J Chem Phys 79(7):3558–3565
Pfeifer P, Avnir D (1983b) Chemistry in nonintegral dimensions between two and three. J Chem Phys 79:3369–3558
Posadas AND, Giménez D, Quiroz R (2003) Multifractal characterization of soil pore systems. Soil Sci Soc Am J 67:1361–1369
Powers MC (1953) A new roundness scale for sedimentary particles. J Sediment Petrol 23(2):117–119
Pyôkàri M (1980) Shape development of trondhjemite pebbles and cobbles on shores in the Southwestern Finnish archipelago. Géogr Phys Quat 34(3):335–350
Pyun S-I, Rhee C-K (2004) An investigation of fractal characteristics of mesoporous carbon electrodes with various pore structures. Electrochim Acta 49:4171–4180
Revil A (2013) Effective conductivity and permittivity of unsaturated porous materials in the frequency range 1 mHz-1GHz. Water Resour Res 49:306–327
Revil A, Koch K, Holliger K (2012) Is it the grain size or the characteristic pore size that controls the induced polarization relaxation time of clean sands and sandstones? Water Resour Res 48:W05602
Rice S, Church M (1998) Grain size along two gravel-bed rivers: statistical variation, spatial pattern and sedimentary links. Earth Surf Processes Landf 23:345–363

Roussillon T, Piegay H, Sivignon I, Tougne L, Lavigne F (2009) Automatic computation of pebble roundness using digital imagery and discrete geometry. Comput Geosci 35(10): 1992–2000
Rogers CDF, Smalley IJ (1993) The shape of loess particles. Naturwissenschaften 80:461–462
Rootare HM, Prenzlow CF (1967) Surface areas from mercury porosimeter measurements. J Phys Chem 71(8):2733–2736
Rubey WW (1933) Settling velocities of gravel, sand, and silt particles. Am J Sci 25(5):325–338
San JM (2010) Multifractal analysis of discretized X-ray CT images for the characterization of soil macropore structures. Geoderma 156:32–42
Saramago B (2010) Thin liquid wetting films. Curr Opin Colloid Interface Sci 15:330–340
Sarkar N, Chaudhuri BB (1992) An efficient approach to estimate fractal dimension of textural images. Pattern Recogn 25(9):1035–1041
Sarkar N, Chaudhuri BB (1994) An efficient differential box-counting approach to compute fractal dimension of image. IEEE Trans Syst Man Cybern 24(1):115–120
Schmitt M, Halisch M, Müller C, Fernandes CP (2016) Classification and quantification of pore shapes in sandstone reservoir rocks with 3-D X-ray micro-computed tomography. Solid Earth 7:285–300
Schwarz G (1962) A theory of the low-frequency dielectric dispersion of colloidal particles in electrolyte solution. J Phys Chem 66:2636–2642
Scott C, Smalley IJ (1991) The original shapes of quartz sand grains. Area 23:353–355
Shao X, Pang X, Hui L, Zhang X (2017) Fractal analysis of pore network in tight gas sandstones using NMR method: a case study from the Ordos Basin, China. Energy Fuels 31(10):10,358–10,368
Shen P, Li K (1994) A new method for determining the fractal dimension of pore structures and its application. In: Proceedings of the 10th offshore South East Asia conference, Singapore, pp 6–9
Shen P, Li K, Jia F (1995) Quantitative description for the heterogeneity of pore structure by using mercury capillary pressure curves. In: International meeting on petroleum engineering. OnePetro
Shi X, Pan J, Hou Q, Jin Y, Wang Z, Niu Q, Li M (2018) Micrometer-scale fractures in coal related to coal rank based on micro-CT scanning and fractal theory. Fuel 212:162–172
Sipos AÁ, Domokos G, Wilson A, Hovius N (2011) A discrete random model describing bedrock erosion. Math Geosci 43:583–591
Sipos AA, Domokos G, Jerolmack DJ (2018) Shape evolution of ooids: a geometric model. Sci Rep 8(1):1–7
Sipos AA, Domokos G, Török J (2021) Particle size dynamics in abrading pebble populations. Earth Surf Dynam 9:235–251
Smalley IJ (1990) Possible formation mechanisms for the modal coarse-silt quartz particles in loess deposits. Quatern Int 7(8):23–28
Smalley IJ (1996) Formation of quartz sand. Nature 211:476–479
Smith DA, Cheung KF (2002) Empirical relationships for grain size parameters of calcareous sand on Oahu, Hawaii. J Coast Res 18(1):82–93
Smith DA, Cheung KF (2003) Settling characteristics of calcareous sand. J Hydraul Eng 129(6):479–483
Sneed E, Folk RL (1958) Pebbles in the lower Colorado River, Texas, a study in particle morphogenesis. J Geol 66:114–150
Sokołowska Z, Sokołowski S (1999) Influence of humic acid on surface fractal dimension of kaolin: analysis of mercury porosimetry water vapour adsorption data. Geoderma 88:233–249
Stallmach F, Vogt C, Kräger J, Helbig K, Jacobs F (2002) Fractal geometry of surface areas of sand grains probed by pulsed field gradient NMR. Phys Rev Lett 88:105505
Strutt (Lord Rayleigh) RJ (1942) The ultimate shape of pebbles, natural and artificial. Proc Math Phys Eng Sci 181(985):107–118
Strutt (Lord Rayleigh) RJ (1944a) Pebbles, natural and artificial, their shape under various conditions of abrasion. Proc Math Phys Eng Sci 182(991):321–335
Strutt (Lord Rayleigh) RJ (1944b) Pebbles of regular shape and their production in experiment. Nature 154(3901):169–171

Sun H, Koch M (1998) Fractal generation of surface area of porous media. Stoch Hydrol Hydraul 12:83–96
Szabó T (2013) A mechanics-based pebble shape classification system and the numerical simulation of the collective shape evolution of pebbles. PhD dissertation, Budapest University of Technology and Economics
Szabó T, Domokos G (2010) A new classification system for pebble and crystal shapes based on static equilibrium points. Cent Eur Geol 53(1):1–19
Szabó T, Domokos G, Grotzinger JP, Jerolmack DJ (2015) Reconstructing the transport history of pebbles on Mars. Nat Commun 6:8366
Toda Y, Toyoda S (1972) Application of mercury porosimetry to coal. Fuel 13:199–201
Tuller M, Or D, Dudley LM (1999) Adsorption and capillary condensation in porous media: liquid retention and interfacial configurations in angular pores. Water Resour Res 35:1949–1964
Turcotte DL (1986) Fractals and fragmentation. J Geophys Res 91:1921–1926
Turcotte DL (1997) Fractals and chaos in geology and geophysics. Cambridge University Press, Cambridge
Tyler SW, Wheatcraft SW (1990) Fractal processes in soil water retention. Water Resour Res 26:1047–1054
Van Dalen G, Koster MW (2012) 2D and 3D particle size analysis of micro-CT images. In: Proceedings, bruker micro-CT user meeting, Brussels, 3–5 April 2012, pp 1–16
Várkonyi P, Domokos G (2006) Static equilibria of rigid bodies: dice, pebbles and the Poincaré-Hopf Theorem. J Nonlinear Sci 16:255–281
Várkonyi P, Domokos G (2011) A general model for collision-based abrasion processes. IMA J Appl Math 76(1):47–56
Várkonyi PL, Laity JE (2012) Formation of surface features on ventifacts: modeling the role of sand grains rebounding within cavities. Geomorphology 139–140:220–229
Várkonyi PL, Domokos G, Sipos AÁ (2008) Pebble shapes and abrasion processes. In: Proceedings of the first workshop on mathematical geosciences, Freiberg, Germany, June 11–12 (2008)
Wadell H (1932) Volume, shape, and roundness of rock particles. J Geol 40:443–451
Wadell H (1935) Volume, shape and roundness of quartz particles. J Geol 43(3):250–280
Wald QR (1990) The form of pebbles. Nature 345:211
Wang S, Li X, Xue H, Shen Z, Chen L (2021) Fractal characteristics of shale pore structure and its influence on seepage flow. R Soc Open Sci 8:202271
Wang X, Jiang Z, Jiang S, Chang J, Zhu L, Li X, Li J (2019) Full-scale pore structure and fractal dimension of the Longmaxi Shale from the Southern Sichuan Basin: investigations using FE-SEM, Gas Adsorption and Mercury Intrusion Porosimetry. Minerals 9:543
Wang Y, Du B, Dou X, Liu J, Shi B, Wang D, Tang H (2007) Study on the pore surface fractal dimension and surface acid–base properties of natural particles around Guanting reservoir. Colloids Surf A: Physicochem Eng Asp 307:16–27
Washburn EW (1921) The dynamics of capillary flow. Phys Rev 17:273–283
Weller A, Ding Y, Zhang Z, Kassab M, Halisch M (2016) Fractal dimension of pore space in carbonate samples from Tushka Area (Egypt). In: International symposium of the society of core analysts, Snowmass, Colorado, USA, 21–26 August 2016. SCA2016: 079
Wentworth CK (1922) The shapes of beach pebbles. US geological survey professional paper 131-C, pp 75–83
Wentworth CK (1923) A method of measuring and plotting the shapes of pebbles. USGS Bull 730-C:91–102
Wiedenmann D, Keller L, Holzer L, Stojadinovic J, Münch B, Suarez L, Furney B, Hagendorfer H, Brünnimann R, Modregger P, Gorbar M, Vogt UF, Züttel A, La Mantia F, Wepf R, Grobety B (2013) Three-dimensional pore structure and ion conductivity of porous ceramic diaphragms. AiChE Lett: Transp Phenom Fluid Mech 59:1446–1457
Willetts BB, Rice MA (1983) Practical representation of characteristic grain shape of sands: a comparison of methods. Sedimentology 30:557–565

Winzer K (2013) On the formation of elliptic stones due to periodic water waves. Eur Phys J B 86(2013):464

Winzer K (2017) The temporal formation and the shape of ellipsoidal stones on the beaches of the oceans. Eur Phys J plus 132:443

Winzer K, Hegerfeldt GC (2021) On the ocean beach—why elliptic pebbles do not become spherical. arXiv:2106.05835 [physics.class-ph] & The European Physical Journal Plus. https://doi.org/10.1140/epjp/s13360-021-01634-2

Wittel FK, Kun F, Herrmann HJ, Kröplin BH, Maloy KJ (2006) Scaling behaviour of fragment shapes. Phys Rev Lett 96:025504

Yang F, Ning Z, Liu H (2014) Fractal characteristics of shales from a shale gas reservoir in the Sichuan Basin, China. Fuel 115:378–384

Yin Y, Qu ZG, Zhang JF (2017) An analytical model for shale gas transport in kerogen nanopores coupled with real gas effect and surface diffusion. Fuel 210:569–577

Yu B, W Liu (2004) Fractal analysis of permeabilities for porous media. AIChE J 50(1):46–57

Zhang B, Li S (1995) Determination of the surface fractal dimension for porous media by Mercury Porosimetry. Ind Eng Chem Res 34(4):1383–1386

Zhang B, Zhu J, He F, Jiang Y (2018) Compressibility and fractal dimension analysis in the bituminous coal specimens. AIP Adv 8:075118

Zhang Z, Weller A (2014) Fractal dimension of pore-space geometry of an Eocene sandstone formation. Geophysics 79(6):D377–D387

Zhao Y, Zhu G, Dong Y, Danesh NN, Chen Z, Zhang T (2017) Comparison of low-field NMR and microfocus X-ray computed tomography in fractal characterization of pores in artificial cores. Fuel 210:217–226

Zhu J, Zhang R, Zhang Y, He F (2019) The fractal characteristics of pore size distribution in cement-based materials and its effect on gas permeability. Sci Rep 9(1):1–12

Zingg T (1935) Beitrag zur Schotteranalyse. Schweizerische Mineralogische Und Petrologische Mitteilungen 15:39–140

Chapter 8
Entropy and Rock Physics

Chapter Highlights. The probabilistic concept of *entropy* as measure of information is introduced in Sect. 8.1 through a classic problem, "how to keep the forecaster honest?". Out of its many applications listed in Table 8.1., the Section briefly discusses "*rock entropy*" and its role in *acoustic scattering*, the *entropy of the Markov Chains* of *sedimentary successions*, and using the changes of entropy for *earthquake prediction* in seismically active regions. Section 8.2 introduces the *maximum entropy principle* and demonstrates its use in three worked-out examples: *maximum entropy* (ME) *pore shape inversion*; a new probabilistic derivation of *Athy's compaction law*; and the *distribution of contact forces in stressed granular media.* A mathematical Appendix deals with the relation between the *Hurst exponent* and the entropy increase in *self-affine random processes.*

8.1 Rock Entropy–or "How to Keep the Forecaster Honest"?

In a historical review of the first thirty years of *Information Theory*, Tribus (1983) emphasized, as a remarkable success of the discipline, that an American Petroleum Geologist, J.P. Dowds had not only used *Markov Chains* and *entropy*, but–based upon such an analysis–he actually purchased the rights to an abandoned oil well in Oklahoma, and christened it "Rock Entropy #1". Actually, John Dowds had been given a hard time by the oil community, as judged from his bitter remark in his important study "*Oil Rocks: Information Theory: Markov Chains: Entropy*" (1969), " … in the report, information theory was mentioned, but I decided it best to avoid unfamiliar words such as *entropy*, *ergodic*, *Markovian*, etc., as these words can cause antagonisms." More than 50 years have passed since those days, and, as we see from Table 8.1, *entropy* has become a basic accessory in the toolbox of the explorationist.

G. Korvin, *Statistical Rock Physics*, Earth and Environmental Sciences Library,
https://doi.org/10.1007/978-3-031-46700-4_8

Table 8.1 Examples for the use of entropy in geosciences

Topic	Authors
Basaltic melts	(Lee et al. 2020)
Blastability	(Zhou and Li 2012)
Channel flow	(Chiu 1987, 1988, 1989; Chin and Murray 1992; Chiu and Said 1995; Cao and Knight 1997; Chiu et al. 2000; Sterling and Knight 2002; Chiu and Tung. 2002; Singh 2011; Luo and Singh 2011; Cui and Singh 2013; Moramarco et al. 2013; Cui and Singh 2014; Bonakdari et al. 2015; Kumbhakar et al. 2016, 2017a; 2017b; 2017c; 2018; Kumbhakar and Ghoshal 2017; Kundu 2017a; 2017b; 2017c; 2017d; Singh et al. 2017; Luo et al. 2018; Zhu 2018; Zhu and Yu 2019)
Colloidal matter	(Manoharan 2015)
Earthquakes	(Benu and Davis 1980; Shen and Mansinha 1983)
Exploration seismics	(Korvin and Petrovics 1975; Korvin 2000; Puryear et al. 2012; Lyu et al. 2020)
Faults	(Liu et al. 2004)
Geochemistry	(Sterligov 2010)
Grain topology	(Edwards and Oakeshott 1989; Fradkov and Udler 1994; Nowak et al. 1998; Fayad et al. 1999; Lazar 2011; Lazar et al. 2012, 2020; Mason et al. 2012; Keller et al. 2014; Meng et al. 2015; Xue et al. 2016; Lutz et al. 2017; Perumal et al. 2017)
Granular matter	(Makse and Kurchan 2002; Bagi 2003; Schröter et al. 2005; Henkes et al. 2007; Briscoe et al. 2008; Herman 2013, Smart 2017)
Hydraulic conductivity	(Sanchez-Vila et al. 2006)
Hydrology	(Singh and Asce 2011)
Hydrothermal systems	(van Ruitenbeek et al. 2020)
Landscape evolution	(Leopold and Langbein 1962)
Markov chains of layers	(Dowds 1969; Gingerich 1969; Miall 1973; Hattori 1976; Power and Easterling 1982; Hota and Maejima 2004; Elfeki and Dekking 2005; Sinha et al. 2015; Szabolcs 2016; Tita and Djomeni 2016)
Minerology	(Timothy 1989)
Modelling	(Woodbury and Ulrych 1993; Elfeki and Dekking 2005; Huang et al 2012; Bianchi et al. 2015))
Permeability	(Journel and Deutsch 1993; Renard and de Marsily 1997)
Polycrystals	(Barmak et al. 2011a, 2011b)
Pore shape spectrum	(Doyen 1987)
Porosity	(Shmerkin 2011; Korvin et al. 2013; Anovitz and Cole 2015)
Remote sensing	(Wu et al. 2022)
Shale compaction	(Korvin 1984)
Soil science	(Oleschko et al. 2002, 2004; Bird et al. 2006; Martinez et al. 2007)
Solute transport	(Kitanidis 1994; Gotovac et al. 2010; Chiogna et al. 2012; Bianchi and Pedretti 2017)

(continued)

Table 8.1 (continued)

Topic	Authors
Unsaturated flow	(Mays et al. 2002)
Wave absorption	(Korvin 1980, 1985)

In Petroleum Exploration, *permeability*–one of the most important property of reservoir rocks–is never "*estimated*" or "*computed*" from well logs, well pressure transients, or rock cuttings, it is always "*predicted*". There is, of course, a hidden caveat in the term: we realize that predictions can go wrong. Soothsaying is a dangerous business. In Dante's *Inferno* (*Canto XX*) the souls of soothsayers who misled their clients have their heads twisted to the rear, so they walk backward. But it is so easy to understand why the diviners used to cheat. Who would dare to upset a mighty Caesar who ordered "Go bid the priests to do present sacrifice/And bring me *their opinions of success*" (Shakespeare: Julius Caesar ii, 2,5–italics mine).

In the early years of *Information Theory*, Good (1952, 1954; see also McCarthy 1956) introduced the "*how to keep the forecaster honest*" paradigm, that is how to design a payoff system which would force the forecaster to give an *unbiased prediction*. Much later it has been proved mathematically (Fischer, 1972; Aczél and Daróczy, 1975) that the only way to do this is intimately connected with the concept of *Shannon entropy*.

Let the probability of the *i*-th possible event be $p_i, i = 1, \ldots, N; \sum p_i = 1; 0 \leq p_i \leq 1$, and suppose the forecaster gets a payoff $f(p_i), i = 1, \ldots, N$ if he predicts the *i*th event, that is his expected payoff is $\langle f \rangle = \sum p_i f(p_i)$.

If we want to keep the forecaster honest, we must select a function $f(p)$ such that for any other probability distribution $\{q_i\}_{i=1}^n$ one should have

$$\sum p_i f(p_i) \geq \sum p_i f(q_i) \tag{8.1.1}$$

that is, the expected payoff is maximal if the forecaster predicts the events according to their correct probability. In a brilliant paper, Pál Fischer (1972) proved that the only function satisfying inequality (8.1.1) is

$$f(p) = const \cdot \ln(p) \tag{8.1.2}$$

that is–apart from a constant factor–the *expected payoff* is the Shannon entropy

$$H = -\sum p_i \ln(p_i) \tag{8.1.3}$$

Putting aside the "forecaster" analogy, we can say that the only reasonable and unbiased quantitative "value" what we can associate with the information about a probability distribution $p_i, i = 1, \ldots, N$ is its entropy, $H = -\sum p_i \ln(p_i)$ (Korvin 2000; Korvin et al. 2013).

Entropy also occurs in a very different context, as measure of the *configurational disorder of random geometrical patterns* (Ziman 1979). In the statistical physics of point patterns *configurational entropy* is defined as $S = k_B \cdot lnW$ where W is the number of different ways the given arrangement can be realized using undistinguishable objects, assuming that all configurations are equally probable (Boltzmann's 1877 equation, $k_B = 1.380649 \times 10^{-23} J/K$ is the Boltzmann constant). If the configurations have different probabilities $p_i, i = 1, \ldots, W$, then combinatorial reasoning and the Stirling's approximation yield $S = -k_B \sum_{i=1}^{W} p_i \ln p_i$ (see Landau and Lifshitz 1980; Korvin 1984).

As an example, in case of irregularly placed objects (points, particles, grains) the distribution of *distances to nearest neighbor* can be used to characterize the randomness of the arrangement (Clark and Evans 1954). It was first proved by Hertz (1909), that if $N \gg 1$ points are distributed on the plain with density ρ;, and for every point $P_i, i = 1, \ldots, N$ its distance to the nearest neighbor is $r_i, i = 1, \ldots, N$, then the expected value of r_i is

$$\langle r_i \rangle = \lim_{N \to \infty} \frac{\sum_{i=1}^{N} r_i}{N} = \frac{1}{2\sqrt{\rho}} \tag{8.1.4}$$

For a *regular lattice*, all distances $\{r_i\}$ are equal, and the Shannon entropy of the distance-to-nearest-neighbor distribution is 0. The more irregular is the lattice, the larger will be the range of values in the set $\{r_i\}$, and the larger will be the Shannon entropy. The Shannon entropy of the distribution of $\{r_i\}$ is a meaningful (and, as we have seen for the forecaster problem, the *only objective*) measure of the irregularity of the point distribution. And, because H only depends on the probabilities, but not on the actual distances, this measure is *scale-free*.

Apart from a multiplying factor, the entropy function $H = -\sum_{i=1}^{N} p_i ln(p_i)$ is the *unique continuous function that satisfies the following three basic properties* (Khinchin 1957; Aczél and Daróczy 1975; Sethna 2017).

Let $p_i, i = 1, \ldots, N; \sum p_i = 1; 0 \leq p_i$ *be an arbitrary probability distribution. Then.*

E1: *It is maximum for equal probabilities*

$$H(p_1, p_2, \ldots, p_N) < H\left(\frac{1}{N}, \ldots, \frac{1}{N}\right) unless\ p_i = \frac{1}{N} for\ i = 1, \ldots, N \tag{8.1.5a}$$

E2: *It is unaffected by states of zero probability*:

$$H(p_1, p_2, \ldots, p_{N-1}, 0) = H(p_1, p_2, \ldots, p_{N-1}) \tag{8.1.5b}$$

E3: *Entropy of a conditional probability* $Prob(A|B)$:

$$\langle H(A|B) \rangle_B = H(AB) - H(B) \tag{8.1.5c}$$

where AB is the Joint Distribution.

There are two *generalized entropy concepts* derived from Shannon's: the *Rényi entropy* (1961), and the *Tsallis entropy* (1988), in special limiting cases these entropies become identical with the *Shannon entropy* (Akopian 2015; Amigó et al., 2018):

(a) If $p(x)$ is a *pdf* (*Probability Density Function*), its *Rényi entropy of order* $\alpha(\alpha \geq 0, \alpha \neq 1)$ is

$$H_\alpha = \frac{1}{1-\alpha} \int_{-\infty}^{\infty} p^\alpha(x)dx \tag{8.1.6}$$

for $\alpha \to 1$ this gives the *Shannon entropy*;

(b) The *Tsallis entropy* of $p(x)$ of order $q, q \neq 1$ is defined as

$$S_q = \frac{1 - \int_{-\infty}^{\infty} p^q(x)dx}{q-1} \tag{8.1.7}$$

for $q \to 1$ it reduces to the *Shannon entropy.*

There are reports on practical applications of these generalized entropies: the *Rényi entropy* has turned up in *hydraulics* (Kumbhakar and Ghoshal 2017; Kumbhakar et al. 2017b), the Tsallis entropy in hydraulics (Cui and Singh 2013, 2014; Singh et al. 2017; Zhu and Yu 2019), and *earthquake prediction* (Varotsos et al. 2018). In (Bonakdari et al. 2015) the two entropies are compared.

In *transfer phenomena*, the heterogeneities of the medium affect transfer properties. In *elastic wave propagation*, for example, the randomness of the medium slows down the propagating wave and causes dissipation of its coherent energy. It is well known, for example, that sound attenuation in crystalline materials is less for a single crystal than for an aggregate (Bradley and Fort 1966).

Ament (1953) studied *sound attenuation in suspensions* and derived for low frequencies the frequency-dependent *absorption coefficient* the expression

$$\alpha(\omega) = \frac{\omega^2}{9\eta} \frac{\Phi(1-\Phi)}{c_0^2} \frac{(\rho_1 - \rho_2)^2}{\varrho_0^2} r^2 \tag{8.1.8}$$

where $\omega = 2\pi f$; η is viscosity, c_0 average velocity, ρ_1, ρ_2, ρ_0 are densities (of the particles, of the water, and their weighted average),$\rho_0 = \Phi\rho_1 + (1-\Phi)\rho_2$; $c_0 = \Phi c_1 + (1-\Phi)$; Φ is volume fraction of the suspended particles. A similar relation was found by Beltzer (1978) for *sound absorption* in *randomly porous solids* with porosityΦ, where again $\alpha(\omega)$ was proportional to the factor $\Phi(1-\Phi)$ that he identified with the *randomness of porosity*. Korvin (1980, 1985) generalized these results for low-frequency sound attenuation in *N*-component rocks composed of different minerals in volume fractions $\Phi_1, \ldots, \Phi_N$; $\sum \Phi_i = 1$, and found that in this case $\alpha(\omega)$ is proportional to the factor

$$\chi = \sum_{i=1}^{N} \Phi_i (1 - \Phi_i) \tag{8.1.9}$$

what he termed *heterogeneity factor*. Similarly to entropy H, χ assumes its maximum at $\{\Phi_i = 1/N, i = 1, \ldots, N\}$, its value being

$$\chi_{max} = (N - 1)/N \tag{8.1.10}$$

Korvin compared the behavior of $\chi(\Phi_1, \ldots, \Phi_N)$ with Shannon's

$$H(\Phi_1, \ldots, \Phi_N) = -\sum_{i=1}^{N} \Phi_i \ln(\Phi_i) \tag{8.1.11}$$

(for which $H_{max} = \ln N$) and found that close to $\{\Phi_i = 1/N, i = 1, \ldots, N\}$ they are related as $\chi_{max} - \chi \approx \frac{2}{N}(H_{max} - H)$.

Byryakovskiy (1968) claimed that for multicomponent rocks $H = -\sum_{i=1}^{N} \Phi_i \ln(\Phi_i)$ is a "criterion for heterogeneity of the rock", in Russian literature it used to be called "*rock entropy*". Actually, in the statistical physics of *disordered materials*, the entropy S of a rock, which is a random aggregate of small grains of different lithology, always consists of two parts:

$$S_{rock} = S_{configurational} + S_{mixture} \tag{8.1.12}$$

(This is the *Flory–Huggins formula*, see Flory 1941; Huggins 1941; Ziman 1979: Sect. 7.2). The second term in this equation, *mixture entropy*, is same as the Shannon entropy (8.1.11) but in itself, without the *configurational term* describing the randomness of geometry, it cannot characterize the heterogeneity of a rock.

In connection with *radiative transfer in the atmosphere*, Casti and Tse (1972: 53) stated, "the *absorption function* … which corresponds to the radiative energy which is input to the atmosphere, but which is neither transmitted nor reflected back out …there is reason to suspect that this function may correspond to the loss of inherent information in the known control input due to interaction with the noisy system … If this corresponence can be made precise, it would seem to be possible to establish a *conservation law for stochastic systems*."

Solute transport in geological formations is controlled by the spatial distribution of *hydraulic conductivity* K, while conductivity itself depends on the variability in the textural properties (porosity, pore size, pore shape) of the sediments.

A recent model study of *solute transport in heterogeneous aquifers* (Bianchi and Pedretti 2017) found that the distribution of arrival times (BCT's, *Break Through Curves*) is determined by two measures of the aquifer's heterogeneity, (a) the variance of the *logarithmic permeability values*, σ_Y^2 (where $Y = \ln K$), and (b) a new *entropy indicator* (H_R) called *relative entropy index*. In the simulation the particles were driven only by advection, and they exited the system when reaching the right boundary.

The relative entropy index H_R was defined as follows. Let the whole aquifer consist of N *hydrofacies* in volumetric proportions $p_{G,n}, n = 1, 2, \ldots, N$. (Different hydrofacies have different values of *permeability* K, or *logarithmic permeability* Y, the index "G" means "global"). The *global entropy* of the *hydrofacies distribution* is

$$H_G = -\sum_{n=1}^{N} p_{G,n} ln p_{G,n} \tag{8.1.13}$$

Consider a randomly selected block of the whole aquifer, of linear size l and calculate its *local entropy*

$$H_L(l) = -\sum_{n=1}^{N} p_{l,n} ln p_{l,n} \tag{8.1.14}$$

where "l" is the linear size of the block; "L" means "local", and $\{p_{l,n}\}, n = 1, \ldots, N$ is the marginal fractional volume distribution of the hydrofacies. Bianchi and Pedretti (2017) fixed $l = 3$ that is $H_L(l)$ was estimated from 27 values. If $\overline{H_L} = \langle H_L(l) \rangle$ is the average of the local entropies over the entire domain, then the *relative entropy index* H_R is defined as:

$$H_R = \overline{H_L}/H_G \tag{8.1.15}$$

Introduce the following notations: L_h *isotropic mean size* of the hydrofacies; K *permeability*; $Y = ln K$; H_R *relative entropy index* (8.1.15); μ_t *mean arrival time* of solute particles; σ_t^2 *variance of the arrival times*; sk_t *skewness* of the arrival time distribution; σ_Y^2 variance of $ln K$, and let the symbol " ~ " indicate the *median value* of the distributions of moments. With these notations, Bianchi and Pedretti (2017) established the following empirical equations. The constants $a, b, \alpha, \beta, \gamma$ are different in each of the following equations:

$$H_R = a \cdot L_h^b \tag{8.1.16a}$$

where the factor a is positive and the exponent b is negative;

$$\tilde{\mu}_t = a H_R^b \tag{8.1.16b}$$

$$\tilde{\sigma_t^2} = a + b \cdot ln H_R \tag{8.1.16c}$$

$$\tilde{sk}_t = a \cdot H_R^b \tag{8.1.16d}$$

$$\tilde{sk}_t = \alpha\left(1 - exp\left[\beta \sigma_Y^2\right]\right) \tag{8.1.16e}$$

where α is positive and β is negative;

$$\widetilde{sk}_t = \alpha\left(1 - exp\left[-\frac{\sigma_Y^2}{\beta}\right]\right)H_R^{-\gamma} \tag{8.1.16f}$$

where $\alpha, \beta, \gamma > 0$;

$$\widetilde{\sigma_t^2} = exp\left[a\left(\sigma_Y^2\right)^b\right] \tag{8.1.16g}$$

$(a > 0, b > 0)$; and

$$\frac{\tilde{\mu}_t}{t_{eq}} = exp\left[a\sigma_Y^2\right] \tag{8.1.16h}$$

where $a > 0$ and t_{eq} is the arrival time of the particles in a *homogeneous field* with constant K equal to the *equivalent conductivity* (K_{eq}) of the *heterogeneous field*. K_{eq} is defined (by Renard and de Marsily 1997; Sanchez-Vila et al. 2006) as

$$K_{eq} = -QL/A\Delta h \tag{8.1.17}$$

with Q the *total flow rate*, L the distance between input and output sides, Δh *differential head*, and A *cross-sectional area.*

Equation 8.1.16a indicates that entropy, as any other extensive rock property, might depend on the *size of the medium* (in this case L_h) from which it is estimated. In other words, it needs a REV, a *Representative Elementary Volume* (or, in 2-D, REA, *Representative Elementary Area*) for its unbiased estimation (Bear 1972; Bear and Bachmat 1984). Martinez et al. (2007) used *entropy dimension* (also called *information dimension*) to find the REA for *multifractal analysis of soil porosity. Information dimension* D_I (Korvin 1992: 173, Eqs. A2.1.5–A2.1.7) is a measure of how the number of bits necessary to describe a fractal point-distribution scales with resolution. In case of a d-dimensional cube of unit size divided to 2^{dk} subcubes of size $r = 2^{-k}$ we need $S(r) = dk$ bits to specify the position of a point hidden in one of the subcubes, that is

$$d = \frac{-S(r)}{log_2 r} \tag{8.1.18}$$

If the fractal is partitioned into boxes B_i of size r, such that a randomly selected point is in B_i with probability p_i, then the average information needed to locate the box containing it is

$$S(r) = -\sum_i p_i log_2 p_i \tag{8.1.19}$$

On the analogy of Eq. (8.1.18), the *information dimension* (or *entropy dimension*) is defined as

$$D_i = \lim_{r \to 0} \frac{\sum_i p_i log_2 p_i}{log_2 r} \tag{8.1.20}$$

In 2D microscopic images of porous rocks, in most cases the image is *binary*, and porosity Φ is understood as the proportion of white pixels on the image. For this case Bird et al. (2006) established the following bounds for the Shannon entropy of the image:

$$2log\left(\frac{L}{\varepsilon}\right) + log\Phi \le -\sum p_i log p_i \le 2log\left(\frac{L}{\varepsilon}\right) \tag{8.1.21}$$

where L is image size, ε pixel size.

Among geologists, the most popular application of entropy is connected with the *Markov chain analysis* of the succession of sedimentary layers (Dowds 1969; Gingerich 1969; Miall 1973; Hattori 1976; Power and Easterling 1982; Hota and Maejima 2004; Elfeki and Dekking 2005; Sinha et al. 2015; Szabolcs 2016; Tita and Djomeni 2016). *Markov chains*, and their *entropy*, as taught in *Probability Theory* (Gagniuc 2017) and as actually used in *sedimentology*, are different in several respects. In the standard sedimentological procedure (Hattori 1976; Power and Easterling 1982; Sinha et al. 2015; Tita and Djomeni 2016) several matrices and kinds of entropies are calculated:

(i) The *transition frequency matrix* $F = F_{ij}$ contains zero elements in the main diagonal $i = j$ that is only *abrupt jumps from one lithology to a different lithology* are recorded;

(ii) The *transition frequency matrix* $F = F_{ij}$ records the number of vertical (upward) transitions between different lithofacies in the stratigraphic column. The row number i refers to the lower ("overlain") facies, the column number j refers to the upper ("overlying") facies;

(iii) The *upward transition probability matrix* $P = P_{ij}$ contains the probabilities of the upward transitions between lithofacies in the succession:

$$P_{ij} = F_{ij}/S_{R_i} \tag{8.1.22a}$$

where S_{R_i} is total number of transitions in the i-th row;

(iv) *The downward transition probability matrix* $Q = Q_{ij}$ contains the probabilities that facies i is overlain by facies j, that is

$$Q_{ji} = F_{ij}/S_{C_j} \tag{8.1.22b}$$

where S_{C_j} is total number of transitions in the j-th column;

(v) The *independent trail matrix* $R = R_{ij}$ contains the probabilities of randomly occurring transitons:

$$R_{ij} = \frac{S_{C_j}}{\left(S_T - S_{R_i}\right)}, \; S_T = \sum_{i=1}^{n}\sum_{j=1}^{n} F_{ij} \qquad (8.1.22c)$$

where S_T is the total number of transitions in the $F = F_{ij}$ matrix;

(vi) The *difference matrix* $D = D_{ij}$ helps to find those transitions that have a probability of occurrence greater (or smaller) than in case of a completely random succession.

$$D_{ij} = P_{ij} - R_{ij} \qquad (8.1.22d)$$

A positive value in D indicates that a particular transition occurs *more frequently* and a negative value indicates that it occurs *less frequently* (Miall 1973);

(vii) The *expected frequency matrix* $E = E_{ij}$ contains the expected number of transitions from lithology i to j,

$$E_{ij} = R_{ij} \times S_{R_i} \qquad (8.1.22e)$$

Hattori (1976) introduced two kinds of entropies with respect to each lithological state; one is *post–depositional entropy* computed from matrix P; the other is *pre–depositional entropy* computed from matrix Q, and also defined *total entropy for the whole sedimentation system*, computed from matrix F.

(viii) *Post-depositional entropy with respect to lithofacies state i* is defined as

$$E_i^{(post)} = -\sum_{j=0}^{n} P_{ij} \log P_{ij} \qquad (8.1.22f)$$

(ix) *Pre-depositional entropy with respect to state i* is defined as

$$E_i^{(pre)} = -\sum_{j=0}^{n} Q_{ij} \log Q_{ij} \qquad (8.1.22g)$$

(x) The *entropy of the whole system* is

$$E^{(system)} = -\sum_{i=1}^{n}\sum_{j=1}^{n} r_{ij} \log r_{ij},$$

$$where\ r_{ij} = \frac{F_{ij}}{S_T}; \; S_T = \sum_{i=1}^{n}\sum_{j=1}^{n} F_{ij} \qquad (8.1.22h)$$

The values of the entropies $E_i^{(post)}$, $E_i^{(pre)}$, $E^{(system)}$ increase with the number of lithological facies. To compensate for this effect, one can use normalized entropies $E_{norm} = E/E_{max}$ where

$$E_{i,max}^{(post)} = E_{i,max}^{(pre)} = -log\left(\frac{1}{n-1}\right);\ E_{max}^{(system)} = -log\left(\frac{1}{n(n-1)}\right) \quad (8.1.22i)$$

Hattori (1976) used cross plots of $E^{(system)}$ versus n (*number of lithologies*), and $E_{norm}^{(post)}$ versus $E_{norm}^{(pre)}$ crossplots, to classify sedimentary cycles based on their *Markov chain entropy*.

The structure of certain *clastic rocks* is well-approximated by jammed, disordered packings of identical spherical grains, so their randomness can be gauged by the *configurational entropy*

$$S \propto \log W \quad (8.1.23)$$

where W is the expected number of *topologically different realizations*, assuming the spheres are undistinguishable and all configurations are equally probable (Vink and Barkema 2002; Briscoe et al. 2010). To count the possible configurations we take recourse of *Graph Theory*. To transform a packing to a *graph*, first we construct the Voronoi cells around each particle, and define real and virtual *Delaunay contacts* when two particles share the same Voronoi side (In *real contact*, the grains physically touch, see Fig. 8.2 in DeGiuli and McElwaine 2011). The vertices of the graphs will correspond to the spheres, with edges between those which are in real Delaunay contact. The classes of topologically equivalent graphs can be found, counted, and the probability of their occurrence determined, by using the algorithms in (Weinberg 1966; McKay 1990), or–more practically–by randomly taking a large number m of clusters of size n from the analyzed packing and count the number of times when a graph belonging to a given class occurs. If clusters belonging to class C_i are observed f_i times, then the probability of this class is

$$p_i \approx f_i/m \quad (8.1.24)$$

where the equation becomes an equality if $m \to \infty$. In case of $\mathcal{N} \gg 1$ classes, the topological entropy can be estimated as

$$H = -\sum_{i=1}^{\mathcal{N}} p_i log p_i \approx -\sum_{i=1}^{\mathcal{N}} \left(\frac{f_i}{m}\right) log\left(\frac{f_i}{m}\right) + O\left(\frac{1}{m}\right) \quad (8.1.25)$$

the precise form of the correction term is derived in (Briscoe et al. 2010: Eq. 22).

There have been many proposals (Sotolongo and Posadas 2004; De Santis et al. 2011; Varotsos et al., 2018; Akopian 2014, 2015; Lopes and Machado 2016; Bressan et al. 2017; Sarlis et al. 2018; Posadas et al. 2021) for using the changes in entropy,

H, for *earthquake prediction* in seismically active regions, from an analysis of past observations of increases in H and a subsequent occurrence of a great earthquake.

If the states of the *seismic system* at time t are given by a probability distribution $P(M, t)$ where M is earthquake magnitude, the Shannon entropy of the system is

$$H(t) = -\int_{M_0}^{\infty} P(M, t) log M dM \tag{8.1.26}$$

According to (Aki 1965; Utsu 1965), if the maximum magnitude is much greater than the minimum magnitude M_0 (in practice, if $M - M_0 \geq 2$), the probability density function for an earthquake distribution with magnitudes M is given by

$$P(M) = b \cdot \ln(10) \cdot 10^{-b(M-M_0)} \tag{8.1.27}$$

where b is the slope in the *Gutenberg-Richter law*

$$\log n(> M) = a - b \cdot M \tag{8.1.28}$$

with $n(> M)$ being the number of earthquakes with magnitude greater than M. Aki (1965) and Utsu (1965) also proved that the b-value figuring in the Gutenberg-Richter law is

$$b = \frac{\log e}{\left[\overline{M} - \left(M_0 - \frac{\Delta M}{2}\right)\right]} \tag{8.1.29}$$

Using this value in Eq. (8.1.26), De Santis et al. (2011) obtained

$$H(t) = log(e \cdot loge) - logb = loge + log\left[\overline{M} - \left(M_0 - \frac{\Delta M}{2}\right)\right] \tag{8.1.30}$$

where $\overline{M}$ is average magnitude in the considered time interval, ΔM is resolution of the determination of magnitudes.

We note that the connection between past and future states of a *random dynamical system*, that is the dependence of $Q = S(t + \tau), \tau > 0$ on $S(t)$, is determined by their *mutual information* μ_I (Fraser and Swinney 1986). The mutual information is given by the *Kullback–Leibler formula* (Jumarie 1990):

$$\mu_I = \sum_{i=1}^{n} \sum_{j=1}^{m} P_{SQ}(s_i, q_j) \cdot log\left(\frac{P_{SQ}(s_i, q_j)}{P_S(s_i) \cdot P_Q(q_j)}\right) \tag{8.1.31}$$

where $P_S(s_i)$ is the probability distribution of the present state, $P_Q(q_j)$ are probabilities of the future state; $P_{SQ}(s_i, q_j)$ is the joint probability distribution of the S and Q states.

8.2 The Maximum Entropy Method

The first scientist to assign probabilities to events was Laplace (1886): he advised that when there is no reason to do otherwise, assign to all outcomes *equal probabilities.* Boltzman (1877) introduced the concept of the *most probable distribution*, using a technique which is similar to the modern *Maximum Entropy* procedure. Planck (1906) maximized the entropy subject to some constraints (this had led him to the so-called *Planck distribution*), his technique has been formalized and made precise by Gibbs (1902). Von Neumann (1927) introduced entropy to *Quantum Mechanics.* Elsasser (1937) stated that the entropy of quantum systems must be maximized. Hartley (1928), Shannon (1948), Shannon and Weaver (1949), and Wiener (1948) worked out the *Theory of Information.* Cox (1946) developed a formal theory of the *algebra of probable inferences.* Jaynes (1957) stated the principle of *maximum entropy* in its most general form, based on Shannon's axioms characterizing the "*uncertainty function*", on Shannon's definition of entropy:

$$H(p) = -\int_{-\infty}^{\infty} p(x) lnp(x) dx$$

where $p(x)$ is a *probability density function*, and on Cox's insight that probability reflects our state of knowledge.

The principle of maximum entropy states that if we are seeking a probability distribution subject to certain constraints, we should select that distribution which satisfies these constraints and has its Shannon-entropy as large as possible. (Without constraints, and for a finite number of outcomes, this reduces to Laplace's principle favoring the *uniform distribution*).

We introduce *maximal entropy* using an *inverse geophysical problem* as example. Make physical measurements of K rock-properties $\sigma_1, \sigma_2, \ldots, \sigma_K$ on the surface of a formation consisting of randomly arranged thin sedimentary layers. We might measure for example the density (by gravity measurement or γ-ray absorption), electric conductivity, hydraulic conductivity, sound speed, etc., of the formation. We shall get K measured values $\sigma_{1,\, meas}, \ldots, \sigma_{K,\, meas}$.

Generally, there are much fewer possible lithologies (sand, shale, dolomite, etc.–let us denote them by $\Lambda_1, \Lambda_2, \ldots, \Lambda_L$) than the number of layers N ($L \ll N$). Suppose that if we know the *fractional thickness distribution* of lithologies$\left\{\Phi_1, \Phi_2, \ldots, \Phi_L | \sum \Phi_i = 1\right\}$, and the *specific values* for each lithology of the measured properties σ: $\left\{\sigma_{k,1}^{(spec)}; \ldots \sigma_{k,L}^{(spec)}\right\}$. $k = 1, 2, \ldots, K$, then we can express theoretically each σ_k with explicit formulae:

$$\sigma_k^{(meas)} = f_k(\Phi_1, \ldots, \Phi_L; \sigma_{k,1}^{(spec)}; \ldots; \sigma_{k,L}^{(spec)}); k = 1, 2, \ldots, K \qquad (8.2.1)$$

In this system of nonlinear equations there are K equations and L unknowns, $\Phi_1, \ldots, \Phi_L$. If $K < L$, there are an infinite number of ways to satisfy (8.2.1)–which solution $\{\Phi_1, \Phi_2, \ldots, \Phi_L\}$ of the nonlinear system (8.2.1), subject the conditions

$$0 \leq \Phi_i; \sum \Phi_i = 1 \tag{8.2.2}$$

is the *most likely*?

If there are N_1 layers of lithology Λ_1, N_2 layers of lithology $\Lambda_2, \ldots, N_L$ layers of lithology Λ_L, such that $\sum_{i=1}^{L} N_i = N$, then $\left\{\Phi_i = \frac{N_i}{N}\right\}, 0 \leq \Phi_i \leq 1; \sum_{i=1}^{L} \Phi_i = 1$ is a *probability distribution*. Every probability distribution represents a certain degree of *randomness*. For example, a distribution $\{\Phi_1 = 1; \Phi_2 = \Phi_3 = \cdots = \Phi_L = 0\}$ is *not random* at all; $\{\Phi_1 = \Phi_2 = \Phi_3 = \cdots = \Phi_L = 1/L\}$ is *the most random;* $\left\{\Phi_1 = 0.25; \Phi_2 = 0.25; \Phi_3 = \Phi_4 = \cdots = \Phi_L = \frac{1}{2(L-2)}\right\}$ is something in between. To characterize the *randomness of the lithology distribution* $\left\{N_i | \sum N_i = N\right\}$ we count how many ways can we classify N objects such that N_1 of them are in group Λ_1, N_2 are in group Λ_2, ..., N_L are in group Λ_L; the larger this number, the more random we consider the distribution. This number N^* is given by the *polynomial coefficient*

$$N^* = \frac{N!}{1!2!\ldots N_{L!}} \tag{8.2.3}$$

It is easier to estimate the logarithm of N^*, recalling that for $n \gg 1$ we have

$$\ln n! = ln1 + \cdots + \ln(n) \approx \int_1^n lnxdx \sim n \cdot \ln(n) \tag{8.2.4}$$

the simplified *Stirling formula*. Thus, we approximate the logarithm of N^* as:

$$\begin{aligned} lnN^* \sim NlnN - \sum_1^L N_i lnN_i = \sum_1^L N_i lnN - \sum_1^L N_i lnN_i = -\sum_1^L N_i ln\frac{N_i}{N} \\ = -N\sum_1^L \frac{N_i}{N} ln\frac{N_i}{N} = N\left(-\sum_1^L \Phi_i ln\Phi_i\right) \end{aligned} \tag{8.2.5}$$

The expression $H = -\sum_1^L \Phi_i ln\Phi_i$ (8.2.6) is the *Shannon-entropy* of the probability distribution $\{\Phi_i\}$. In the *Maximum Entropy Technique we accept that particular solution of the system (8.2.1 subject to conditions 8.2.2) which results in the most random lithology, in the sense that its entropy is maximal.*

We shall discuss three applications, from different parts of rock physics (Korvin 2021).

8.2.1 *Maximum Entropy (ME) Pore Shape Inversion*

Doyen (1987) used the ME method to determine the *pore and crack distribution* in igneous rocks from *hydraulic conductivity* and/or *dc electric conductivity* values measured at a series of confining pressures. His rock model consisted of a random distribution of spherical pores connected with each others by tubes of elliptic cross-section ("throats"). This model has been later applied to a wide range of problems in (Akbar et al., 1994; Korvin et al. 2014).

At a given reference pressure $P = P_0$ the pores' radii r are distributed according to the *probability density function* (pdf) $n(r)$; the throat-length l is distributed according to the *pdf* $n(l)$; the throats have elliptic cross-section with semi-axes b and c. The bivariate *pdf* of the throat-shape at reference pressure is $n(\alpha,c)$ where α is the aspect ratio $\alpha = b/c$. Using *Elasticity Theory* (Zimmerman 1991) to compute the deformation of voids under pressure P, the *pdf*'s for $r(P)$, $c(P)$, $l(P)$, $a(P)$ can be determined at any pressure steps from their values at reference pressure. The *pdf* $n(\alpha,c)$ satisfies the *self-consistency equation* (Kirkpatrick 1973) for all pressure steps P:

$$\sum_{\alpha,c} n(\alpha, c)\frac{g_e^*(P) - g_e(P, \alpha, c, l)}{g_e(P, \alpha, c, l) + \left(\frac{Z}{2} - 1\right)g_e^*(P)} \tag{8.2.6a}$$

$$\sum_{\alpha,c} n(\alpha, c)\frac{g_h^*(P) - g_h(P, \alpha, c, l)}{g_h(P, \alpha, c, l) + \left(\frac{Z}{2} - 1\right)g_h^*(P)} = 0 \tag{8.2.6b}$$

where Z is the average *pore coordination number* (number of throats incident to a given pore); $g_e^*(P)$ and $g_h^*(P)$ are the measured bulk electric- and hydraulic conductances at pressure P, and $g_e = g_e(P, \alpha, c, l)$ and $g_h = g_h(P, \alpha, c, l)$ are the theoretical conductances of an originally (α,c)-shaped throat of length l subjected to pressure P.

Elasticity Theory also provides the theoretical value $\beta_p(r, \Phi)$ for the compressibility of a rock at pressure P containing a volume-fraction Φ of r-sized pores; and an expression for its expected value $\beta_P(\{n(r)\},\Phi) := \langle\beta_P(\mathrm{r}, \Phi)\rangle_{n(r)}$. Similarly, for a volume fraction Φ of throats of length l and cross-sectional shape (a,c) both $\beta_P(\alpha, c, l, \Phi)$ and its expected value can be theoretically determined.

Subjecting the rock to pressure P, the pore parameters change as $r \to r(P), l \to l(P), c \to (P), \alpha \to \alpha(P)$, and the measured bulk compressibility becomes $\beta(P)$. Assuming that the compressibilities due to pores and throats are independent and additive, one gets

$$\beta_p\big(\{n(r)\}, \Phi_p, P\big) + \beta_l(\{n(\alpha, c)\}, \Phi_l, l, P) = \beta(P) \tag{8.2.7}$$

Let $V(r)$ be the volume of a pore of radius r, and $V(a,c,l)$ the volume of an (α,c)-shaped tube of length l. Then $n(r)$ and $n(\alpha,c)$ satisfy the additional constraints:

$$\sum\nolimits_r n(r) = 1, \sum\nolimits_{\alpha,c} n(\alpha, c) = 1, N_p \sum\nolimits_r n(r)V(r)$$

$$+ N_t \sum\nolimits_{\alpha,c} n(\alpha.c) V(\alpha.c, l) = \Phi_p + \Phi_t = \Phi \tag{8.2.8}$$

where N_p and $N_t = N_p \frac{Z}{2}$ are the numbers of pores, respectively throats, in a unit volume of rock. By simple geometry, N_p, N_t, Φ_p, Φ_t, Z, l can be expressed in terms of *n(r)*,*n*(α, *c*), Φ. As there are much more possible values of *n(r)* and *n(a,c)* than pressure steps *P*, we assume that *n(r)* and *n(a,c)* are *maximum entropy solutions*, that is

$$-\sum_{r} n(r) \log n(r) = \max; \quad -\sum_{\alpha,c} n(\alpha, c) \log n(\alpha, c) = \max \tag{8.2.9}$$

subject to the conditions

$$\sum_{\alpha,c} n(\alpha, c) \frac{g_e^*(P) - g_e(P, \alpha, c, l)}{g_e(P, \alpha, c, l) + \left(\frac{Z}{2} - 1\right) g_e^*(P)} = 0 \tag{8.2.10a}$$

$$\sum_{\alpha,c} n(\alpha, c) \frac{g_h^*(P) - g_h(P, \alpha, c, l)}{g_h(P, \alpha, c, l) + \left(\frac{Z}{2} - 1\right) g_h^*(P)} = 0 \tag{8.2.10b}$$

$$\beta_p\big(\{n(r)\}, \Phi_p, P\big) + \beta_l(\{n(\alpha, c)\}, \Phi_l, l, P) = \beta(P) \tag{8.2.10c}$$

$$\sum\nolimits_r n(r) = 1, \sum\nolimits_{\alpha,c} n(\alpha, c) = 1, N_p \sum\nolimits_r n(r) V(r)$$
$$+ N_t \sum\nolimits_n (\alpha.c) V(\alpha.c, l) = \Phi_p + \Phi_t = \Phi \tag{8.2.10d}$$

The constraining Eqs. (8.2.10a–8.2.10d) can be written in the common general form

$$\sum_{r,\alpha,c} n(r.\alpha, c) A(r, \alpha, c, P) = 0 \text{ for } P = P_1, P_2, \ldots, P_N \tag{8.2.11}$$

where $n(r, \alpha.c)$, satisfying $n(r, \alpha, c) \geq 0$; $\sum_{r,\alpha,c} n(r, \alpha, c) = 1$ is the joint pdf of r, α, c at reference pressure. Its Shannon entropy is

$$S[n(r, \alpha, c)] = -\sum_{r,\alpha,c} n(r, \alpha, c) log n(r, \alpha, c) \tag{8.2.12}$$

Using Lagrange multipliers, we first determine the vector which minimizes the function

$$f\left(\lambda_{P_1}, \ldots, \lambda_{P_N}\right) = log\left[\sum_{r,\alpha,c} exp\left\{-\sum_{k=1}^{N} \lambda_{P_k} A(r, \alpha, c, P_k)\right\}\right] \tag{8.2.13}$$

and then find the ME solution for the *pdf*:

$$n(r, \alpha, c) = \frac{exp\left\{-\sum_{k=1}^{N} \lambda_{P_k} A(r, \alpha, c, P_k)\right\}}{\sum_{r,\alpha,c} exp\left\{-\sum_{k=1}^{N} \lambda_{P_k} A(r, \alpha, c, P_k)\right\}} \tag{8.2.14}$$

8.2.2 *Shale Compaction and Maximal Entropy*

Athy's (1930) law, expressing the decrease of porosity as function of depth in thick pure shale columns states that:

$$\rho(z) = \rho_0 + (\rho_\infty - \rho_0)\big[1 - \exp(-kz)\big] \tag{8.2.15}$$

where ρ_0 is surface density, ρ_∞ the upper limit of shale density, z depth and k is a constant. An equivalent form of this law describes the decrease of porosity with depth:

$$\Phi(z) = \Phi_0 \exp(-kz) \tag{8.2.16}$$

where $\Phi(z)$ is the porosity at depth z,Φ_0 is the porosity at the surface. Assuming all pores are of the same volume, the porosity of a rock is proportional to the number of pores in a unit volume of the rock, and Athy 's law states that the pores in compacted shales are distributed in such a manner that their number in a unit volume of rock exponentially decreases with depth. Korvin (1984) noted that there are several analogies of this rule from different branches of physics. The most familiar one is the *barometric equation of Boltzmann*, which gives the density *n(z)* of the air at an *altitude h* as

$$n(h) = n(0) \cdot exp\left[-\frac{mgh}{kT}\right] \tag{8.2.17}$$

where m is the mass of a single gas molecule, g gravity acceleration, k Boltzmann's constant, T absolute temperature. In statistical physics (Landau and Lifshitz 1980) the proof of (8.2.17) is based on the assumption that the gas particles move independently of each other and the whole system tends toward its most probable (*maximum entropy*) state.

The decrease in porosity of shales during compaction is due to the expulsion of water with some mechanical rearrangement of clay particles towards a more dense system of packing. Following Litwiniszyn's model (1974), we shall consider the compaction of shales as an *upward migration of pores*. Consider a rectangular prism of unit cross-section of the compacted shale down to the basement at depth Z_0 and suppose in its present state its *mean porosity* is Φ i.e. it contains a fractional volume ΦZ_0 of fluid and a volume $(1 - \Phi)Z_0$ of solid clay particles. Assume that the system's evolution is *ergodic*, i.e. it tends towards the *maximum-entropy final state*. Neglecting the actual history of how the clay particles had been deposited into the water, we assume that at time $t = 0$ the initial condition was that a prism of water of unit cross-section, height ΦZ_0 and density ρ_1 was overlain by a prism of solid clay particles of height $(1-\Phi)Z_0$ and densityρ_2, $\rho_1 < \rho_2$. Suppose the prism of water can be divided into $\mathfrak{N}$ particles (water-filled pores), each of volume ΔV, all of which start to migrate upwards independently of each other, until the final (*maximum entropy*) state is reached. The initial potential energy of the system is

$$E = \mathfrak{N} \cdot \Delta V \cdot g(\rho_2 - \rho_1)(1-\Phi)Z_0 \tag{8.2.18}$$

the initial porosity and particle number are connected by

$$\mathfrak{N} = \frac{Z_0\Phi}{\Delta V} \tag{8.2.19}$$

We use a coordinate system (x, y, z). z pointing downwards, and divide the prism H into N slabs of equal thickness $\Delta z = Z_0/\Delta z$. Denote the i-th slab by $\gamma_i (i = 0, 1, \ldots, N-1)$ where

$$\gamma_i = \{(x, y, z)|x\epsilon[0, 1], y\epsilon[0, 1]; i \cdot \Delta z \le z < (i+1) \cdot \Delta z\}(i = 0, 1, \ldots./, N-2) \tag{8.2.20a}$$

$$\gamma_{N-1} = \{(x, y, z)|x\epsilon[0, 1], y\epsilon[0, 1]; (N-1) \cdot \Delta z \le z \le Z_0\} \tag{8.2.20b}$$

We divide the entire prism H into $N^* = Z_0/\Delta V$ non-overlapping small cubes. We have $\mathfrak{N} \ll N^*$ if Φ is sufficiently small (by 8.2.19). We rank the N^* possible positions, called "states", of a pore into N groups: a pore is said to belong to the group γ_i if and only if its centre (x, y, z) lies within the slab γ_i. This implies every group γ_i contains $G = \Delta z/\Delta V$ states. Suppose that N_i pores are found in state γ_i. The numbers N_i satisfy two constraints, the *conservation of pore-particle number*, and the *conservation of total potential energy*:

$$\sum_{i=0}^{N-1} N_i = \mathfrak{N} \tag{8.2.21a}$$

$$\sum_{i=0}^{N-1} \varepsilon_i N_i = E \tag{8.2.21b}$$

where E is the total energy (see Eq. 8.3.2.18); ε_i is the potential energy of a single pore particle in group γ_i, due to *buoyancy*:

$$\varepsilon_i = g(\rho_2 - \rho_1) \cdot \Delta V \cdot i \cdot \Delta z \tag{8.2.22}$$

The set of numbers $\{N_i\}$ determine the *macroscopic* distribution of pores inside H. We determine next the *thermodynamical probability* Γ of the macroscopic state $\{N_i\}$, i.e. how many microscopic realizations of this state exists. If the groups γ_i are independent and Γ_i denotes the thermodynamic probability of the i-th group. then

$$\Gamma = \prod_{i=0}^{N-1} \Gamma_i \text{where } \Gamma_i = \binom{G}{N_i} = \frac{G!}{N_i!(G - N_i)!} \tag{8.2.23}$$

Apart from a constant factor, the entropy of the distribution {Ni} is the logarithm of Γ:

$$S = log\Gamma = \sum_{i=0}^{N-1} log\Gamma_i \tag{8.2.24}$$

Since the numbers G, N_i, G$-N_i$ are large, Stirling's formula applies and, apart from a constant factor and an additive term

$$S = -\sum_{i=1}^{N-1} N_i log\frac{eG}{N_i} \tag{8.2.25}$$

Denote the average number of pore particles in group γ_i by $\overline{n}_i$, then $\overline{n}_i = N_i/G$ and

$$S = -G\sum_{i=1}^{N-1} \overline{n}_i log\frac{e}{\overline{n}_i} \tag{8.2.26}$$

while the constraints (8.2.21a, b) become

$$G \cdot \sum_{i=0}^{N-1} \overline{n}_i = \mathfrak{N}, G \cdot \sum_{i=0}^{N-1} \overline{n}_i \varepsilon_i = E \tag{8.2.27}$$

After a very long time the pore particles will be distributed such that the entropy (8.2.25) is maximal. To find this state we maximize the entropy (8.2.25) subject to the constraints (8.2.27a, b). Introducing Lagrange multipliers α, β

$$\frac{\partial}{\partial\overline{n}_i}\left(S + \alpha\frac{\mathfrak{N}}{G} + \beta\frac{E}{G}\right) = 0 \ (i = 0, 1, \ldots, N - 1) \tag{8.2.28}$$

that is

$$G \cdot (-log\overline{n}_i + \alpha + \beta\varepsilon_i) = 0 \ (i = 0, 1, \ldots, N - 1) \tag{8.2.29}$$

whence

$$\overline{n}_i = exp(\alpha + \beta\varepsilon_i) \tag{8.2.30}$$

A long computation gives

$$exp\ \alpha = \frac{\Phi}{1-\phi}, \beta = -\frac{1}{E} \tag{8.2.31}$$

and

$$\overline{n}_z = \frac{\Phi}{1-\phi} exp\left[-\frac{z}{\left(1-\frac{\Phi}{(1-\phi)Z_0}\right)}\right] \tag{8.2.32}$$

The expected number of pores inside a slab (z, $z + dz$) of the prism H is $N_z = \frac{dz}{\Delta V}\overline{n}_z$, their total volume is $N_z \cdot \Delta V = dz \cdot \overline{n}_z$, that is the porosity of this slab is $\Phi_z = (N_z \cdot \Delta V) / dz$, and the porosity-depth dependence of compacted shale is expressed by the simple rule

$$\Phi(z) = \frac{\Phi}{1-\phi} \cdot exp\left[-\frac{z}{(1-\phi)Z_0}\right] \tag{8.2.33}$$

Identifying the first factor in (8.2.33) with the surface porosity Φ_0 of the shale: $\Phi_0 = \Phi/(1-\Phi)$ and, solving for Φ, (8.2.33) becomes in terms of surface porosity

$$\Phi(z) = \Phi_0 \cdot exp\left[-\frac{(1+\Phi_0)z}{Z_0}\right] \tag{8.2.34}$$

an equation that reproduces Athy's compaction law and qualitatively expresses the known *inverse dependence of compaction rate on basin depth* Z_0 (Nafe and Drake 1957).

8.2.3 Contact Force Distribution in Granular Media

Bagi (2003) determined the probability density functions of *contact force components* corresponding to the principal directions of average stress in *random granular assemblies*, by maximizing the statistical entropy, and proved that the distribution of force magnitude is exponential. She considered a body consisting of randomly packed, cohesion-less grains. If external stresses are acting on this body's boundary, contact forces will arise between neighboring grains and the average stress of the assembly, $\bar{\sigma}_{ij}$, depends on the contact forces f_i between neighbor grains and on the vectors l_i pointing from the center of the first grain to the center of the second

$$\bar{\sigma}_{ij} = \frac{1}{V}\sum_{c=1}^{M} l_i^c f_j^c \tag{8.2.35}$$

(M is the total number of contacts; the index "c" runs over all contacts; i, j = 1, 2, 3 show the vector components; V is total volume). The axes of the system of coordinates $\left(x_1, x_{2,} x_3\right)$ are selected to be parallel to the principal directions of $\bar{\sigma}_{ij}$ hence, $\bar{\sigma}_{ij} = 0$ if $i \neq j$. Consider Eq. (8.2.35) at first only for the case i = j = 1:

$$\bar{\sigma}_{11} = \frac{1}{V}\sum_{c=1}^{M} l_1^c f_1^c \tag{8.2.36}$$

Since, when two grains are in contact, it is arbitrary which one is the 'first' and which is 'second', we can always make the sign of f_1^c positive, that is the possible values of f_1 lie in (0, + ∞). We discretize this domain to equal (small) intervals: a particular f_1^c can fall into the 1st interval where the medium value is F_{11}, into the 2nd interval where the medium value is F_{12}, etc. The distribution of the f_1-component can be described by $p_1^{(1)}$, $p_2^{(1)}$, ..., $p_k^{(1)}$, ... these are the probabilities that f_1^c falls into the 1st, 2nd, ... kth,... interval. Next, consider all those contacts where the f_1^c contact force component falls into a chosen k-th interval (that is $f_1^c \doteq F_{1k}$). In these contacts, focus on the branch vectors. Denote the average of the l_1^c components of these vectors as $\bar{l}_{1k}^{(1)}$, where the superscript "(1)" expresses that the grouping was done according to the f_1-components. Applying the above notations, Eq. (8.2.36) becomes

$$\bar{\sigma}_{11} = \frac{1}{V}\sum_{k=1}^{\infty}\left(M \cdot p_k^{(1)}\right)\bar{l}_k^{(1)} F_{1k}^k \tag{8.2.37}$$

For the case i = 2, j = 1 we similarly get

$$\bar{\sigma}_{21} = \frac{1}{V}\sum_{k=1}^{\infty}\left(M \cdot p_k^{(1)}\right)\bar{l}_{2k}^{(1)} F_{1k}^k \tag{8.2.38}$$

and with similar reasoning we get nine equations, summarized as

$$\bar{\sigma}_{ij} = \frac{1}{V}\sum_{k=1}^{\infty}\left(M \cdot p_k^{(j)}\right)\bar{l}_{ik}^{(j)} F_{jk}^k \tag{8.2.39}$$

Suppose the geometry and average stress are given and we are searching for the *distribution* of the *i*-th contact force component $\{f_i\}$. The unknown $\{p_1, p_2, \ldots, p_k, \ldots\}$ probabilities satisfy

$$\sum_{k=1}^{\infty} p_k = 1 \tag{8.2.40}$$

and

$$\bar{\sigma}_{ij} = \frac{M}{V} \sum_{k=1}^{\infty} p_k \bar{l}_{ik}^{(j)} F_{jk} \tag{8.2.41}$$

We need to make an assumption on the $\bar{l}_{ik}^{(j)}$ values. Suppose that they are constant:

$$\bar{l}_{i1}^{(j)} = \bar{l}_{i2}^{(j)} = \bar{l}_{i3}^{(j)} = \cdots := \bar{l}_{i}^{(j)} \tag{8.2.42}$$

The average stress then can be expressed as

$$\bar{\sigma}_{ij} = \frac{M}{V} \bar{l}_i \sum_{k=1}^{\infty} p_k F_{jk} \tag{8.2.43}$$

We are looking for that particular distribution which has the highest entropy:

$$-\sum\nolimits_{k=1}^{\infty} p_k log p_k = max, \tag{8.2.44}$$

subject to the conditions

$\sum_{k=1}^{\infty} p_k = 1$ and $\tilde{\sigma}_{ij} = \frac{M}{V} \sum_{k=1}^{\infty} p_k \bar{l}_{ik}^{(j)} F_{jk}$ (see 8.2.40 and 8.2.41).

Introduce the Lagrange-multipliers α and β associated with the respective constraints (8.2.40 and 8.2.41). Easy algebra yields the solution:

$$p_k = \frac{1}{exp(1+\alpha)} \cdot exp\left(-\beta \bar{l}_i^{(j)} F_{jk}\right) \tag{8.2.45}$$

where the Langrange-multipliers can be computed from the following system of non-linear equations (using the given data $\bar{\sigma}_{ij}$, $\bar{l}_i$, V and M):

$$\sum_{k-1}^{\infty} F_{jk} exp\left(-1 - \beta \bar{l}_i^{(j)} F_{jk} - \alpha\right) - \frac{V}{M \bar{l}_i^{(j)}} \bar{\sigma}_{ij} = 0 \tag{8.2.46}$$

$$\sum_{k=1}^{\infty} exp\left(-1 - \beta \bar{l}_i^{(j)} F_{jk} - \alpha\right) - 1 = 0 \tag{8.2.47}$$

Equation (8.2.45) expresses that the *contact force components are exponentially distributed*, provided that (8.2.42) is an acceptable simplification.

8.3 Appendix: Relation Between the Hurst Exponent and the Entropy of a Self-affine Random Process

It has been reported (Fig. 8.3 in Martinez et al. 2007) that the entropy estimated from increasing volumes of random spatial data first increased, and has only become stationary when we reached the *Representative Elementary Area,* or *Volume* (REA, or REV). In this Appendix we investigate whether we experience a similar increase in entropy when we estimate it from increasingly longer temporal sections $t\epsilon[0, T]$ of a *self-affine random function* $x(t)$ with Hurst exponent α. (For *self-affine functions* and their *Hurst exponent* see Korvin 1992, 2022. We use the physicists' notation α for the Hurst exponent, as e.g. in Barabási and Stanley 1995, rather than "H", because H is used in this book for the Shannon entropy.)

After some preparations, we shall prove the following.

Theorem *Let* $x(t)$ *be a self-affine random function in* $(0, T)$ *which has* 0 *mean value and a scatter growing with T as* $\sigma = \langle x^2(t)\rangle^{1/2}\Big|_{t\epsilon[0,T]} \propto T^{\alpha}$ *as* $T \to \infty$. *Suppose the shape of the pdf of* $x(t)$ *in* $(0, T)$ *scales with* σ *according to Eq.* (8.3.10). *Then the Shannon entropy* $H(T)$ *of the amplitude distribution of* $x(t)$ *computed from the set of values* $\{(x(t)|t \in (0, T))\}$ *grows with* T *as*

$$H(T) = const + \alpha \cdot lnT \ as \ T \to \infty \tag{8.3.1}$$

Shannon defined the *entropy* of a continuous *pdf* (Probability Distribution Function) as:

$$H(p) = -\int_{-\infty}^{\infty} p(x)ln\ p(x)dx \tag{8.3.2}$$

if this integral exists.

For some distributions, *H(p)* can be computed in closed form (Shannon and Weaver 1949; Goldman 1955). For the Gaussian (normal) *pdf* with mean μ and variance σ^2:

$$\begin{aligned} H = &-\int_{-\infty}^{\infty} \frac{1}{\sqrt{2\pi}\cdot\sigma} exp\left[-\left(\frac{1}{2}\right)\left[\frac{x-\mu}{\sigma}\right]^2\right] \\ &\cdot\left(-ln\sqrt{2\pi}\cdot\sigma - (\frac{1}{2})\left[\frac{x-\mu}{\sigma}\right]^2\right) \\ &= \frac{1}{2}\left[1 + ln\left(2\pi\sigma^2\right)\right] \end{aligned} \tag{8.3.3}$$

For a *uniform distribution*, defined as $p(x) = \frac{1}{b-a}\ if\ x\epsilon[a, b], 0\ otherwise$, one has

$$H = -\frac{1}{b-a}\int_a^b ln\left(\frac{1}{b-a}\right)dx = \ln(b-a) \qquad (8.3.4)$$

Consider now a *self-affine random function* $x(t)$, with zero mean and Hurst exponent α, that is, letting angular brackets denote expected value, suppose that

$$\sigma = \langle x^2(t)\rangle^{1/2}\Big|_{t\epsilon(0,T)} \approx \left(\frac{1}{T}\int_0^T x^2(t)dt\right)^{1/2} \propto T^\alpha \qquad (8.3.5)$$

For a sampled version of the self-affine function, that is when $f_i = f(i\cdot\Delta)$; $(i = 0, 1, \ldots, N)$, the average variance of the functional values observed in $(0, N)$ scales as:

$$\langle f_i^2\rangle^{1/2} \approx \sqrt{\frac{1}{N+1}\sum_{k=0}^N f_k^2} \propto N^\alpha; 0 \le i \le N \qquad (8.3.6)$$

If the amplitudes $x(t)$ are *normally distributed*, from Eqs. (8.3.3 and 8.3.5) the Shannon entropy of the distribution of *x(t)* will grow, as function of observation length T:

$$\begin{aligned} H(T) &= \frac{1}{2}\left[1 + ln\left(2\pi\sigma^2\right)\right] = \frac{1}{2}\left[1 + ln\left(2\pi\cdot const\cdot T^{2\alpha}\right)\right] \\ &= const + \alpha\cdot lnT\ if\ T \to \infty \end{aligned} \qquad (8.3.7)$$

For a uniformly distributed variable in $(-a, a)$ we have $\frac{1}{2a}\int_{-a}^{a} x^2 dx = \frac{a^2}{3} = \sigma^2$, that is $a = \sigma\sqrt{3}$. Consequently, if the amplitudes of the self-affine, random *x(t)* are uniformly distributed, their average *range* will grow with T when $x(t)\epsilon\left(-const\cdot\sqrt{3}T^\alpha, +const\cdot\sqrt{3}T^\alpha\right)$, and by Eq. (8.3.4)

$$H(T) = \ln(const\cdot T^\alpha) \approx const + \alpha\cdot lnT\ if\ T \to \infty. \qquad (8.3.8)$$

We prove now a general Theorem, from which Eqs. (8.3.7) and (8.3.8) would also follow.

Theorem *Let x(t) be a self-affine random function with zero mean and Hurst exponent* α*, that is*

$$\sigma = \langle x^2(t)^{1/2}\Big|_{t\in(0,T)} \approx \left(\frac{1}{T}\int_0^T x^2(t)dt\right)^{\frac{1}{2}} = \sigma_0 T^\alpha \tag{8.3.9}$$

Assume that for $t \in [0, 1]$ *the pdf of* $x(t)$, $p(x)$, *only depends on* x *and on the single parameter* σ_0, *which is by Eq.* (8.3.9) *the scatter of* $x(t)$ *in* $[0, 1]$. *Assume further, that* $p(x) \equiv F(x, \sigma)$ *satisfies the scaling law*

$$F(x, \lambda\sigma) = F\Big(\frac{x}{\lambda}, \sigma\Big) \cdot \frac{1}{\lambda}\, for\ all\, \lambda > 0 \tag{8.3.10}$$

Denote by $p_T(x)$ the Pdf of $x(t)$ for $t \in (0, T)$. then the Shannon Entropy of $p_T(x)$ increases with time T as

$$H(p_T(x)) = const + \alpha \cdot lnT \tag{8.3.11}$$

Proof We apply the scaling law (8.3.10) with $\lambda = T^\alpha$:

$$H(p_T(x)) = -\int_{-\infty}^{\infty} p_T(x)ln[p_T(x)]dx = -\int_{-\infty}^{\infty} F(x, T^\alpha\sigma_0)ln\big[F(x, T^\alpha\sigma_0)\big]dx =$$

$$-\frac{1}{T^\alpha}\int_{-\infty}^{\infty} F\Big(\frac{x}{T^\alpha}, \sigma_0\Big)\Big\{lnF\Big(\frac{x}{T^\alpha}, \sigma_0\Big) - ln(T^\alpha)\Big\}dx =$$

$$-\int_{-\infty}^{\infty} F(y, \sigma_0)ln[F(y, \sigma_0]dy + \alpha lnT$$

where $y = \frac{x}{T^\alpha}$.

Note: It is straightforward to check that the *pdf* of the normal distribution with zero mean, and of the uniform distribution in $(-a, a)$, both satisfy the scaling rule (8.3.10), that is Eqs. (8.3.7) and (8.3.8) are special cases of the general law (8.3.11).

As found by (Scafetta et al. 2003) the fulfillment of *both* scaling laws $\sigma = \langle x^2(t)\rangle^{1/2}\Big|_{t\in(0,T)} = \sigma_0 T^\alpha$ (8.3.9), and $F(x, \lambda\sigma) = F\big(\frac{x}{\lambda}, \sigma\big)\cdot\frac{1}{\lambda}$ (8.3.10) are necessary to get $H(T) = const + \alpha T^\alpha$. If only Eq. (8.3.9) holds but the *pdf* is not scaling as (8.3.10), one can only derive an *upper bound* for *H(T),* by using the following extremum property of the normal distribution (Goldman 1955: 129–134; Shannon and Weaver 1949: Sect. 20): *For all pdf's with variance* σ^2 *one has*

$$H(p) \le \frac{1}{2}\big[1 + ln(2\pi\sigma^2)\big] \tag{8.3.12}$$

with Equality if and only if P is Gaussian with variance σ^2.

By Eq. (8.3.9), *for all self-affine functions satisfying* $\sigma = \langle x^2(t)\rangle^{1/2}\Big|_{t\epsilon(0,T)} = \sigma_0 T^\alpha$, *independently of whether the pdf is scaling like (8.3.10), the entropy cannot grow faster with T than* $(const + \alpha \cdot lnT)$.

References

Aczél J, Daróczy Z (1975) On measures of information and their characterization. Academic Press, New York

Agnieszka H (2013) Shear-jamming in two-dimensional granular materials with power-law grain-size distribution. Entropy 15:4802–4821

Akbar N, Mavko G, Nur A, Dvorkin J (1994) Seismic signatures of reservoir properties and pore fluid distribution. Geophysics 59(8):1222–1236

Aki K (1965) Maximum likelihood estimate of *b* in the formula *log (N)=a−bm* and its confidence limits. Bull Earthq Res Inst Tokyo Univ 43:237–239

Akopian ST (2014) Seismic systems of Japan: entropy and monitoring of the Tohoku earthquake, March 11, 2011. Seism Instrum 50(4):347–368

Akopian ST (2015) Open dissipative seismic systems and ensembles of strong earthquakes: energy balance and entropy funnels. Geophys J Int 201:1618–1641

Ament WS (1953) Sound propagation in gross mixtures. J Acoust Soc Am 25(4):638

Amigó J, Balogh SG, Hernández S (2018) Brief review of generalized entropies. Entropy 20:813

Anovitz LM, Cole DR (2015) Characterization and analysis of porosity and pore structures. Rev Mineral Geochem 80:61–164

Athy LI (1930) Compaction and oil migration. Bull Am Ass Petrol Geol 14:25–35

Bagi K (2003) Statistical analysis of contact force components in random granular assemblies. Granular Matter 5:45–54

Barabási AL, Stanley HE (1995) Fractal concepts in surface growth. Cambridge U. Press, Cambridge

Barmak K, Eggeling E, Emelianenko M, Epshteyn Y, Kinderlehrer D, Sharp R, Ta'asan S (2011) An entropy based theory of the grain boundary character distribution. Discret Contin Dynam Syst 30(2):427

Barmak K, Eggeling E, Emelianenko M, Epshteyn Y, Kinderlehrer D, Sharp R, Ta'asan S (2011) Critical events, entropy, and the grain boundary character distribution. Phys Rev B 83(13):134117

Bear J (1972) Dynamics of fluids in porous media. Dover Publ. Inc., New York

Bear J, Bachmat Y (1984) Transport in porous media. Basic equations. In: Bear J, Corapeioglu MY (eds) Fundamentals of transport phenomena in porous media. NATO ASI series, M. Nijhoff, Dordrecht, pp 3–61

Beltzer A (1978) The influence of random porosity on elastic wave propagation. J Sound Vibr 58(2):251–256

Benu JB, Davis RO (1980) Maximum entropy and the magnitude distribution. Bull Seism Soc Am 70:1823–1831

Bianchi M, Pedretti D (2017) Geological entropy and solute transport in heterogeneous porous media. Water Resour Res 53(6):4691–4708

Bianchi M, Kearsey T, Kingdon A (2015) Integrating deterministic lithostratigraphic models in stochastic realizations of subsurface heterogeneity. Impact on predictions of lithology, hydraulic heads and groundwater fluxes. J Hydrol 531:557–573

Bin L, Qi J, Sinha S, Li J, Marfurt KJ (2020) Improving fault delineation using maximum entropy multispectral coherence. Interpretation/november 2020:T835–T850

Bird N, Díaz MC, Saa A, Tarquis AM (2006) Fractal and multifractal analysis of pore-scale images of soil. J Hydrol 322:211–219

Boltzmann L (1877) Über die Beziehung zwischen dem zweiten Hauptsatze der mechanischen Warmetheorie und der Wahrscheinlichkeitsrechnung, respective den Satzen über das Warmgleichsgewicht. Wiener Berichte 76:373–395

Bonakdari H, Sheikh Z, Tooshmalani M (2015) Comparison between Shannon and Tsallis entropies for prediction of shear stress distribution in open channels. Stoch Environ Res Risk Assess 29:1–11

Borka S (2016) Markov chains and entropy tests in genetic-based lithofacies analysis of deep-water clastic depositional systems. Open Geosci 8:45–51

Bradley JJ, Fort AN Jr (1966) Internal friction in rocks. In: Clark SP Jr (ed) Handbook of physical constants. Geol Soc Am Memoir 97:175–193

Bressan G, Barnaba C, Gentili S, Rossi R (2017) Information entropy of earthquake populations in northeastern Italy and western Slovenia. Phys Earth Planet Int 271:29–46

Byryakovskiy LA (1968) Entropy as criterion of heterogeneity of rocks. Soviet Geol 3:135–138

Cao S, Knight DW (1997) Entropy-based design approach of threshold alluvial channels. J Hydraul Res 35(4):505–524

Casti J, Tse E (1972) Optimal linear filtering theory and radiative transfer: comparisons and interconnections. J Math Anal Appl 40:45–54

Chin CL, Murray DW (1992) Variation of velocity distribution along nonuniform open-channel flow. J Hydraul Eng 118(7):989–1001

Chiogna G, Hochstetler DL, Bellin A, Kitanidis PK, Rolle M (2012) Mixing, entropy and reactive solute transport. Geophys Res Lett 39:L20405

Chiu CL (1987) Entropy and probability concepts in hydraulics. J Hydraul Eng 113(5):583–599

Chiu CL (1988) Entropy and 2-D velocity distribution in open channels. J Hydraul Eng 114(7):738–756

Chiu CL (1989) Velocity distribution in open channel flow. J Hydraul Eng 115(5):576–594

Chiu CL, Said CAA (1995) Maximum and mean velocities and entropy in open-channel flow. J Hydraul Eng 121:26–35

Chiu CL, Tung NC (2002) Maximum velocity and regularities in open-channel flow. J Hydraul Eng 128(4):390–398

Chiu CL, Jin W, Chen YC (2000) Mathematical models of distribution of sediment concentration. J Hydraul Eng 1:16–23

Christopher B, Song C, Wang P, Makse HA (2008) Entropy of jammed matter. Phys Rev Letters 101:188001

Christopher B, Song C, Wang P, Makse HA (2010) Jamming III: Characterizing randomness via the entropy of jammed matter. Physica A 389(19):3978–3999

Clark PJ, Evans FC (1954) Distance to nearest neighbor as a measure of spatial relationships in populations. Ecology 35:445–453

Cox RT (1946) Probability, frequency and reasonable expectations. Am J Phys 14:1–14

Cui H, Singh VP (2013) Suspended sediment concentration in open channels using Tsallis entropy. J Hydrol Eng 19:966–977

Cui H, Singh VP (2014) One dimensional velocity distribution in open channels using Tsallis entropy. J Hydrol Eng 2014:290–298

De Santis A, Cianchini G, Favali P, Beranzoli L, Boschi E (2011) The Gutenberg-Richter law and entropy of earthquakes: two case studies in central Italy. Bull Seism Soc Am 101:1386–1395

DeGiuli E, McElwaine J (2011) Laws of granular solids: geometry and topology. Phys Rev E 84(4):041310

Dowds JP (1969) Oil rocks: information theory: Markov Chains: entropy. Quart Col School Mines 64:275–293

Doyen PE (1987) Crack geometry of igneous rocks: a maximum entropy inversion of elastic and transport properties. J Geophys Res 92(B8):8169–8181

Edwards SF, Oakeshott RBS (1989) Theory of powders. Physica A 157(3):1080–1090

Elfeki AMM, Dekking FM (2005) Modelling subsurface heterogeneity by coupled Markov chains: directional dependency, walther's law and entropy. Geotech Geol Eng 23(6):721–756
Elsasser WM (1937) On quantum measurements and the role of uncertainty relations in statistical mechanics. Phys Rev 52:987–999
Elvis TD, Djomeni AL (2016) Application of Markov chain and entropy analysis to lithologic succession-an example from the post-rift cretaceous and tertiary formations, Kombe-Nsepe area, Douala Subbasin, Cameroon. Int J Sci Res Publ 6(5):692–709
Fayad W, Thompson CV, Frost HJ (1999) Steady-state grain-size distributions resulting from grain growth in two dimensions. Scr Mater 409(10):1199–1204
Fischer P (1972) On the inequality $\Sigma p_i f(p_i) \geq \Sigma p_i f(q_i)$. Metrika 18:199–208
Flory PJ (1941) Thermodynamics of high polymer solutions. J Chem Phys 9:660–661
Fradkov V, Udler D (1994) Two-dimensional normal grain growth: topological aspects. Adv Phys 43(6):739–789
Fraser A, Swinney H (1986) Independent coordinates for strange attractors from mutual information. Phys Rev A 33(2):1134–1140
Gabor K, Sterligov B, Oleschko K, Cherkasov S (2013) Entropy of shortest distance (ESD) as pore detector and pore-shape classifier. Entropy 15:2384–2397
Gagniuc PA (2017) Markov chains: from theory to implementation and experimentation. John Wiley & Sons, NJ, USA
Gibbs JW (1902) Elementary principles of statistical mechanics. Yale University Press, New Haven
Gingerich PD (1969) Markov analysis of cyclical alluvial sediments. J Sediment Petrol 39:330–332
Goldman S (1955) Information theory. Prentice Hall, New York
Good IJ (1952) Rational decisions. J Roy Stat Soc Ser B 14:107–114
Good IJ (1954) Uncertainty and business decisions. Liverpool University Press, Liverpool
Gotovac H, Cvetkovic V, Andricevic R (2010) Significance of higher moments for complete characterization of the travel time probability density function in heterogeneous porous media using the maximum entropy principle. Water Resour Res 46:W05502
Hartley RVL (1928) Transmission of information. Bell System Technical Journal 3: July 1928
Hattori I (1976) Entropy in Markov Chains and discrimination of cyclic patterns in lithologic successions. Math Geol 8(4):477–497
Hertz P (1909) Über den gegenseitigen durchschnittlichen Abstand von Punkten, die mit bekannter mittlerer Dichte im Raume angeordnet sind. Math Annalen 64:387–398
Hota RN, Maejima W (2004) Comparative study of cyclicity of litofacies in Lower Gondwana formations of Talchir basin, Orissa, India: a statistical analysis of subsurface logs. Gondwana Res 7:353–362
Huang L, Ritzi RW, Ramanathan R (2012) Conservative models: parametric entropy versus temporal entropy in outcomes. Ground Water 50(2):199–206
Huggins ML (1941) Solutions of long chain compounds. J Chem Phys 9:440
Jaynes ET (1957) Information theory and statistical physics Pt. I. Phys Rev 106:620–630; Pt. II., Phys Rev 108:171–190. repr. In: Statistical Physics, Brandeis Summer Inst., 1962, W.A. Benjamin, New York, pp 181–218
Journel AG, Deutsch CV (1993) Entropy and spatial disorder. Math Geol 25(3):329–355
Jumarie G (1990) Relative information. Theories and applications. Springer-Verlag, Berlin–Heidelberg
Keller T, Cutler B, Lazar EA, Yauney G, Lewis DJ (2014) Comparative grain topology. Acta Mater 66:414–423
Khinchin AI (1957) Mathematical foundations of information theory. Dover, New York
Kirkpatrick S (1973) Percolation and conduction. Rev Mod Phys 45(4):574–588
Kitanidis PK (1994) The concept of the dilution index. Water Resour Res 30(7):2011–2026
Korvin G (1980) Effect of random porosity on random elastic wave attenuation. Geophys Trans 26:43–56
Korvin G (1984) Shale compaction and statistical physics. Geophys J R Astron Soc 78:35–50
Korvin G (1985) A few unsolved problems of applied geophysics. Geophys Trans 31(4):373–389

Korvin G (1992) Fractal models in the earth sciences. Elsevier, Amsterdam

Korvin G (2000) The value of information in the interactive, integrative and evolutionary world model: a case history. Humanomics 16(1):15–24

Korvin G, Petrovics I (1975) Seismic data processing using a reduced number of bits. Geophys Trans 23:47–69

Korvin G, Oleschko K, Abdulraheem A (2014) A simple geometric model of sedimentary rock to connect transfer and acoustic properties. Arab J Geosci 7(3):1127–1138

Korvin G (2021) Statistical rock physics. In: Daya Sagar B, Cheng Q, McKinley J, Agterberg F (eds) Encyclopedia of mathematical geosciences. Encyclopedia of earth sciences series. Springer, Cham

Korvin G (2022) Rescaled range analysis. In: Daya Sagar BS, Cheng Q, McKinley J, Agterberg F (eds) Encyclopedia of mathematical geosciences. Encyclopedia of earth sciences series. Springer, Cham

Kumbhakar M, Ghoshal K (2017) One-dimensional velocity distribution in open channels using Renyi entropy. Stoch Environ Res Risk Assess 31:949–959

Kumbhakar M, Kundu S, Ghoshal K, Singh VP (2016) Entropy-based modeling of velocity lag in sediment-laden open channel turbulent flow. Entropy 18(9):318

Kumbhakar M, Ghoshal K, Singh VP (2017a) Derivation of Rouse equation for sediment concentration using Shannon entropy. Physics A 465:494–499

Kumbhakar M, Ghoshal K, Singh VP (2017b) Renyi entropy and random walk hypothesis to study suspended sediment concentration. J Hydrol Eng 22(8):04017027

Kumbhakar M, Kundu S, Ghoshal K (2017c) Hindered settling velocity in particle-fluid mixture: a theoretical study using the entropy concept. J Hydraul Eng 143(11):06017019

Kumbhakar M, Kundu S, Ghoshal K (2018) An explicit analytical expression for bed-load layer thickness based on maximum entropy principle. Phys Lett A 382:2297–2304

Kundu S (2017) Derivation of hunt equation for suspension distribution using Shannon entropy theory. Physica A 488:96–111

Kundu S (2017) Prediction of velocity-dip-position over entire cross section of open cannel flows using entropy theory. Environ Earth Sci 76(10):363

Kundu S (2017) Prediction of velocity-dip-position at the central section of open channels using entropy theory. J Appl Fluid Mech 10(1):221–229

Kundu S (2017) Derivation of different suspension equations in sediment-laden flow from Shannon entropy. Stoch Env Res Risk Assess 32:563–576

Landau LD, Lifshitz EM (1980) Statistical physics. Pt.1. (Vol. 5 of Course of Theoretical Physics). Pergamon Press, Oxford, pp 106–114

Laplace PS (1886) Theorie analytique les probabilities 1812–1820, 1st edn. 1812, 2nd edn. 1814, 3rd edn with suppl. Paris, 1820. repr. In: Œuvres complètes de Laplace, vol 7, Gauthier-Villars, Paris

Lazar EA, Mason JK, MacPherson RD, Srolovitz DJ (2012) Complete topology of cells, grains, and bubbles in three-dimensional microstructures. Phys Rev Lett 109(9):95505

Lazar EA, Mason JK, MacPherson RD, Srolovitz DJ (2020) Distribution of topological types in grain-growth microstructures. Phys Rev Lett 125(1):015501

Lazar EA (2011) The evolution of cellular structures via curvature flow. PhD Thesis, Princeton University.

Lee SK, Mosenfelder JL, Park SY, Lee AC, Asimow PD (2020) Configurational entropy of basaltic melts in earth's mantle. Proc Natl Acad Sci USA 117(36):21938–21944

Leopold Luna B, Langbein WB (1962) The concept of entropy in landscape evolution. Theoretical papers in the hydrologic and geomorphic sciences. Geological survey professional paper 500-a. United States Government Printing Office, Washington D.C.

Litwiniszyn J (1974) Stochastic methods in the mechanics of granular bodies. International Centre for Mechanical Sciences, Udine. Courses and Lectures no. 93. Springer-Verlag, Wien

Liu KD, Cao QK, Pang YJ (2004) A method of fault diagnosis based on unascertained set. Acta Automatic Sinica 30(5):747–756

Lopes AM, Machado JAT (2016) Integer and fractional-order entropy analysis of earthquake data series. Nonlinear Dyn 84:79–90
Luo H, Singh VP (2011) Entropy theory for two-dimensional velocity distribution. J Hydrol Eng 16:303–315
Luo H, Singh V, Schmidt A (2018) Comparative study of 1D entropy-based and conventional deterministic velocity distribution equations for open channel flows. J Hydrol 563:679–693
Lutz FH, Mason JK, Lazar EA, MacPherson RD (2017) Roundness of grains in cellular microstructures. Phys Rev E 96(2):023001
Makse HA, Kurchan J (2002) Testing the thermodynamic approach to granular matter with a numerical model of a decisive experiment. Nature 415(6872):614–617
Manoharan VN (2015) Colloidal matter: packing, geometry, and entropy. Science 349(6251):1253751
Mason JK, Lazar EA, MacPherson RD, Srolovitz DJ (2012) Statistical topology of cellular networks in two and three dimensions. Phys Rev E 86:051128
Matthias S, Goldman DI, Swinney HL (2005) Stationary state volume fluctuations in a granular medium. Phys Rev E 71:030301
Mays DC, Faybishenko BA, Finsterle S (2002) Information entropy to measure temporal and spatial complexity of unsaturated flow in heterogeneous media. Water Resour Res 38(12):1313
McCarthy J (1956) Measures of the value of information. Proc Natl Acad Sci USA 42:654–655
McKay BD (1990) Nauty user's guide, version 1.5, Tech Rep TR-CS-90–02. Australian National University, Canberra
Meng L, Wang H, Liu G, Chen Y (2015) Study on topological properties in two-dimensional grain networks via large-scale Monte Carlo simulation. Comput Mater Sci 103:165–169
Miall AD (1973) Markov chain analysis applied to an ancient alluvial plain succession. Sedimentol 20:347–364
Moramarco T, Corato G, Melone F, Singh VP (2013) An entropy-based method for determining the flow depth distribution in natural channels. J Hydrol 497:176–188
Myron T (1983) Thirty years of information theory. In: Levine RD, Tribus M (eds) Maximum Entropy Formalism. MIT Press, Cambridge, Mass, pp 1–14
Nafe JE, Drake CL (1957) Variation with depth in shallow and deep water marine sediments of porosity, density and the velocities of compressional and shear waves. Geophysics 22:523–552
Nowak ER, Knight JB, Ben-Naim E, Jaeger HM, Nagel SR (1998) Density fluctuations in vibrated granular materials. Phys Rev E 57(2):1971
Oleschko K, Figuerora-Sandoval B, Korvin G, Martinez Menes M (2004) Agroecometry: a toolbox for the design of virtual agriculture. Agricultura, sociedad y desarollo 1(4):53–71 (In English and Spanish)
Oleschko K, Korvin G, Figuerora B (2002) Entropy based triangle for designing sustainable soil management. 17th World Congress on Soil Science, Bangkok, Thailand
Perumal R, Amos PK, Selzer M, Nestler B (2017) Phase-field study on the formation of first-neighbour topological clusters during the isotropic grain growth. Comput Mater Sci 140:209–223
Planck M (1906) Theorie der Warmestrahlung. J.A. Barth, Leipzig
Posadas A, Morales J, Posadas-Garzon A (2021) Earthquakes and entropy: characterization of occurrence of earthquakes in southern Spain and Alboran sea. Chaos 31:043124
Power DW, Easterling GR (1982) Improved methodology for using embedded Markov chains to describe cyclical sediments. J Sedim Petrol 53:913–923
Puryear CI, Portniaguine ON, Cobos CM, Castagna JP (2012) Constrained least-squares spectral analysis: application to seismic data. Geophysics 77(5):V143–V167
Renard P, de Marsily G (1997) Calculating equivalent permeability: a review. Adv Water Resour 20(5–6):253–278
Rényi A (1961) On measures of entropy and information. In: Proceedings of the Berkeley symposium on mathematical statistics and probability. vol 1, pp 547−561

San José MF, Caniego FJ, García-Gutiérrez C, Espejo R (2007) Representative elementary area for multifractal analysis of soil porosity using entropy dimension. Nonlinear Process Geophys 14(4):503–511

Sanchez-Vila X, Guadagnini A, Carrera J (2006) Representative hydraulic conductivities in saturated groundwater flow. Rev Geophys 44:RG3002

Sarlis NV, Skordas ES, Varotsos PA (2018) A remarkable change of the entropy of seismicity in natural time under time reversal before the super-giant M9 Tohoku earthquake on 11 March 2011. Europhys Lett 124:29001–29008

Scafetta N, Grigolini P, Hamilton P, West BJ (2003) Nonextensive diffusion entropy analysis and teen birth phenomena. In: Gell-Mann M, Tsallis C (eds) Nonextensive entropy –interdisciplinary applications. Oxford University Press, New York

Sethna James P (2017) Statistical mechanics. Entropy, order parameters, and complexity. Clarendon Press, Oxford

Shannon CE (1948) A mathematical theory of communication. Bell Syst Tech J 27(379–423):623–656

Shannon CE, Weaver W (1949) The mathematical theory of communication. U. of Illinois Press, Urbana, IL

Shen PY, Mansinha L (1983) On the principle of maximum entropy and the earthquake frequency-magnitude relation. Geophys J Int 74(3):777–785

Shmerkin P (2011) Porosity, dimension, and local entropies: a survey. *arXiv preprint*: 1110.5682

Silke H, O'Hern CS, Chakraborty B (2007) Entropy and temperature of a static granular assembly: an ab initio approach. Phys Rev Lett 99(3):038002

Singh VP (2011) Derivation of power law and logarithmic velocity distributions using the Shannon entropy. J Hydrol Eng 16(5):478–483

Singh VP, Asce F (2011) Hydrological synthesis using entropy theory: review. J Hydrol Engin 16:421–433

Singh VP, Sivakumar B, Cui HJ (2017) Tsallis entropy theory for modelling in water engineering: a review. Entropy 19:641

Sinha SK, Das S, Sahoo SR (2015) Application of Markov Chain and entropy function for cyclicity analysis of a lithostratigraphic sequence–a case history from the Kolhan Basin, Jharkhand, Eastern India. J Geol & Geophysxof Geology & Geophysics 4:1–7

Smart AG (2017) A thermodynamic theory of granular material endures. Phys Today 70(9):20

Sotolongo O, Posadas A (2004) Fragment-asperity interaction model for earthquakes. Phys Rev Lett 92(4):048501

Sterligov B (2010) *Analyse probabiliste des relations spatiales entre les gisements aurifères et les structures crustales: developpement méthodologique et applications à l'Yennisei Ridge (Russie).* Ph.D. Thesis, Lomonosov State University, Moscow, and Institut des Sciences de la Terre d'Orleáns.

Sterling M, Knight D (2002) An attempt at using the entropy approach to predict the transverse distribution of boundary shear stress in open channel flow. Stoch Environ Res Risk Assess 16:127–142

Timothy JB (1989) Dependence of entropy on volume for silicate and oxide minerals: a review and a predictive model. Am Miner 74:5–13

Tsallis C (1988) Possible generalization of Boltzmann-Gibbs statistics. J Stat Phys 52:479

Utsu T (1965) A method for determining the value of b in a formula $log\ n = a–bm$ showing the magnitude-frequency relation for earthquakes. Geophys Bull Hokkaido Univ Jpn 13:99–103

van Ruitenbeek FJA, Goseling J, Bakker WH, Hein KAA (2020) Shannon entropy as an indicator for sorting processes in hydrothermal systems. Entropy 22:656

Varotsos PA, Sarlis NV, Skordas ES (2018) Tsallis entropy index q and the complexity measure of seismicity in natural time under time reversal before the M9 Tohoku earthquake in 2011. Entropy 20(10):757–773

Vink RLC, Barkema GT (2002) Configurational entropy of network-forming materials. Phys Rev Lett 89:076405

Von Neumann J (1927) Thermodynamik quantenmechanischer Gesamtheiten. Göttinger Nachr: 273–291

Weinberg L (1966) On the maximum order of the automorphism group of a planar triply connected graph. SIAM J Appl Math 14(4):729–738

Wiener N (1948) Cybernetics. MIT Press, Cambridge

Woodbury AD, Ulrych TJ (1993) Minimum relative entropy: forward probabilistic modeling. Water Resour Res 29(8):2847–2860

Wu P, Mu R, Deng Y (2022) Robust crater detection algorithm based on maximum entropy threshold segmentation. Int J Aerosp Eng 2022:4872248

Xue W, Wang H, Liu G, Meng L, Xiang S, Ma G, Li W (2016) Matrix description of the complete topology of three-dimensional cells. Sci Rep 6:25877

Zhou J, Li X (2012) Integrating unascertained measurement and information entropy theory to assess blastability of rock mass. J Cent South Univ (Changsha, China) 19:1953−1960

Zhu Z, Yu J (2019) Estimating the bed-load layer thickness in open channels by Tsallis entropy. Entropy 21:123

Zhu Z (2018) A simple explicit expression for the flocculation dynamics modeling of cohesive sediment based on entropy considerations. Entropy 20:845

Ziman JM (1979) Models of disorder. The theoretical physics of homogeneously disordered systems. Cambridge University Press, Cambridge

Zimmerman RW (1991) Compressibility of sandstones. Elsevier, Amsterdam

Chapter 9
Effective Properties of Rocks

Chapter Highlights *Effective Medium Theory* (EMT) is a method to estimate the macroscopic properties of a composite material based upon the relative fractions and specific properties of its constituent parts. Its history, and the mathematical treatment of its different versions (*Maxwell's Approximation*, the *Self Consistent Method*, and *Differential Effective Method*) are given in Sect. 9.1. Section 9.1.1 illustrates the use of EMT through (Kirkpatrick 1973) determination of the *effective conductivity* of a *network of random conductors*. *Differential Effective Medium* (DEM) approximation (Bruggeman 1935) is treated in Sect. 9.2, in the context of electric conductivity and elasticity. In Sect. 9.3 we discuss *variational inequalities* (Hashin and Shtrikman in Mech Phys Solids 1963, and in Phys Rev 1963), and a *generalized mean value theorem* (Korvin 1978, 1982), for effective rock properties, and present an example (Korvin 2012) for the anisotropic case. Section 9.4 reviews attempts to theoretically derive *Archie's law* $\sigma_{rock} = \sigma_{fluid}\Phi^m$ from basic principles, and to determine the *cementation exponent m*.

9.1 The Effective Medium (EM) Approximation

In a recent study modeling *organic-rich shales with different maturity levels* (Qin et al. 2014) the authors used four different EM techniques (EMTs): an anisotropic version of the *Differential Effective Medium* (DEM) method (Nishizawa 1982) to generate the clay-fluid composite; then the *Anisotropic Self-Consistent Approximation* (ASCA) (Hornby et al. 1994) to include quartz and calcite in the clay-fluid composites. The cracked kerogen's effective elastic properties were calculated using the *Self-Consistent Approximation* (SCA) (Mavko et al. 2009); and the kerogen inclusions' proper alignment along the shale bedding plane was assured by an anisotropic DEM method.

According to Choy (2015), EMT "defines averages, which one hopes will be representative of the system and be connected with experimental measurements". These methods have been applied to elastic properties, thermal properties, magnetic properties, electrical properties, transport properties, etc., in all fields of science and

G. Korvin, *Statistical Rock Physics*, Earth and Environmental Sciences Library,
https://doi.org/10.1007/978-3-031-46700-4_9

technology (Table 9.1),where they are variously called *homogenization*, *upscaling*, or *effective medium methods* (Sahimi 1995; Choy 2015; Torquato 2002; Jurgawczynski 2007; Edwards and Berg 2021).

The first attempts to describe the *optical properties* of composite materials in terms of their specific densities were due to (Lorentz 1880; Lorenz 1880) who described the *refractive index* of a medium as function of its density. They considered the propagation of light through the medium in the *dilute limit*, where the opaque, non-transparent particles were small compared to the distance between them. Lord Rayleigh (1892) extended these results to obstacles whose dimensions were no longer negligible with respect to their spacing. He assumed spherical or cylindrical objects with uniform conductance arranged in a "rectangular or square" (cubic) order, embedded in a

Table 9.1 Some important applications of the DEM (*Differential Effective Medium*) method

Application	Authors
Acoustic velocities in patchy and partially saturated porous rock	Berryman et al. (2002a, b)
Anisotropic DEM method	Nishizawa (1982)
Anisotropic seismic properties of partially molten rocks	Mainprice (1997)
DEM lies between the Hashin-Shtrikman bounds for spherical inclusions	McLaughlin (1977)
Dielectric propoerties	Norris et al. (1990)
Elastic constants for calcite and muscovite mixtures	Bjarne et al. (2011)
Elastic moduli of a cracked solid	Henyey and Pomphrey (1982)
Elastic modulus of carbonate rocks	Haidar et al. (2018)
Elastic properties of mixtures of calcite and muscovite	Almqvist et al. (2011)
Elastic properties of high-porosity marine calcareous sediments	Ruiz and Dvorkin (2007)
Elasticity in nonclastic rocks	Ruiz and Dvorkin (2010)
Elastic properties of cracked rocks	Mavko et al. 1998; Bjarne et al. (2011), Haidar et al. (2018)
Electric conductivity of composites	Bruggeman (1935)
Granular effective medium approximation	Yonezawa and Cohen (1983); Allo (2019)
Organic-rich shales	Qin et al. (2014)
Permeability of 3-D fractured rock	Ebigbo et al. (2016)
Pore types in carbonate reservoirs	Candikia et al. (2017); Rosid et al. (2017)
Randomly oriented spheroid inclusions	Norris (1985)
Resistivity in marine sediments	Yonezawa and Cohen (1983)
Seismic velocities in geological materials	Bruner (1976)

conducting medium. He derived the effective transport coefficient of this two-phase medium as

$$\frac{D_{eff} - D_2}{D_{eff} + mD_2} = \frac{\nu_1}{\nu_2}\frac{D_1 - D_2}{D_1 + D_2} \tag{9.1.1}$$

if $\nu_1 \ll \nu_2$, where D_i, ν_i refer to the transport coefficient and volume fraction of the two phases, respectively (we use the modern notations of Tjaden et al. 2016; Edwards and Berg 2021). The variable m is a shape factor, which is an inevitable ingredient of every effective medium solution. In Rayleigh's formula (4.1.1), $m = 1$ for cylinders and $m = 2$ for spheres. The generalization of Rayleigh's equation was the *Maxwell–Garnett formula* (Garnett 1904) for the effective dielectric constant, D_{eff}, of a composite medium. It assumed that *spherical inclusions* of material 1 are embedded into material 0, and stated that

$$\frac{D_{eff} - D_0}{D_{eff} + 2D_0} = \nu_1\frac{D_1 - D_0}{D_1 + 2D_0} \tag{9.1.2}$$

Note that on the RHS of (9.1.2) ν_1 is the *volume fraction* of the inclusions. The equation also holds for N *constituents*:

$$\frac{D_{eff} - D_0}{D_{eff} + 2D_0} = \sum\nolimits_{i=1}^{N} \nu_i\frac{D_i - D_0}{D_i + 2D_0} \tag{9.1.3}$$

but it is only accurate at small values of ν_i. Bruggeman (1935) further developed the Rayleigh-Maxwell–Garnett formulae, from a cubic-lattice arrangement of the inclusions to their arbitrarily random placements. He needed three assumptions: (i) The two phases are homogeneous and isotropic; (ii) the particle sizes of the phases are small in comparison to the overall sample size; (iii) the phases are randomly distributed within the sample. He applied an *iterative embedding* procedure during which the embedding medium had been gradually filled up with increasing amounts of the two phases. In the limit when the embedding medium is completely replaced by the two kinds of inclusions, one gets

$$\nu_1 \cdot \frac{D_1 - D_{eff}}{D_1 + mD_{eff}} + \nu_2 \cdot \frac{D_2 - D_{eff}}{D_2 + mD_{eff}} = 0 \tag{9.1.4}$$

and for an arbitrary number of phases

$$\sum_{i=1}^{N} \nu_i \cdot \frac{D_i - D_{eff}}{D_i + mD_{eff}} = 0 \tag{9.1.5}$$

(Choy 2015; Tjaden et al. 2016; Edwards and Berg 2021). For *porous media*, transport through one of the phases is often negligible (e.g. *d.c. current* will flow in brine-filled pores between grains, but not through the grains themselves). In this case Eq. (9.1.4) reduces to

$$D_{eff} = \frac{\nu}{\tau} \tag{9.1.6}$$

where D and ν are the *transport coefficient* and *volume fraction of the transporting phase*. τ is the so-called "*tortuosity factor*", given by

$$\tau = \begin{cases} \nu^{-1/2} & for\ spherical\ inclusions \\ \nu^{-1} & for\ cylindrical\ inclusions \end{cases} \tag{9.1.7}$$

The EMT developed by Kirkpatrick (1971, 1973) was also formulated in the spirit of Bruggeman's work. Kirkpatrick considered a *network of random resistors*, arranged as a square (2D) or cubic (3D) lattice, such as shown in Fig. 9.1. Earlier, Landauer (1952) had already attempted to predict the resistance of binary mixtures of conducting materials using simple *mixing laws* (see Sect. 9.3.2). However, his results were inapplicable to the limiting case of insulating-conducting mixtures, the most common case in rock physics. Kirkpatrick (1971, 1973) presented a solution to this problem, using the so-called *Effective Medium Approximation* (EMA).

Before discussing in details Kirkpatrick's (1971, 1973) celebrated solution for the effective conductivity of a random conductor network (Sect. 9.1.1), we briefly review some further *inclusion-based effective medium models* applicable for *hydraulic flow* in *fracture networks* (Sævik et al. 2013, 2014; Ebigbo et al. 2016).

Usually, a *fracture* is defined as an *oblate spheroidal inclusion* within a rock matrix. Its two equal semi-axis lengths are the fracture radius R, while the length of its short semi-axis is one half of the maximum aperture, that is $h/2$. The *aspect ratio* of the fracture is then

$$\alpha = h/2R(\alpha \ll 1) \tag{9.1.8}$$

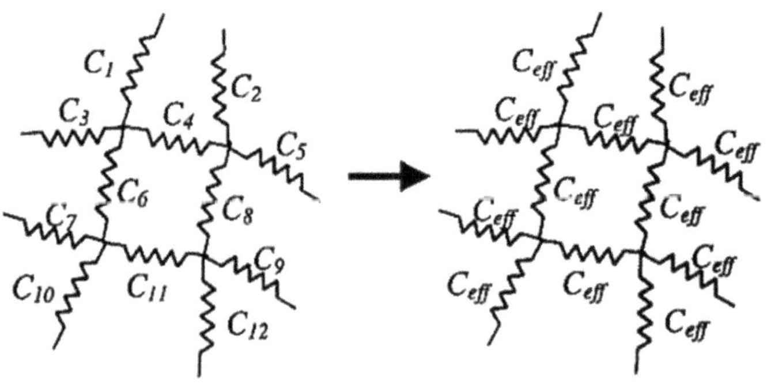

Fig. 9.1 Substitution of a random network of conductances by a *homogenized network* of the same topology. (After Lock 2001: Fig. 6.2.1; reproduced with the kind permission of Dr. Peter Lock)

There are other models (e.g. Mourzenko et al. 2011), where fractures are *flat discs* of radius R with an aperture α_f. By equating the volumes of the two kind of fractures we find

$$\alpha_f = \frac{4}{3} R\alpha \tag{9.1.9}$$

A *fracture set* is a group of fractures with the same radius. The density, ρ_i, of fracture set i is the number of fracture centres, N_f, per volume:

$$\rho_i = \frac{N_{f,i}}{L^3} \tag{9.1.10}$$

where $L > 2R$ is system size. Denote by ϕ_i the total volume occupied by a fracture set per unit volume of rock mass. When all aspect ratios are equal,

$$\phi_i = \frac{4}{3}\pi\alpha R_i^3 \rho_i, \quad \phi = \sum_{i=1}^{n} \phi_i \tag{9.1.11}$$

if there are n fracture sets. We shall also need *fracture density*, defined *as*

$$\varepsilon_i = \varrho_i R_i^3 \tag{9.1.12}$$

Let each fracture set have a permeability K_f, for simplicity *computed with the* "cubic law" (Zimmerman and and Bodvarsson 1996) as:

$$K_f = \alpha_f^2/12 \tag{9.1.13}$$

Finally, let κ be the *permeability ratio,*

$$\kappa = \frac{K_m}{K_f} (\kappa \ll 1) \tag{9.1.14}$$

where K_m is the permeability of the background rock matrix.

With these conventions and notations, let us first apply the *Hashin–Shtrikman* (H–S) *upper bound* (Hashin and Shtrikman 1961,1962a, 1962b, 1962c; Watt et al. 1976; see Sect. 9.3.1) to find the *largest posible effective permeability*. According to (Zimmerman 1989), in case of a monodisperse, isotropic fracture network

$$K \le K_{HS}^{+} = K_f + \frac{3K_f(K_m - K_f)(1-\phi)}{3K_f + (K_m - K_f)\phi} \tag{9.1.15}$$

A simple model of fractured rocks, with zero matrix permeability ($K_m = 0$) is that of Snow (1969; further developed by Oda 1995). In this model

$$K_{Snow} = \frac{2}{3}\rho\pi R^2 K_f \alpha_f \tag{9.1.16}$$

Using Eq. (9.1.9) to change α_f to α, and Eq. (9.1.11) for relative fracture volume, Eq. (9.1.16) becomes

$$K_{Snow} = \frac{2}{3}\phi K_f \tag{9.1.17}$$

On the other hand, for $K_m = 0$ we have

$$K_{HS}^{+}(K_m = 0) = \frac{2\phi K_f}{3 - \phi} \tag{9.1.18}$$

and, comparing (9.1.17) with (9.1.18) for $\phi \ll 1$, we find that the Snow model (1969) yields about the same upper bound as HS: $K \leq K_{HS}^{+}(K_m = 0) \approx K_{Snow} = \frac{2}{3}\phi K_f$ for

$$K_m = 0, \phi \ll 1 \tag{9.1.19}$$

as first pointed out by (Ebigbo et al. 2016).

As noted by (Zimmerman 1989: 221), "the general approach to determining the effective properties of a composite material consisting of inclusions of any shape is first to calculate the effect of an infinitesimally small concentration of inclusions, and then use some approximate method to account for finite concentration." For rock permeability, the *dilute-limit solution* had been found by Fricke, as early as 1924, in the analogue context of *electrical conductivity*, for *randomly oriented spheroidal inclusions*. Fricke's solution, rewritten for permeability, is:

$$K = K_m(1 + \beta\phi) \tag{9.1.20}$$

with

$$\beta = \frac{r-1}{3}\left[\frac{4}{2 + (r-1)M} + \frac{1}{1 + (r-1)(1-M)}\right] \tag{9.1.21}$$

where K is the effective permeability of the rock mass, $r = K^{-1}$, and

$$M = \frac{2\Theta - sin2\Theta}{2tg\Theta sin^2\Theta} \tag{9.1.22}$$

with

$$\Theta = \arccos\alpha \tag{9.1.23}$$

For *flat spheroids* (Sævik et al. 2013, 2014) $\alpha \ll 1$, that is $M \approx \alpha\frac{\pi}{2} \ll 1$and in Eq. (9.1.21) $(1 - M) \approx 1$. Assuming $r \gg 1$, Eq. (9.1.21) becomes

$$\beta = \frac{4r}{3\left(2 + \frac{\pi}{2}\alpha r\right)} = \frac{8}{3\kappa\left(4 + \pi\frac{\alpha}{\kappa}\right)} \tag{9.1.24}$$

The coefficient β in Eq. (9.1.20) has two asymptotic regimes as function of α:

$$\beta = \begin{cases} \frac{8}{3\pi\alpha} \; for \; \alpha \gg \kappa \\ \frac{2}{3\kappa} \; for \; \alpha \ll \kappa \end{cases} \tag{9.1.25}$$

Following are some of the *Effective Medium solutions* for the rock permeability problem (Sævik et al. 2013).

Maxwell Approximation: In Maxwell's model, the overall effect of the inclusions on the pressure field and on the permeabilities summed over all fracture sets is assumed to be the same as the effect of a *large sphere of permeability K*, representing the *effective medium* inside the same rock matrix where the inclusions are (Sevostianov 2014; Lutz and Zimmerman 2016):

$$\sum\nolimits_{i=1}^{n} \beta_i \phi_i = \beta_{sphere} \tag{9.1.26}$$

β_{sphere} can be calculated from Eq. (9.1.21) using $M_{sphere} = \frac{2}{3}$; $r_{sphere} = \frac{K}{K_m}$. Rearrangement yields

$$K = K_m \cdot \frac{\left(1 + \frac{2}{3}\sum_1^n \phi_i \beta_i\right)}{1 - \frac{1}{3}\sum_1^n \phi_i \beta_i} \tag{9.1.27}$$

Using (9.1.24) for β, and substituting

$$\phi_i = \frac{4}{3}\pi\alpha\varepsilon_i \tag{9.1.28}$$

yield Maxwell's solution in the form

$$K - K_m = \frac{32}{9}\left(\frac{1}{3}K + \frac{2}{3}K_m\right)\sum_1^n \frac{\varepsilon_i}{\left(\frac{4\kappa}{\pi\alpha} + 1\right)} \tag{9.1.29}$$

The Self Consistent Method:

Sævik et al. (2013) solved the problem using the *asymmetric and symmetric versions of the Self-Consistent Method*. In both cases they started out from the *dilute limit solution* $K = K_m(1 + \beta\phi)$ (9.1.20), with $\beta = \frac{4r}{3\left(2+\frac{\pi}{2}\alpha r\right)} = \frac{8}{3\kappa\left(4+\pi\frac{\alpha}{\kappa}\right)}$ (9.1.24). Note that the use of (9.1.24) implicitly presupposes that $K_{f,i} \gg K$. The *Asymmetric Self-Consistent Method* resulted in the expression

$$K - K_m = \frac{32}{9} K \sum_{i=1}^{n} \frac{\varepsilon_i}{\frac{4K}{\pi \alpha K_{f,i}} + 1} \tag{9.1.30}$$

In the alternative, *Symmetric Self-Consistent Method* the rock matrix where the fractures are embedded also contains a set of *spherical inclusions*. The corresponding equation for the effective permeablity is.

$$K - K_m = \frac{32}{9} \left(\frac{2}{3} K + \frac{1}{3} K_m \right) \sum_{i=1}^{n} \frac{\varepsilon_i}{\frac{4K}{\pi \alpha K_{f,i}} + 1} \tag{9.1.31}$$

In the *Differential Effective Medium* (DEM) method the fractures are added in infinitesimal increments, and in each step the results are *homogenized*. This leads to to the following *differential equation* for the evolving effective permeability as function of fracture porosity (McLaughlin 1977; Zimmerman 1996; Sævik et al. 2013):

$$\frac{dK}{d\phi} = K \sum_{i=1}^{n} \beta_i \hat{\phi}_i \tag{9.1.32}$$

For

$$\kappa \ll 1, \alpha \ll 1 \tag{9.1.33}$$

the equation becomes

$$\frac{dK}{d\varepsilon} = K \sum_{i=1}^{n} \frac{\hat{\varepsilon}_i}{\frac{4K}{\pi \alpha K_{f,i}} + 1} \tag{9.1.34}$$

where $\hat{\varepsilon}_i$ is analogous to $\widehat{\phi_i}$. For *monodisperse, isotropic fracture networks*, (9.1.34)can be integrated which yields (Sævik et al. 2013):

$$\frac{4}{\pi \alpha K_f} (K - K_m) + ln \left(\frac{K}{K_m} \right) = \frac{32}{9} \varepsilon \tag{9.1.35}$$

9.1.1 Kirkpatrick's Random Conductor Network

The best example for the use of *Effective Medium Theory* is Kirkpatrick's (1973) classic study of random conductors placed on the bonds of a regular lattice (Fig. 9.1). Apply an external field along the vertical axis so that the voltages increase by a constant amount V_m per row of nodes. The average effect of the random conductors can be described by an *effective medium*, which is a hypothetical homogeneous medium in which the total field inside is equal to the external field. Consider an *effective field* which is made up of equal conductances, C_m, connecting nearest neighbors on a *square-* (in 2-D), or (in 3-D) *cubic mesh.* To find C_m, we require that the local perturbations in this medium due to the voltages, induced when individual conductances C_{ij} and C_m are interchanged, should average to zero. Consider a conductance C_{AB} between randomly picked neighboring nodes *A, B* in the direction of the external field. Using Kirchoff's laws it can be shown (Kirkpatrick 1973; Schlueter 1995; Lock 2001; Jurgawczynski 2007; Edwards and Berg 2021) that if we change C_m to C_{AB}, the arising voltage, V_0, induced between A and B will be

$$V_0 = V_m(C_m - C_{AB})/\left[C_{AB} + \left(\frac{Z}{2} - 1\right)C_m\right] \tag{9.1.36}$$

where V_m is the constant increase in voltages per row, Z is *coordination number* of the mesh. If the conductivities C_{ij} are distributed according to the *pdf* (*probability density function*) (C), the requirement that the average of V_0 should vanish yields an equation for the effective conductivity C_m:

$$\langle V_0 \rangle = \int f(C)\frac{V_0(C)(C_m - C)}{C + \left(\frac{Z}{2} - 1\right)C_m}dC = 0 \tag{9.1.37}$$

which can be used to determine the effective conductance C_m, for any regular infinite lattice and any distribution of conductances.

Kirkpatrick derived Eq. (9.1.37) using the assumption that the coordination numbers Z at each node are identical. Koplik (1982) deduced the equation for *irregular networks*, when Z is *the mean co-ordination number* for all nodes in the network.

For a binary distribution of conductivities, Eq. (9.1.37) has a simple *algebraic solution* (Edwards and Berg 2021). Let the distribution be

$$f(C) = p\delta(C - \alpha) + (1 - p)\delta(C - \beta) \tag{9.1.38}$$

where $\delta(x)$ is the *Dirac delta function*, $p \in (0, 1)$ is the probability that a given bond has conductivity α, and $1 - p$ is the probability that it has conductivity β. Substituting Eq. (9.1.38) into (9.1.37), we obtain

$$p\frac{C_m - \alpha}{\alpha + \left(\frac{Z}{2} - 1\right)C_m} + (1 - p)\frac{C_m - \beta}{\beta + \left(\frac{Z}{2} - 1\right)C_m} \tag{9.1.39a}$$

that yields a quadratic equation

$$\left(\frac{Z}{2} - 1\right)C_m^2 + \left\{\beta\left[\frac{Z}{2}(p - 1) + 1\right] + \alpha\left[1 - \frac{Z}{2}p\right]\right\}C_m - \alpha\beta = 0 \tag{9.1.40a}$$

It has two real roots, positive and negative, the *physical value* of C_m is provided by the positive root.

For the case of $N > 2$ different conductivities $\{\sigma_1, \cdots, \sigma_N\}$ distributed with equal probabilities $p_1 = \cdots = p_N = 1/N$, instead of Eq. (9.1.39a) we have

$$\sum_{i=1}^{N} \frac{C_{eff} - \sigma_i}{\left(\frac{Z}{2} - 1\right)C_{eff} + \sigma_i} = 0 \tag{9.1.39b}$$

Assuming that the effective conductivity C_{eff} is a monotonically inceasing function of the coordination number Z (Darnowski et al. 2019), the two extreme cases $Z = 2$ (*series arrangement of conductors*), and $Z \to \infty$ (*parallel arrangement of conductors*) yield two natural, *lower,* resp. *upper bounds* for the effective conductivity:

$$\left[\frac{1}{N}\sum_{i=1}^{N} \frac{1}{\sigma_i}\right]^{-1} \leq C_{eff}(Z) \leq \frac{1}{N}\sum_{i=1}^{N} \sigma_i \tag{9.1.41}$$

When estimating the *effective elastic modulus* of a composite rock using EMT (Mavko et al. 1998), a similar, more general, estimation is used for the lower and upper bounds:

$$\frac{1}{M_R} = \sum_{i=1}^{N} \frac{f_i}{M_i} \leq M_{eff} \leq M_V = \sum_{i=1}^{N} f_i M_i \tag{9.1.42a}$$

where

$$M_R = \left\{\sum_{i=1}^{N} \frac{f_i}{M_i}\right\}^{-1} \tag{9.1.42b}$$

is the *Reuss elastic modulus* (Reuss 1929),

$$M_V = \sum_{i=1}^{N} f_i M_i \tag{9.1.42c}$$

is the *Voigt elastic modulus* (Voigt 1928).

M_i are elastic moduli of the minerals, f_i are the mineral fractions. In practice (Haidar et al. 2018), petrophysicists often use M_{VRH}, the "*Voigt-Reuss-Hill modulus*", which is the average of the two extreme moduli (Hill 1952, 1963a, b; Chung 1963):

$$M_{VRH} = \frac{M_V + M_R}{2} \tag{9.1.43}$$

Some of the efforts to find closed-form solutions to Kirkpatrick's equation $\langle V_0 \rangle = \int f(C) \frac{V_0(C)(C_m - C)}{C + (\frac{Z}{2} - 1)C_m} dC = 0$ (9.1.37) for *continuous probabity distribution functions*

$f(C)$were reviewed in (Lock 2001; Jurgawczynski 2007). If $f(C)$is *uniform* in the interval $C_1 \leq C \leq C_2$, that is

$$f(C) = \begin{cases} \frac{1}{C_2 - C_1} & if \quad C\epsilon\left[C_{1,}, C_2\right] \\ 0 & otherwise \end{cases} \tag{9.1.44}$$

then from(9.1.37) we find

$$\frac{Z}{2} C_{eff} \cdot ln\left\{1 + \frac{C_2 - C_1}{\left(\frac{Z}{2} - 1\right)C_{eff} + C_1}\right\} = C_2 - C_1 \tag{9.1.45}$$

(David et al. 1990: Eq. 4), an equation which can be solved numerically for C_{eff}.

Schlueter (1995) solved Eq. (9.1.37) analytically for a *log-uniform distribution function*

$$f(C) = \begin{cases} (2Cln\Omega)^{-1} & forC\epsilon\left(\Omega^{-1}, \Omega\right) \\ 0 & otherwise \end{cases} \tag{9.1.46}$$

for three special geometric arrangement: (i) for $Z = 2$ (*conductors in series*); (ii) for $Z = \infty$ (*parallel conductors*); and (iii) $Z = 6$ (*isotropic cubic lattice*), and found:

(i) for $Z = 2$

$$C_{eff} = \frac{2\Omega\ln\Omega}{\Omega^2 - 1} \tag{9.1.47a}$$

(ii) for $Z = \infty$

$$C_{eff} = \frac{\Omega^2 - 1}{2\Omega\ln\Omega} \tag{9.1.47b}$$

(iii) for $Z = 6$

$$C_{eff} = \frac{\Omega^{4/3} - 1}{2\left(\Omega - \Omega^{1/3}\right)} \tag{9.1.47c}$$

From the last equation, for $Z = 6$ and $\Omega \gg 1$,

$$C_{eff} \propto \frac{1}{2}\Omega^{1/3} \tag{9.1.47d}$$

For the general case of a *continuous distribution of conductivities* $f(C)$ in (4.1.37), replace the integration with summation *over all bonds of the random network* to get the nonlinear equation

$$\sum_{i=1}^{M} \frac{C_{eff} - C_i}{\left(\frac{z}{2} - 1\right)C_{eff} + C_i} = 0 \tag{9.1.48}$$

Denote the function on the LHS by

$$K\left(C_{eff}\right) := \sum_{i=1}^{M} \frac{C_{eff} - C_i}{\left(\frac{z}{2} - 1\right)C_{eff} + C_i}, \tag{9.1.49}$$

("*Kirkpatrick's function*"). It is easy to check that.

$$\text{for } C_{eff} = 0 \text{ we have } K(0) = -M \tag{9.1.50a}$$

and

$$\text{for } C_{eff} \to \infty \text{ we have } K\left(C_{eff}\right) \to K(\infty) = \frac{M}{\left(\frac{z}{2} - 1\right)} \tag{9.1.50b}$$

if $Z \neq 2$. Thus, if $Z > 2$, by the *Intermediate Value Theorem* (Rudin 1976: 42, 93) there must exist at least one positive value of C_{eff} which is a root of (9.1.48), i.e. satisfies $K\left(C_{eff}\right) = 0$. To see that this root is *unique*, take the derivative of the *Kirkpatrick's function*:

$$\frac{dK\left(C_{eff}\right)}{dC_{eff}} = \sum_{i=1}^{M} \frac{\frac{Z}{2} \cdot C_i}{\left\{\left(\frac{Z}{2} - 1\right)C_{eff} + C_i\right\}^2} > 0 \tag{9.1.51}$$

As the derivative is positive, $K\left(C_{eff}\right)$ is increasing in $C_{eff} \epsilon [0, \infty]$, and the root is unique.

The *Effective Medium Method* of Kirkpatrick (1973) can be modified and extended to *anisotropic lattices*. Bemasconi (1974) developed an *Effective Medium Approximation* for square lattices with *anisotropic binary distribution of conductivities*. Toledo et al. (1992) made the technique applicable to any regular lattice.

In these methods, the *anisotropic conductor network* is substituted by an *equivalent anisotropic effective medium* where all bonds parallel with a particular direction ν have the same effective conductivity C_{eff}^{ν}. Applying the same lines of reasoning as in the isotropic case, Kirchoff's Laws yield

$$\left\langle \frac{C_{eff}^{\nu} - C_k}{S_\nu\left(C_{eff}^{\nu_1}, \cdots, C_{eff}^{\nu_d}\right) + C_k} \right\rangle = 0 \tag{9.1.52}$$

that is

$$\sum_{k=1}^{n} \frac{C_{eff}^{\nu} - C_k}{S_\nu\left(C_{eff}^{\nu_1}, \cdots, C_{eff}^{\nu_d}\right) + C_k} = 0 \tag{9.1.53a}$$

where d is the number of different directions, S_ν is a complicated function which depends on the effective conductivities in different directions, and on network topology. In the simplest case of a *cubic network* Bernasconi (1974) approximated S_ν as

$$S_\nu(C_{eff}^{\nu=1}, C_{eff}^{\nu=2}, C_{eff}^{\nu=3}) \approx C_m^{\nu=1} \cdot \frac{tan^{-1}\left\{\sqrt{\left[C_{eff}^{\nu=1}C_{eff}^{\nu=2} + C_{eff}^{\nu=2}C_{eff}^{\nu=3} + C_{eff}^{\nu=3}C_{eff}^{\nu=1}\right]}/C_{eff}^{\nu=1}\right\}}{tan^{-1}\left\{C_{eff}^{\nu=1}/\sqrt{\left[C_{eff}^{\nu=1}C_{eff}^{\nu=2} + C_{eff}^{\nu=2}C_{eff}^{\nu=3} + C_{eff}^{\nu=3}C_{eff}^{\nu=1}\right]}\right\}} \tag{9.1.53b}$$

It is easy to check, that for an *isotropic cubic mesh of* conductors the system of Eqs. (9.1.53a, b) reduces to Kirkpatrick's isotropic equation (9.1.48).

The anisotropic EM method has also proved to be useful for estimating the permeability of *anisotropic fracture networks* (Harris 1990, 1992; Hestir and Long, 1990).

A simple case of the *EM* approximation (Stroud 1998; Ghanbarian et al. 2014; Korvin 2021) concerns the random mixture of two types of grains, present in relative volume fractions p and $1 - p$, which are characterized by conductivities σ_1 and σ_2. To calculate the *effective conductivity* σ_{eff} of this composite, imagine that each grain is embedded in a homogeneous *effective medium,* whose conductivity σ_{eff} is determined *self-consistently*. If the grains are spherical, the electric field inside a grain is

$$E_i = E_0 \frac{3\sigma_{eff}}{\sigma_i + 2\sigma_{eff}} (i = 1 \text{ or } i = 2) \tag{9.1.54}$$

where E_0 is the electric field far from the grain. *Self-consistency* requires that the *average electric field* within the grain should equal E_0, that is

$$pE_0 \frac{3\sigma_{eff}}{\sigma_1 + 2\sigma_{eff}} + (1 - p)E_0 \frac{3\sigma_{eff}}{\sigma_2 + 2\sigma_{eff}} = E_0 \tag{9.1.55}$$

leading to the quadratic equation

$$p\frac{\sigma_1 - \sigma_{eff}}{\sigma_1 + 2\sigma_{eff}} + (1 - p)\frac{\sigma_2 - \sigma_{\sigma_{eff}}}{\sigma_2 + 2\sigma_{eff}} = 0 \tag{9.1.56}$$

The equation has two solutions for σ_{eff}, one physically acceptable, the other unphysical.

In *Percolation Theory* the *transport properties* (such as conductivity) grow as a power-function of $|p - p_c|$ above the *percolation threshold*, p_c. For $p > p_c$ the electrical conductivity σ of a random network grows as

$$\sigma \propto (p - p_c)^t, p > p_c \tag{9.1.57}$$

where p is the occupation probability for a site or bond, and the scaling exponent t is equal to 2 in 3-D. In the *Effective Medium (EM)* approach (Stroud 1998; Ghanbarian et al. 2014) a randomly selected part of the heterogeneous medium is replaced by the homogeneous material. Replacing the actual conductivity by an *effective* one perturbs the local current. The effective conductivity is determined from the requirement that the average of these perturbations should be zero. For the (bond) percolation model of electric conductivity in rocks this gives (in case of two components)

$$\sigma(p)/\sigma_b = 1 - \frac{1-p}{1-\frac{2}{Z}} \tag{9.1.58}$$

where σ_b is the electrical conductivity of the *brine*, and the grains are *non-conducting*, Z is the coordination number of the lattice. As in n-dimensional lattices the average coordination number Z, and percolation threshold $p_c$are approximately related as

$$Zp_c \approx d/(d-1) \tag{9.1.59}$$

(Vyssotsky et al. 1961; Korvin 1992: 22), in 2D one has $2/Z \approx p_c$, and the relation (9.1.58) can be interpreted as a *percolation scaling rule*

$$\frac{\sigma(p)}{\sigma_b} \approx \frac{p-p_c}{1-p_c} \tag{9.1.60}$$

for $p > p_c$.

9.2 Differential Effective Medium (DEM) Approximation

The DEM approach was invented by Bruggeman (1935) in the context of *electric conductivity*. He considered a composite with two constituents in volume fractions $x = v_1$ and $y = v_2 = 1 - x$; where the type-1 material of specific resistivity σ_1 is the *host*, and the type-2 material of specific resistivity σ_2 is the *inclusion*. He proved that if the value of the *effective conductivity* $\sigma^*_{DEM}(y)$ is known, then its value after adding a further infinitesimal amount dy of type-2 material satisfies the differential equation

$$(1-y)\frac{d}{dy}\left[\sigma^*_{DEM}(y)\right] = \frac{\sigma_2 - \sigma^*_{DEM}(y)}{\sigma_2 + 2\sigma^*_{DEM}(y)} \cdot \left[3\sigma^*_{DEM}(y)\right] \qquad (9.2.1)$$

The equation can be integrated analytically. Starting with $\sigma^*_{DEM}(y=0) = \sigma_1$ one finds

$$\left(\frac{\sigma_2 - \sigma^*_{DEM}(y)}{\sigma_2 - \sigma_1}\right) = \left(\frac{\sigma_1}{\sigma^*_{DEM}(y)}\right)^{1/3} \qquad (9.2.2)$$

The basic assumption of *Differential Effective Medium* (DEM) theory (Bruggeman 1935; Cleary et al. 1980; Walsh 1980; Norris 1985; Avellaneda 1987) is that a composite material can be modeleded by making *infinitesimal changes* in an already existing composite. In many cases this leads to analytical results, its numerical computation is easy, it is applicable for the estimation of several kinds of rock properties (see selected applications in Table 9.1). It has been proved (McLaughlin 1977; Berryman et al. 2002a, b; Almquist et al. 2011) that the results satisfy theoretically established physical and mathematical constraints, such as the *Hashin-Shtrikman bounds* (Hashin and Shtrikman 1961, 1962a, b, c—see Sect. 9.3.1), or the *Reuss and Voigt inequalities* (Voigt 1928; Reuss 1929—see Eqs. 9.1.42a, b, c).

The *elastic properties* of porous or fractured rocks are commonly modeled by the *DEM* approach (Table 9.1). For example, a two-phase porous or cracked rock can be built up by incrementally adding small fractions of pores or cracks to the rock matrix. As we insert new inclusions into the background model, its elastic properties continuously change as:

$$\left.\begin{aligned} (1-y)\tfrac{dK^*(y)}{dy} &= [K_i - K^*(y)]P^{*i} \\ (1-y)\tfrac{d\mu^*(y)}{dy} &= [\mu_i - \mu^*(y)]Q^{*i} \end{aligned}\right\} \qquad (9.2.3a,b)$$

(Mavko et al. 1998; Berryman et al. 2002a, b), where the index *i* refers to the *inclusio*n; yis volume fraction of the inclusion; K^*(y) is the effective bulk modulus of the DEM model; P^{*i} is a *shape factor* for an inclusion of material 2 in a background medium with effective moduli K* and μ^*; μ^*(y) is the effective shear modulus of the DEM model; Q^{*i} is shape factor for an inclusion of material 2 in a background medium with effective moduli K^* and μ^*.

The DEM Eqs. (9.2.3a, b) are *coupled* because $K^*(y)$ and $\mu^*(y)$ are included in *both shape factors* (also called *polarization factors*) P^{*i} and Q^{*i}. These factors generally depend on the bulk and shear moduli of both the inclusion and the host medium, and on the shapes of the inclusions. The shape factors in (Berryman 1980a, b; Mavko et al. 1998) had been computed based on the theoretical results of (Eshelby 1957; Walsh 1980; Wu 1966).

For example, (Berryman et al. 2002a, b)considered *penny-shaped cracks*, in this case

$$P^{*i} = \frac{K^* + \frac{4}{3}\mu_i}{K_i + \frac{4}{3}\mu_i + \pi\alpha\gamma^*} \tag{9.2.4a}$$

$$Q^{*i} = \frac{1}{5}\left[1 + \frac{8\mu^*}{4\mu_i + \pi\alpha(\mu^* + 2\gamma^*)} + 2\frac{K_i + \frac{2}{3}(\mu_i + \mu^*)}{K_i + \frac{4}{3}\mu_i + \pi\alpha\gamma^*}\right] \tag{9.2.4b}$$

where $\alpha (0 < \alpha < 1)$ is the *oblate spheroidal crack*'s *aspect ratio,*

$$\gamma^* = \mu^*\left[\frac{(3K^* + \mu^*)}{(3K^* + 4\mu^*)}\right] \tag{9.2.4c}$$

and the superscript $*$ identifies constants of the *matrix material* when the inclusion's volume fraction is y.

In the study of (Ruiz and Dvorkin 2010) ellipsoidal inclusions are used, the inclusions are filled with sea-water under in situ conditions, the matrix properties are those of *pure calcite*. The shape factors P and Q are given by

$$P = \frac{1}{2}T_{iijj};\ Q = \frac{1}{2}\left(T_{ijij} - \frac{1}{3}T_{iijj}\right) \tag{9.2.5a, b}$$

where the tensor T relates the uniform far-field strain to the strain within the ellipsoidal inclusion (Wu 1966).

Ruiz and Dvorkin (2010) posed the question" Can a theoretical inclusion model—specifically, the differential effective medium (DEM) model—match experimental velocity data in rocks that are not necessarily made of inclusions, such as clastics?". Their answer is affirmative, "it is indeed possible in some cases by using an almost constant inclusion aspect ratio (AR) within wide ranges of porosity and mineralogy".

The Eqs. (9.2.5a, b) are numerically integrated starting from inclusion volume fraction $y = 0$ where the initial values are

$$K^*(0) = K_m \text{ and } \mu^*(0) = \mu_m \tag{9.2.6a, b}$$

which are the values for the homogeneous solid constituent. Integration proceeds from $y = 0$ to the desired highest value $y = y_{max}$ as new inclusions are slowly introduced into the solid. The method needs as input the *shape factors*, the *elastic moduli of the inclusions* and the maximal fraction of inclusions. For the shape factors one also needs the *aspect ratios*. In the model of (Ruiz and Dvorkin 2010) there are three ranges of aspect ratio values, for the *interparticle pores*, for the *stiff* pores and for the *cracks*.

It had been recentlly noted, that the results depend on the way as the inclusions had been gradually added. Edigbo et al. (2016) noted "the predictions of the differential method depend slightly on the integration path"; Haidar et al. (2018) similarly state "In the DEM method, the effective moduli depend on the path taken to construct the final composite". To avoid this problem, Edigbo et al. (2016) selected a protocol (based on Sævik et al. 2013) where "the infinitesimal amount of each fracture set added, at each step, is proportional to the final proportion of the set's volume fraction, that is

$$(\phi_i/\phi)_j = (\phi_i/\phi)_{j+1} = \widehat{\phi}_i \tag{9.2.7}$$

In the solution of Haidar et al. (2018) there is a control step where one calculates the reference velocity V_P^{ref} of the composite using the parameters provided by the DEM equations and compares it with the measured P-wave velocity. The reference value V_P^{ref} serves to decide whether stiff pores or cracks are to be added to the rock matrix. If $V_P^{ref} < V_P^{meas}$ then stiff pores are added, if $V_P^{ref} > V_P^{meas}$ then cracks are added to the matrix. When the process is completed, the effective elastic moduli can be calculated. This controlled introduction of crack (or soft) porosity into a material containing spherical (or stiff) porosity is similar to the *porosity distribution model* of Mavko and Jizba (1991).

In an interesting application (Bjarne et al. 2011) of *DEM* to compute the elastic constants for *calcite* and *muscovite* mixtures, the background medium was *anisotropic*. For anisotropic elastic media *DEM* is computed as

$$\frac{dC^{DEM}}{dV} = \frac{1}{1-V}\left(C_i - C^{DEM}\right)A_i \tag{9.2.8}$$

where C^{DEM} is the elastic tensor of the effective media, V is the volume fraction of the inclusions, C_i is the elastic tensor of the inclusion and A_i, which relates the strain inside the inclusion with that of the background matrix, is defined as

$$A_i = \left[I + G\left(C_i - C^{DEM}\right)\right]^{-1} \tag{9.2.9}$$

Here I is the 4th-rank unit tensor, G is a symmetric tensorial Green's function, C_i is the elastic modulus of the inclusion, the tensor C^{DEM} is updated in each inclusion step. Inclusions are ellipsoids with given aspect ratio ($a \geq b \geq c$).

There are several other examples for anisotropic DEM. In Qin and co-workers' (2014) modeling study of *organic-rich shales* they used an anisotropic version of the DEM method (Nishizawa 1982). Almqvist et al. (2011) calculated the elastic properties of rock aggregates using a DEM assuming an anisotropic background medium with *triclinic or higher symmetry*, into which ellipsoidal inclusions were added incrementally. Mainprice (1997) who modeled the anisotropic seismic properties of *partially molten rocks at mid-ocean ridges*, also assumed an anisotropic background medium. An important technical tool in these studies is Barnett's (1972) method of computing the derivatives of anistropic elastic Green's functions.

9.3 Bounds for the Effective Properties

9.3.1 Variational Bounds

When asked by the editors of the *Current Contents*' "This week's Citation Classics" (on 11 February 1980), how their paper (Hashin and Shtrikman 1963) has become one of the most frequently cited and widely used articles by geoscientists, engineers, and physicists (see Table 9.2), Dr. Zvi Hashin (1980) recalled:

> When we finished this work in 1961 we submitted it for publication to a prestigious applied mechanics journal in the US. It was ignominiously rejected. The reviewer, described to us as an 'outstanding authority,' had classified the paper as 'ramblings.' We subsequently sent it to the journal of the *Mechanics and Physics of Solids* which accepted it at once. This may appear funny today, but we failed to appreciate the humor of the situation in 1962.
>
> I believe the reason for frequent citation of the paper is that it established permanent results for general two-phase materials, namely, best bounds in terms of easily available information for arbitrary phase geometry.

Brown (1955) had shown that *phase volume fractions* and *specific material properties* of the phases cannot determine unambigously the effective properties of a two-phase composite, because the effective properties also depend on *phase geometry*. If we only have this *limited information*, *bounds* are needed to delimit the possible range of variation of the effective properties.

Table 9.2 Some important applications of the Hashin-Shtrikman Bounds

Application	Authors
Debye characteristic temperature of cubic crystals	Shukla and Padial (1973)
Elastic moduli of inhomogeneous systems	Walpole (1966a, b)
Elastic moduli of polycrystals and rocks	Hill (1952), Beran and Molyneux (1966), Kumazawa (1969), Miller (1969), McCoy (1970), Thomsen (1972), Lakkad et al. (1973), Corson (1974c), Watt and Peselnick (1980)
Elastic properties of reinforced solids	Hill (1963a)
Electrical conductivity and permittivity in saturated porous media	Elsayed (1974), Brovelli and Cassiani (2010)
Electrical conductivity in a partially molten rock	Waff (1974), Park and Ducea (2003), Park (2004)
Magnetic properties	Elsayed (1974)
Mechanical properties of nonlinear isotropic solids	Ponte Castañeda (1991, 1992)
Thermal properties of porous media	Woodside and Messmer (1961), Corson (1974d), Elsayed (1974), Berryman (2005)
Thermoelastic properties	Lakkad et al. (1973)

We shall briefly discuss the four most frequently used bounds, the *Reuss bound* B_R, the *Voigt bound* B_V, the lower and upper Hashin-Shtrikman bounds B_{HS^-} and B_{HS^+} and the so-called *Voigt-Reuss-Hill average*, defined as $\langle \varepsilon \rangle_{VRH} = \frac{1}{2}[B_R(\varepsilon) + B_v(\varepsilon)]$ (Voigt 1928; Reuss 1929; Hill 1952; Chung 1963). For any effective petrophysical property ε and volume fractions $v_1, v_2 \epsilon [0, 1]$; $v_1 + v_2 = 1$ these five quantities *usually* satisfy

$$B_R(\varepsilon, v_1) \le B_{HS^-}(\varepsilon, v_1) \le \langle \varepsilon, v_1 \rangle_{VRH} \le B_{HS^+}(\varepsilon, v_1) \le B_V(\varepsilon, v_1) \tag{9.3.1}$$

with equalities only for $v_1 = 0$ or $v_1 = 1$. "Usually" here means that for some values of specific properties and certain ranges of v_1 the *Voigt-Reuss-Hill average* can step out from the range of the HS bounds, as shown by Watt et al. (1976). Thomsen (1972) checked the performance the *Voigt-Reuss-Hill average* on a large collection of polycrystalline data and concluded that "the VRH average has no more experimental support than it has theoretical justification" (*op c*it.: 318).

To gain a feeling of the magnitude of the deviations of the VRH average from the HS bounds, and from the Reuss and Voigt bounds in Eq. (9.3.1), consider the study of Chang and Graham (1975), who determined the single-crystal elastic constants of *cassiterite* (SnO_2). They computed P-wave and S-wave velocites using VRH average elastic constants and obtained $V_P = 7.06 \pm 0.23\text{km/s}$ and $V_S = 3.82 \pm 0.26\text{km/s}$. When using the arithmetic mean of the HS values, this resulted in $V_P = 7.07 \pm 0.07\text{km/s}$ and $V_S = 3.83 \pm 0.07\text{km/s}$ where the uncertainties are connected in the first case with the distance between the VRH value and the Reuss and Voigt values, in the second case with the distance between the upper and lower HS values.

We already mentioned (Eqs. 9.1.41 and 9.1.42b) that for an N-component conducting material where each component has the same volume fraction, the *effective conductivity* satisfies the inequalities

$$\left[\frac{1}{N}\sum\nolimits_{i=1}^{N}\frac{1}{\sigma_i}\right]^{-1} \le C_{eff}(Z) \le \frac{1}{N}\sum\nolimits_{i=1}^{N}\sigma_i \tag{9.3.2}$$

and similarly, the *effective elastic modulus* of an N-component composite rock obeys

$$\frac{1}{M_R} = \sum\nolimits_{i=1}^{N}\frac{f_i}{M_i} \le M_{eff} \le M_V = \sum\nolimits_{i=1}^{N} f_i M_i \tag{9.3.3}$$

where M_i are the *elastic moduli* of the minerals, f_i are the mineral fractions. The lower bounds in these inequalities are the *Reuss bound* (Reuss 1929), the upper bounds are called *Voigt bounds* (Voigt 1928). For electric properties these bounds are sometimes called *lower*, resp. *upper Wiener bounds* to honor the early work of Otto Wiener (1912).

Following the treatment of (Hill 1963a, summarized in Watt et al. 1996), in the elastic case these bounds can be derived from basic principles.

Consider a finite, 2-component elastic body inside which the stress and strain fields are related as

$$\tau_{ij}(\boldsymbol{x}) = c_{ijkl}(\boldsymbol{x}) \cdot e_{kl}(\boldsymbol{x}); \tag{9.3.4}$$

$$e_{ij}(\boldsymbol{x}) = s_{ijkl}(\boldsymbol{x}) \cdot \tau_{kl}(\boldsymbol{x}) \tag{9.3.5}$$

where s_{ijkl} is the *elastic compliance tensor*, and

$$c_{ijkl}s_{klmn} = \frac{1}{2}\{\delta_{im}\delta_{jn} + \delta_{in}\delta_{jm}\} \tag{9.3.5}$$

In case of two isotropic phases mixed in volume fractions v_1 and v_2, the average stress and strain are

$$\left.\begin{aligned} \langle E\rangle &= v_1\langle E_1\rangle + v_2\langle E_2\rangle \\ \langle \tau\rangle &= v_1\langle \tau_1\rangle + v_2\langle \tau_2\rangle \end{aligned}\right\} \tag{9.3.6a, b}$$

where averages are volumetric averages, for example, $\langle E_1\rangle = V_1^{-1}\int_{V_1} E_1(\boldsymbol{x})dV$. Using (9.3.3), and (9.3.4)

$$\left.\begin{aligned} \langle E\rangle &= v_1 S_1\langle \tau_1\rangle + v_2 S_2\langle \tau_2\rangle \\ \langle \tau\rangle &= v_1 C_1\langle E_1\rangle + v_2 C_2\langle E_2\rangle \end{aligned}\right\} \tag{9.3.7a, b}$$

Suppose the average strain in each phase depends on the average strain in the composite as

$$\left.\begin{aligned} \langle E_1\rangle &= A_1\langle E\rangle \\ \langle E_2\rangle &= A_2\langle E\rangle \end{aligned}\right\} \tag{9.3.8}$$

with unknown A_1, A_2. From (9.3.7a, b)

$$v_1 A_1 + v_2 A_2 = I \tag{9.3.9}$$

Average stress and average strain are related as $\langle \tau\rangle = C^*\langle E\rangle$ (9.3.10), where $C^*$is the *effective modulus*:

$$C^* = v_1 C_1 A_1 + v_2 C_2 A_2 \tag{9.3.10}$$

Analogously, we get: $\langle \tau_i\rangle = B_i\langle \tau\rangle$; $v_1 B_1 + v_2 B_2 = I$; $\langle E\rangle = S^*\langle \tau\rangle$ and

$$S^* = v_1 S_1 B_1 + v_2 S_2 B_2 \tag{9.3.11}$$

If the average stress or strain in either constituent is known,

$$\left.\begin{aligned} C^* &= C_2 + v_1(C_1 - C_2)A_1 \\ S^* &= S_2 + v_1(S_1 - S_2)B_1 \end{aligned}\right\} \tag{9.3.12}$$

To get the *Voigt average*, assume that the *strain is uniform* everywhere in the composite ($A_1 = A_2 = I$); and to get the *Reuss average*, assume an *equal stress* throughout the composite ($B_1 = B_2 = I$).

The mathematically correct demonstration of the bounding properties of the two averages was given by Hill (1952), who (1963a) also generalized the bounds for N components as

$$M_R^* = \left(\sum_{i=1}^{M} \frac{v_i}{M_i}\right)^{-1} \le M^* \le \sum_{i=1}^{M} v_i M_i = M_V^* \tag{9.3.13}$$

where the modulus M is either K or μ.

Hill (1952) proposed that either the arithmetic mean

$$\frac{M_V^* + M_R^*}{2} \tag{9.3.14a}$$

or the geometric mean

$$(M_R^* M_V^*)^{1/2} \tag{9.3.14b}$$

would be a fair approximation of the effective modulus. The arithmetic mean, called the Voigt-Reuss-Hill (VRH) average is still widely used, in spite of its known shortcomings (Thomsen 1972: 318). The geometric mean was used by Kumazawa (1969). Shukla and Padial (1973) applied the harmonic means of the Voigt and Reuss bounds:

$$\left\{\frac{1}{2}M_V^{-1} + \frac{1}{2}M_R^{-1}\right\}^{-1} = 2M_V M_R/(M_V + M_R) \tag{9.3.14c}$$

for the calculation of the *Debye characteristic temperature of cubic crystals*. Korvin (1978, 1982) generalized these ideas and suggested the effective modulus M_{eff} must necessarily be given by the expression

$$M_R \le M_{eff} = \left\{\frac{1}{2}M_V^t + \frac{1}{2}M_R^t\right\}^{1/t} \le M_V \tag{9.3.15}$$

for some real parameter t, $-\infty \le t \le \infty$.

Note the special cases: for $t = -\infty$ $M_R = M_{eff}$; $t = -1$ yields the harmonic mean (9.3.14c); $t = 0$ reproduces (with L'Hospital's rule) the geometric mean (9.3.14b); $t = 1$ gives the VRH average (9.3.14.a); $t = \infty$ leads to $M_V = M_{eff}$. The *generalized mean value* $\{v_1 M^t + v_2 M^t\}^{1/t}$ will be discussed in Sect. 9.3.2.

In the electric case for a porous medium filled with conducting fluid, the Reuss and Voigt bounds are named *Wiener bounds* (Wiener 1912). These are obtained considering an equivalent medium of the same porosity, where the solid and fluid phases are arranged in layers. The two bounds are obtained when the flux is perpendicular or parallel to the layers, respectively. In the first case, the porous medium can be substituted by an electric circuit composed of *resistances in series*, in the second case of *resistances in parallel*. The circuits in series and in parallel correspond to the *maximum and minimum resistance*, respectively.

Hashin and Shtrikman (1962 a, b, c) used a *variational approach* for isotropic granular porous materials and derived upper and lower bounds for the *effective permittivity*. For the upper bound they used a model of spherical grains, each coated by a shell of fluid. The fluid forms an *interconnected percolation cluster*, while the grains do not touch. In the model for the lower bound the two materials are interchanged. For a fully saturated porous rock with porosity Φ , the HS bounds for permittivity are:

$$\left.\begin{aligned} \varepsilon_{HSL} &= \varepsilon_s + \frac{\Phi}{\left(\varepsilon_f - \varepsilon_s\right)^{-1} + \frac{1-\Phi}{3\varepsilon_s}} \\ \varepsilon_{HSU} &= \varepsilon_s + \frac{\Phi}{\left(\varepsilon_s - \varepsilon_f\right)^{-1} + \frac{\Phi}{3\varepsilon_f}} \end{aligned}\right\} \tag{9.3.16}$$

(Brovelli and Cassiani 2010),

where the subscripts mean: HSL = Hashin-Shtrikhman Lower. HSU = HS Upper, f = fluid, s = solid. The subscripts “HSL” and “HSU” refer to the case $\varepsilon_f > \varepsilon_s$, in the opposite case (dry rock, or sweet water-, or oil-bearing porous rocks) the bounds are inverted. To relate bulk permittivity with *bulk conductivity*, Brovelli and Cassiani (2010: Eq. 6) deduced from Archie’s law that for $\varepsilon_f \gg \varepsilon_s$ one has $\frac{\sigma_b}{\sigma_f} = \Phi^m \approx \frac{\varepsilon_b}{\varepsilon_f}$ (4.3.16) where m is the so-called *cementation exponent* (see Sect. 9.4).

The Hashin–Shtrikman bounds for for the bulk and shear moduli of a porous rock are given by

$$K_{HS}^{\pm} = K_2 + \frac{\Phi}{(K_1 - K_2)^{-1} + (1-\Phi)\left(K_2 + \frac{4}{3}\mu_2\right)^{-1}} \tag{9.3.17}$$

$$\mu_{HS}^{\pm} = \mu_2 + \frac{\Phi}{(\mu_1 - \mu_2)^{-1} + \frac{2(1-\Phi)(K_2+2\mu_2)}{5\mu_2\left(K_2+\frac{4}{3}\mu_2\right)}} \tag{9.3.18}$$

When $K_2 > K_1$, the equations provide the *upper bounds* K_{HS}^{+} and μ_{HS}^{+}. The lower bounds K_{HS}^{-} and μ_{HS}^{-} are obtained by interchanging the índices in Eqs. Eqs. (9.3.17 and 9.3.18). For an N-component rock the corresponding equations can be found in (Watt et al. 1976: 546, Eqs. 49–52).

Hashin and Shtrikman’s original results (1962a, b, c) were only valid under restrictions on the relative sizes of K_1, K_2, μ_1, μ_2. These conditions have been later removed in the more careful derivation of the bounds by Walpole (1966a, b).

Hashin and Shtrikman (1963) proved that the bounds on the bulk modulus (9.3.17) are *optimal*, given only the phase volume fractions and the moduli of the individual phases, by constructing two model geometries for which the effective bulk modulus could be exactly computed, and they agreed with the theoretical bounds. Apparently, the optimality of the bounds (4.3.18) for *shear modulus* has not been demonstrated as yet.

There have been many efforts to tighten the HS bounds by including geometry information in the form of *many-point correlation functions*. Lakkad et al. (1973) obtained improved bounds on the effective elastic and thermoelastic properties using 2-point correlation functions. (Beran and Molyneux 1966; Miller 1969; McCoy 1970) incorporated 3-point correlations into the bounds, Corson (1974a, b, c, d) simplified their technique and applied it to *elastic moduli* and *thermal conductivity*. Elsayed (1974) experimented with 4- and 5-point correlations to get *improved bounds* for thermal, electric, and magnetic properties.

In a recent paper devoted to the effect of two-point correlations on effective elasticity we are warned that "the two-point correlation functions are not in general sufficient to estimate accurately the response of composites, especially when their underlying phases exhibit infinite contrast, e.g., porous materials" (Zerhouni et al. 2021: Abstract). Two-point correlations certainly cannot improve the HS bounds in case of *anisotropic microstructures*, because 2-point correlations have been already incorporated in the general expression of these bounds (Willis 1977). Torquato (1997) claims that for a random two-phase composite an *exact expansion of the effective moduli* would involve *the infinite set of n-point correlations*.

9.3.2 The General Mixture Rule

Suppose we are given a composite material of volume V consisting of two phases of the respective volume fractions P, Q; $P + Q = V$, and suppose both constituents are uniformly distributed within the total volume. Suppose g is some strictly positive physically measurable property that assumes the values $g_1 > 0$ and $g_2 > 0$ respectively, for the two constituents, and a value $\overline{g}$ for the composite. Suppose, further, that the value of $\overline{g}$ is unambiguously determined by the volume fractions P, Q and the specific properties g_1 and g_2:

$$\overline{g} = M\left(g_{1,}g_2, P.Q\right) \tag{9.3.19}$$

Korvin (1978, 1982)has shown, using the *Theory of Functional Equations* (Aczél 1961), that, if a set of physically plausible conditions are met, the only possible functional form of $M\left(g_{1,}g_2, P.Q\right)$ is the "general mixture rule":

$$M\left(g_{1,}g_2, P.Q\right) = \left\{\Phi g_1^t + (1-\Phi)g_2^t\right\}^{1/t} \tag{9.3.20}$$

for some real t, $t \neq 0$, or

$$M\left(g_{1,}g_2, P.Q\right) = g_1^{\Phi} g_2^{1-\Phi} \tag{9.3.21}$$

Here

$$\Phi = \mathrm{P}/(P + Q) \tag{9.3.22}$$

We expect that the function $M\left(g_{1,}g_2, P.Q\right)$ should satisfy the following set of conditions:

Condition 1: *reflexivity*

$$M(g_1, g_1, P, Q) = g_1 \mathrm{forall} P, Q \mathrm{suchthat} P + Q > 0 \tag{9.3.23a}$$

Condition 2: *idempotency* (Fuchs 1950).

$$M(g_1, g_2, P, 0) = g_1 \mathrm{forall} P > 0 \tag{9.3.23b}$$

$$M(g_1, g_{21}, 0, Q) = g_2 \mathrm{forall} Q > 0 (9.3.23c)$$

Conditon 3: *homogeneity (of 0-th order) with respect to the volume fractions.* The physical property $\overline{g}$ of the composite does not depend on the actual values of the volume fractions, only on their ratio:

$$M(g_1, g_1, P, Q) = M(g_1, g_1, \lambda P, \lambda Q) if P + Q > 0, \lambda > 0 \tag{9.3.23d}$$

Condition 4: *internity*. The property $\overline{g}$ for the composite lies between the specific values g_1 and g_2 of the constituents. If $g_1 < g_2$, $P + Q > 0$, then

$$M(g_1, g_2, 1, 0) \leq M(g_1, g_2, P, Q) \leq M(g_1, g_2, 0, 1) \tag{9.3.23e}$$

Condition 5: *bi-symmetry* (this concept is due to Aczél 1946). Given two composites, the first consisting of P_1 resp. Q_1 parts of materials of g_1 resp. g_2 properties, the second of P_2 resp. Q_2 parts of materials of G_1 resp. G_2 properties, the following two expressions for the measured property $\overline{g}$ of the *four-component aggregate* must be equal:

$$\begin{aligned} &M(M(g_1, g_2, P_1, Q_1); M(G_1, G_2, P_2, Q_2); P_1 + Q_1; P_2 + Q_2) = \\ &M(M(g_1, G_1, P_1, P_2); M(g_2, G_2, Q_1, Q_2); P_1 + P_2; Q_1 + Q_2) \end{aligned} \tag{9.3.23f}$$

Condition 6: *monotonicity with respect to the volume fractions*

$$If g_1 < g_2 P + Q_1 > 0; Q_1 < Q_2 then M(g_1, g_2, P, Q_1) \leq M(g_1, g_2, P, Q_2) \tag{9.3.23g}$$

Condition 7: *monotonicity with respect to the physical properties*

$$If\, P + Q > 0, g_2 < g_3 then M(g_1, g_2, P, Q) < M(g_1, g_3, P, Q) \quad (9.3.23h)$$

Condition 8: *homogeneity (of first order) with respect to the physical properties.* For all P, Q, λ such that $P + Q > 0, \lambda > 0$ we have

$$M\left(\lambda g_1, \lambda g_2, P, Q\right) = \lambda M(g_1, g_2, P, Q) \quad (9.3.23i)$$

According to the theorem of De Finetti, Kitagawa and Aczél (Aczél 1961) any function $M(g_1, g_2, P, Q)$ satisfying the Conditions 1 – 7 (i.e., Eqs. 4.3.23a – i) can be described in the general form:

$$M(g_1, g_2, P, Q) = f^{-1}\left(\frac{Pf(g_1) + Qf(g_2)}{P + Q}\right) \quad (9.3.24)$$

that is by Eq. (9.3.22) we have

$$M(g_1, g_2, P, Q) = f^{-1}[\Phi f(g_1) + (1 - \Phi) f(g_2)] \quad (9.3.25)$$

where f is some arbitrary, continuous, strictly monotone function (f is the so-called *Kolmogorov-Nagumo function* (Kolmogorov 1930; Nagumo 1930) corresponding to the given mean $M(g_1, g_2, P, Q)$). The Kolmogorov-Nagumo function f corresponding to $M(g_1, g_2, P, Q)$ is only determined up to constant factors. Introducing the notation

$$\mathfrak{M}_f(g_1, g_2, \Phi, 1 - \Phi) = f^{-1}[\Phi f(g_1) + (1 - \Phi) f(g_2)] \quad (9.3.26)$$

we have the following theorem of Jessen (Hardy et al. 1934): *In order that* $\mathfrak{M}_\phi(g_1, g_2, \Phi, 1 - \Phi) = \mathfrak{M}_\psi(g_1, g_2, \Phi, 1 - \Phi)$(4.3.27) *for all* g_1, g_2 *and all* $\Phi\epsilon[0, 1]$ *it is necessary and sufficient that* $\psi = a\phi + b$ (4.3.27) where a, b are constants, $a \neq 0$.

We introduce a further notation:

$$M_t(g_1, g_2, \Phi, 1 - \Phi) = \left[\Phi g_1^t + (1 - \Phi) g_2^t\right]^{1/t} (g_1, g_2 > 0; t \neq 0) \quad (9.3.27a)$$

$$M_0(g_1, g_2, \Phi, 1 - \Phi) = g_1^\Phi g_2^{1-\Phi} g_1, g_2 > 0 \quad (9.3.27b)$$

The following important characterization of the means $M_t$is due to Nagumo, De Finetti and Jessen (De Finetti 1931; Hardy et al., 1934): *Suppose that* $\phi(x)$ *is continuous in* $(0, \infty)$ *and that*:

$$\mathfrak{M}_\phi\left(\lambda g_1, \lambda g_2, \Phi, 1 - \Phi\right) = \lambda \mathfrak{M}_\phi(g_1, g_2, \Phi, 1 - \Phi) \quad (9.3.28)$$

for all positive λ, g_1, g_2 *and for all* $\Phi\epsilon[0, 1]$. *Then*

$$\mathfrak{M}_\phi(g_1, g_2, \Phi, 1-\Phi) = M_t(g_1, g_2, \Phi, 1-\Phi) \quad (9.3.29)$$

for some real t.

The expression $M_t(g_1, g_2, \Phi, 1-\Phi)$is called the *generalized weighted mean of t-th order* of g_1 and g_2 (Beckenbach and Bellman 1961). On strength of the above-stated three theorems the following conclusion can be made: *if the expression* $M(g_1, g_2, P, Q)$ *defining the effective physical parameter of a two-phase composite satisfies Conditions* 1 – 8 (Eqs. 9.3.23.a – i), $P + Q > 0, g_1 > 0, g_2 > 0$ *then necessarily* $\overline{g} = M(g_1, g_2, P, Q) = \left[\Phi g_1^t + (1-\Phi)\right]^{1/t}$ (*for some* $t \neq 0$) *or* $\overline{g} = g_1^\Phi g_2^{1-\Phi}$.

In case of sound speeds in fluid-filled sedimentary rocks the above general rule contains, in particular, the following widely used velocity formulae:

– for $t = -2$ *Wood equation* (Wood 1955)

$$\frac{1}{\overline{v}^2} = \frac{\Phi}{v_1^2} + \frac{1-\Phi}{v_2^2} \quad (9.3.30a)$$

– for $t = -1$ *time-average equation* (Wyllie et al. 1956)

$$\frac{1}{\overline{v}} = \frac{\Phi}{v_1} + \frac{1-\Phi}{v_2} \quad (9.3.30b)$$

– for $t = 0$ "*vugular carbonate*" formula (Meese and Walther 1967)

$$\overline{v} = v_1^\Phi v_2^{1-\Phi} \quad (9.3.30c)$$

– for $t = 1$ *average velocity* formula (Berry 1959)

$$\overline{v} = \Phi v_1 + (1-\Phi)v_2 \quad (9.3.30d)$$

For geophysical-physical applications the following property of the generalized mean M_t is especially important: *If* g_1 *and* g_2 *are positive,* $\Phi \neq 0$, $\Phi \neq 1$, $g_1 \neq g_2$ *then* $M_t(g_1, g_2, \Phi, 1-\Phi)$ *is a strictly monotone function of t in* $(-\infty, \infty)$.

The assertion easily follows from Jensen's inequality, see (Beckenbach and Bellman 1961). For the above-quoted velocity rules, this theorem implies that:

$$v_{Wood} < v_{wyllie} < v_{Meese-Walther} < v_{average} \quad (9.3.31)$$

Korvin (1978: Figs. 1, 2) found that measured *sandstone* and *carbonate* P-wave velocity vs. porosity values are best fitted by the M_t curve corresponding to $t = -0.6$, that is to the formula

$$\overline{v} = \left\{\Phi v_{fluid}^{-0.6} + (1-\Phi)v_{matrix}^{-0.6}\right\}^{1/(-0.6)} \tag{9.3.32}$$

9.3.3 *Bounding the Electric Anisotropy of Thinly Laminated Sand-Shale*

Korvin (2012) studied the *electric resistivity anisotropy* of thinly laminated *sand-shale formations*, for the case when both sandstone and shale are *electrically anisotropic*, and derived simple upper- and lower bounds for the possible maximal and minimal values for the *coefficient of resistivity anisotropy* in such formations.

The introduction of *induction logging tools with multi-directional coils* (Kriegshäuser et al. 2000; Clavaud et al. 2005) has made possible to independently measure horizontal and vertical effective resistivities ρ_h and ρ_v in wells and to derive from them vertical and horizontal shale resistivities ρ_{sh_v} and ρ_{sh_h}, and a single resistivity value ρ_{sd} for the sandstone which is considered isotropic. We shall consider now the more general case when *both sandstone and shale are electrically anisotropic*.

Suppose we have a horizontal stack of inherently anisotropic shale layers of horizontal resistivity ρ_{sh_h} and vertical resistivity ρ_{sh_v} alternating with anisotropic sandstone layers of horizontal resistivity ρ_{sd_h} and vertical resistivity ρ_{sh_v} ρ_{sd_v}. The volume fractions of shale and sand, respectively, are V_{sh} and V_{sd} with $0 \le V_{sh} \le 1$, $0 \le V_{sd} \le 1$; in the absence of any further lithology $V_{sh} + V_{sd} = 1$. The two volume fractions are assumed as known, because they can be estimated from *Gamma Ray log*, *SP log*, or *porosity crossplots*.

The assumption, that both sand and shale are electrically anisotropic, generalizes the model of Klein et al. (1997) who assumed isotropy for both sand and shale, and it also improves upon published models (Clavaud et al. 2005; Minh et al. 2008) where only the shale is taken as anisotropic but sand is assumed isotropic.

The *Klein equations* (Klein et al. 1997) which are based on Kirchoff's laws for laminated composites (Maxwell 1891; Grant and West 1965; Mei and Vernescu 2010), express the effective horizontal resistivity ρ_h and effective vertical resistivity ρ_v of a stack of sand/shale layers in terms of specific resistivities:

$$\left.\begin{array}{ccccc} V_{sh}\rho_{sh} & + & V_{sd}\rho_{sd} & = & \rho_v \\ \frac{V_{sh}}{\rho_{sh}} & + & \frac{V_{sd}}{\rho_{sd}} & = & \frac{1}{\rho_h} \end{array}\right\} \tag{9.3.33a, b}$$

From (9.3.33a),

$$\rho_{sh} = \frac{\rho_v - V_{sd}\rho_{sd}}{V_{sh}} \tag{9.3.34}$$

where the sand resistivity ρ_{sd} is obtained by solving the quadratic equation

$$A\rho_{sd}^2 + B\rho_{sd} + C = 0, \tag{9.3.35}$$

with

$$\begin{aligned} A &= V_{sd} \\ B &= \rho_h\left(V_{sh}^2 - V_{sd}^2 - \frac{\rho_v}{\rho_h}\right) = \rho_h\left(V_{sh}^2 - V_{sd}^2 - \lambda^2\right) \\ C &= \rho_h \rho_v V_{sd} \end{aligned} \tag{9.3.36a–c}$$

Here $\lambda = \sqrt{\rho_v/\rho_h}$ (which is always greater than 1) is the *anisotropy coefficient* of the layered structure. Clavaud et al. (2005) assumed that the sand is isotropic with resistivity ρ_{sd}, but the shale layers are inherently anisotropic with two different resistivities ρ_{sh_v} and ρ_{sh_h}.

In this case Kirchoff 's rules give

$$\left.\begin{aligned} V_{sh}\rho_{sh\_v} + V_{sd}\rho_{sd} &= \rho_v \\ \frac{V_{sh}}{\rho_{sh\_h}} + \frac{V_{sd}}{\rho_{sd}} &= \frac{1}{\rho_h} \end{aligned}\right\} \tag{9.3.37}$$

If $\lambda_{sh} = \sqrt{\rho_{sh\_v}/\rho_{sh\_h}}$ is known from some independent measurement, Eqs. (9.3.37a, b) can be written as

$$\left.\begin{aligned} V_{sh}\lambda_{sh}^2\rho_{sh\_h} + V_{sd}\rho_{sd} &= \rho_v \\ \frac{V_{sh}}{\rho_{sh\_h}} + \frac{V_{sd}}{\rho_{sd}} &= \frac{1}{\rho_h} \end{aligned}\right\} \tag{9.3.38}$$

that is

$$\rho_{sh\_h} = \frac{\rho_v - V_{sd}\rho_{sd}}{V_{sh}\lambda_{sh}^2} \tag{9.3.39}$$

leading to

$$A\rho_{sd}^2 + B\rho_{sd} + C = 0 \tag{9.3.40}$$

where the coefficients are

$$\begin{aligned} A &= V_{sd} \\ B &= \rho_h\left(V_{sh}^2\lambda_{sh}^2 - V_{sd}^2 - \lambda^2\right) \\ C &= \rho_h \rho_v V_{sd} \end{aligned} \tag{9.3.41}$$

Korvin (2012) considered the most general case. He realized that a reasonably strong ($\lambda_{sd} = \sqrt{\rho_{sd\_v}/\rho_{sd\_h}}$ between 1 and 2) *inherent electrical anisotropy* can develop in a *shale-free sandstone*, especially if it is *hydrocarbon bearing* (Anderson et al.1994; Jing et al. 2002; Kennedy and Herrick 2004).

In such cases sand anisotropy cannot be excluded, and a more general set of equations must be used than Clavaud's or Klein's, namely

$$\left.\begin{array}{l} V_{sh}\rho_{sh\_v} + V_{sd}\rho_{sd\_v} = \rho_v \\ \frac{V_{sh}}{\rho_{sh\_h}} + \frac{V_{sd}}{\rho_{sd\_h}} = \frac{1}{\rho_h} \end{array}\right\} \tag{9.3.42}$$

Using *known anisotropy values* (measured on cores or obtained from logs in nearby thick shale and sand) $\lambda_{sh} = \sqrt{\rho_{sh\_v}/\rho_{sh\_h}}$ and $\lambda_{sd} = \sqrt{\rho_{sd\_v}/\rho_{sd\_h}}$, Eqs. (9.3.42a, b) become:

$$\left.\begin{array}{l} V_{sh}\lambda_{sh}^2\rho_{sh\_h} + V_{sd}\lambda_{sd}^2\rho_{sd\_h} = \rho_v \\ \frac{V_{sh}}{\rho_{sh\_h}} + \frac{V_{sd}}{\rho_{sd\_h}} = \frac{1}{\rho_h} \end{array}\right\} \tag{9.3.43}$$

From Eq. (9.3.43a):

$$\rho_{sh\_h} = \frac{\rho_v - V_{sd}\lambda_{sd}^2\rho_{sd\_h}}{V_{sh}\lambda_{sh}^2} \tag{9.3.44}$$

that is

$$A\rho_{sd\_h}^2 + B\rho_{sd\_h} + C = 0, \tag{9.3.45a}$$

wherefrom

$$\rho_{sd\_h} = \frac{-B \pm \sqrt{B^2 - 4AC}}{2A}, \tag{9.3.45b}$$

With

$$\begin{array}{l} A = V_{sd}\lambda_{sd}^2 \\ B = \rho_h\left(V_{sh}^2\lambda_{sh}^2 - V_{sd}^2 - \lambda^2\right) \\ C = \rho_h\rho_v V_{sd} \end{array} \tag{9.3.46a, b, c}$$

For $\lambda_{sd} = 1$, Eqs. (9.3.43a, b) reduce to Clavaud's equations, and when both $\lambda_{sd} = \lambda_{sh} = 1$ we get back Klein's equations. To recognize the pioneering role of these authors, we shall call Eqs. (9.3.43a, b) *Generalized Klein-Clavaud Equations.*

As seen from Eqs. (9.3.43–9.3.46), in the general case of parallel, intrinsically anisotropic sand and shale layers, the anisotropic sand- and shale resistivitivities can be obtained in three steps:

(1) computing A, *B*, *C*; (2) solving the quadratic Eq. (4.3.45a) for ρ_{sd_h}; (3) then computing ρ_{sh_h} using Eq. (9.3.44). The specific vertical resistivities are obtained as $\rho_{Sh\_v} = \lambda^2{}_{sh}\rho_{sh\_h}$, $\rho_{sd\_v} = \lambda^2{}_{sd}\rho_{sd\_h}$.

For these calculations we need the following input data:

- three values inferred from well log measurements: V_{sh} (shale volume), ρ_h (horizontal resistivity of the formation, parallel with the bedding), ρ_v (vertical resistivity of the formation, perpendicular to the bedding);
- two computed values: $V_{sd} = 1 - V_{sh}$ (sand volume), and $\lambda = \sqrt{\rho_v/\rho_h}$ (formation anisotropy);
- core-derived or defaulted specific anisotropy values: λ_{sd} and λ_{sh}.

The independent parameters which are needed for Eqs. (9.3.46) span a *5-dimensional space.*

(V_{sh}, ρ_h, ρ_v, λ_{sd}, λ_{sh}), where they satisfy the obvious contraints that $0 \leq V_{sh} \leq 1$; ρ_h and ρ_v are positive real numbers; and none of the specific anisotropies λ_{sd} and λ_{sh} is less than one. To see why specific anisotropies cannot be less than one, consider the two cases presented in Figs. 9.2 and 9.3.

Figure 9.2 shows the vertical and horizontal resistivities of a formation, computed by the *Generalized Klein-Clavaud Eqs.* (9.3.43a, b), as function of shale volume. The specific sand- and shale resistivities used for the calculation are contained in Table 9.3. The plot presents a reasonable situation: for all shale volumes one has $\rho_v > \rho_h$ that is $\lambda = \sqrt{\rho_v/\rho_h} > 1$, as it should be. In Fig. 9.3, on the other hand, for two ranges of V_{sh} (very small and very large shale volumes) there arises a "nonphysical" case: $\rho_v < \rho_h$ that is $\lambda = \sqrt{\rho_v/\rho_h} < 1$. As seen in Table 9.2, in case of Fig. 9.2 both specific anisotropies are greater than one, while when constructing Fig. 9.3, both were, *unphysically*, less than one.

A little algebra shows that the simultaneous fulfillment of the two inequalities

$$\frac{\rho_{sh\_v}}{\rho_{sh\_s}} = \lambda_{sh}^2 \geq 1; \ \frac{\rho_{s\_v}}{\rho_{sd\_s}} = \lambda_{sd}^2 \geq 1 \tag{9.3.47}$$

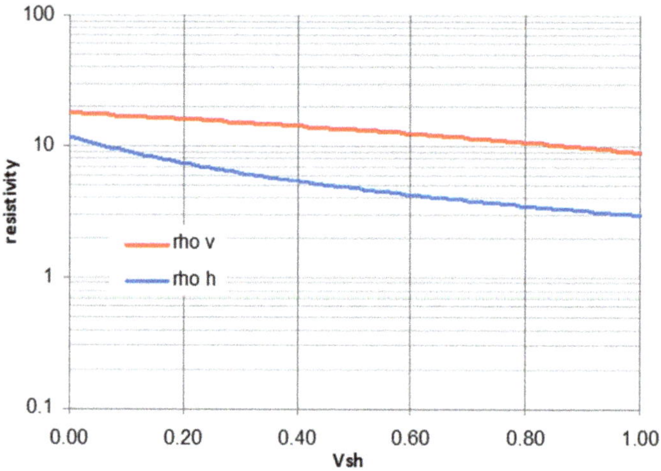

Fig. 9.2 Example for a "physical" situation. (From Korvin 2012, reproduced with the kind permission of the *Society of Petrophysicists and Well Log Analysts*)

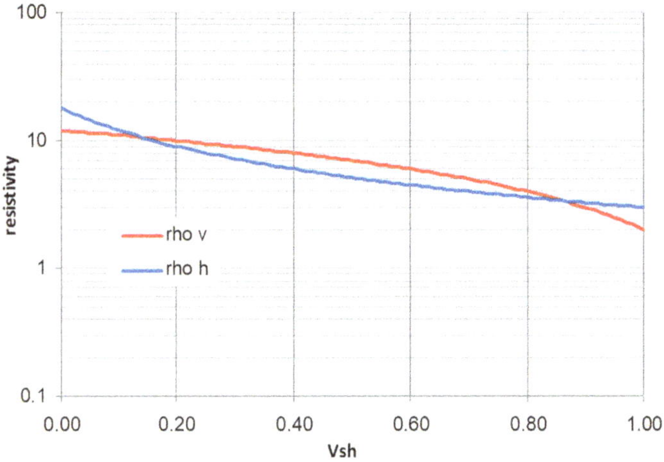

Fig. 9.3 A “nonphysical” case. (From Korvin 2012, reproduced with the kind permission of the *Society of Petrophysicists and Well Log Analysts*)

guarantees that if the solutions ρ_v and ρ_h of the *Generalized Klein-Clavaud Equations* are positive and real, then they satisfy $\lambda = \sqrt{\rho_v/\rho_h} > 1$. Indeed, from Eqs. (9.3.43a, b) and (9.3.47)

$\frac{1}{\rho_h} = \frac{V_{sh}}{\rho_{sh\_h}} + \frac{V_{sd}}{\rho_{sd\_h}} > \frac{V_{sh}}{\rho_{sh\_v}} + \frac{V_{sd}}{\rho_{sd\_v}}$, i.e. $\rho_h < \left\{\frac{V_{sh}}{\rho_{sh\_v}} + \frac{V_{sd}}{\rho_{sd\_v}}\right\}^{-1} < V_{sh}\rho_{sh\_v} + V_{sd}\rho_{sd\_v} = \rho_v$.

In the last step we used *Jensen's theorem* (Beckenbach and Bellman 1961; Bullen 2003) according to which *for any two unequal positive numbers their weighted harmonic mean is less than their arithmetic mean.*

Some further numerical experimentation in the *parameter space* (V_{sh}, ρ_h, ρ_v, λ_{sd}, λ_{sh}) reveals that Conditions (9.3.47a, b), in themselves, still do not guarantee that the quadratic equation $A\rho_{sd\_h}^2 + B\rho_{sd\_h} + C = 0$ (where the coefficients are computed from Eqs. (9.3.46a,b,c) would have a real positive solution for ρ_{sd_h}. The question naturally arises: is there a way to characterize those points (V_{sh}, ρ_h, ρ_v, λ_{sd}, λ_{sh}) of the parameter space for which the *Generalized Klein-Clavaud Equations* have physically meaningful (real and positive) solutions for ρ_{sd_h} and ρ_{sh_h}? Korvin (2012) proved the following two theorems:

Theorem 1 *The overall anisotropy of the layered sand/shale formation satisfies the inequalities.*

$$(V_{sh}\lambda_{sh} + V_{sd}\lambda_{sd})^2 \leq \lambda^2 \leq V_{sh}^2\lambda_{sh}^2 + V_{sd}^2\lambda_{sd}^2 + 2V_{sh}V_{sd}\max\left(\lambda_{sd/sh}^2, \lambda_{sh/sd}^2\right) \quad (9.3.48)$$

where $\lambda = \sqrt{\rho_v/\rho_h}$, $\lambda_{sh} = \sqrt{\rho_{sh\_v}/\rho_{sh\_h}}$, $\lambda_{sd} = \sqrt{\rho_{sd\_v}/\rho_{sd\_h}}$, and we introduced the “*cross-anisotropies*” $\lambda_{sh/sd} = \sqrt{\rho_{sh\_v}/\rho_{sd\_h}}$, $\lambda_{sd/sh} = \sqrt{\rho_{sd\_v}/\rho_{sh\_h}}$.

Table 9.3 Parameters used for constructing Figs. 9.2 and 9.3 (From Korvin 2012)

Figure #	V_{sh}	ρ_{sd_h} Ωm	ρ_{sd_v} Ωm	λ_{sd}	ρ_{sh_h} Ωm	ρ_{sh_v} Ωm	λ_{sh}
9.2	0 to 1	12	18	$\sqrt{1.5} = 1.22$	3	9	$\sqrt{3} = 1.73$
9.3	0 to 1	3	2	0.82	18	12	0.82

Theorem 2 *If the parameters* $(V_{sh}, V_{sd}, \lambda, \lambda_{sd}, \lambda_{sh})$ *satisfy.*

$$(V_{sh}\lambda_{sh} + V_{sd}\lambda_{sd})^2 \le \lambda^2 \tag{9.3.49}$$

then the generalized Klein-Clavaud Equations (165a,b) *have physically meaningful (real and positive) solutions* $\rho_{sh\_h}, \rho_{sh\_v}, \rho_{sd\_h}, \rho_{sd\_v}$.

The lower bound in Inequality (9.3.48) *can be used in the numerical or graphical interpretation of triaxial induction logs to exclude such "nonphysical" cases when the graphical or numerical solutions would result in negative, or complex-valued specific resistivities, or in unrealistic formation anisotropies for which* $\lambda = \sqrt{\rho_v/\rho_h} < 1$.

The inequalities (9.3.48) give interesting Hashin–Shtrikman type bounds (Hashin and Shtrikman 1963b; Racherla and Castañeda 2008; Mei and Vernescu 2010) for the resistivity anisotropy of TI (*transversely isotropic*) materials. The classical Hashin–Shtrikman variational bounds for the effective electric conductivity of polycrystals cannot be applied for the present case because they relate to a *statistically homogeneous and isotropic* (rather than laminated, that is *transversely isotropic*) assembly of randomly oriented single anisotropic crystals. I hope one of my Readers will succeed to improve the upper bound in Eq. (9.3.48), which does not seem optimal.

9.4 The Mysterious Cementation Exponent

The electric conductivity of a salt-water-filled rock sample can be expressed with high accuracy by means of a very simple empirical law (Archie's Law, 1942):

$$\sigma_{rock} = \sigma_{fluid}\Phi^m \tag{9.4.1}$$

where $\sigma_{rock}, \sigma_{fluid}, \Phi, m$ are, respectively, the *dc-conductivities of the fluid-filled rock and of the fluid, the porosity, and a constant called "cementation exponent"*. The exponent m is close to 2 for almost all rock types, it is between 1.5 and 2, except for "problematic reservoirs" (Worthington 2011). There is ample experimental evidence for the law, but since 1942 nobody could derive it satisfactorily from basic laws of Physics. According to (Thovert et al. 1990), "it is well verified experimentally though it does not have any theoretical justification."

Here we review some attempts based on *Effective Medium Theory*, *Granular Effective Medium Theory* (a version of *Differential Effective Medium Theory*), *Fractal Random Walks*, *Generalized Mean Values* to prove Archie's law, and to show that $m \sim 2$.

9.4.1 Effective Medium Theory Solution

Suppose that, as in Kirkpatrick's (1973) EMT solution (Fig. 9.1) the network can be substituted by a topologically identical "*effective*" network where all bonds have the same conductivity g_m. To find g_m we assume that the local electric field generated when a bond between A and B is substituted by its original value g_0 averages to zero. (A and B are neighboring vertices, the bond between them is in the direction of the external field). Let V_m be the constant voltage increase per row, V_0 the local disturbance created, then Kirchoff's equations give

$$V_0 = \frac{V_m(g_m - g_0)}{\left[g_0 + \left(\frac{Z}{2} - 1\right)g_m\right]} \tag{9.4.2}$$

(Z is the coordination number of the lattice). For a random distribution of conductances $\{g_m\}$ with PDF (*probability density function*) $f(g)$

$$\langle V_0 \rangle = \int \frac{dg f(g)(g_m - g)}{\left[g + \left(\frac{Z}{2} - 1\right)g_m\right]} = 0 \tag{9.4.3}$$

For a binary-valued conductivity distribution $\left\{g_{fluid} = 1;\, g_{grain} = \alpha \ll 1\right\}$ the PDF is

$$f(g) = \Phi\delta(g - 1) + (1 - \Phi)\delta(g - \alpha) \tag{9.4.4}$$

Equation (9.4.3) becomes quadratic, and for $\alpha \to 0$ its physically meaningful solution is $g_m \approx 1 - \frac{1-\Phi}{1-\frac{2}{Z}}$ which is linear in Φ rather than a power function required by Archie's Law.

9.4.2 Granular Effective Medium Theory Solution

This approach (Yonezawa and Cohen 1983) applies *Granular Effective Medium Theory* (Allo 2019). It is a version of DEM (Differential EMT): we start from saline water (sea water) and replace at every step small fractions of water with an amount Δq_k of rock pieces. In each step we iteratively use EMT for the 2-component medium. Denote by $\sigma^{(k)}$ the mixture's conductivity at the kth step. By EMT:

$$\frac{\sigma^{(k)} - \sigma^{(k+1)}}{\sigma^{(k)} + \left(\gamma^{-1} - 1\right)\sigma^{(k+1)}}(1 - \Delta q_k) + \Delta q_k \frac{\sigma_2 - \sigma^{(k+1)}}{+\left(\gamma^{-1} - 1\right)\sigma^{(k+1)}} = 0 \tag{9.4.5}$$

where σ_2 is rock conductivity, $\gamma = 2/Z$, Z is mean coordination number of the pack of solid grains. If $\Delta q_k \ll 1$, Φ_k denotes water fraction in step k, $\sigma^{(k+1)} \approx$

$\sigma^{(k)}\left[1+\Delta q_k \gamma^{-1}\cdot\frac{\sigma_2-\sigma^{(k)}}{\sigma_2+(\gamma^{-1}-1)\sigma^{(k)}}\right]$ and $\Delta q_k \Phi_k = \Phi_k - \Phi_{k+1}$. For $\Delta q_k \to 0$ we get the ODE (*Ordinary Differential Equation*)

$$\frac{d\sigma^{(k)}}{\sigma^{(k)}} = -\frac{d\Phi_k}{\Phi_k}\gamma^{-1}\cdot\frac{\sigma_2-\sigma^{(k)}}{\sigma_2+\left(\gamma^{-1}-1\right)\sigma^{(k)}} \tag{9.4.6}$$

with the *boundary condition*

$$\sigma^{(k)} = \sigma_1 for\ \Phi_k = 1 \tag{9.4.7}$$

whose solution for $\sigma_2 \to 0$ is $\sigma = \sigma_1 \Phi^m$ with $m = \frac{\gamma^{-1}}{\gamma^{-1}-1} = \frac{Z}{Z-2}$. This reproduces Archie's Law, but not with $m = 2$, unless $Z = 4$.

9.4.3 *Fractal Random Walk Model*

There have been several attempts to derive Archie's Law from the rules of *Fractal Random Walk* (Korvin 1992). Roy and Tarafdar (1997) started out from Einstein's equation

$$\sigma \propto n_c D_e \tag{9.4.8}$$

with $n_c \propto \Phi$ being charge carrier density, D_e diffusivity. For classic, *nonfractal random walk* $\langle r^2\rangle \propto D_e t$. If porosity is *fractally scaling* with dimension D in the range $[l, L]$; $l \ll L$, then the *fractal walk* in the pore space is *anomalous*: $\langle r^2\rangle \propto t^{2/D_w}$; $D_w > 2$ as long as $\langle r^2\rangle^{1/2} < L$, that is up to a time $t' \propto \left(\frac{L}{l}\right)^{D_w}$. For $t \gg t'$ diffusion becomes normal, with $D_e \propto \left(\frac{L}{l}\right)^2 \cdot \frac{1}{t'} \propto \left(\frac{L}{l}\right)^{2-D_w}$. But in a rock of fractal porosity, porosity scales as $\Phi \propto \left[\left(\frac{L}{l}\right)^D \times l^3\right] \times L^{-3} = \left(\frac{L}{l}\right)^{D-3}$ and Einstein's Eq. (9.4.8) gives Archie's $\sigma \propto \Phi^m$ with

$$m = 1 + \frac{2-D_w}{D-3} \tag{9.4.9}$$

($D \neq 3$; $D < 3$; see Korvin 1992: 287).

9.4.4 *Using Generalized Mean Values*

Korvin's (1978, 1982) *Generalized Mean Value formalism* (see Sect. 9.3b) suggests a *generalized Kirchoff's Law* for the conductivity of a 2-component composite:

$$\sigma = \left\{\lambda\sigma_1^t + (1-\lambda)\sigma_2^t\right\}^{1/t} \tag{9.4.10}$$

where λ, $1-\lambda$ are the respective volume ratios, $t\epsilon(-\infty, \infty)$ is a real parameter. Letting $\sigma_2 \rightarrow 0$ and matching the result with *Archie's Law* $\sigma = \sigma_1\lambda^m$ we obtain $\lambda^{1/t} = \lambda^m$, that is

$$1/t = m \tag{9.4.11}$$

References

Aczél J (1946) The notion of mean values. Nor. Vidensk. Selsk. Forh. 19:83–86
Aczél J (1961) Vorlesungen über Funktionalgleichungen und ihre Anwendungen. VEB Deutscher Verlag der Wissenschaften, Berlin
Allo F (2019) Consolidating rock-physics classics: A practical take on granular effective medium models. The Leading Edge 38:334–340
Almqvist BSG, Mainprice D, Madonna C, Burlini L, Hirt AM (2011) Application of differential effective medium, magnetic pore fabric analysis, and X-ray microtomography to calculate elastic properties of porous and anisotropic rock aggregates. J Geophys Res 116:B01204
Anderson B, Bryant I, Lüling M, Spies B, Helbig K (1994) Oilfield anisotropy: Its origins and electrical characteristics. Oilfield Rev:48–56
Archie GE (1942) The electrical resistivity log as an aid in determining some reservoir characteristics. Pet Trans AIME 146:54–62
Avellaneda M (1987) Iterated homogenization, differential effective medium theory and applications. Commun Pure Appl Math 40:527–554
Barnett DM (1972) The precise evaluation of derivatives of the anistropic elastic Green's functions. Phys Status Solidi C 49:741–748
Beckenbach EF, Bellman R (1961) Inequalities. Springer Verlag, Berlin-Göttingen-Heidelberg
Bemasconi J (1974) Conduction in anisotropic disordered systems: Effective medium theory. Phys Rev B 9:4575–4579
Berry JE (1959) Acoustic velocity in porous media. J Pet Technol 11:262–270
Berryman JG (1980) Long-wavelength propagation in composite elastic media. I Spherical inclusions. J Acoust Soc Am 68:1809–1819
Berryman JG (1980) Long-wavelength propagation in composite elastic media II Ellipsoidal inclusions. J Acoust Soc Am 68:1820–1831
Berryman JG (2005) Thermal conductivity of porous media. Appl Phys Lett 86:032905
Berryman JG, Berge PA, Bonner BP (2002) Estimating rock porosity and fluid saturation using only seismic velocities. Geophysics 67:391–404
Berryman James G, Pride Steven R, Wang Herbert F (2002) A differential scheme for elastic properties of rocks with dry or saturated cracks. Geophys J Int 151(2):597–611
Brovelli A, Cassiani G (2010) A combination of the Hashin-Shtrikman bounds aimed at modelling electrical conductivity and permittivity of variably saturated porous media. Geophys J Int 180(1):225–237
Brown WF, Jr (1955) Solid mixture permittivities. J Chem Phys 23:1514–1517
Bruggeman DAG (1935) Berechnung verschiedener physikalischer Konstanten von heterogenen Substanzen. I. Dielektrizitätskonstanten und Leitfähigkeiten der Mischkörper aus isotropen Substanzen. Annalen der Physik 416(7):636–664
Bruner WM (1976) Comment on "Seismic velocities in dry and saturated cracked solids" by Richard J. O'Connell and Bernard Budiansky". J Geophys Res 81:2573–2576

Bullen PS (2003) Handbook of means and their inequalities. Kluwer Academic Publishers, Dordrecht-Boston-London
Candikia YN, Rosid MS, Haidar MW (2017) Determining pore type in carbonate reservoir. AIP Conf Proc 1862:030185
Chang E, Graham EK (1975) The elastic constants of cassiterite SnO_2, and their pressure and temperature dependence. J Geophys Res 80:2595–2599
Choy TC (2015) Effective medium theory, Chapter 1. Essentials, 2nd edn. Oxford University Press, Oxford, pp 1–26
Chung DH (1963) Elastic moduli of single crystal and polycrystalline MgO. Phil Mag 8:833–841
Chung DH (1974) General relationships among sound speeds,1. New experimental information. Phys Earth Planet Interiors 8:113–120
Clavaud J-B, Nelson R, Guru UK, Wang (2005) Field example of enhanced hydrocarbon estimation in thinly laminated formation with a triaxial array induction tool: a laminated sand-shale analysis with anisotropic shale. In: SPWLA annual logging symp., New Orleans, Louisiana
Cleary MP, Chen I-W, Lee S-M (1980) Self-consistent techniques for heterogeneous media. ASCE J Eng Mech 106:861–887
Corson PB (1974a) Correlation functions for predicting properties of heterogeneous materials. I. Experimental measurement of spatial correlation functions in multiphase solids. J Appl Phys 45(7):3159–3164
Corson PB (1974b) Correlation functions for predicting properties of heterogeneous materials. II. Empirical construction of spatial correlation functions for two-phase solids. J Appl Phys 45(7):3165–3170
Corson PB (1974c) Correlation functions for predicting properties of heterogeneous materials. III. Effective elastic moduli of two-phase solids. J Appl Phys 45(7):3171–3179
Corson PB (1974d) Correlation functions for predicting properties of heterogeneous materials. IV. Effective thermal conductivity of two-phase solids. J Appl Phys 45(7):3180–3182
Darnowski P, Furmański P, Domański R (2019) Relation between thermal conductivity and coordination number for fibre-reinforced composite with random distribution of fibres. Arch Thermodyn 40(1):21–48
David C, Gueguen Y, Pampoukis G (1990) Effective medium theory and network theory applied to the transport properties of rock. J Geophys Res 95:6993–7005
De Finetti B (1931) Sul concetto di media. G Ist Ital Attuari 2:369–396
Ebigbo A, Lang PS, Paluszny A, Zimmerman RW (2016) Inclusion-based effective medium models for the permeability of a 3D fractured rock mass. Transport in Porous Media:137–158
Edwards J, Berg P (2021) Pore-network models and effective medium theory: a convergence analysis. arXiv:2109.07
Elsayed MA (1974) Bounds for effective thermal, electrical, and magnetic properties of heterogeneous materials using high order statistical information. J Math Phys 15(11):2001–2015
Eshelby JD (1957) The determination of the elastic field of an ellipsoidal inclusion, and related problems. Proc R Soc London Ser A 241:376–396
Fricke H (1924) A mathematical treatment of the electric conductivity and capacity of disperse systems. Phys Rev 24:575–587
Fuchs L (1950) On mean systems. Acta Math 1:303–320
Garnett JCM (1904) XII. Colours in metal glasses and in metallic films. Philos Trans R Soc A 203:285–420
Ghanbarian Behzad, Hunt Allen G, Ewing Robert P, Skinner Thomas E (2014) Universal scaling of the formation factor in porous media derived by combining percolation and effective medium theories. Geoph Res Lett 41(11):3884–3890
Grant FS, West GF (1965) Interpretation theory in applied geophysics. McGraw-Hill Book Co, New York
Haidar MW, Wardhana R, Iksan M, Rosid MS (2018) Optimization of rock physics models by combining the differential effective medium (DEM) and adaptive Batzle-Wang methods in "R" field, East Java. Sci Bruneiana 17(2):23–33

Hardy GH, Littlewood JE, Polya G (1934) Inequalities. Cambridge Univ. Press, London
Harris CK (1990) Application of generalised effective-medium theory to transport in porous media. Transp Porous Med 5:517–542
Harris CK (1992) Effective-medium treatment of flow through anisotropic fracture systems-improved permeability estimates using a new lattice mapping. Transp Porous Med 9:287–295
Hashin Z (1980) This Week's Citation Classic: Hashin Z & Shtrikman S. A variational approach to the theory of the elastic behaviour of multiphase materials. J Mech Phys Solids 11:127–40, 1963. Curr Contents 11:344
Hashin Z, Shtrikman S (1961) Note on a variational approach to the theory of composite elastic materials. J Franklin Inst 271:336–341
Hashin Z, Shtrikman S (1962) On some variational principles in anisotropic and nonhomogeneous elasticity. J Mech Phys Solids 10:335–342
Hashin Z, Shtrikman S (1962) A variational approach to the theory of elastic behavior of polycrystals. J Mech Phys Solids 10:343–352
Hashin Z, Shtrikman S (1962) A variational approach to the theory of the effective magnetic permeability of multiphase materials. J Appl Phys 33:3125–3131
Hashin Z, Shtrikman S (1963) A variational approach to the elastic behavior of multiphase materials. J Mech Phys Solids 11:127–140
Hashin Z, Shtrikman S (1963) Conductivity of polycrystals. Phys Rev 130(1):129–133
Henyey F, Pomphrey N (1982) Self-consistent elastic moduli of a cracked solid. Geophys Res Lett 9:903–906
Hestir K, Long JCS (1990) Analytical expressions for the permeability of random two-dimensional Poisson fracture networks based on regular lattice percolation and equivalent media theories. J Geophys Res 95:21,565–21,581
Hill R (1952) The elastic behavior of a crystalline aggregate. Proc Phys Soc London Sect A 65:349–354
Hill R (1963) Elastic properties of reinforced solids: Some theoretical principles. J Mech Phys Solids 11:357–372
Hill R (1963) New derivations of some elastic extremum principles. Progress in Applied Mechanics. MacMillan, The Prager Anniversary Volume. New York, pp 99–106
Hornby BE, Schwartz LM, Hudson JA (1994) Anisotropic effective medium modeling of the elastic properties of shales. Geophysics 59(10):1570–1583
Jing XD, Al-Harthy S, King S (2002) Petrophysical properties and anisotropy of sandstones under true-triaxial stress conditions. Petrophysics 43:358–362
Jurgawczynski M (2007) Predicting absolute and relative permeabilities of carbonate rocks using image analysis and effective medium theory. Doctor of Philosophy Thesis of the University of London, and the Diploma of Imperial College
Kennedy D, Herrick D (2004) Conductivity anisotropy in shale-free sandstone. Petrophysics 45:38–58
Kirkpatrick S (1971) Classical transport in disordered media: Scaling and effective medium theories. Phys Rev Lett 27:1722–1725
Kirkpatrick S (1973) Percolation and conduction. Rev Mod Phys 45(4):574–588
Klein JD, Martin PR, Allen DF (1997) The petrophysics of electrically anisotropic reservoirs. Log Analyst 38(3)
Kolmogorov A (1930) Sur la notion de la moyenne. Atti R. Accad. Naz. Lincei 12:388
Koplik J (1982) Creeping flow in two-dimensional networks. J Fluid Mech 119:219–247
Korvin G (1978) The hierarchy of velocity formulae: Generalized mean value theorems. Acta Geod Geoph et Mont Acad Sci Hung 13(1–2):211–222
Korvin G (1982) Axiomatic characterization of the general mixture rule. Geoexploration 19(4):267–276
Korvin G (1992) Fractal models in the earth sciences. Elsevier, Amsterdam
Korvin G (2012) Bounds for the resistivity anisotropy in thinly-laminated sand-shale. Petrophysics 53(1):14–21

Korvin G (2021) Statistical rock physics. In: Daya Sagar B, Cheng Q, McKinley J, Agterberg F (eds) Encyclopedia of mathematical geosciences. Encyclopedia of earth sciences series. Springer, Cham
Kriegshäuser B, Fanini O, Forgang S, Itskovich G, Rabinovich M, Tabarovsky L, Yu L, Epov M, VD Horst J (2000) A new multicomponent induction logging tool to resolve anisotropic formations. In: SPWLA 40th Logging Symp
Kumazawa M (1969) The elastic constant of polycrystalline rocks and nonelastic behavior inherent in them. J Geophys Res 74:5311–5320
Lakkad SC, Miatt BB, Parsons B (1973) Elastic and thermoelastic properties of a statistically isotropic heterogeneous medium. J Phys D: Appl Phys 6(13):1566
Landauer R (1952) The electrical resistance of binary metallic mixtures. J Appl Phys 23:779–784
Lock PA (2001) Estimating the permeability of reservoir sandstones using image analysis of pore structure. Doctoral dissertation, Department of Earth Science and Engineering, Imperial College
Lorentz HA (1880) Ueber die Beziehung zwischen der Fortpflanzungsgeschwindigkeit des Lichtes und der Körperdichte. Annalen der Physik 245(4):641–665
Lorenz L (1880) Ueber die Refractionsconstante. Annalen der Physik 247(9):70–103
Lutz MP, Zimmerman RW (2016) Effect of the interphase zone on the conductivity or diffusivity of a particulate composite using Maxwell's homogenization method. Int J Eng Sci 98(SI):51–59
Mainprice D (1997) Modeling the anisotropic seismic properties of partially molten rocks found at mid-ocean ridges. Tectonophysics 279:161–179
Mavko G, Jizba D (1991) Estimating grain-scale fluid effects on velocity dispersion in rocks. Geophysics 56:1940–1949
Mavko G, Mukerji T, Dvorkin J (1998) The rock physics handbook: tools for seismic analysis in porous media. Cambridge University Press, Cambridge
Mavko G, Mukerji T, Dvorkin J (2009) The rock physics handbook. Cambridge University Press
Maxwell JC (1891) A treatise on electricity and magnetism. London: Clarendon, London (Repr. edn. by New York: Dover 1954)
McCoy JJ (1970) On the displacement field in an elastic medium with random variations in material properties. In: Recent advances in engineering science (Ed. by A. C. Eringen). Gordon and Breach, New York, pp 235–254
McLaughlin RA (1977) A study of the differential scheme for composite materials. Int. J. Eng. Sci. 15:237–244
Meese AD, Walther HC (1967) An investigation of sonic velocities in vugular carbonates. In: 8th SPWLA Symp., Denver
Mei Chiand C, Vernescu Bogdan (2010) Homogenization methods for multiscale mechanics. World Scientific, Singapore
Miller MN (1969) Bounds for effective bulk modulus of heterogeneous materials. J Math Phys 10:2005–2013
Minh CC, Clavaud JB, Sundararaman P, Froment S, Caroli E, Billon O, Davis G, and Fairbairn R (2008) Graphical analysis of laminated sand-shale formations in the presence of anisotropic shales1. Petrophysics. SPWLA J Form Eval Reserv Descr 149(05)
Mourzenko VV, Thovert JF, Adler PM (2011) Permeability of isotropic and anisotropic fracture networks, from the percolation threshold to very large densities. Phys Rev E 84(3):036307
Nagumo M (1930) Ūber eine Klasse der Mittelwerte. Jpn J Math: Trans Abstr 7:71–79
Nishizawa O (1982) Seismic velocity anisotropy in a medium containing oriented cracks. J Phys Earth 30(4):331–347
Norris A (1985) A differential scheme for the effective moduli of composites. Mech Mater 4:1–16
Norris AN, Sheng P, Callegari AJ (1990) Effective-medium theories for two-phase dielectric media. J Appl Phys 57:1990–1996
Oda M (1995) Permeability tensor for discontinuous rock masses. Géotechnique 35(4):483–495
Park SK (2004) Mantle heterogeneity beneath eastern California from magnetotelluric measurements. J Geophys Res 109:B09406

Park SK, Ducea MN (2003) Can in situ measurements of mantle electrical conductivity be used to infer properties of partial melt? J Geophys Res 108(B5):2270
Ponte Castañeda P (1991) The effective mechanical properties of nonlinear isotropic solids. J Mech Phys Solids 39:45–71
Ponte Castañeda P (1992) Bounds and estimates for the properties of nonlinear heterogeneous systems. Phil Trans R Soc London A 340(1659):531–567
Qin X, Han DH, Zhao L (2014) Rock physics modeling of organic-rich shales with different maturity levels. SEG Tech Program Expand Abstr 2014:2952–2957
Racherla V, Castañeda PP (2008) Linear comparison estimates for the effective resistivity of three-dimensional nonlinear polycrystals. Proc Royal Soc A 464(2097):2391–2410
Rayleigh L (1892) LVI. On the influence of obstacles arranged in rectangular order upon the properties of a medium. Philos Mag 34(211):481–502
Reuss A (1929) Berechnung der Fliessgrenze von Mischkristallen auf Grund der Plastizitätsbedingung für Einkristalle. Z Angew Math Mech 9:49–58
Rosid MS, Wahyuni SD, Haidar MW (2017) Carbonate reservoir characterization with pore type inversion using Differential Effective Medium (DEM) model at "X" Field East Java. AIP Conf Proc 1862:030179
Roy S, Tarafdar S (1997) Archie's law from a fractal model for porous rocks. Phys Rev B 55(13):8038
Rudin W (1976) Principles of mathematical analysis. McGraw-Hill, New York
Ruiz F, Dvorkin J (2007) Rock physics model for deep-sea shallow calcareous sediment with porous grains. SEG/San Antonio 2007 Annual Meeting Technical Program. Expanded Abstracts
Ruiz F, Dvorkin J (2010) Predicting elasticity in nonclastic rocks with a differential effective medium model. Geophysics 75(1):E41–E53
Schlueter EM (1995) Predicting the transport properties of sedimentary rocks from microstructure. Ph.D. Thesis, University of California, Berkeley
Sævik PN, Berre I, Jakobsen M, Lien M (2013) A 3D computational study of effective medium methods applied to fractured media. Transp Porous Media 100(1):115–142
Sævik PN, Jakobsen M, Lien M, Berre I (2014) Anisotropic effective conductivity in fractured rocks by explicit effective medium methods. Geophys Prospect 62(6):1297–1314
Sahimi M (1995) Flow and transport in porous media and fractured rock. VCH Publishers, Weinheim, Germany
Sevostianov I (2014) On the shape of effective inclusion in the Maxwell homogenization scheme for anisotropic elastic composites. Mech Mater 75:45–59
Shukla MM, Padial NT (1973) Calculation of the Debye characteristic temperature of cubic crystals. Rev Brasil Fis 3:39–45
Snow DT (1969) Anisotropic permeability of fractured media. Water Resour Res 5(6):1273–1289
Stroud D (1998) The effective medium approximations: Some recent developments. Superlattices Microstruct 23(3/4):567–573
Thomsen L (1972) Elasticity of polycrystals and rocks. J Geophys Res 77(2):315–327
Thovert JF, Wary F, Adler PM (1990) Thermal conductivity of random media and regular fractals. J Appl Phys 68:3872–3883
Tjaden B, Cooper SJ, Brett DJ, Kramer D, Shearing PR (2016) On the origin and application of the Bruggeman correlation for analysing transport phenomena in electrochemical systems. Curr Opin Chem Eng 12:44–51
Toledo PG, Davis HT, Scriven LE (1992) Transport properties of anisotropic media: effective medium theory. Chem Eng Sci 47:391–405
Torquato S (1997) Effective stiffness tensor of composite media—I. Exact series expansions. J Mech Phys Solids 45(9):1421–1448
Torquato S (2002) Random heterogeneous materials: microstructure and macroscopic properties. Springer-Verlag, Princeton
Voigt W (1928) Lehrbuch der Kristallphysik. Teubner, Leipzig
Waff HS (1974) Theoretical considerations of the electrical conductivity in a partially molten mantle and implications for geothermometry. J Geophys Res 79(26):4003–4010

Walpole LJ (1966) On bounds for the overall elastic modulus of inhomogeneous systems-I. J Mech Phys Solids 14:151–162
Walpole LJ (1966) On bounds for the overall elastic moduli of inhomogeneoussystems-II. J Mech Phys Solids 14:289–301
Walsh JB (1980) Static deformation of rock. ASCE J Eng Mech 106:1005–1019
Watt JP, Peselnick L (1980) Clarification of the Hashin-Shtrikman bounds on the effective elastic moduli of polycrystals with hexagonal, trigonal and tetragonal symmetries. J Appl Phys 51(3):1525–1531
Watt JP, Davies GF, O'Connell RJ (1976) The elastic properties of composite materials. Rev Geophys 14(4):541–563
Wiener OH (1912) Die Theorie des Mischkörpers für das Feld der stationären Strömung. 1 Die Mittelwertsätze für Kraft, Polarisation und Energie. Leipzig, B.G. Teubner
Willis J (1977) Bounds and self-consistent estimates for the overall properties of anisotropic composites. J Mech Phys Solids 5(3):185–202
Wood AB (1955) A textbook of sound. Macmillan, New York
Woodside W, Messmer JH (1961) Thermal conductivity of porous media. I: unconsolidated sands. J Appl Phys 32(9):1688–1699
Worthington Paul F (2011) The petrophysics of problematic reservoirs. J Petr Technol 63(12):88–97
Wu T (1966) The effect of inclusion shape on the elastic moduli of a two-phase material. Int J Solids Struct 2:1–8
Wyllie MRJ, Gregory AR, Gardner LW (1956) Elastic wave velocities in heterogeneous and porous media. Geophysics 21:41–70
Yonezawa F, Cohen MH (1983) Granular effective medium approximation. J Appl Phys 54(6):2895–2899
Zerhouni O, Brisard S, Danas K (2021) Quantifying the effect of two-point correlations on the effective elasticity of specific classes of random porous materials with and without connectivity. Int J Eng Sci 166:103520
Zimmerman RW (1989) Thermal conductivity of fluid-saturated rocks. J Petrol Sci Eng 3(3):219–227
Zimmerman RW (1996) Effective conductivity of a two-dimensional medium containing elliptical inhomogeneities. Proc R Soc A Math Phys Eng Sci 452:1713–1727
Zimmerman RW, Bodvarsson GS (1996) Hydraulic conductivity of rock fractures. Transp Porous Media 23(1):1–30

Chapter 10
Markov Random Fields and Random Walks

Chapter Highlights *Markov Random Fields*, which are spatial generalizations of *Markov Chains* in time, are mainly used in *geophysical inverse problems* and *geologic image processing* (Table 10.1). *Discrete-time* and *continuous-time Random Walk models* (Sect. 10.2, and Table 10.2) are among the most popular petrophysical simulation tools. Three important applications are discussed in details: *anomalous diffusion*, the *NMR response of fluids in porous rocks*, and an attempt (of Ioannidis et al. 1997) to derive the formation factor F in Archie's equation. Mathematical details (such as: *Derivation of the Gibbs Distribution*; *Proof of the Hammersley-Clifford Theorem*; *Polya's Theorem and its proof*; and *Partial Differential Equations governing Continuous Time Random Walks*) are discussed in appendices.

10.1 Markov Random Fields

When we model rocks under *ambient conditions*, it is natural to assume that the spatial distribution of material properties inside any volume $\mathcal{V}$ of the rock is independent from the distribution in any other *far-away region* $\mathcal{V}'$. In many cases we go to the extreme, and assume that the probability measures describing the distributions inside $\mathcal{V}$ have a certain 'spatial Markov property', that is the probability distributions inside $\mathcal{V}$ only depend on the distributions on and around the boundary $\partial\mathcal{V}$ but are independent of the distributions farther outside.

In the wide range of applications of MRF (*Markov Random Field*), see Table 10.1, in most cases we work with 2-D or 3-D *images* where it is assumed that in any pixel (voxel) the value of a variable is only related to pixels in the neighborhood (Salomão and Remacre 2001). In MRF-based procedures we make sequential changes in the image, according to a transition probability, until a prescribed objective function measuring the success of simulation is optimized. The simulation steps are usually governed by some modification of the algorithm of (Metropolis et al. 1953). In this simple case the model is said to be a MRF if, for the neighborhood ∂_s of each pixel s, the following relationship is valid

G. Korvin, *Statistical Rock Physics*, Earth and Environmental Sciences Library,
https://doi.org/10.1007/978-3-031-46700-4_10

$$P(X_s = x_s|X_r = x_r, r \neq s) = P(X_s = x_s|X_r = x_r, r \in \partial_s) \tag{10.1.1}$$

In a frequently used hierarchical neighborhood structure introduced by (Derin and Elliot 1987), and applied by Uçan's group (Uçan et al. 2000, 2004; Uçan and Albora 2009) for the *separation of regional and residual gravity- and geomagnetic anomalies* the territories sharing a common boundary are considered *neighbors* (Steinsland 2003: Fig. 31). In the *graph* representation of this neighborhood structure *vertices* correspond to the regions, those representing neighboring regions are joined by *edges*. Further details for the graph representation of the neighborhood structures found in 2- or 3-D physical space can be found in (Rue and Held 2005).

A subset of vertices of a graph is called *clique* if every pair of its *vertices* is joined by an *edge*. Mathematically, cliques are defined as (Majoros 2007): *Let* $G = (V, E)$ *be an undirected graph with a set of vertices* V *and set of edges* E. $C \subseteq V$ *is a clique if for all* $(a, b) \in C, a \neq b$ *we have* $(a, b) \in E$. *Singleton sets* $\{v\}, v \in V$ *are also considered cliques.*

A general definition of MRF is that of Majoros (2007), according to which a discrete-valued Markov Random Field (MRF) is a model $\mathcal{M} = (\{\alpha\}, \{X\}, P_{\mathcal{M}}, G)$, where

$\{\alpha\}$	is a finite *alphabet* (i.e. set of colors, properties, labels, etc.)
$\{X\}$	is a set of variables randomly taking values from $\{\alpha\}$;
$P_{\mathcal{M}}$	is the probability distribution describing the random assignment of values from $\{\alpha\}$ *to* $\{X\}$;
$G(\{X\}, E)$	is an *undirected graph* with set of vertices $\{X\}$, and set of edges E describing the *dependence relations* between the variables in $\{X\}$ in the following way:

Let $X_i \in \{X\}$. *Denote by* $\mathcal{N}_G(X_i)$ *the set of neighbors of* X_i, *that is*

$$\mathcal{N}_G(X_i) = \left\{X_j \middle| j \neq i; \left(X_i; X_j\right) \in E\right\}. \tag{10.1.2}$$

Then we require that $P_{\mathcal{M}}$ *satisfy*

$$P_{\mathcal{M}}\left(\{X_i\}\middle|\left\{X_{k\neq i}\right\}\right) = P_{\mathcal{M}}\left(X_i|\mathcal{N}_g(X_i)\right). \tag{10.1.3}$$

It has been shown by Besag (1974) that the only probability distribution for this MRF model in which the conditional probabilities $P_{\mathcal{M}}(X_i|\mathcal{N}_g(X_i))$ satisfy Eq. (10.1.3), and the joint probability distribution is consistent and unique, is provided by the following Theorem, a consequence of the *Hammersley-Clifford Theorem* (see Sect. 10.1.1).

Theorem *Consider the random allocation of properties from* α *on the vertices* X *of the graph* $G(V, E)$. *Suppose* $P(\boldsymbol{x}) > 0$ *for all joint value assignments* $\boldsymbol{x}$ *to the variables in X. Then the likelihood of* $\boldsymbol{x}$ *under model* $\mathcal{M}$ *is given by:*

$$P_{\mathcal{M}}(\boldsymbol{x}) = \frac{1}{Z}e^{Q(\boldsymbol{x})}, \tag{10.1.4}$$

with the normalization factor

$$Z = \sum_{x'} e^{Q(x')}; \tag{10.1.5a}$$

and $Q(\boldsymbol{x})$ has a unique expansion given by

$$\begin{aligned} Q(\boldsymbol{x}) &= Q(x_0, x_1, \cdots, x_{n-1}) \\ &= \sum_{0 \le i < n} \Phi_i(x_i) + \sum_{0 \le i < j < n} x_i x_j \Phi_{i,j}(x_i, x_j) \\ &+ \cdots + x_0 x_1 \cdots x_{n-1} \Phi_{0,1,\ldots,n-1}(x_0, x_1, \ldots, x_{n-1}), \end{aligned} \tag{10.1.5b}$$

where any Φ term whose indices do not correspond to a clique must be zero.

Denoting by $\mathcal{C}$ the set of cliques of the graph $G(X, E)$, Eq. (10.1.4) can be written as

$$P_{MRF}(\boldsymbol{x}) = \frac{1}{Z} \exp\left[\sum_{c \in \mathcal{C}} \Phi_c(\boldsymbol{x})\right]. \tag{10.1.6}$$

The MRF likelihood functions (10.1.4) or (10.1.6) resemble the *Boltzmann* (also called "*Gibbs*") distribution function, defined in *Statistical Mechanics as*

$$P_{Boltzmann}(\boldsymbol{x}) = \frac{1}{Z} \exp\left[-\frac{E(\boldsymbol{x})}{kT}\right]. \tag{10.1.7}$$

This distribution gives the probability that a particular molecular configuration $\boldsymbol{x}$ would occur in an ideal gas at temperature T. Here, $k = 1.38 \times 10^{-23}$ is the *Boltzmann constant*, the normalizing factor Z is called the *partition function*, *E(**x**)* is the *Gibbs free energy* (Landau and Lifshitz 1980: Sect. 28). A simple case of this distribution will be derived in Sect. 10.1.1.

In real-world problems, the *partition function* is *exponentially complex* (Forbes and Peyrard 2003; Wang et al. 2017) and its exact computation is near impossible. According to the Hammersley-Clifford theorem, the *local conditional probabilities* can be written in the form (Besag 1986; Geman and Geman 1984)

$$P\left(x_j \boldsymbol{X}_{\partial_j}\right) = \frac{P\left(x_j, \boldsymbol{X}_{\partial_j}\right)}{\sum_{x'_j \in L} P\left(x'_j, \boldsymbol{X}_{\partial_j}\right)} = \frac{\exp\left[-U\left(x_j, \boldsymbol{X}_{\partial_j}\right)\right]}{\sum_{x'_j \in L} \exp\left[-U\left(x'_j, \boldsymbol{X}_{\partial_j}\right)\right]} \tag{10.1.8}$$

where L is the set of "labels" that can be assigned to the vertices.

While, according to Wang et al. (2017) this equation "is the cornerstone of the simulation of an MRF using a classical Gibbs sampler", to find the local energy function $U\left(x_j, \boldsymbol{X}_{\partial_j}\right)$ is a delicate problem. In the simple case of the *Ising model*

(Ising 1925; Cipra 1987) which is a *binary-valued MRF*, the *local conditional probability* depends on the values assigned to the neighborhood $\{v_i\}$ and on the attraction parameter, as

$$P(X_i = x_i | X_r = x_r, r \neq i) = \frac{\exp(-\beta x_i \sum v_i}{1 + \exp(-\beta \sum v_i)} \qquad (10.1.9)$$

(Salomão and Remacre 2001).

In the *Potts model* (Potts 1952; Koller and Friedman 2009), which extends Ising's model and uses more than two labels, the local energy function is selected as

$$U\left(x_j, \boldsymbol{X}_{\partial_j}\right) = V_j\left(x_j\right) + \sum_{i \in \partial_j} V_{i,j}(x_i, x_j), \qquad (10.1.10)$$

where the potential functions are

$$V_{i,j}\left(x_i, x_j\right) = \begin{cases} 0 \text{ if } x_i = x_j \\ \beta \text{ if } x_i \neq x_j \end{cases}, \qquad (10.1.11)$$

$\beta > 0$ is sometimes called the *granularity coefficient* because it introduces a "smoothness" constraint. The role of β has been much studied for simple models. For the *Ising model* (an MRF model with two values: spin ↑ or ↓), there exists a critical value β_c such that for $\beta > \beta_c$ a phase transition occurs (Kindermann and Snell 1980). For the Potts model Forbes and Preyrard (2003) found in image processing experiments that for small values of β the realizations look "noisy", for large values of β the simulations lead to mono-color images, no matter how multi-colored the initial image had been.

For voxelized models (Wang et al. 2017) Eqs. (10.1.10–10.1.11) imply that two neighboring voxels have the lowest possible pairwise potential (zero) if they have the same label.

Strauss (1975, 1977), who worked with MRF models on multi-colored lattices, used the equation:

$$P(X = w) \propto \exp\left[-\sum_{0 \le k < l \le N} \sum \beta_{kl} n_{kl}\right] \qquad (10.1.12)$$

where n_{kl} is the number of pairs of neighbors having the colors (k, l) and β_{kl} is a factor expressing *repulsion* between two different colors ($\beta_{kl} = \beta_{lk}$). The *transition probability* is given by the relation:

$$P(x_i = k) \propto \exp\left[-\sum_{l \neq k} \beta_{kl} u_i(l)\right] \qquad (10.1.13)$$

where $u_i(l)$ represents the number of neighbors of i having color l. For *gray-level* images, the repulsion between neighbors is usually taken as proportional to the difference between their *gray scales*.

Note that the potential (10.1.11) is *isotropic* because it does not take into account the local information about *orientation*. Suppose we simulate with MRF the distribution of *lithologies* and want to encourage voxels to have the same consistent lithology in *dipping layers* characterized with unit normal vectors $\{\boldsymbol{n}_j\}$, $j \in \mathcal{D}$. Denote the unit orientation vector between two neighboring voxels v_i, v_j as $\boldsymbol{\psi}_{ij} := \frac{\overrightarrow{v_i v_j}}{|\overrightarrow{v_i v_j}|}$. Then (Wang et al. 2017) the potential function has the *anisotropic* form

$$V_{i,j}^{aniso}(x_i, x_j) = \begin{cases} 0 & \text{if} \quad x_i = x_j \\ \beta\left[2 - |\boldsymbol{n}_i \cdot \boldsymbol{\psi}_{ji}| - |\boldsymbol{n}_j \cdot \boldsymbol{\psi}_{ij}|\right] & \text{if} \quad x_i \neq x_j \end{cases}. \tag{10.1.14}$$

By the *Hammmersley-Clifford Theorem* the joint probability distribution of an MRF is a *Gibbs distribution*

$$P_G(z) = Z^{-1} \exp[-E(z)] \tag{10.1.15a}$$

with

$$E(z) = \sum_C V_C(z_C), \tag{10.1.15b}$$

where the summation is over cliques. To circumvent the difficulties in computing the partition function Z, Besag (1975) approximated the distribution (10.1.15a) on finite set of sites S with the *pseudo-likelihood*

$$\mathcal{PL}(z) = \prod_{i \in S} P_G(z_i \rfloor z_{N(i)}), \tag{10.1.16a}$$

where for a given value z_i of variable Z_i

$$P_G(z_i \rfloor z_{N(i)}) = \frac{\exp\left[-\sum_{C, i \in C} V_C(z_C)\right]}{\sum_{z_i'} \exp\left[-\sum_{C, i \in C} V_C(Z_C')\right]}, \tag{10.1.16b}$$

the sums in the exponentials are over cliques that contain the vertex i. The outer sum in the denominator is over all possible values forZ_i. If C contains i and z_i' then Z_C' denotes values $\{z_i', z_j, j \in, j \neq i\}$ for sites in C.

Another useful equation proposed by Besag (1972, 1974, 1986) expresses the relationship between the probabilities of two configurations X and X':

$$\frac{P(X)}{P(X')} = \prod_{i=1}^{n} \frac{P(x_i | x_1, \ldots, x_{i-1}, x_{i+1}', \ldots, x_n')}{P(x_i' | x_1, \ldots, x_{i-1}, x_{i+1}', \ldots, x_n')}. \tag{10.1.17}$$

We regret that an important recent paradigm, *Hidden Random Field Models* could not be dealt with in this introductory treatment. The Reader is referred to the excellent papers (Zhang et al. 2001; Forbes and Preyrard 2003; Wang et al. 2017). We also recommend to look up (Guyon and Hardouin 2002) who introduced, on a meteorological example, the concept of *Markov Chain Markov Field* (MCMF) to study the *time dynamics* of Markov Fields. In the model there are two kinds of potentials, the time-delay potentials (a *Markov Chain*), and an instantaneous spatial interaction potential which defines a *Markov Field* on the finite number of sites conditionally to the past.

10.1.1 Mathematical Appendices

Derivation of the Gibbs Distribution

In Sect. 8.2.2. ("Shale Compaction and Maximal Entropy") we derived the probability distribution

$$\{p_k = \exp(\alpha - \beta\varepsilon_k)\}_{k=0}^{N};$$

subject to

$$\sum_{k=0}^{N} \exp(\alpha - \beta\varepsilon_k) = 1, \tag{10.1.18}$$

and

$$\sum_{k=0}^{N} \varepsilon_k \exp(\alpha - \beta\varepsilon_k) = \langle E\rangle \tag{10.1.19}$$

(where $\langle E\rangle$ is *expected energy*) from *maximum entropy principles*. Here we solve the system of Eqs. (10.1.18–10.1.19) for α and β in the special case $N = \infty$ and $\varepsilon_k = k \cdot \varepsilon$; $\varepsilon > 0, \varepsilon \ll 1$.

With the notation $A = \exp(\alpha)$:

$$A \sum_{k=0}^{\infty} \exp(-\beta k\varepsilon) = 1; \tag{10.1.20}$$

$$A \sum_{k=0}^{\infty} k\varepsilon \cdot \exp(-\beta k\varepsilon) = \langle E\rangle. \tag{10.1.21}$$

In (10.1.20) we must have $\beta > 0$ for the convergence of the sum. We solve this nonlinear system of equations by *successive approximations*. From Eq. (10.1.20):

$$1 = A\sum_{k=0}^{\infty}\exp(-\beta k\varepsilon) = \frac{A}{1-\exp(-\beta\varepsilon)} \approx \frac{A}{1-(1-\beta\varepsilon)} = \frac{A}{\beta\varepsilon}, \qquad (10.1.22)$$

and the first approximation of A is $A \approx \beta\varepsilon$. In Eq. (10.1.21), observe that

$$\frac{\partial}{\partial\beta}\exp(-\beta k\varepsilon) = -\varepsilon k \cdot \exp(-\beta k\varepsilon).$$

So,

$$\begin{aligned}\frac{\langle E\rangle}{A} &= \sum_{k=0}^{\infty} k\varepsilon \cdot \exp(-\beta k\varepsilon) \\ &= -\sum_{k=0}^{\infty}\frac{\partial}{\partial\beta}\exp(-\beta k\varepsilon) = -\frac{\partial}{\partial\beta}\left(\sum_{k=0}^{\infty}\exp(-\beta k\varepsilon)\right) \\ &= -\frac{\partial}{\partial\beta}\left(\frac{1}{1-\exp(-\beta\varepsilon)}\right) \approx -\frac{\partial}{\partial\beta}\frac{1}{\beta\varepsilon} \\ &= \frac{1}{\varepsilon\beta^2} = \frac{1}{(\varepsilon\beta)\beta} \approx \frac{1}{A\beta}, \end{aligned} \qquad (10.1.23)$$

that is

$$\beta \approx \frac{1}{\langle E\rangle}. \qquad (10.1.24)$$

Substituting back to (10.1.20):

$$1 = A\sum_{k=0}^{\infty}\exp(-\beta k\varepsilon) = A,$$

that is

$$A = \left\{\sum_{k=0}^{\infty}\exp\left(-\frac{\varepsilon_k}{\langle E\rangle}\right)\right\}^{-1},$$

and we get

$$p_i = \exp\left(-\frac{\varepsilon_i}{\langle E\rangle}\right)\cdot\left\{\sum_{k=0}^{\infty}\exp\left(-\frac{\varepsilon_k}{\langle E\rangle}\right)\right\}^{-1} \qquad (10.1.25)$$

which is the *Gibbs distribution*.

Proof of the Hammersley-Clifford Theorem (After Grimmett 1973, the proof is not reproduced in its full generality).

Preliminaries: We shall need some definitions from the foundations of *Combinatorial Theory* (Rota 1964; Bender and Goldman 1975).

A *partially ordered* set P is a set, for certain pairs of elements of which there exists a relation $a \leq b$ (pronounced "*a is less than or equal to b*" or "*a precedes b*"). "Partial" here indicates that not every pairs of elements are comparable, that is there may exist pairs $\{p, q\}$ such that neither relation $p \leq q$ or $q \leq p$ holds. The relation $\leq$ in a partially ordered set P satisfies for all $p, q, r \in P$ the following three properties (Davey and Priestley 2002):

Reflexivity:

$$p \leq p \tag{10.1.26a}$$

Antisymmetry:

$$\text{if } p \leq q \text{ and } q \leq p \text{ then } p = q \tag{10.1.26b}$$

Transitivity:

$$\text{if } p \leq q \text{ and } q \leq r \text{ then } p \leq r. \tag{10.1.26c}$$

If $p, r;\ p \leq r$ are elements of the partially ordered set P, the *segment* $[p, r]$ is the set of all elements q which are *between* p and r, that is, satisfy $p \leq q \leq r$. A partially ordered set is *locally finite* if every segment in it is finite.

The *Möbius function* $\mu(x, y), x, y \in P$, is an integer-valued function defined by the equations

$$\left.\begin{array}{ll} \mu(x, z) = 0 & when\ x \nleq z \\ \sum_{y:x \leq y \leq z} \mu(x, y) = \delta(x, z) & when\ x \leq z \end{array}\right\}; \tag{10.1.27}$$

(Bender and Goldman 1975: Eq. 2), where

$$\delta(x, z) = \left.\begin{array}{l} 1\ if\ x = z \\ 0\ if\ x \neq z \end{array}\right\} \tag{10.1.28}$$

is the *Kronecker delta*.

The *Möbius Inversion Formula* (Rota 1964: Proposition 2): *Let $f(x)$ be a real-valued function defined for $x \in P$ where P is a locally finite partially ordered set. Suppose there exists an element $\xi \in P$ with the property that $f(x) = 0$ unless $x \geq \xi$. Define $g(x)$ as*

$$g(x) = \sum_{y \leq x} f(y) \tag{10.1.29}$$

Then

$$f(x) = \sum_{y \leq x} g(y)\mu(y, x). \tag{10.1.30}$$

Colored graphs: Denote by Λ the set of *vertices* (also called "*sites*", or "*nodes*") of a finite undirected graph $\mathcal{G}$ with no *loops* or multiple *edges*. Two vertices are considered *neighbors* if they are connected by an edge. We shall denote by capital letters the subsets of Λ with at least two elements, and use lower case letters for single vertices. Instead of $A \bigcup B$ and $A \setminus B$ we simply write $A + B$ and $A - B$. A set $\mathcal{K} \subseteq \Lambda$ of vertices is a *clique* if $\mathcal{K}$ contains only a single vertex or any two of its vertices are neighbours.

Let us now assign *properties*, called '*colors*' to the vertices in Λ. At any vertex $x \in \Lambda$ we can select from a countable set of colors $\{\omega_1, \omega_2, \ldots\}$ to color x. There is one specific color, called *black*. Write Ω for the set of all colorings of Λ, and $\mathcal{F}$ for the collection of all subsets of Ω. For each coloring $\omega \in \Omega$ we denote by $L(\omega)$ the set of sites of Λ which are *not colored black* by ω. Denote by Ω' the set of those colorings of Λ which blacken every site of Λ except the sites of a clique, formally

$$\Omega' = \{\omega \in \Omega | L(\omega) is a clique\}. \tag{10.1.31}$$

For any coloring $\omega \in \Omega$ and any set of vertices $A \subseteq \Lambda$ the notation ωA represents that particular coloring of Λ which assigns the color ω to each site of A and makes black all vertices outside A. The coloring which blackens all sites in Λ will be denoted by β. With these definitions we obviously have

$$\omega = \omega L(\omega) \text{ and } \beta = \omega\emptyset, \tag{10.1.32}$$

where $\varnothing$ is the empty set.

Markov and Gibbs random fields: If P is a probability measure on $(\Omega, \mathcal{F})$, then the triple $(\Omega, \mathcal{F}, P)$ is called *a random field on the domain* Λ. There are two specially important random fields among those that can be defined on Λ: the *Markov field* and the *Gibbs field*.

(A) The random field $(\Omega, \mathcal{F}, P)$ is a *Markov field* if it satisfies the following two conditions:

Condition 1. (*Positivity*). Every coloring has positive probability:

$$P(\omega) > 0 for all \omega \in \Omega; \tag{10.1.33}$$

Condition 2. (*Markov property*). The random colour of the site x, given the colours of every other site, depends only upon the colours of the *neighbours* of x: for any coloring $\omega \in \Omega$, set of vertices $A \subseteq \Lambda$ and pair $\{x, y\}$ of non-neighboring sites such that $x \notin A$ and $y \in A$,

$$\frac{P(\omega(A+x))}{P(\omega A)+P(\omega(A+x))} = \frac{P(\omega(A-y+x))}{P(\omega(A-y))+P(\omega(A-y+x))} \tag{10.1.34}$$

(B) The random field $(\Omega, \mathcal{F}, P)$ is a *Gibbs field* if there exists a *potential function* $V : \Omega' \to \mathbb{R}$ such that

$$P(\omega) = Z^{-1}\exp\left(\sum_{K\subseteq L(\omega)} V(\omega K)\right) \text{ for all } \omega \in \Omega, \tag{10.1.35}$$

where the summation is over the cliques contained in $L(\omega)$, and Z is a normalization factor to make

$$\sum_{\omega\in\Omega} P(\omega) = 1. \tag{10.1.36}$$

The *Hammersley-Clifford theorem* (also called the *Fundamental theorem of random fields*) connects these two, apparently quite different concepts. It proves that the *Markov Random Field* and *Gibbs Field* are equivalent with regard to the same graph. As stated and proved by Grimmett (1973).

Theorem *The random field $(\Omega, \mathcal{F}, P)$ is a Gibbs field if and only if it is a Markov field. If $(\Omega, \mathcal{F}, P)$ is a Markov field, then its potential function is given by*

$$V(\omega) = \sum_{A\subseteq L(\omega)} (-1)^{|L(\omega)-A|} \log P(\omega A), \textit{ for all } \omega \in \Omega', \tag{10.1.37}$$

where the summation is over all subsets of the clique $L(\omega)$.

Proof of the Theorem Suppose first that $(\Omega, \mathcal{F}, P)$ is a Markov field, and define a potential function $V : \Omega \to \mathbb{R}$ by

$$V(\omega) = \sum_{B\subseteq L(\omega)} (-1)^{|L(\omega)-B|} \log P(\omega B), \text{ for all } \omega \in \Omega. \tag{10.1.38}$$

Let $\omega \in \Omega$ and $A \subseteq L(\omega)$. Then, by (10.1.38),

$$V(\omega A) = \sum_{B\subseteq A} (-1)^{|A-B|} \log P(\omega B). \tag{10.1.39}$$

We show that $V(\omega A) = 0$ unless A is a clique. Indeed, otherwise A would contain two sites x, y which are not neighbors. Then, by (10.1.39)

$$
\begin{aligned}
V(\omega A) = &\sum_{\substack{B \subseteq A \\ x, y \in B}} (-1)^{|A-B|} \log P(\omega B) + \sum_{\substack{B \subseteq A \\ x \in A,\, y \notin B}} (-1)^{|A-B|} \log P(\omega B) \\
&+ \sum_{\substack{B \subseteq A \\ x \notin A,\, y \in B}} (-1)^{|A-B|} \log P(\omega B) + \sum_{\substack{B \subseteq A \\ x, y \notin\in B}} (-1)^{|A-B|} \log P(\omega B) \\
&= \sum_{B \subseteq A-x-y} (-1)^{|A-B|} \log\left(\frac{P(\omega(B+y+x))}{P(\omega(B+y))} \Big/ \frac{P(\omega(B+x))}{P(\omega B)} \right) = 0
\end{aligned}
\tag{10.1.40}
$$

by the condition (10.1.34).

The set of all subsets of $L(\omega)$ is *partially ordered by set-theoretical inclusion*, and its *Möbius function* μ is given by

$$
\mu(B, A) = (-1)^{|A-B|} \text{ for all } A, B \subseteq L(\omega) \text{ such that } B \subseteq A \tag{10.1.41}
$$

(Rota 1964: 345 Corollary "*Principle of Exclusion-Inclusion*"). Writing Eq. (10.1.39) in the form

$$
V(\omega A) = \sum_{B \subseteq A} \mu(B, A) \log P(\omega B), \text{ for all } A \subseteq L(\omega), \tag{10.1.42}
$$

and applying the *Möbius Inversion Theorem* yield

$$
\log P(\omega A) = \sum_{B \subseteq A} V(\omega B), \text{ for all } A \subseteq L(\omega), \tag{10.1.43}
$$

Hence, by (10.1.40)

$$
P(\omega) = P(\beta) \exp\left(\sum_{K \subseteq L(\omega)} V(\omega K) \right), \text{ for all } \omega \in \Omega, \tag{10.1.44}
$$

since, by (10.1.39)

$$
V(\beta) = \log P(\beta). \tag{10.1.45}
$$

Thus $(\Omega, \mathcal{F}, P)$ is a *Gibbs field* with *potential function* given by (10.1.36). (Grimmett 1973 mentioned without proof that this potential function is unique.)

On the other other hand, let $(\Omega, \mathcal{F}, P)$ be a *Gibbs field* with potential function V. Clearly

$$P(\omega) > 0, \text{ for all } \omega \in \Omega. \tag{10.1.46}$$

Let $\in \Omega$, and let $A \subseteq \Lambda$ and x, y be a pair of non-neighboring sites such that $x \notin A$ and$y \in A$. It is sufficient to show that the Markov property (10.1.34) holds if $A + x \subseteq L(\omega)$ (since if $x \notin L(\omega)$ then $\omega(A + x) = \omega A$, and if $z \in A + x - L(\omega)$ then $\omega(A + x) = \omega(A - z + x)$ and $\omega A = \omega(A - z)$). Then, by (10.1.35)

$$\begin{aligned}\log P(\omega(A+x)) - \log P(\omega A) &= \sum_{K\subseteq A+x} V(\omega K) - \sum_{K\subseteq A} V(\omega K) \\ &= \sum_{\substack{K \subseteq A+x \\ x \in K}} V(\omega K) = \sum_{\substack{K \subseteq A-y+x \\ x \in K}} V(\omega K) \\ &= \sum_{K\subseteq A-y+x} V(\omega K) - \sum_{K\subseteq A-y} V(\omega K) \\ &= \log P(\omega(A-y-x)) - \log P(\omega(A-y)),\end{aligned} \tag{10.1.47}$$

and we deduce, using (10.1.34), that $(\Omega, \mathcal{F}, P)$ is a Markov field.

Historical Notes (Grimmett and Welsh 2007): The theory of probability measures that satisfy certain "spatial Markov properties" had been developed around 1970 (Dobrushin 1968; Averintsev 1970; Spitzer 1971). The Hammersley and Clifford Theorem (1971, unpublished but available on the Web, see also Clifford 1990) generalized these results for an arbitrary discrete network and stated that a positive measure defined on a network is a Markov field if and only if it has a Gibbsian representation in terms of some potential function. Later authors simplified and clarified the proof (Grimmett 1973; Preston1973; Besag 1974; Geman and Geman 1984). Geoffrey Grimmett was a student of Hammersley, his proof, reproduced here, is based on an ingenious application of the *inclusion–exclusion principle*. Another student of Hammersley (Moussouris 1974) showed by counter-examples that the positivity condition (10.1.33) cannot be relaxed. The terminology of the proof related to *color*s has proved very convenient for dealing with variables with countable (but more than two) possible values, such as facies (Strauss 1975, 1977).

10.2 Random Walk Models

10.2.1 History and Applications

The classical theory of *random walks* (Spitzer 1964; Lovász 1996; Lawler 2010) provided a combinatorial and probabilistic treatment of random walks on *infinite grids*. The first significant result in the field was due to Polya (1921) who proved that *if a walker randomly walks on a d-dimensional grid, then* (*with probability* 1) *it returns to the starting point infinitely often if* $d = 2$, *but only a finite number of times if* $d \geq 3$. (We shall prove *Polya's Theorem* in Sect. 10.2.2.) According to "one possibly apocryphal story told about Pólya—(related in Holtz 2014)—"he really liked to walk and, while on one of his walks, kept bumping into the same couple. The couple finally got annoyed and asked him if he was stalking them. Pólya assured them he wasn't... and started thinking about the mathematical implications of this question. He subsequently discovered that a random walk in dimensions 1 and 2—but not higher—is recurrent, i.e., keeps coming back to the same point with probability 1".

Gaussian random walk is a mathematical model of the *Brownian motion* of small particles in liquids. For this motion the change of the position $X(t)$ of the randomly moving particle $\Delta X(t) = X(t + \Delta t) - X(t)$ obeys a Gaussian distribution with mean $\langle \Delta X(t) \rangle = 0$ and variance

$$\langle [\Delta X(t)]^2 \rangle^{1/2} \propto \Delta. \tag{10.2.1}$$

According to Lévy's definition (1965, English translation in Metzler and Klafter 2000), "the stochastic process which we will call linear Brownian motion is a schematic representation which describes well the properties of real Brownian motion, observable on a small enough, but not infinitely small scale, and which assumes that the same properties exist on whatever scale". Brownian motion has been named after the Scottish botanist Robert Brown (1773–1858) who in 1828 discovered the erratic movement of pollen grains in water (Brown 1828). Perrin (1909) repeated Brown's experiment with refined tools and succeeded to trace the path of a few pollen particles.

The mathematical studies of *continuous random walks* started in 1905 with an amusing correspondence in *Nature* between Prof. Karl Pearson and Lord Rayleigh. In his *Letter to Nature* Pearson asked "A man starts from a point O and walks l yards in a straight line; he then turns through any angle whatever and walks another l yards in a second straight line. He repeats this process n times. I require the probability that after these n stretches he is at a distance between r and $r + \delta r$ from his starting point O". The answer came in the next issue of the journal from Lord Rayleigh. He pointed out that he had already solved the problem some 25 years ago (Rayleigh 1880; see also Rayleigh 1945: vol. 1, Sect. 42.a), and "if n be very great, the probability sought is $\frac{2}{n} e^{-r^2/n} r dr$".

The amazingly wide range of applications of random walks in Earth Sciences is illustrated by the examples in Table 10.2.

Tobochnik (1990) describes the usual workflow of a random walk study as follows. (i) One decides which are the *non-conducting obstacles* into which the walker may not step; (ii) one puts down a walker randomly somewhere in the *conducting region* and allows it to exercise a random walk with fixed step length in random directions along coordinate axes; or in uniformly random directions; or in randomly selected directions among a finite number of possibilities; (iii) the step is allowed if the walker does not step through, or into a nonconducting obstacle, in this case the time is incremented by one unit; (v) otherwise, the walker is returned to its original position. This is repeated with many other independent walkers and the mean-square displacements from the original position are averaged to find $\langle R^2 \rangle = \langle R(t)^2 \rangle$. Selection of the step size is crucial: it must be small enough so that the walker could navigate through the narrow crevices (for porous materials also called "throats" or "necks") between the obstacles, and at the same time it must be large enough to sample a large part of the configuration space within affordable computer time. In the long-time limit

$$\frac{\langle R(t)^2 \rangle}{t} \to 2dD \tag{10.2.2}$$

where D is the *diffusion constant*, d is the dimension. In *electric conductivity* applications, by Einstein's (1905, 1906) equation

$$\sigma = const \cdot \Phi D \tag{10.2.3}$$

where σ is *conductivity*, Φ *porosity*.

Tobochnik (1990) recognized that this algorithm is inefficient, because the walker spends too much time in large pores without even approaching any of the obstacles. He solved this problem with a very nice *Monte Carlo algorithm* based on the *distribution of first passage times of random walkers on d-dimensional hyperspheres*. First, we find the largest d-dimensional sphere centered on the present position of the walker, which has no common points with any non-conducting obstacles. Then we move the walker immediately to a random position on the surface of this sphere. (An efficient algorithm for randomly selecting points in and on hyperspheres is due to Luban and Staunton 1988). The time is increased by an amount randomly picked from the *distribution of first-passage times of a random walk from the origin to the surface of a sphere.*

The *first-passage time distribution* $\rho(a, t)dt$ is defined as the probability of first reaching the surface of a d-dimensional sphere of radius a in a time between t and $t + dt$. In order to determine $\rho(a, t)dt$ we calculate $P(r, t)$, which is the probability that a particle starting at the origin will be at a distance r from the origin at time t, having never gone a distance greater than a. This is done by solving the diffusion equation

$$D\nabla^2 \mathrm{P}(\mathbf{r}, \mathrm{t}) = \frac{\partial P(\boldsymbol{r}, t)}{\partial t} \tag{10.2.4}$$

with initial condition

$$\mathrm{P}(\mathbf{r}, 0) = \delta(\boldsymbol{r}) = \delta(r)/r, \tag{10.2.5a}$$

and absorbing boundary condition

$$P(a, t) = 0. \tag{10.2.5b}$$

The solution in 2-D is

$$P(r, t) = \sum_{n=1}^{\infty} \frac{J_0(j_n r/a)}{\pi a^2 [J_1(j_n)]^2} \exp\left[-(j_n/a)^2 Dt\right], \tag{10.2.6}$$

where J_0 and J_1 are Bessel functions; j_n is the nth zero of $J_0(z)$.

In 3-D:

$$P(r, t) = \sum_{n=1}^{\infty} \frac{n \cdot \sin(n\pi r/a)}{2a^2 r} \exp\left[-(n\pi/a)^2 Dt\right]. \tag{10.2.7}$$

The probability $P(t)$ of *not being absorbed at time t* is the integral of $P(r, t)$ over $r < a$, that is:

in 2-D:

$$P(t) = \sum_{n=1}^{\infty} \frac{2}{j_n J_1(j_n)} \exp\left[-j_n^2 Dt/a^2\right]; \tag{10.2.8}$$

in 3-D:

$$P(t) = 2 \sum_{n=1}^{\infty} (-1)^{n+1} \exp\left[-(n\pi/a)^2 Dt\right]. \tag{10.2.9}$$

The probablity of *being absorbed in the time interval* $(t, t + dt)$ is obtained by differentiating Eqs. (10.2.8–10.2.9), because

$$\rho(a, t)dt = -\frac{\partial P(t)}{\partial t} dt. \tag{10.2.10}$$

We get:

in 2-D:

$$\rho(a, t)dt = 2D \sum_{n=1}^{\infty} \frac{j_n}{J_1(j_n)} \cdot \exp\left[-\frac{j_n^2 Dt}{a^2}\right] \frac{dt}{a^2}, \tag{10.2.11}$$

in 3-D:

$$\rho(a, t)dt = 2D \sum_{n=1}^{\infty} (-1)^{n+1} (n\pi)^2 \exp\left[-(n\pi)^2 \frac{Dt}{a^2}\right] \frac{dt}{a^2}, \tag{10.2.12}$$

In both dimensions $\rho(a,t)dt$ only depends on the ratios t/a^2 and dt/a^2. In 2-D the computation of the intricated expression (10.2.11) can be simplified by using the approximations (Abramowitz and Stegun 1965):

$$j_n \approx (n - 0.25)\pi; \qquad (10.2.13a)$$

$$J_1(z) \approx \left(\frac{2}{\pi z}\right)^{\frac{1}{2}} \cos\left(z - \frac{3\pi}{4}\right). \qquad (10.2.13b)$$

Theoretically, the *Mean Squared Displacement* (MSD) $\langle R^2(t)\rangle$ for a typical *diffusive random walk* should be linearly related to time (or number of steps) as in (10.2.1, 10.2.2). However, many examples have been found both in physical experiments and random-walk simulations where the MSD is not linear in time but scales according to some other power-law. Such processes are known as *anomalous diffusion* (or *non-Fickian*) (Berkowitz and Scher 1995; Weeks et al. 1996; De W. Van Siclen 1999b; Metzler 2000; Metzler and Klafter 2000, 2004; Shlesinger 2003; Dentz et al. 2004; Cortis and Berkowitz 2004, 2005; Berkowitz et al. 2006; Codling et al. 2008; Gouze et al. 2008; Bijeljic et al. 2011, 2013; Kang et al. 2014; Noetinger et al. 2016; Abou-Saleh et al. 2019; Yang and Wang 2019). Codling et al. (2008) consider the following five cases for $\langle R^2(t)\rangle \propto t^{\mu}$:

(1) $\mu = 0$ *stationarity*, no movement over the period of observation;
(2) $\mu = 1$ this is the *standard relation* between MSD and time for diffusive processes;
(3) $0 < \mu < 1$ *sub-diffusion*, the MSD increases slower than in case of standard diffusion. This might occur if waiting times are included between steps (Weeks et al. 1996) or the spatial domain is constricted, e.g. by barriers (Coscoy et al. 2007).
(4) $1 < \mu < 2$ *super-diffusion*, the MSD increases at a faster rate than in standard diffusion. This typically occurs when the step lengths in the walk have a heavy-tailed probability distribution with infinite variance (*Lévy walks*, *Lévy flights*: Lévy 1954; Mandelbrot 1977; Barthelemy et al. 2008).
(5) $\mu = 2$ *ballistic* or *wavelike*, the MSD increases quadratically with t as in a wave process.

Several explanations have been put forward for anomalous diffusion. (In these studies the terms *walker, tracer, particle, contaminant, and solute* are used interchangeably.) DeW. Van Siclen (1999b, 2021) claims that macroscopic properties such as transport coefficients should be determined for length scales not smaller than the correlation length ξ of the inhomogeneous medium. "For displacements < ξ, the walker diffusion is anomalous rather than Gaussian due to the heterogeneity of the composite at length scales less than ξ".

Bijeljic et al. (2013) wonder "how much geometric complexity is needed to induce anomalous transport, and whether anomalous transport can be persistently observed in porous media with simple pore space structure." Yang and Wang (2019) answer this question and claim the *ubiquity of anomalous transport in porous media*. They write "Anomalous transport in porous media is commonly believed to be induced by the highly complex pore space geometry. However, this phenomenon is also observed in porous media with rather simple pore structure". They found by *Lattice Boltzmann simulations* (see Sect. 11.2) that the solute transport process might be anomalous even in simple structures at high Péclet numbers. (The Péclet number is defined as

$$Pe = \frac{v \cdot L_{char}}{D} = \frac{\textit{convective transport}}{\textit{diffusive transport}}, \tag{10.2.14}$$

see Rapp 2017: Eq. 9.18).

The *Continuous Time Random Walk* (CTRW) model has been proposed to explain and simulate *anomalous*, i.e., *non-Fickian*, transport in heterogeneous media (Scher and Lax 1973; comprehensive review: Berkowitz et al. 2006; numerical implementation: Cortis and Berkowitz 2005). In this type of walk we allow the walker (i) to jump according to a *probability distribution function* (pdf) $\Psi(x, \tau)$ a length x, or (ii) spend a random *retention time* τ at a given location. Jump directions have a uniform random distribution. If the length of jumps and the retention times are independent, we have

$$\Psi(x, \tau) = p(x)\psi(\tau) \tag{10.2.15}$$

where $p(x)$ is the pdf of jump lengths, and $\psi(\tau)$ is pdf of the retention times. When both $p(x)$ and $\psi(\tau)$ have finite first and second moments, the CTRW and RW are similar. If either of the pdf-s has infinite variance (such as Cauchy, or power-law distributions) then the CTRW yields a different physical picture, and the random paths are referred to as *Lévy walks* (infinite variance of (τ)) or *Lévy flights* (infinite variance of $p(x)$). Gautestad (2013: Fig. 1) presents a simulated Lévy flight where the jump lengths were defined as

$$l = \alpha(1 - RND)^{-(\beta-1)}, \tag{10.2.16}$$

with $= 1$, $\beta = 2$, and *RND* a uniformly distributed random number in (0, 1).

To show the strength and beauty of *Random Walk modeling*, I briefly discuss two characteristic rock-physical applications (Korvin 2021). The first is a study of *Nuclear Magnetic Relaxation* (NMR) in brine-filled porous rocks (Jin et al. 2009), the other a random-walk determination of the electric conductivity in rocks (Ioannidis et al. 1997).

The *random-walk* simulation of the NMR response of fluids in porous rocks can be applied on *grain-based* and *voxel-based representations* of the rock. In the grain-based approach a spherical grain pack is the input, where the solid surface is analytically defined; in the voxel-based approach the input is a computer-generated 3D image of the porous media. The implementation of random walk is the same in both cases.

For fluids in porous media, total magnetization originates from *diffusion* and *relaxation processes*. Three processes are involved in the relaxation of spin magnetization (only transverse relaxation is considered): (1) bulk fluid relaxation, with characteristic time T_{2B} due to dipole–dipole interactions between spins within the fluid; (2) surface relaxation, characterized by effective relaxation time T_{2S}, due to the additional interaction of protons at the pore-grain interface with paramagnetic impurities in the grains, and the motion of water molecules in a layer near the pore-grain interface; and (3) relaxation due to background magnetic field heterogeneities, T_{2D}. Surface morphology has a strong effect on the relaxation time T_{2S} which, in the fast-diffusion limit ($\frac{\rho R}{D_B} < 1$) is inversely proportional to specific surface area (Brownstein and Tarr 1979):

$$\frac{1}{T_{2S}} = \rho \frac{S}{V}, \tag{10.2.17}$$

where ρ is *surface relaxivity*, D_B diffusion coefficient of the bulk fluid, S is pore surface area, and V pore volume. The decay rate T_2 of the magnetization is given by

$$\frac{1}{T_2} = \frac{1}{T_{2B}} + \frac{1}{T_{2S}}. \tag{10.2.18}$$

Far away from a boundary, T_2 is equal to the bulk fluid's transverse relaxation time, T_{2B}. For spins close to a pore boundary, the magnetization decay is locally enhanced by surface relaxation. Random-walk simulations of NMR generally assume (Jin et al. 2009) that the spin is "killed" with a probability p when it reaches a boundary, and relate this probability to surface relaxivity ρ as

$$p = 0.96A\frac{\rho \Delta r}{D_B} \tag{10.2.19}$$

where Δr is the displacement that the spin achieves during the interval $[t, t + \Delta t]$ when it reaches the boundary, and A is an empirical factor of order 1 (usually set to 2/3). In more realistic models (see Refs. [1, 5, 7, 13, 15, 16, 18] in Jin et al. 2009) of *local surface relaxation* the spin is not "killed", but its magnetization is decreased by a factor

$$(1 - p) \approx \exp[-\Delta t(p/\Delta t)], \tag{10.2.20}$$

so that the total relaxation rate becomes $1/T_2 = 1/T_{2B} + p/\Delta t$. Equivalently, the relaxation time used in each step of the random walk can be written as

$$\frac{1}{T_2} = \frac{1}{T_{2B}} + \chi \frac{3.84\,\rho}{\Delta r} \tag{10.2.21}$$

where in the grain-based model, $\chi = 1$ when the walker is located within a step of the relaxing boundary, and 0 otherwise.

The starting points of walkers are randomly picked in the pore-space occupied by fluids. Each walker is allowed to walk for a sufficiently long time, whenever it attempts to step out of the pore space into the matrix, its magnetization decays with a rate determined by Eq. (10.2.20). At each time-step Δt, a random vector $\vec{n} = (n_x, n_y, n_z)$ is generated, whose independent components are normally distributed with unit variance and zero mean. The walker is then spatially displaced by $\Delta\vec{r} = \sqrt{(6D_B\Delta t)}\frac{\vec{n}}{\|\vec{n}\|}$. The time step Δt is dynamically changed along the walk so that it is smaller than any magnetic-field pulse duration, and the amplitude of the total displacement, $\Delta\vec{r}$, is much shorter than the relevant geometric length scales (e.g., pore throats, wetting film thickness, etc.). It is assumed that the pores are fully brine-saturated, the pulse sequence is of the *Carr–Purcell–Meiboom–Gill* (CPMG) type (Carr and Purcell 1954; Meiboom and Gill 1958) with an inter-echo spacing of 1 ms, the parameters were $T_{2B} = 2800$ ms, $D_B = 2\,\mu\text{m}^2/\text{ms}$, $\rho = 0.005$. One analytically treatable case, a *cubic packing of spheres*, a *random packing of monodisperse and polydisperse spheres*, and a digitized micro X-ray-CT image were used as test cases. The theoretical case was a single spherical pore of radius R where the magnetization decay in the fast diffusion limit ($\rho R/D_B < 1$) is given by (Hidajat et al. 2002)

$$\frac{M(t)}{M_0} = \exp\left[-t\left(\frac{1}{T_{2B}} + \rho\frac{S}{V}\right)\right], \tag{10.2.22}$$

M_0 being the initial transverse magnetization (equal to 1 in this study). Simulations were performed for the case $\frac{\rho R}{D_B} = 0.02$.

Grain-pack based simulations were made for the cubic packing of spheres, and for random packing of monodisperse and polydisperse spheres. The voxel-based approach was run on the micro X-ray-CT image of a Fontainebleau sandstone sample with 21.5%. porosity, the image consisted of 512 ×512 × 512 voxels, with 5.68 μm voxel size.

In another random-walk model (Ioannidis et al. 1997) they attempted to find the *formation factor F* in *Archie's equation*. Instead of solving Laplace's equation these authors exploited the formal analogy between random walks and the solutions of the Laplace equation (Kim and Torquato 1990). They determined the *mean square displacement* as function of time for a large number of random walkers. A single walker starts at a random void voxel (i_0, j_0, k_0). It can take random steps to any neighboring void, without penetrating solid grains. The time required for each step is, by Einstein's rule

$$t_{v\to v} = \frac{(\partial D)^2}{\sigma_0}, \tag{10.2.23}$$

where σ_0 is conductivity of the pore fluid (set to 1), ∂D is distance travelled in one step in a 3D lattice of spacing δ, so that it can be δ, $\delta\sqrt{2}$ or $\delta\sqrt{3}$. To avoid premature disappearance of walkers, periodic boundary conditions are used. The random walker travels a large number of steps (~10^6) and there are many (~10^3) independent walkers. The average displacement after time t will be $\langle R^2 \rangle = \left(\frac{\sigma}{\Phi_a}\right) t$ where $\sigma (= \sigma_0/F)$ is the effective electrical conductivity, and Φ_a is the fraction of voxels accessible for the walker. For such rocks (carbonates, shales) where solid-void, void-solid, solid–solid transitions are also allowed, Einstein's formula changes to

$$t_{s \to s} = \frac{(\partial D)^2}{\sigma_s} = \frac{(\partial D)^2}{\sigma_0} F_s \tag{10.2.24}$$

(F_s is the formation factor of the matrix). Steps between solid and void are only allowed with a given probability. In the model they used $p_{v \to s} = \frac{1}{1+F_s}$ and $p_{s \to v} = \frac{F_s}{1+F_s}$. The time to cross an interface is

$$\tau = \left(\frac{2}{1+F_s}\right) \cdot \frac{(\partial D)^2}{4\sigma_0} F_s, \tag{10.2.24}$$

and

$$\begin{aligned} t_{s \to v} &= \frac{t_{s \to s}}{4} + \tau, \\ t_{v \to s} &= \frac{t_{v \to v}}{4} + \tau. \end{aligned} \tag{10.2.25}$$

For a recent, very nice derivation of *Archie's law*, and the cementation exponent m using *random walk theory* (the "*Walker Diffusion Method*" of De W. Van Siclen1999a) and *fractal scaling*, see De W. Van Siclen (2021).

10.2.2 Mathematical Apppendices

Polya's Theorem and Its Proof

Polya's Theorem (Polya 1921; Lovász 1996; Lawler 2010; Durrett 2019) deals with a classical problem of random walks on the d-dimensional grid of integers $\mathbb{Z}^d$, where the walker makes a random move of length one at equidistant time instants in one of the lattice directions. In 1-D, assume the walker starts at x, then in n steps it will get to

$$S_n = x + \sum_{i=1}^{n} X_i,$$

where $X_i = \pm 1$ are independent random variables with

$$P(X_i = +1) = P(X_i = -1) = 1/2.$$

Assume that $x = 0$, that is the walker starts from the origin. After an odd number of steps the walker is at an odd integer, after an even number of steps it is at an even integer. It is easy to check that

$$\langle S_n \rangle = \left\langle \sum_{i=1}^{n} X_i \right\rangle = 0; \tag{10.2.26}$$

$$\langle S_n^2 \rangle = \left\langle \left(\sum_{i=1}^{n} X_i \right) \left(\sum_{j=1}^{n} X_i \right) \right\rangle = n + \sum_{i \neq j} \sum \langle X_i X_j \rangle = n. \tag{10.2.27}$$

Consider Polya's problem first in 1-D. The probability of getting back to the origin in an odd number of steps is zero, getting back in $2n$ steps is

$$P(S_{2n} = 0) = \binom{2n}{n} 2^{-2n} = \frac{(2n)!}{(n!)^2} 2^{-2n}. \tag{10.2.28}$$

Using Stirling's formula

$$n! \sim \sqrt{2\pi} \cdot n^{n+1/2} e^{-n} \tag{10.2.29}$$

for $n \gg 1$ we get

$$P(S_n = 0)|_{n \to \infty} \sim \frac{\sqrt{2}}{\lambda n^{1/2}}, \tag{10.2.30a}$$

with

$$\lambda = \sqrt{2\pi}. \tag{10.2.30b}$$

Denote by $\chi(E)$ the indicator function of the *random event E*. This is a *random variable*: $\chi(E) = 1$ if E occurs, $\chi(E) = 0$ if E does not occur. The number of times the walker visits the origin is $= \sum_{n=0}^{\infty} \chi(S_{2n} = 0)$, the expected number of such visits is

$$\langle V \rangle = \sum_{n=0}^{\infty} \langle \chi(S_{2n} = 0) \rangle = \sum_{n=0}^{\infty} P(S_{2n} = 0).$$

By (10.2.30a, 10.2.30b)

$$P(S_{2n} = 0) \propto 1/\sqrt{n}$$

that is

$$\langle V \rangle = \sum_{n=0}^{\infty} P(S_{2n} = 0) \propto \sum_{n=0}^{\infty} \frac{1}{\sqrt{2n}} = \infty. \tag{10.2.31}$$

Denote by q the probability that the walker ever returns to the origin. We shall prove that $q = 1$. Otherwise (indirect assumption) we would have $q < 1$. Compute for this case the distribution of the number of visits. For example,

$$P(V = 1) = 1 - q, \tag{10.2.32}$$

because $V = 1$ if and only if the walker starts from the origin and never returns there again. Similarly,

$$P(V = k) = q^{k-1}(1 - q), k = 1, 2, \ldots, \tag{10.2.33}$$

that is

$$\langle V \rangle = \sum_{k=1}^{\infty} kP(V = k) = \sum_{k=1}^{\infty} kq^{k-1}(1 - q) = (1 - q)\frac{d}{dq}\sum_{0}^{\infty} q^k = \frac{1}{1 - q} < \infty. \tag{10.2.34}$$

This contradicts (10.2.31), thus we have proved *Polya's Theorem* for 1-D walks: *A random walk on* $\mathbb{Z}^d(d = 1)$ *with probability one returns to the origin infinitely many times.*

The discussion of the problem for walks on $\mathbb{Z}^d(d \geq 2)$ needs more sophisticated tools (see Lawler 2010: 15–17 and Exercise 1.17 on p. 48), we give here only a heuristic scaling argument. Consider a random walker on the grid $\mathbb{Z}^d = \{x_1, \ldots, x_d\}$ where x_i are integers. Again, $\boldsymbol{S}_n = \boldsymbol{x}+\boldsymbol{X}_1+\cdots+\boldsymbol{X}_n$ is the position of the walker after n steps, and for convenience assume that $\boldsymbol{S}_0 = 0 = (0, 0, \ldots, 0)$. The increments $\boldsymbol{X}_i$ are d-dimensional vectors of unit length, with a single non-zero component ± 1. We have

$$\boldsymbol{X}_i \cdot \boldsymbol{X}_i = 1 \tag{10.2.35a}$$

and if $j \neq k$ then

$$\boldsymbol{X}_j \cdot \boldsymbol{X}_k = \begin{cases} +1 \; \textit{with probability} \; 1/2d \\ -1 \; \textit{with probability} \; 1/2d \\ 0 \qquad \textit{otherwise} \end{cases} \tag{10.2.35b}$$

From (10.2.35a, 10.2.35b)

$$\langle \boldsymbol{X}_i \cdot \boldsymbol{X}_i \rangle = 1;\; if\;\, j \neq k \langle \boldsymbol{X}_j \cdot \boldsymbol{X}_k \rangle = 0;\; if\;\, \boldsymbol{S}_0 = \boldsymbol{0}\, then\, \langle \boldsymbol{S}_n \rangle = \boldsymbol{0} \tag{10.2.36}$$

Similarly to the 1-D case (see 10.2.27), we have

$$\langle |\boldsymbol{S}_n|^2\rangle = \langle \boldsymbol{S}_n \cdot \boldsymbol{S}_n\rangle = \left\langle \sum_{j=1}^{n} \boldsymbol{X}_j \cdot \sum_{k=1}^{n} \boldsymbol{X}_k \right\rangle = n. \tag{10.2.37}$$

What is the probability that in an *even* number of steps n the walker returns to the origin? (For an *odd* number of steps $n = 2k+1$ the probability is obviously zero.) By Eq. (10.2.37) the typical distance of $\boldsymbol{S}_n$ from the origin 0 is $\sim \sqrt{n}$. In d-dimensions the number of lattice points in a sphere of radius $R = \sqrt{n}$ around the origin is about

$$Vol^{(d)}(R) = \frac{\pi^{d/2}}{\Gamma\left(\frac{d}{2}+1\right)} \cdot R^d = const \cdot \left(\sqrt{n}\right)^d. \tag{10.2.38}$$

The probability that the walker will "land" *on any particular point* after n steps will thus decrease with n as $\sim n^{-d/2}$.

If V is the number of visits to the origin, then—as in the 1-D case—

$$q_d = \sum_{n=0}^{\infty} P(S_{2n} = 0) = \frac{1}{1-q_d} \tag{10.2.39}$$

where q_d is the probability that the d-dimensional walk returns to the origin. Since $P(S_{2n} = 0) \sim c/n^{d/2}$, we have

$$\langle V\rangle = \sum_{n=0}^{\infty} P(S_{2n} = 0) = \begin{cases} < \infty \; if \; d \geq 3 \\ \infty \;\; if \; d = 2 \end{cases}. \tag{10.2.40}$$

Thus, we have proved Polya's Theorem: *Suppose S_n is a simple random walk on F^d with $S_0 = 0$. If $d = 1, 2$ then the random walk is recurrent, i.e. with probability one it returns to the origin infinitely often. If $d \geq 3$ the random walk is transient that is with probability one it returns to the origin only finitely often. Also, the event that it would never return, has a positive probability:*

$$P\{S_{2n} \neq 0 \; for \; all \; n > 0\} > 0 \; if \; d \geq 3. \tag{10.2.41}$$

Classic PDEs (Partial Differential Equations) of Continuous Time Random Walks (CTRWs).

For Readers' convenience we introduce here (in alphabetic order) those *Partial Differential Equations* (PDEs) of random walks which are the most frequently quoted and used in petrophysics studies.

Advection-Diffusion Equation (*ADE*) (Berkowitz et al. 2002, 2006).

According to Berkowitz et al. (2006), "the ADE, and variants thereof, have continued to be used to this day as the principal means for considering and quantifying tracer transport in porous media". This equation states

$$\frac{\partial c(\boldsymbol{s},t)}{\partial t} = -\boldsymbol{v}\cdot\nabla c(\boldsymbol{s},t) + \boldsymbol{D}\cdot\nabla\nabla c(\boldsymbol{s},t) \tag{10.2.42}$$

(Berkowitz et al. 2006: Eq. 74; Codling et al. 2008: Eq. 2.7), where $\boldsymbol{v}$ is fluid velocity, $\boldsymbol{D}$ is dispersion. Berkowitz et al. (2002) derived ADE from the *Master Equation*

$$\frac{\partial C(\boldsymbol{s},t)}{\partial t} = -\sum_{\boldsymbol{s}'} w(\boldsymbol{s}',\boldsymbol{s}) + \sum_{\boldsymbol{s}'} w(\boldsymbol{s},\boldsymbol{s}')C(\boldsymbol{s},t) \tag{10.2.43}$$

(Berkowitz et al. 2006: Eq. 1; Oppenheim et al. 1977; Shlesinger 1996), where $C(\boldsymbol{s},t)$ is *normalized particle concentration*, $\sum_{\boldsymbol{s}'} w(\boldsymbol{s},\boldsymbol{s}')$ is *rate of particle transfer* from $\boldsymbol{s}$ to $\boldsymbol{s}'$. They used Taylor expansions of the normalized particle concentration$C(\boldsymbol{s},t)$, and of $w(\boldsymbol{s},\boldsymbol{s}')$, which is the rate of particle transfer between $\boldsymbol{s}$ and $\boldsymbol{s}'$.

The classical derivation of the ADE for porous rocks (Bear 1972) is based on the following assumptions (Berkowitz et al. 2006); (i) the rock is fully saturated; (ii) Darcy's law applies; (iii) the transport of the tracer is the sum of an *advective part* and a *dispersion part*; (iv) the dispersion obey's *Fick's Law*; (v) the diffusion coefficient $\boldsymbol{D}$ is sum of a molecular diffusion part D_m and a part depending on velocity: in 1-dimension $D = D_m + \alpha v$ where α is *dispersivity;* (vi) the *transport velocity* and *fluid velocity* are equal; (vii) the variations in fluid velocity can be neglected.

CTRW PDE (*Continuous Random Walk Partial Differential Equation*) *and its Connection with ADE* (*Advection–Dispersion Equation*) (Dentz et al. 2004; Cortis and Birkholzer 2008).

The CTRW PDE which governs the spatial and temporal development of a passive tracer $c(\boldsymbol{x},\tau)$ is (Dentz et al. 2004):

$$u\tilde{c}(x,u) - c_0(x) = -\tilde{M}(u)\partial_x\left[\tilde{c}(x,u) - \alpha\partial_x\tilde{c}(x,u)\right], \tag{10.2.44}$$

where the tilde indicates Laplace Transform $\tilde{f}(u) = \mathcal{L}f(\tau) = \int_0^\infty f(\tau)\exp(-u\tau)d\tau$ and u is the Laplace variable; $c_0(x)$ is the initial condition; α is local dispersivity; $\tilde{M}(u)$ is a memory function constructed from the *pdf* (*probability density function*) of the retention times, $\psi(u)$, as

$$\tilde{M}(u) = u\cdot\frac{\tilde{\psi}(u)}{1-\tilde{\psi}(u)}. \tag{10.2.45}$$

In Eq. (10.2.44) length and time are nondimensional, the nondimensional transport velocity is made unity. Equation (10.2.44) becomes the classical *ADE* (*Advection–Dispersion Equation*) when $\tilde{M}(u) = 1$, that is $M(\tau) = \delta(\tau)$, which corresponds to an exponentially decaying pdf for the retention times: $\tilde{\psi}(u) = \frac{1}{1+u}$, that is to $\psi(\tau) = e^{-\tau}$.

Diffusion Equation

Consider a diffusion process, let t be time, ρ density, $\boldsymbol{v}$ velocity, $\boldsymbol{j}$ flux, ∇ gradient. The *Equation of Continuity* requires $\frac{d\rho}{dt} = -\nabla \boldsymbol{j} = -\nabla \rho\ \mathbf{v}$. As *flux* is proportional to the density gradient: $\boldsymbol{j} = -\nabla\rho\ (D > 0)$, we get the *Diffusion Equation*:

$$\frac{d\rho}{dt} = D\nabla^2\rho. \tag{10.2.46}$$

Consider a random walker starting at $t = 0$ from point $(0, 0)$ on the 2-D square lattice $F^2 = \{i, j\}$, where the walker gets to (i, j) at n steps with probability $P_{i,j}(n)$. The probabilities satisfy the *recursion relation*

$$\begin{aligned} P_{i,j}(n) - P_{i,j}(n-1) = \frac{1}{4}\big[\big(P_{i-1,j}(n-1) - 2P_{i,j}(n-1) + P_{i+1,j}(n-1)\big) \\ +\big(P_{i,j-1}(n) - 2P_{i,j}(n-1) + P_{i,j+1}(n-1)\big)\big]. \end{aligned} \tag{10.2.47}$$

Let $\Delta t \ll 1$, $\Delta x \ll 1$, and express time and position in terms of them as $t = n \cdot \Delta t$; $x = i \cdot \Delta x$; $y = j \cdot \Delta y$.

Then

$$P(n) - P(n-1) = P\left(\frac{t}{\Delta t}\right) - P\left(\frac{t - \Delta t}{\Delta t}\right) \approx \Delta t \cdot \frac{dP(t)}{dt}; \tag{10.2.48a}$$

$$P_{i-1} - 2P_i + P_{i+1} \approx (\Delta x)^2 \frac{d^2P(x)}{dx^2}. \tag{10.2.48b}$$

Combining Eq. (10.2.47) with (10.2.48a, 10.2.48b) yields the *Diffusion Equation* for a *Continuous-Time Random Walk* (CTRW):

$$\frac{dP(t)}{dt} = \frac{(\Delta x)^2}{2d \cdot \Delta t} \cdot \frac{\mathrm{d}^2\mathrm{P(x)}}{dx^2} = D \cdot \frac{\mathrm{d}^2\mathrm{P(x)}}{dx^2}, \tag{10.2.49}$$

where

$D = \frac{(\Delta x)^2}{2d \cdot \Delta t}$ is the *diffusion coefficient*, d is the dimension of the lattice. This derivation of the diffusion equation, through difference equations, was due to Lord Rayleigh (1880, 1905, 1945; see also Woynar 2013).

Diffusion Equation with Absorption (Tobochnik 1990). Denote by $P(r, t)$ the probability that a particle starting at $t = 0$ from the origin will be at a distance r from the origin at time t, having never gone a distance greater than a. This probability is solution to the diffusion equation:

$$D\nabla^2 P(r, t) = \frac{\partial P(r, t)}{\partial t} \tag{10.2.50}$$

with initial condition

$$P(\boldsymbol{r}, 0) = \delta(\boldsymbol{r}) = \delta(r)/r \tag{10.2.51}$$

and *absorbing boundary condition*

$$P(a, t) = 0. \tag{10.2.52}$$

In 2-D the solution is

$$P(r, t) = \sum_{n=1}^{\infty} \frac{J_0(\frac{j_n r}{a})}{\pi a^2 [J_1(j_n)]^2} e^{-(j_n/a)^{2Dt}}, \tag{10.2.53}$$

where J_0 and J_1 are Bessel functions, j_n is the nth zero of $J_0(z)$. In 3-D:

$$P(r, t) = \sum_{n=1}^{\infty} \frac{n \cdot \sin(\frac{n\pi r}{a})}{\pi a^2 r} e^{-(n\pi/a)^2} \cdot Dt. \tag{10.2.54}$$

Diffusion Equation with Absorption and Back-reflection (Nan et al. 2019) and *Calculation of the Mean Penetration Depth (MPD).*

Consider a *Brownian motion* in a finite porous material starting at $\boldsymbol{x}_0 = \boldsymbol{x}'$ on its outer boundary plane, then proceeding in the void medium until the walker hits a solid surface at a random position, $\boldsymbol{x}^*$. The random *depth of the hit point* relative to the outer boundary plane (i.e. $z(\omega)$, ω is the ensemble index) is called the *penetration depth of the Brownian motion,* denoted by $l(\boldsymbol{x}', \omega)$. The mean penetration depth (MPD) is the *ensemble mean* of the penetration depths of such walks starting from an area Ω on the outer boundary (termed "*departure region*"), i.e.,

$$MPD = \frac{1}{S} \int_{\boldsymbol{x}' \in \Omega} l\left(\boldsymbol{x}', \omega\right) d\boldsymbol{x}', \tag{10.2.55}$$

where $\langle \cdot \rangle$ stands for expectation with respect to the *ensemble index* ω, and S is the area of Ω.

The importance of the MPD and other measures of absorption of the Brownian motion in the void space of porous rocks is due to their intimate connection with *hydraulic permeability*. Torquato (1990) documented that

$$k \le \eta \Phi \overline{\tau}, \tag{10.2.56}$$

where η is the diffusion coefficient of the Brownian motion, $\overline{\tau}$ is mean survival time, and Φ is porosity. Hwang et al. (2000) found a simple relation

$$k = (MPD)^2 \tag{10.2.57}$$

between permeability and *Mean Penetration Depth*. Simonov and Mascagni (2004) included porosity to formula (10.2.57) and claimed

$$k = C_{SM}\Phi(MPD)^2 \tag{10.2.58}$$

C_{SM} is an empirical factor. Nan et al. (2019) derived the theoretical values of the empirical constant C_{SM} using Green's functions for pore shapes of simplified geometries.

In fact, the Brownian motion in the void space of a rock, is a special kind of diffusion process that is reflected back to void space at the outer boundary and is "killed" (absorbed) when hitting the interface of grains. If the starting point is $\boldsymbol{x}_0 = \boldsymbol{x}'$, the spatial density $c(\boldsymbol{x}|\boldsymbol{x}')$ satisfies the PDE

$$\Delta c(\boldsymbol{x}|\boldsymbol{x}') + \delta(\boldsymbol{x} - \boldsymbol{x}'), \tag{10.2.59a}$$

$$c(\boldsymbol{x}|\boldsymbol{x}') = 0\, E\, \boldsymbol{x} \in \boldsymbol{\Gamma} \tag{10.2.59b}$$

(absorption at the void/solid interfaces);

$$\frac{\partial c(\boldsymbol{x}|\boldsymbol{x}')}{\partial \boldsymbol{n_e}} = 0\,\text{if}\, \boldsymbol{x} \in \boldsymbol{\Omega} \tag{10.2.59c}$$

(reflection at the depature region of the outer boundary), where $\boldsymbol{x}' = (x', y', z') \in \boldsymbol{\Omega}$; $\boldsymbol{n_e}$ is the vector normal to the outer boundary.

The solution $c(\boldsymbol{x}|\boldsymbol{x}')$ of Eqs. (10.2.59a–10.2.59c) is the Green's function of the *Poisson equation*

$$\Delta u(\boldsymbol{x}) = -f(\boldsymbol{x}) \tag{10.2.60a, b}$$

subject to a Dirichlet boundary condition at $\boldsymbol{\Gamma}$; and *Neumann boundary condition at* $\boldsymbol{\Omega}$. In the knowledge of $c(\boldsymbol{x}|\boldsymbol{x}')$ the *probability distribution* of the hit point x^* at the inner boundary can be computed (Deaconu and Lejay 2006) for some ideal, simplified pore geometries using known analytical solutions for the Green function (Cole et al. 2011):

$$P(\boldsymbol{x}^*|\boldsymbol{x}_0 = \boldsymbol{x}') = -\frac{\partial}{\partial \boldsymbol{n}} c(\boldsymbol{x}^*|\boldsymbol{x}'),\ \boldsymbol{x}^* \in \boldsymbol{\Gamma} \tag{10.2.61}$$

where $\boldsymbol{n}$ is normal to $\boldsymbol{\Gamma}$ and points into the solid. The mean penetration depth of point $\boldsymbol{x}'$ is

$$\langle l(\boldsymbol{x}', \omega)\rangle = -\int_{\Gamma} z\left(\frac{\partial c(\boldsymbol{x}^*|\boldsymbol{x}')}{\partial \boldsymbol{n}}\right)\Bigg|_{x\in\Gamma} d\boldsymbol{x}, \tag{10.2.62}$$

and a further averaging over the departure region $\boldsymbol{\Omega}$ yields the MDP:

$$MDP = \frac{1}{S} \int_{\boldsymbol{x}' \in \boldsymbol{\Omega}} \langle l(\langle l(\boldsymbol{x}', \omega) \rangle, \omega) \rangle d\boldsymbol{x}' = -\frac{1}{S} \int_{\boldsymbol{x}' \in \boldsymbol{\Omega}} \int_{\boldsymbol{\Gamma}} z \frac{\partial}{\partial \boldsymbol{n}} c(\boldsymbol{x} | \boldsymbol{x}') \Big|_{\boldsymbol{x} \in \boldsymbol{\Gamma}} d\boldsymbol{x}, \quad (10.2.63)$$

where S is the area of $\boldsymbol{\Omega}$.

In (Cole et al. 2011) the Green's functions $c(\boldsymbol{x}|\boldsymbol{x}')$ are available for a few simple pore shapes: (a) inside two parallel planes; (b) tubes with square cross section; (c) circular tubes. Denote by φ the *isoperimetric ratio* $\varphi = l_c^2/A$ to indicate how a shape of perimeter l_c and area A differs from the circle. By a known geometric theorem (Berger 2010) one always has $\varphi \geq 4\pi$ with equality for the circle. For the parallel planes $\varphi = \infty$, for the square cross section $\varphi = \frac{(4d)^2}{d^2} = 16$; for equilateral triangle $\varphi = \frac{(3d)^2}{\left(\frac{\sqrt{3}d^2}{4}\right)} = 12\sqrt{3} \approx 20.8$; for the ellipse with semi-axes $a, b, a \geq b$ we have $4\pi \leq \varphi < \infty$, in particular $\varphi \approx 15.0$ in case $a = 2b$. For parallel planes, a distance d from each other, the Green's function is

$$c(x, z | x', z' = 0) = \frac{2}{\pi} \sum_{n=1}^{\infty} \frac{1}{n} e^{-n\pi z/d} \sin\left(\frac{n\pi x}{d}\right) \sin\left(\frac{npx'}{d}\right), \quad (10.2.64)$$

that is by Eq. (10.2.63) we have

$$MPD = 7\zeta(3) \frac{d}{\pi^3} \approx 0.2714d \quad (10.2.65)$$

(Nan et al. 2019: Eq. 5), where $\zeta(\cdot)$ is the *zeta function*.

For straight tubes of square cross section the Green's function is

$$\begin{aligned} & c(x, y, z | x', y', z' = 0) \\ & = \frac{4}{\pi d} \sum_{m=1}^{\infty} \sum_{n=1}^{\infty} \frac{\exp\left[-\left(\pi \sqrt{n^2 + m^2} z\right)/d\right]}{\sqrt{n^2 + m^2}} \cdot \sin\left(\frac{m\pi x}{d}\right) \sin\left(\frac{m\pi x'}{d}\right) \quad (10.2.66) \end{aligned}$$

which yields

$$\begin{aligned} MDP &= \frac{16d}{\pi^5} \sum_{m=1}^{\infty} \sum_{n=1}^{\infty} \frac{1}{n^2 (n^2 + m^2)^{3/2}} [1 - (-1^m)][1 - (-1^n)]^2 \\ &\approx \frac{16d}{\pi^5} \cdot 3.2535 = 0.1701d; \quad (10.2.67) \end{aligned}$$

(Nan et al. 2019: Eq. 8).

If the cross sections are circular, with radius R, we have

$$c\left(r, zr', z' = 0\right) = \frac{1}{\pi R} \cdot \sum_{m=1}^{\infty} \frac{1}{\beta_m J_1^2(\beta_m)} \cdot \exp(-)\frac{\beta_m z}{R} J_0\left(\frac{\beta_m r}{R}\right) J_0\left(\frac{\beta_m r'}{R}\right), \tag{10.2.68}$$

and

$$MDP = 4R \sum_{m=1}^{\infty} \frac{1}{\beta_m^3} \approx 0.3235R \tag{10.2.69}$$

(Nan et al. 2019: Eq. 10), where J_0 and J_1 are Bessel functions of zeroth and first orders of the first kind, and β_m is the m-th root of J_0.

Fractional Differential Equations of Random Walks (Metzler and Klafter 2000, 2004; Berkowitz et al. 2006).

In their important review paper Berkowitz et al. (2006) convincingly demonstrated that "fractional equations have come of age as a complementary tool in the description of anomalous transport processes". (They referred to works such as Benson et al. 2000; Metzler and Klafter 2000, 2004; Baeumer et al. 2001; Schumer et al. 2003; Zhang et al. 2005 and others.) In such equations the differential operator either in time or space or both can be of *fractional order*.

The *temporal FD*E can be written as (Metzler and Klafter 2000; Berkowitz et al. 2006: Eqs. 107–108)

$$\frac{\partial c(s,t)}{\partial t} = -\frac{\partial^{1-\beta}}{t^{1-\beta}}\left\{\boldsymbol{v}'(s) \cdot \nabla c(s,t) - \nabla \cdot \left[\boldsymbol{D}'(s) \cdot \nabla(s,t)\right]\right\} \tag{10.2.70}$$

where the fractional operator is defined by

$$\frac{\partial^{-\gamma}}{t^{-\gamma}} := \frac{1}{\Gamma(\gamma} \int_0^t \frac{c\left(s,t'\right)}{(t-t')^{1-\gamma}} dt'. \tag{10.2.71}$$

The primes on the symbols $\boldsymbol{v}'$ and $\boldsymbol{D}'$ indicate that these quantities do not have their usual physical dimensions, for example the dimension of $\boldsymbol{v}'$ is $[Length]/\left[Time^{\beta}\right]$. The Laplace transform of the fractional operator in Eq. (10.2.70) satisfies

$$\mathcal{L}\left[\frac{\partial^{-\gamma}}{t^{-\gamma}} c(s,t); u\right] = u^{-\gamma} \tilde{c}(s,u). \tag{10.2.72}$$

Metzler et al. (1998) developed an FDE, valid under special conditions (Berkowitz et al. 2006), where *spatial fractional operators* are used:

$$\frac{\partial}{\partial t}c(s,t) + \boldsymbol{v}\cdot\nabla c(s,t) = D^{\varsigma}\nabla^{\xi}c(s,t) \tag{10.2.73}$$

(Berkowitz et al. 2006: Eq. 111), here D is a *generalized diffusion parameter*.

Telegraph Equation (Codling et al. 2008).

Consider a large number of walkers moving on an infinite line at constant speed v. At location x and time t there are $\alpha(x,t)$ walkers moving to the left, and $\beta(x,t)$ of them moving to the right. The population density is

$$p(x,t) = \alpha(x,t) + \beta(x,t). \tag{10.2.74}$$

Denote by $\tau \ll 1$ a time step, and assume that at each time step every walker can either change direction (with probability $r = \lambda\tau$) and moves a distance δ in this direction; or (with probability $q = 1 - r = 1 - \lambda\tau$) keeps going in the old direction making a distance δ. The events of changing directions is a *Poisson process* of rate λ. After a time step the number densities of the walkers moving right and left, respectively, are

$$\alpha(x,t+\tau) = (1-\lambda\tau)\alpha(x-\delta,t) + \lambda\tau\beta(x-\delta,t) \tag{10.2.75a}$$

$$\beta(x,t+\tau) = \lambda\tau\alpha(x+\delta,t) + (1-\lambda\tau)\beta(x-\delta,t). \tag{10.2.75b}$$

Expanding α and β into Taylor series and letting $\tau \to 0$, $\delta \to 0$ in such a way that $\delta/\tau = v$, yield

$$\begin{aligned}\frac{\partial\alpha}{\partial t} &= -v\frac{\partial\alpha}{\partial x} + \lambda(\beta-\alpha);\\ \frac{\partial\beta}{\partial t} &= v\frac{\partial\beta}{\partial x} - \lambda(\beta-\alpha).\end{aligned} \tag{10.2.76a, b}$$

Using that $\alpha + \beta = p$, a long but elementary calculation (Codling et al. 2008: Eqs. 2.21–2.27) yields the PDE

$$\frac{\partial^2 p}{\partial t^2} + 2\lambda\frac{\partial p}{\partial t} = v^2\frac{\partial^2 p}{\partial x^2}, \tag{10.2.77}$$

with initial conditions for $p(x,0)$ and $\frac{\partial}{\partial t}p(x,0)$. The full solution is found in Morse and Feschbach (1953). An equation of this form was named the *Telegraphist's Equation* by Lord Kelvin who derived it in 1855 to describe signals propagating across the Transatlantic cable (Thomson 1854–1855). It was recognized by (Goldstein 1951; Kac 1974) that the equation can be applied for the above-described random walk

Table. 10.1 Some important applications of random Markov fields

Application	Authors
Estimation of pressure–temperature paths	Kuwatani et al. (2012)
Geomagnetic and gravity anomaly separation	Uçan and Albora (2009), Uçan et al. (2000, 2004)
Inversion of seismic amplitude	Tjelmeland et al. (2019)
Joint inversion of seismic and electromagnetic data	Chen and Hoversten (2012)
Land-cover mapping	Tolpekin and Stein (2009)
LIDAR data	Reitberger et al. (2009)
Lithology/fluid prediction	Rimstad and Omre (2010), Ulvmoen and Omre (2010)
Meteorology	Guyon and Hardouin (2002)
Modeling geological structures	Norberg et al. (2002)
Partial melt mapping	Kuwatani et al. (2014)
Remote sensing	Chen et al. (2019)
SAR speckle reduction	Xie et al. (2002)
Satellite imagery	Solberg et al. (1996)
Seismic horizon tracking	Abbas and Swoboda (2019)
Seismic monitoring of hydrocarbon reservoir	Räss et al. (2019)
Simulation of petroleum reservoirs	Salomão and Remacre (2001)
Stratigraphy	Li et al. (2016)
3D segmentation	Wang et al. (2017)
Tunnel construction	Wang et al. (2016)

model (now called *Goldstein-Kac model*). More recently, Shlesinger (2003) derived Eq. (10.2.77) using a random walk with *coupled space–time memory*.

Appendix

See Tables 10.1 and 10.2.

Table. 10.2 Some applications of random walk models

Application	Authors
Anomalous diffusion	Metzler and Klafter (2000)
Anomalous transport in soil and sand	Cortis and Berkowitz (2004)
Anomalous contaminant transport	Cortis and Berkowitz (2005)
Bacterial migration in porous media	Duffy et al. (1995)
Conductivity of porous media	Ioannidis et al. (1997)
Contaminant transport	Baeumer et al. (2001)
Diffusion and trapping in random media	Dentz et al. (2012)
Diffusion in crystalline rock	Sardini et al. (2003, 2007)
Dispersion in a porous medium	Le Borgne et al. (2011), Dentz et al. (2018)
Double porosity fractured media	Noetinger and Estebenet (2000), Noetinger et al. (2001a, b)
Effective conductivity	Shuler and Mohanty (1981), Kim and Torquato (1992)
Effective permeability	McCarthy (1990, 1991, 1993a, b)
Effective properties of porous media	Simonov and Mascagni (2004)
Emulsion transport in porous media	Cortis and Ghezzehei (2007)
Fracture networks	Berkowitz and Scher (1995), Landereau et al. (2001)
Groundwater modelling	Kinzelbach and Uffink (1991)
Heat transfer in porous media	Emmanuel and Berkowitz (2007)
Hydraulic pumping tests	Cortis and Knudby (2006)
Hydrodynamic transport	Noetinger et al. (2016), Soulaine et al. (2021)
Mass transfer in porous media	Hoteit et al. (2002)
NMR	Schwartz and Banavar (1989), Wilkinson et al. (1991), Sen et al. (1994), Bergman et al. (1995), Øren et al. (2002), Valckenborg et al. (2002), Olayinka and Ioannidis (2004), Arns et al. (2005), Regier and Schuchman (2005), Toumelin et al. (2003, 2007), Valfouskaya et al. (2006), Jin et al. (2009), Néel et al. (2011, 2014) Guillon et al. (2013, 2014), Fleury et al. (2015)
NMR in carbonate rocks	Ramakrishnan et al. (1999)
Non-Fickian transport in geological formations	Berkowitz et al. (2006), Gouze et al. (2008), Bijeljic et al. (2011, 2013), Kang et al. (2014), Noetinger et al. (2016), Abou-Saleh et al. (2019)

(continued)

Table. 10.2 (continued)

Application	Authors
Nuclide transport in fractured rock	Wu et al. (2017)
Percolation conductivity	Tobochnik (1990)
Permeability	Torquato(1990), Hwang et al. (2000), Nan et al. (2019)
Pollutant transport	Kinzelbach (1988), Zimmermann et al. (2001)
Random conductance model	Biskup (2011)
Residence times of solutes in fractured media	Robinet et al. (2007)
Solute transport	Bodin et al. (2003), Schumer et al. (2003), Haggerty et al. (2004)
Solute transport in fracture network	Delay and Bodin (2001), Delay et al. 2002, 2005; Bodin et al. 2007; Cortis and Birkholzer (2008), Geiger et al. (2010), Bechtold et al. (2011), Stalgorova and Babadagli (2012), Cvetkovic et al. (2014); Bodin (2015)
Solute transport in karstic systems	Anwar et al. (2007)
Stream transport of solutes	Boano et al. (2007)
Tracers in porous media	Bijeljic and Blunt (2006)
Transport in fault zones	O'Brien et al. (2003a, b)
Transport of biocolloids	Cortis et al. (2006)
Transport properties of composite materials	De W. Van Siclen (1999b)
Wave propagation	Korvin and Lux (1972)

References

Abbas A, Swoboda P (2019) Bottleneck potentials in Markov random fields. In: Proceedings of the IEEE/CVF international conference on computer vision, pp 3175–3184

Abou-Saleh K, Dweik J, Haidar Y, Ghaddar A (2019) Solving diffusion time in heterogeneous microscale rock matrix by 3D computations: non-Fickian dispersion observed in porous media. J Geosci Environ Protect 7:42–52

Abramowitz M, Stegun IA (eds) (1965) Handbook of mathematical functions. Dover, New York

Anwar S, Cortis A, Sukop MC (2007) Lattice Boltzmann Simulation of solute transport in heterogeneous porous media with conduits to estimate macroscopic continuous time random walk model parameters. Progr Comput Fluid Dyn 8:213–221

Arns CH, Sheppard AP, Sok RM, Knackstedt MA (2005) NMR petrophysical predictions on digitized core materials. In: SPWLA 46th annual logging symposium, New Orleans, Lousiana, USA, 2005: Paper MMM

Averintsev MB (1970) On a method of describing complete parameter fields. Problemy Peredachi Informatsii 6:100–109

Baeumer B, Benson DA, Meerschaert MM, Wheatcraft SW (2001) Subordinated advection-dispersion equation for contaminant transport. Water Resour Res 37(6):1543–1550

Barthelemy P, Bertolotti J, Wiersma DA (2008) Lévy flight for light. Nature 453:495–498

Bear J (1972) Dynamics of fluids in porous media. Elsevier, New York

Bechtold M, Vanderborght J, Ippisch O, Vereecken H (2011) Efficient random walk particle tracking algorithm for advective-dispersive transport in media with discontinuous dispersion coefficients and water content. Water Resour Res 47(10):W1052

Bender EA, Goldman JR (1975) On the applications of Möbius inversion in combinatorial analysis. Am Math Mon 82(8):789–803

Benson DA, Wheatcraft SW, Meerschaert MM (2000) The fractional-order governing equation of Lévy motion. Water Resour ReS 36(6):1413–1423

Berger M (2010) Geometry revealed: a Jacob's ladder to modern higher geometry. Springer-Verlag, Berlin-Heidelberg

Bergman DJ, Dunn K-J, Schwartz LM, Mitra PP (1995) Self-diffusion in a periodic porous medium: a comparison of different approaches. Phys Rev E 51:3393–3400

Berkowitz B, Scher H (1995) On characterization of anomalous-dispersion in porous and fractured media. Water Resour Res 3:1461–1466

Berkowitz B, Klafter J, Metzler R, Scher H (2002) Physical pictures of transport in heterogeneous media: advection-dispersion, random walk and fractional derivative formulations. Water Resour Res 38(10):1191

Berkowitz B, Cortis A, Dentz M, Scher H (2006) Modeling non-Fickian transport in geological formations. Rev Geophys 44(2): Article no 608

Besag JE (1972) Nearest-neighbor systems and the auto-logistic model for binary data. J R Stat Soc B 34:75–83

Besag JE (1974) Spatial interaction and the statistical analysis of lattice systems. J R Stat Soc B 36:192–221

Besag J (1975) Statistical analysis of non-lattice data. The Statistician 24:179–195

Besag J (1986) On the statistical analysis of dirty pictures. J R Stat Soc 48(3):259–302

Bijeljic B, Blunt MJ (2006) Pore-scale modeling of transverse dispersion in porous media. Water Resources Res 43(12)

Bijeljic B, Mostaghimi P, Blunt MJ (2011) Signature of non-Fickian solute transport in complex heterogeneous porous media. Phys Rev Lett 107:204502

Bijeljic B, Raeini A, Mostaghimi P, Blunt MJ (2013) Predictions of non-Fickian solute transport in different clases of porous media using direct simulation on pore-scale images. Phys Rev E 87:013011

Biskup M (2011) Recent progress on the random conductance model. Probab Surv 8:294–373

Boano F, Packman AI, Cortis A, Revelli R, Ridolfi L (2007) A continuous time random walk approach to the stream transport of solutes. Water Resour Res 43:W10425

Bodin J (2015) From analytical solutions of solute transport equations to multidimensional time-domain random walk (TDRW) algorithms. Water Resour Res 51(3):1860–1871

Bodin J, Delay F, de Marsily G (2003) Solute transport in a single fracture with negligible matrix permeability: 2. Mathematical formalism. Hydrogeol J 11434–454

Bodin J, Porel G, Delay F, Ubertosi F, Bernard S, de Dreuzy J-R (2007) Simulation and analysis of solute transport in 2D fracture/pipe networks. J Contam Hydrol 89(1–2):1–28

Brown R (1828) A brief account of microscopical observations in the months of June, July and August, 1827, on the particles contained in the pollen of plants; and the general existence of active molecules in organic and inorganic bodies. Philos Mag 4:161–173

Brownstein KR, Tarr CE (1979) Importance of classical diffusion in NMR studies of water in biological cell. Phys Rev A 19(6):2446–2453

Carr HY, Purcell EM (1954) Effects of diffusion on free precession in nuclear magnetic resonance experiments. Phys Rev 94:630–638

Chen J, Michael Hoversten G (2012) Joint inversion of marine seismic AVA and CSEM data using statistical rock-physics models and Markov random fields. Geophysics 77(1):R65–R80

Chen Z, Pan X, Chen X, Yang X, Xin X, Su L (2019) An object-based Markov random field model with anisotropic penalty for semantic segmentation of high spatial resolution remote sensing imagery. Remote Sens 11(23):2878

Cipra BA (1987) An introduction to the Ising model. Amer Math Monthly 94:937–959

Clifford P (1990) Markov random fields in statistics. In: Grimmett GR, Welsh DJA (eds) Disorder in physical systems: a volume in honour of John M. Hammersley. Oxford University Press, Oxford

Codling Edward A, Plank MJ, Benhamou S (2008) Random walk models in biology. J R Soc Interface 5:813–834

Cole KD, Beck JV, Haji-Sheikh A, Litkouhi B (2011) Heat conduction using Green's functions, 2nd edn. CRC Press, Boca Raton

Cortis A, Berkowitz B (2004) Anomalous transport in "classical" soil and sand columns. Soil Sci Soc Am J 68:1539–1548

Cortis A, Berkowitz B (2005) Computing '"anomalous"' contaminant transport in porous media: the CTRW MATLAB toolbox. Ground Water 43(6):947–950

Cortis A, Birkholzer J (2008) Continuous time random walk analysis of solute transport in fractured porous media. Water Resour Res 44:W0641

Cortis A, Ghezzehei TA (2007) On the transport of emulsions in porous media. J Colloid Interface Sci 313(1):1–4

Cortis A, Knudby C (2006) A continuous time random walk approach to transient flow in heterogeneous porous media. Water Resour Res 42(10):W10201

Cortis A, Harter T, Hou L, Atwill ER, Packman A, Green P (2006) Transport of *Cryptosporidium parvum* in porous media: long-term elution experiments and continuous time random walk filtration modeling. Water Resour Res 42(12):W12S13

Coscoy S, Huguet E, Amblard F (2007) Statistical analysis of sets of random walks: how to resolve their generating mechanism. Bull Math Biol 6:2467–2492

Cvetkovic V, Fiori A, Dagan G (2014) Solute transport in aquifers of arbitrary variability: a time-domain random walk formulation. Water Resour Res 50(7):5759–5773

Davey BA, Priestley HA (2002) Introduction to lattices and order, 2nd edn. Cambridge University Press, New York

Deaconu M, Lejay A (2006) A random walk on rectangles algorithm. Methodol Comput Appl Probab 8:135–151

Delay F, Bodin J (2001) Time domain random walk method to simulate transport by advection-dispersion and matrix diffusion in fracture networks. Geophys Res Lett 28(21):4051–4054

Delay F, Porel G, Sardini P (2002) Modelling diffusion in a heterogeneous rock matrix with a time-domain Lagrange method and an inversion procedure. CR Geoscience 334:967–973

Delay F, Ackerer P, Danquigny C (2005) Simulating solute transport in porous or fractured formations using random walk particle tracking: a review. Vadose Zone J 4(2):360–379

Dentz M, Cortis A, Scher H, Berkowitz B (2004) Time behavior of solute transport in heterogeneous media: Transition from anomalous to normal transport. Adv Water Resour 27:155–173

Dentz M, Gouze P, Russian A, Dweik J, Delay F (2012) Diffusion and trapping in heterogeneous media: an inhomogeneous continuous time random walk approach. Adv Water Resour 49:13–22

Dentz M, Icardi M, Hidalgo JJ (2018) Mechanisms of dispersion in a porous medium. J Fluid Mech 841:851–882

Derin H, Elliot AH (1987) Modeling and segmentation of noisy and textured images using Gibbs random field. IEEE Trans Pattern Anal Mach Intell 9:39–55

De W Van Siclen C (1999a) Walker diffusion method for calculation of transport properties of composite materials. Phys Rev E 59(3):2804–2807

De W Van Siclen C (1999b) Anomalous walker diffusion through composite systems. J Phys a: Math Gen 3:5763–5771

De W Van Siclen C (2021) Random walker derivation of Archie's law. arXiv: 2103.14099 [cond-mat.stat-mech]

Dobrushin PL (1968) The description of a random field by means of conditional probabilities and conditions of its regularity. Theory Probab Appl 13(2):197–224

Duffy KJ, Cummings PT, Ford RM (1995) Random walk calculations for bacterial migration in porous media. Biophys J 68(3):800–806

Durrett R (2019) Probability: theory and examples, 5th edn. Cambridge University Press, Cambridge

Einstein A (1905) Über die von der molekularkinetischen Theorie der Wärme geforderte Bewegung von in ruhenden Flüssigkeiten suspendierten Teilchen. Ann Phys 17:549–560
Einstein A (1906) Zur Theorie der Brownschen Bewegung. Ann Phys 19:371–381
Emmanuel S, Berkowitz B (2007) Continuous time random walks and heat transfer in porous media. Transp Porous Media 67(3):413–430
Fleury M, Bauer D, Néel M (2015) Modeling of super-dispersion in unsaturated porous media using NMR propagators. Microporous Mesoporous Mater 205:75–78
Forbes F, Peyrard N (2003) Hidden Markov random field model selection criteria based on mean field-like approximations. IEEE Trans Pattern Anal Mach Intell 25:1089–1101
Gautestad Arild O (2013) Lévy meets Poisson: a statistical artifact may lead to erroneous recategorization of Lévy walk as Brownian motion. Am Nat 181(3):440–450
Geiger S, Cortis WA, Birkholzer JT (2010) Upscaling solute transport in naturally fractured porous media with the continuous time random walk method. Water Resour Res 46(12):W12530
Geman S, Geman D (1984) Stochastic relaxation, Gibbs distributions and the Bayesian restoration of images. IEEE Trans PAMI 6:721–741
Goldstein S (1951) On diffusion by discontinuous movements, and on the telegraph equation. J Mech Appl Math 6:129–156
Gouze P, Melean Y, Le Borgne T, Dentz M, Carrera J (2008) Non-Fickian dispersion in porous media explained by heterogeneous microscale matrix diffusion. Water Resour Res 44(11):W11416–W11435
Grimmett GR (1973) A theorem about random fields. Bull Lond Math Soc 5(1):81–84
Grimmett G, Welsh D (2007) John Michael Hammersley 21 March 1920–2 May 2004. Biogr Mems Fell R Soc 53:163–183
Guillon V, Fleury M, Bauer D, Néel MC (2013) Superdispersion in homogeneous unsaturated porous media using NMR propagators. Phys Rev E 87(4)
Guillon V, Bauer D, Fleury M, Néel MC (2014) Computing the longtime behaviour of NMR propagators in porous media using a pore network random walk model. Transp Porous Media 101(2):251–267
Guyon X, Hardouin C (2002) Markov chain Markov field dynamics: models and statistics. Stat A J Theor Appl Stat 36(4):339–363
Haggerty R, Harvey CF, Schwerin CF, Meigs LC (2004) What controls the apparent timescale of solute mass transfer in aquifers and soils? A comparison of experimental results. Water Resour Res 40:W01510
Hammersley JM, Clifford P (1971) Markov fields on finite graphs and lattices. Unpublished. http://www.statslab.cam.ac.uk/~grg/books/hammfest/hamm-cliff.pdf
Hidajat I, Singh M, Cooper J, Mohanty KK (2002) Permeability of porous media from simulated NMR response. Transp Porous Media 48(2):225–247
Holtz O (2014) My random walks with Pólya and Szegő. Institute for Advanced Study. The Institute Letter, Summer. https://www.ias.edu/publications/institute-letter/institute-letter-summer-2014
Hoteit H, Mose R, Younes A, Lehmann F, Ackerer P (2002) Three-dimensional modeling of mass transfer in porous media using the mixed hybrid finite elements and the random-walk methods. Math Geol 34(4):435–456
Hwang C-O, Given JA, Mascagni M (2000) On the rapid estimation of permeability for porous media using Brownian motion paths. Phys Fluids 12(7):1699–1709
Ioannidis MA, Kwiecen MJ, Chatzis I (1997) Electrical conductivity and percolation aspects of statistically homogeneous porous media. Transp Porous Media 29(1):61–83
Ising E (1925) Beitrag zur Theorie des Ferromagnetismus. Zeitschrift Für Physik A Hadrons Nuclei 31:253–258
Jin G, Carlos T-V, Emmanuel T (2009) Comparison of NMR simulations of porous media derived from analytical and voxelized representations. J Mag Reson 200:313–320
Kac M (1974) A stochastic model related to the telegraphers equation. Rocky Mt J Math 4:497–509

Kang PK, Anna P, Nunes JP, Bijeljic B, Blunt MJ, Juanes R (2014) Pore-scale intermittent velocity structure underpinning anomalous transport through 3-d porous media. Geophys Res Lett 410(17):6184–6190
Kim IC, Torquato S (1990) Determination of the effective conductivity of heterogeneous media by Brownian motion simulation. J Appl Phys 68:3892–3903
Kim IC, Torquato S (1992) Effective conductivity of suspensions of overlapping spheres. J Appl Phys 71(6):2727–2735
Kindermann R, Snell JL (1980) Markov random fields and their applications. American Mathematical Society, Providence
Kinzelbach W (1988) The random walk method in pollutant transport simulation. In: Custodio E, Gurgui A, Ferreira JPL (eds) Groundwater flow and quality modelling. NATO ASI series, vol 224. Springer, Dordrecht, pp 227–245
Kinzelbach W, Uffink G (1991) The random walk method and extensions in groundwater modelling. In: Bear J, Corapcioglu MY (eds) Transport processes in porous media. NATO ASI series, vol 202. Springer, Dordrecht
Koller D, Friedman N (2009) Probabilistic graphical models: principles and techniques. MIT Press, Cambridge
Korvin G (2021) Statistical rock physics. In: Daya Sagar B, Cheng Q, McKinley J, Agterberg F (eds) Encyclopedia of mathematical geosciences. Encyclopedia of earth sciences series. Springer, Cham
Korvin G, Lux I (1972) An analysis of the propagation of sound waves in porous media by means of the Monte Carlo method. Geophys Trans 21(3–4):91–106
Kuwatani T, Nagata K, Okada M, Toriumi M (2012) Precise estimation of pressure-temperature paths from zoned minerals using Markov random field modeling: theory and synthetic inversion. Contrib Mineral Petrol 163(3):547–562
Kuwatani T, Nagata K, Okada M, Toriumi M (2014) Markov random field modeling for mapping geofluid distributions from seismic velocity structures. Earth, Planets Space 66(1):1–9
Landau LD, Mikhailovich LE (1980) Statistical physics. Course of theoretical physics, vol 5, 3rd ed. Pergamon Press, Oxford
Landereau P, Noetinger B, Quintard M (2001) Quasi steady two equation models for transport in fractured porous media. Adv Water Resour 24(8):863–876
Lawler GE (2010) Random walk and the heat equation. Student mathematical library, vol 55. American Mathematical Society, Providence, Rhode Island
Le Borgne T, Bolster D, Dentz M, de Anna P, Tartakovsky A (2011) Effective pore-scale dispersion upscaling with a correlated continuous time random walk approach. Water Resour Res 47(12):W12538
Lévy P (1954) Théorie de l'Addition des Variables Aléatoires. Gauthier-Villars, Paris
Lévy P (1965) Processus stochastiques et mouvement Brownien. Gauthier-Villars, Paris
Li Z, Wang X, Wang H, Liang RY (2016) Quantifying stratigraphic uncertainties by stochastic simulation techniques based on Markov random field. Eng Geol 201:106–122
Lovász L (1996) Random walks on graphs: a survey. In: Miklós D et al (eds) Combinatorics: Paul Erdős is eighty, vol 2, Budapest. János Bolyai Math Soc 353–397
Luban M, Staunton LP (1988) An efficient method for generating a uniform distribution of points within a hypersphere. Comput Phys 2(6):55–60
Majoros WH (2007) Conditional random fields. Online supplement to: Methods for computational gene prediction. Cambridge University Press
Mandelbrot B (1977) The fractal geometry of nature. Freeman, New York
McCarthy JF (1990) Effective permeability of sandstone-shale reservoirs by a random walk method. J Phys A Math Gener 23(9):L445
McCarthy JF (1991) Analytical models of the effective permeability of sand-shale reservoirs. Geophys J Int 105(2):513–527
McCarthy JF (1993a) Continuous-time random walks on random media. J Phys A Math Gener 26(11):2495–2503

McCarthy JF (1993b) Reservoir characterization: efficient random-walk methods for upscaling and image selection. In: SPE Asia pacific oil and gas conference, 8–10 February, Singapore: 25334
Meiboom S, Gill D (1958) Modified spin-echo method for measuring nuclear relaxation times. Rev Sci Instrum 29(8):688–691
Metropolis N, Rosenbluth AW, Teller AH, Teller E (1953) Equations of state calculations by fast computing machines. J Chem Phys 21:1087–1092
Metzler R, Klafter J, Sokolov IM (1998) Anomalous transport in external fields: continuous time random walks and fractional diffusion equations extended. Phys Rev E 58(2):1621–1633
Metzler R (2000) Generalized Chapman-Kolmogorov equation: a unifying approach to the description of anomalous transport in external fields. Phys Rev E 62(5):6233–6245
Metzler R, Klafter J (2000) The random walk's guide to anomalous diffusion: a fractional dynamics approach. Phys Rep 339:1–77
Metzler R, Klafter J (2004) The restaurant at the end of the random walk: recent developments in the description of anomalous transport by fractional dynamics. J Phys A Math Gen 37(31):R161
Morse PM, Feshbach H (1953) Methods of theoretical physics. McGraw-Hill, New York, NY
Moussouris J (1974) Gibbs and Markov random fields with constraints. J Statist Phys 10:11–33
Nan T, Wu J, Li K, Jiang J (2019) Permeability estimation based on the geometry of pore space via random walk on grids. Geofluids. Article ID 924020
Néel MC, Rakotonasyl SH, Bauer D, Joelson M, Fleury M (2011) All order moments and other functionals of the increments of some non-Markovian processes. J Stat Mech Theory Exp 2011:P02006
Néel MC, Bauer D, Fleury M (2014) Model to interpret pulsed-field-gradient NMR data including memory and superdispersion effects. Phys Rev E 89(6)
Noetinger B, Estebenet T (2000) Up-scaling of double porosity fractured media using continuous-time random walks methods. Transp Porous Media 39(3):315–337
Noetinger B, Estebenet T, Landereau P (2001a) A direct determination of the transient ex-change term of fractured media using a continuous time random walk method. Transp Porous Media 44(3):539–557
Noetinger B, Estebenet T, Quintard M (2001b) Up scaling of fractured media: equivalence between the large scale averaging theory and the continuous time random walk method. Transp Porous Media 43(3):581–596
Noetinger B, Roubinet D, Russian A, Le Borgne T, Delay F, Dentz M, de Dreuzy J-R, Gouze Ph (2016) Random Walk methods for modeling hydrodynamic transport in porous and fractured media from pore to reservoir scale. Transp Porous Med 115:345–385
Norberg T, Rosén L, Baran A, Baran S (2002) On modelling discrete geological structures as Markov random fields. Math Geol 34:63–77
O'Brien GS, Bean CJ, McDermott F (2003a) Numerical investigations of passive and reactive flow through generic single fractures with heterogeneous permeability. Earth Planet Sci Lett 213(3–4):271–284
O'Brien GS, Bean CJ, McDermott F (2003b) A numerical study of passive transport through fault zones. Earth Planet Sci Lett 214(3–4):633–643
Olayinka S, Ioannidis MA (2004) Time-dependent diffusion and surface-enhanced relaxation in stochastic replicas of porous rock. Transp Porous Media 54(3):273–295
Oppenheim I, Shuler KE, Weiss GH (1977) Stochastic processes in chemical phyics: the master equation. MIT Press, Cambridge, Mass.
Øren PE, Antonsen F, Rueslåtten HG, Bakke S (2002) Numerical simulations of NMR responses for improved interpretations of NMR measurements in reservoir rocks. In: SPE annual technical conference and exhibition, San Antonio, Texas, 2002: SPE 77398
Pearson K (1905) The problem of the random walk. Nature 1905(July 27) 72:294
Perrin J (1909) Mouvement Brownien et réalité moléculaire. Ann Chim Phys VIII(18):5–114
Polya G (1921) Über eine Aufgabe der Wahrscheinlichkeitsrechnung betreffend die Irrfahrt im Strassennetz. Math Annalen 84:149–160

Potts RB (1952) Some generalized order-disorder transformations. Math Proc Cambridge Philos Soc 48(1):106–109

Preston CJ (1973) Generalized Gibbs states and Markov random fields. Adv Appl Probab 5(2):242–261

Ramakrishnan TS, Schwartz LM, Fordham EJ, Kenyon WE, Wilkinson DJ (1999) Forward models for nuclear magnetic resonance in carbonate rocks. Log Anal 40(4):260–270

Rapp BE (2017) Microfluidics: modelling, mechanics and mathematics. Elsevier, Amsterdam

Rayleigh (J.W. Strutt) (1880) On the resultant of a large number of vibrations of the same pitch and of arbitrary phase. Philos Mag 10:73

Rayleigh (J.W. Strutt) (1905) The problem of the random walk. Nature 72:31

Rayleigh (J.W. Strutt) (1945) The theory of sound, volume 1, section 42a. Second edn., revised and enlarged. Dover Publications, New York

Räss L, Kolyukhin D, Minakov A (2019) Efficient parallel random field generator for large 3-D geophysical problems. Comput Geosci 131:158–169

Regier M, Schuchmann HP (2005) Monte Carlo simulations of observation time-dependent self-diffusion in porous media models. Transp Porous Media 59(1):115–126

Reitberger J, Schnörr C, Krzystek P, Stilla U (2009) 3D segmentation of single trees exploiting full waveform LIDAR data. ISPRS J Photogramm Remote Sens 64:561–574

Rimstad K, Omre H (2010) Impact of rock-physics depth trends and Markov random fields on hierarchical Bayesian lithology/fluid prediction. Geophysics 7:R93–R108

Robinet JC, Sardini P, Delay F, Hellmuth KH (2007) The effect of rock matrix heterogeneities near fracture walls on the residence time distribution (RTD) of solutes. Transp Porous Media 72(3):393–408

Rota GC (1964) On the foundations of combinatorial theory, I. Theory of Mobius functions. Zeitschr Wahrsch Theorie Verw Geb 2:340–368

Rue H, Held L (2005) Gaussian Markov random fields: theory and applications. CRC Press, Boca Raton

Salomão MC, Remacre AZ (2001) The use of discrete Markov random fields in reservoir characterization. J Petrol Sci Eng 32(2–4):257–264

Sardini P, Delay F, Hellmuth K-H (2003) Interpretation of out-diffusion experiments on crystalline rocks using random walk modelling. J Contam Hydrol 61:339–350

Sardini P, Robinet J-C, Siitari-Kappi M (2007) Direct simulation of heterogeneous diffusion and inversion procedure applied to an out-diffusion experiment. Test case of Palmetto granite. J Contam Hydrol 93:21–37

Scher H, Lax M (1973) Stochastic transport in a disordered solid. I. Theory. Phys Rev B 7(10):4491

Schumer R, Benson DA, Meerschaert MM (2003) Fractal mobile/immobile solute transport. Water Resour Res 39(10):1296

Schwartz LM, Banavar JR (1989) Transport properties of disordered continuum systems. Phys Rev B 39:11965–11970

Sen PN, Schwartz LM, Mitra PP, Halperin BI (1994) Surface relaxation and the long-time diffusion coefficient in porous media: periodic geometries. Phys Rev B 49(1):215–225

Shlesinger MF (1996) Random processes. In: Encyclopedia of applied physics, vol 16. Wiley, Hoboken, N.J.

Shlesinger MF (2003) Supra-diffusion. In: Ranagarajan G, Ding M (eds) Processes with long-range correlations. Springer, Berlin, pp 139–147

Shuler KE, Mohanty U (1981) Random walk properties from lattice bond enumeration: anisotropic diffusion in lattices with periodic and randomly distributed scatterers. Proc Natl Acad Sci USA 78(11):6576–6578

Simonov NA, Mascagni M (2004) Random Walk algorithms for estimating effective properties of digitized porous media. Monte Carlo Methods and Appl. 10(3–4):599–608

Solberg AHS, Taxt T, Jain AK (1996) A Markov random field model for classification of multisource satellite imagery. IEEE Trans Geosci Remote Sens 34:100–113

Soulaine C, Girolami L, Arbaret L, Roman S (2021) Digital Rock Physics: computation of hydrodynamic dispersion. Oil & Gas Sci Technol-Revue d'IFP Energies Nouvelles 76:51

Spitzer F (1964) Principles of random walk. Graduate texts in mathematics. Springer New York, New York

Spitzer F (1971) Markov random fields and Gibbs ensembles. Am Math Mon 78:142–154

Stalgorova E, Babadagli T (2012) Field-scale modeling of tracer injection in naturally fractured reservoirs using the random-walk particle-tracking simulation. SPE J 17(2):580–592

Steinsland I (2003) Parallel sampling of GMRFs and geostatistical GMRF models. Norges Teknisk-Naturvitenskapelige Universitet Preprint Statistics no. 7/2003

Strauss DJ (1975) Analyzing binary lattice data with the nearest-neighbor property. J Appl Prob 12:702–712

Strauss DJ (1977) Clustering on coloured lattices. J Appl Prob 14:135–143

Thomson W (Lord Kelvin) (1854–1855) On the theory of the electric telegraph. Proc R Soc Lond 7:382–399

Tjelmeland H, Luo X, Fjeldstad T (2019) A Bayesian model for lithology/fluid class prediction using a Markov mesh prior fitted from a training image. Geophys Prospect 67:609–623

Tobochnik J (1990) Efficient random walk algorithm for computing conductivity in continuum percolation systems. Comput Phys IEEE Comput Sci Eng 4(2):181–184

Tolpekin VA, Stein A (2009) Quantification of the effects of land-cover-class spectral separability on the accuracy of Markov-random-field-based superresolution mapping. IEEE Trans Geosci Remote Sens 47:3283–3297

Torquato S (1990) Relationship between permeability and diffusion-controlled trapping constant of porous media. Phys Rev Lett 64(22):2644–2646

Toumelin E, Torres-Verdín C, Chen S (2003) Modeling of multiple echo-time NMR measurements for complex pore geometries and multiphase saturations. SPE Reservoir Eval Eng 6(4):234–243

Toumelin E, Torres-Verdín C, Sun B, Dunn K-J (2007) Random-walk technique for simulating NMR measurements and 2D NMR maps of porous media with relaxing and permeable boundaries. J Magn Reson 188:83–96

Uçan ON, Muhittin Albora A (2009) Markov random field image processing applications on ruins of the Hittite Empire. Near Surface Geophys 7(347):111–122

Uçan ON, Sen A, Albora AM, Ozmen A (2000) A new gravity anomaly separation approach: differential Markov random field (DMRF). Electron Geosci 5:1–13

Uçan ON, Muhittin Albora A, Aydoğan D (2004) On the use of Markov Random Field in geophysical applications: Gelibolu Peninsula. İstanbul Üniv Müh Fak Yerbilimleri Dergisi 17(1):35–46

Ulvmoen M, Omre H (2010) Improved resolution in Bayesian lithology/fluid inversion from prestack seismic data and well observations: Part 1-methodology. Geophysics 75:R21–R35

Valckenborg RME, Huinink HP, Sande JJvd, Kopinga K (2002) Random-walk simulations of NMR dephasing effects due to uniform magnetic-field gradients in a pore. Phys Rev E 65:021306

Valfouskaya A, Adler PM, Thovert J-F, Fleury M (2006) Nuclear magnetic resonance diffusion with surface relaxation in porous media. J Colloid Interface Sci 2951:188–201

Wang H, Wellmann JF, Li Z, Wang X, Liang RY (2017) A segmentation approach for stochastic geological modeling using hidden Markov random fields. Math Geosci 49:145–177

Wang X, Li Z, Wang H, Rong Q, Liang RY (2016) Probabilistic analysis of shield-driven tunnel in multiple strata considering stratigraphic uncertainty. Struct Saf 62:88–100

Weeks ER, Urbach JS, Swinney HL (1996) Anomalous diffusion in asymmetric random walks with a quasigeostrophic flow example. Physica D 97:291–310

Wilkinson DJ, Johnson DL, Schwartz LM (1991) Nuclear magnetic relaxation in porous media: the role of the mean lifetime $\tau(\rho, D)$. Phys Rev B 44:4960–4973

Woynar R (2013) Random walk, diffusion and wave equation. Acta Phys Pol, B 44(5):1067–1084

Wu Y, Liu Q, Chan AHC, Liu H (2017) Implementation of a time-domain random-walk method into a discrete element method to simulate nuclide transport in fractured rock masses. Geofluids 2017:5940380

Xie H, Pierce LE, Ulaby FT (2002) SAR speckle reduction using wavelet denoising and Markov random field modeling. IEEE Trans Geosci Remote Sens 40:2196–2212

Yang XR, Wang Y (2019) Ubiquity of anomalous transport in porous media: numerical evidence, continuous time random walk modelling, and hydrodynamic interpretation. Sci Rep 9:4601

Zhang X, Crawford JW, Deeks LK, Stutter MI, Bengough AG, Young IM (2005) A mass balance based numerical method for the fractional advection-dispersion equation: theory and application. Water Resour Res 41:W07029

Zhang Y, Brady M, Smith S (2001) Segmentation of brain MR images through a hidden Markov random field model and the expectation-maximization algorithm. IEEE Trans Med Imaging 20:45–57

Zimmermann S, Koumoutsakos P, Kinzelbach W (2001) Simulation of pollutant transport using a particle method. J Comput Phys 173(1):322–347

Chapter 11
Thermodynamic Algorithms

Chapter Highlights Four important computational tools are reviewed, each of them is based on *thermodynamic principles*. *Simulated Annealing* (Sect. 11.1) is presented through two applications: *Reconstructing 3D pore space from 2D sections*, and *multiphase fluid distribution in porous rocks*. *Lattice Boltzman Methods* (LBMs), recently widely used in *Digital Rock Physics*, are treated in Sect. 11.2, including their theoretical derivation, computational aspects, and generalizations. Their applications include simulation of *multiphase flow*, and *Knudsen flow in tight shales*. The powerful method of RNG (*Renormalization Group*, Sect. 11.3) has been borrowed from the *Theory of Phase Transitions*. After a theoretical introduction it is applied to determine the *critical point of failure*, and *time of failure*, of stressed rocks; to describe the *erosion of tilted landscapes*, and for *permeability upscaling*. Section 11.4 is devoted to *discrete scale invariance*, and its use for *earthquake prediction* and *rock fragmentation*.

11.1 Simulated Annealing

Politis et al. (2008b) aptly summarized the *SA* (*Simulated Annealing*) method by quoting the Apostle's message to the Thessalonians ("Test them all; hold on to what is good, reject every kind of evil"—1 Thessalonians 5: 21–22, New International Version), then paraphrasing it as "One could summarize the method in one sentence: try all that matter, keep the best and stay away from local traps." *Simulated Annealing* (*SA*) is an *optimization algorithm* devised by Kirkpatrick and coworkers (1982, 1983), who found an analogy between the thermodynamics of the *annealing* of metals (i.e. the metallurgic technology of alternating *heating* and *controlled cooling* of a metal to reduce its defects), and the optimization problem of finding the *global minimum* of a multi-parameter *objective function* (also called *cost-*, *penalty-*, *misfit-*, or *energy-function*). If we carefully cool a crystalline material through its melting point, ordered, low-lattice-energy crystals can be formed. In this physical process the lattice energy $E(\boldsymbol{m}) = E(m_1, m_2, \ldots, m_n)$ plays the role of the *objective function* to be minimized, the free parameters of the system $\{m_1, m_2, \ldots, m_n\}$ correspond to the *state space*

G. Korvin, *Statistical Rock Physics*, Earth and Environmental Sciences Library,
https://doi.org/10.1007/978-3-031-46700-4_11

variables. Based on this idea, and using the *Metropolis algorithm* (Metropolis et al. 1953) to perturb the parameters as if they were material molecules randomly vibrating under the effect of heat, Kirkpatrick et al. solved the famous, notoriously difficult ("*NP-hard*", Garey and Johnson 1979) problem of finding the *optimal route* for a '*travelling salesman*' (Lawler et al. 1985) applying this scheme, and found the optimal solutions for up to 6,000 sites (Kirkpatrick et al. 1983). Actually, the SA algorithm was also discovered, around the same time and independently, by Pincus (1970), and Cerny (1982, 1985).

In *SA* a random perturbation is applied to a system's *current configuration*, so that a *trial configuration* is obtained. Let E_c and E_t be the energy levels of the current and trial configurations. If $E_c \geq E_t$ then the trial configuration is unconditionally accepted and it becomes the current configuration. On the other hand, if $E_c < E_t$ then the trial configuration is accepted with a probability given by

$$P(\Delta E) = \exp[-\Delta E/k_B T]$$

where $\Delta E = E_t - E_c$, k_B is the Boltzmann constant and T is temperature. This step helps the system to jump out from local minima. After a sufficient number of iterations at a fixed temperature, the system approaches equilibrium, where the free energy reaches its minimum value. By gradually decreasing T and repeating the simulation process (starting out every time from the equilibrium state found for the previous T value), new lower energy configurations are achieved.

The three basic steps of the *Metropolis algorithm* are:

(i) If the system is in state $E(\boldsymbol{m}) = E(m_1, m_2, \ldots, m_n)$, perturb m_1, keep all other parameters fixed. Denote the new state by $\boldsymbol{m}'$.

(ii) If $E(\boldsymbol{m}') \leq E(\boldsymbol{m})$ then $\boldsymbol{m}'$ is accepted as the new state. If $E(\boldsymbol{m}') > E(\boldsymbol{m})$ then $\boldsymbol{m}'$ is accepted with probability

$$P(\boldsymbol{m} \to \boldsymbol{m}') = \exp\left[-\frac{E(\boldsymbol{m}') - E(\boldsymbol{m})}{T}\right]. \qquad (11.1.1)$$

Each component $m_i, i = 1, \ldots, n$ of $\boldsymbol{m}$ is perturbed the same way, one after the other, and the changes are accepted or rejected as in Eq. (11.1.1).

(iii) The steps (i)–(ii) are repeated several times *at the same temperature*. Then temperature is slightly decreased.

An other way of updating the trial configuration, and accepting or rejecting the change, is provided by the *heat-bath algorithm* (Creutz 1983; Rebbi 1984; Rothman 1986). In this algorithm *generation* and *acceptance* are done as follows.

(*Generation*): Assume that coordinate $m_i, i = 1, \ldots, n$ of $\boldsymbol{m}$ has k allowable values $\{\mu_1, \ldots, \mu_k\}$. Compute the energy functions $E(\mu_j) = E(m_1, \ldots, m_{i-1}, m_i', m_{i+1}, \ldots, \mu_k)$ with $m_i' = \mu_j$ and (*acceptance*): *randomly choose* the new model parameter according to the probability distribution

$$P(\mu_j) = \exp\left[\frac{E(\mu_j)}{T}\right] \bigg/ \sum_{i=1}^{k} \exp\left[\left[\frac{E(\mu_i)}{T}\right]\right]. \tag{11.1.2}$$

This is done for each coordinate in turn at the same temperature several times, then (iii) T is decreased as in the Metropolis algorithm, and the *generation* and *acceptance* steps are repeated.

The probability of acceptance in (11.1.1) follows from the assumed *Boltzmann distribution* of the energies in the system (Ingber 1993). In a system of units where $k_B = 1$, the probability of obtaining a new state with energy $E(\boldsymbol{m}')$ on the condition that the present state is $E(\boldsymbol{m})$ is

$$P[E(\boldsymbol{m}) \to E(\boldsymbol{m}')] = \frac{\exp[E(\boldsymbol{m}')/T]}{\exp[E(\boldsymbol{m}')/T] + \exp[E(\boldsymbol{m})/T]} = \frac{1}{1 + \exp[\Delta E/T]}$$
$$\approx \exp[-\Delta E/T] \text{ with } \Delta E = E(\boldsymbol{m}') - E(\boldsymbol{m}). \tag{11.1.3}$$

It was shown by Hammersley and Handscomb (1964) that if the following two conditions hold, then at any temperature T after (infinitely) many iterations the system will tend towards *thermal equilibrium.*

Condition (A): *any state of the system can be reached from any other state using the allowed rules of moves between states*;

Condition (B): *in any iteration 'move' there must be non-zero probability of staying in the current state.*

In thermal equilibrium at temperature T, the states $\{\omega\}$ of a large system, consisting of replicas of our considered system, are distributed according to the Boltzmann distribution

$$P_B(\omega) = \frac{\exp[-E(\omega)/T]}{Z(T)}; \tag{11.1.4}$$

where the normalizing constant $Z(T)$ is the *partition function*

$$Z(T) = \sum_{\omega} \exp[-E(\omega)/T]. \tag{11.1.5}$$

Suppose the energy function $E(\omega)$ has only one global minimum E_0, attained for state ω_0, and that there are only a finite number of states. Denote by p_0 the probability of the event that the system is in the optimal state ω_0, and by p_ω that the system is in some other state ω, $\omega \neq \omega_0$ with energy E_ω. We have

$$\frac{p_0}{p_\omega} = \exp\left[-\frac{E_0 - E_\omega}{T}\right] \geq 1, \tag{11.1.6}$$

and

$$\lim_{T \to 0} \frac{p_0}{p_\omega} = \infty. \tag{11.1.7}$$

Since

$$p_\omega \le 1 \text{ for all } \omega \tag{11.1.8}$$

and

$$\sum_\omega p_\omega = 1, \tag{11.1.9}$$

it follows from Eq. (11.1.7) that

$$p_0 \to 1 \text{ as } T \to 0. \tag{11.1.10}$$

This shows that the system can be optimized if we attain equilibrium at $T \approx 0$.

The best technique of cooling, that is the controlled way of letting T $\to$ 0 to assure ergodicity, convergence, and an affordable computer time, is a highly debated point among theorists and practitioners of SA (Geman and Geman 1984; Mitra et al. 1986; Laarhoven and Aarts 1987; Szu and Hartley 1987; Andresen et al. 1988; Hajek 1988; Nulton and Salamon 1988; Salamon et al. 1988; Matsuba 1989; Ingber 1993; Billings 1994; Nourani and Andresen 1998; Ben-Ameur 2004; Mahdi et al. 2017; Peprah et al. 2017; Khairuddin and Zainuddin 2019).

Here are a few examples for the proposed cooling schedules.

Geometric schedules:

(Stefaniuk et al. 2016; Zhang et al. 2020):

$$T_N = \lambda^N T_0, \tag{11.1.11a}$$

λ $(0 < \lambda < 1)$is a tunable parameter;

(Knight et al. 1990):

$$T_N = 0.999^N T_0; \tag{11.1.11b}$$

(Cabrera et al. 2014; Peprah et al. 2017; Mahdi et al. 2017; Khairuddin and Zainuddin 2019):

$$T_{k+1} = 0.99 T_k \text{ i.e. } T_k = T_0 0.99^k; \tag{11.1.11c}$$

Logarithmic schedules:

(Ingber 1993):

$$T(N) = \frac{T_0}{\ln N}, \tag{11.1.12a}$$

(Geman and Geman 1984; Mitra et al. 1986):

$$T_k = \frac{T_0 \log 2}{\log(k+1)}, \tag{11.1.12b}$$

(Mitra et al. 1986):

$$T_m = \frac{\gamma}{\log(m + m_0 + 1)}; \; m = 0, 1, 2, \ldots, \text{ where } 1 \leq m_0 < \infty, \tag{11.1.12c}$$

and γ will be defined later (in Eq. 11.1.17i).

Fast cooling schedules:

Fast Annealing: (Szu and Hartley 1987):

$$T(k) = \frac{T_0}{k}, \tag{11.1.13a}$$

Very Fast Simulated Annealing (*VFSA*) (Zhang and Toksöz 2012):

$$T_k = T_0 \exp\left[-dk^{\beta}\right]. \tag{11.1.13b}$$

Cooling with constant thermodynamic speed: This technique had been developed by Nulton and Salamon (1988) in an attempt to lower the temperature at an *optimum rate*. In their cooling process a constant difference is kept between the *non-equilibrium mean value* $\langle \mathrm{E} \rangle$ of the system and the theoretical mean value at equilibrium $\langle \mathrm{E} \rangle_{eq}$ at the same temperature T, measured in units of the *SD* (*standard deviation*) of

$$\mathrm{E} : \frac{\langle \mathrm{E} \rangle - \langle \mathrm{E} \rangle_{eq}}{\sigma_E(T)} = v = const. \tag{11.1.14a}$$

Thermodynamic relaxation at temperature T is governed by the differential equation

$$\frac{dE}{dt} = -\frac{\langle \mathrm{E} \rangle - \langle \mathrm{E} \rangle_{eq}}{\varepsilon(T)}, \tag{11.1.14b}$$

where "time" t is defined as number of steps in the cooling process, $\varepsilon(T)$ is the *relaxation time* of the system at temperature T. It is related to *heat capacity* as

$$C(T) \text{ as } C(T) = \frac{\sigma_E^2(T)}{T^2}. \tag{11.1.14c}$$

From Eqs. (11.1.14a–11.1.14c) we get

$$\frac{dT}{dt} = -\frac{vT}{\varepsilon(T)\sqrt{C(T)}}. \tag{11.1.14d}$$

We can find $\sigma_E(T)$, $\varepsilon(T)$ and $C(T)$ in the following steps (Andresen et al. 1988; Nulton and Salamon 1988; Mosegaard and Sambridge 2002):

(i) Prescribe a *constant difference* δE between two successive energy levels of the objective function, E_{k+1} and E_k;

(ii) Starting from the first iteration, count the number of successful transitions n_{ij} between states $E \in (E_i + \delta E)$ and $E \in \left(E_j + \delta E\right)$, and form the *transition frequency matrix*

$$\boldsymbol{Q} = \left\{Q_{ij}\right\} = \left\{\frac{n_{ij}}{\sum_j n_{ij}}\right\}; \tag{11.1.15a}$$

(iii) Form a temperature-dependent matrix $\boldsymbol{G}(T)$ by (iii.a) multiplying those elements of $\boldsymbol{Q}$ for which $i < j$ with the Boltzmann factors $\exp\left[-\frac{E_j - E_i}{T}\right]$; and (iii.b) adjusting the diagonal terms Q_{ii} so as to keep the sum of rows to stay unity;

(iv) $\boldsymbol{G}(T)$ is a *stochastic matrix*, so its largest eigenvalue is 1 (Perron-Frobenius Theorem: MacCluer 2000) with the corresponding normalized eigenvector $\boldsymbol{p} = \{p_i\}$. The variance $\sigma_E^2(T)$ is estimated as

$$\sigma_E^2(T) = \langle E^2(T) \rangle - \langle E(T) \rangle^2 \tag{11.1.15b}$$

where

$$\langle E^n \rangle = \sum_i (E_i)^n p_i; \tag{11.1.15c}$$

(v) The *heat capacity* is provided by Eq. (11.1.14c);

(vi) If the second largest eigenvalue of $\boldsymbol{G}(T)$ is $\lambda_2(T)$, the relaxation time is estimated as

$$\varepsilon(T) = -\frac{1}{\log(\lambda_2)} \tag{11.1.15d}$$

(using a Theorem in Levin and Peres 2017: 193, Eq. 13.29);

(vii) Because the number of iterations is limited by the available computer resources (Jakobsen et al. 1987; Andresen et al. 1988; Koren et al. 1991) the thermodynamic speed v must be selected with initial experimentations.

In D-dimensional *Boltzmann Annealing* (BA, see Ingber 2000) the new point x_{k+1} is selected according to the Gaussian probability distribution

$$g(\Delta x) = (2\pi T)^{-D/2} \exp\left[-(\Delta x)^2/2T\right] \tag{11.1.16a}$$

where $\Delta x = x_{k+1} - x_k$.

It can be shown heuristically (Szu and Hartley 1987; Ingber 2000) that the cooling schedule

$$T(k) = \frac{T_0}{\log k} \tag{11.1.16a}$$

suffices to obtain a *global minimum* of $E(x)$ if T_0 is large enough and $\Delta x = x_{k+1} - x_k$ is bounded. To prove that in this case every state x_k is *visited infinitely often*, we should show that the probability of *not reaching* one of the states is zero, that is

$$\prod_{k=1}^{\infty}(1 - g_k) = 0. \tag{11.1.16b}$$

Recalling the theory of *infinite products* (Knopp 1954), this is equivalent to

$$\sum_{k=1}^{\infty} g_k = \infty. \tag{11.1.16c}$$

For *Boltzmann Annealing*, using the cooling schedule (11.1.12a), from (11.1.16a)

$$g_k = \left(\frac{\log k}{2\pi T_0}\right)^{D/2} \cdot \exp\left[-\frac{(\Delta x)^2}{2T_0} \cdot \log k\right].$$

There exist T^*, k_0 such that for $T_0 \geq T^*$, $k \geq k_0$ we have

$$\left(\frac{\log k}{2\pi T_0}\right)^{D/2} \geq 1 \text{ and } \frac{(\Delta x)^2}{2T_0} \leq 1, \tag{11.1.16d,e}$$

so that with some constant $\alpha > 0$:

$$\sum_{k=1}^{\infty} g_k = \alpha + \sum_{k=k_0}^{\infty} g_k \geq \alpha + \sum_{k=k_0}^{\infty} \exp(-\log k) = \alpha + \sum_{k=k_0}^{\infty} \frac{1}{k} = \infty. \tag{11.1.16f}$$

Mitra et al. (1986) dealt with the convergence of SA in a mathematically more precise manner, based on the theory of *non-stationary Markov chains*. They presented an annealing schedule for which the Markov chain of state-to-state changes is *strongly ergodic* and the algorithm *converges to a global optimum*. For the statement of their main result, we must introduce a few concepts. The SA process can be represented by a *Markov chain* whose transition probabilities are specified by the *accept* and *generate functions*. Let G be a *directed graph* corresponding to this Markov chain, whose set of vertices correspond to the set of states S of the annealing problem, and there is an edge $i \to j$ between two different vertices if j can be generated started from i. Two such vertices are called *neighbors*, the set of neighbors of i is $N(i)$, we assume $i \notin N(i)$ that is, starting from i we do not recur to it in one step. A given site $j \in N(i)$ can be generated from i with the probability

$$Prob(i \to j) = g(i,j)/g(i) \tag{11.1.17a}$$

where the normalizing function $g(i)$ assures that

$$\frac{1}{g(i)} \cdot \sum_{j \in N(i)} g(i,j) = 1. \tag{11.1.17b}$$

We assume that G is *connected*. A vertex $i \in S$ is a *local maximum* if

$$E(j) \leq E(i) \text{ for all } j \in N(i). \tag{11.1.17c}$$

Denote the set of all points which are local maxima by S_m, and let $d(i,j)$ be the *distance* between any two different vertices defined as the number of edges in the *shortest path* between i and j. The *radius* r of the graph G is defined as

$$r = \min_{i \in (S - S_m)} \max_{j \in S} d(i,j) \tag{11.1.17d}$$

the vertex $\hat{i}$ where the minimum of $d(i,j)$ is attained is the *center of the graph.*

We characterize the local slope of the energy function by a *Lipschitz-like constant*; and we also determine the lower bound for the generating probabilities. (Recall, that a function $f : \mathbb{R} \to \mathbb{R}$ is *Lipschitz continuous* if there exists a real constant L such that for all $x_1, x_2 \in \mathbb{R}$ we have $|f(x_1) - f(x_2)| \leq L \cdot |x_1 - x_2|$; Searcóid 2006: 154). The Lipschitz constant in our case is given by

$$L = \max_{i \in S} \max_{j \in N(i)} |E(j) - E(i)|; \tag{11.1.17f}$$

while the lower bound on generation probabilities is

$$w = \min_{i \in S} \min_{j \in N(i)} \frac{g(i,j)}{g(i)}. \tag{11.1.17a,g}$$

Assume that

$$w > 0. \tag{11.1.17h}$$

Theorem (Mitra et al. 1986). *If* (a) *the energy function E is bounded, and* (b) *satisfies a Lipschitz-like condition with Lipschitz-constant L; if* (c) *the corresponding directed graph G is connected; if* (d) *the generation probabilities satisfy the inequalities*

(11.1.17g) *with* $w > 0$ (11.1.17h), *and* (e) *if* γ *is any real number satisfying*

$$\gamma \geq r \cdot L, \tag{11.1.17i}$$

then the annealing schedule with cooling function

$$T_m = \frac{\gamma}{\log(m + m_0 + 1)};\ m = 0, 1, 2, \ldots, \text{ where } 1 \leq m_0 < \infty \tag{11.1.12c}$$

assures that the individual state probabilities converge either to a positive value or to 0, depending upon whether the given state is of globally minimal energy or not.

Further theoretical details of the SA algorithm in its original and modified versions can be found in the literature compiled in Table 11.1. The Reader is advised to study the theory of SA, and its applications (summarized in Table 11.2) before applying this subtle and tricky technique.

Because of space limitations, we can only discuss here two characteristic applications (Korvin 2021).

(a) *Reconstructing 3D pore space from 2D sections*

In this technique, statistical information, obtained from binary images of 2-D thin-sections, is used to create a 3D representation of the porous structure. The 3D model should honor the statistics of the thin sections, and is used to determine the *topology* and *effective physical properties* of the rock. Yeong and Torquato (1998a, b) solved this task with *Simulated Annealing*,

The statistics they used are the first two moments of the *characteristic function* $\chi(\vec{r})$ ($\chi(\vec{r})$ is unity if $\vec{r}$ lies in the pore space, and zero otherwise): the porosity, $\Phi = \chi(\vec{r})$, and the normalized autocorrelation function,

$$\rho(\vec{u}) = \frac{\left[\chi(\vec{r}) - \Phi\right] \cdot \left[\chi(\vec{r} + \vec{u}) - \Phi\right]}{\Phi(1 - \Phi)}.$$

The *ACF*is fitted by the stretched exponential model $\exp[-(u/r_0)^n]$, where $u = |\vec{u}|$, r_0 is a characteristic length scale (*correlation length*) and n is an adjustable parameter. In each iteration step i we compute the "energy" function

$$E_i(\rho) = \sum_k \left[\rho_i(u_k) - \rho_{ref}(u_i)\right]^2$$

which is to be minimized, where $\{u_i\} = \{0, \Delta u, 2\Delta u, \ldots\}$, $\rho_i(u)$ is the actually observed *ACF*in this iteration step, $\rho_{ref}(u)$ is the required *ACF*of the field to be reconstructed. Starting from a random configuration with porosity Φ, the values of two voxels of different phases are exchanged at each iteration step. Thus Φ remains constant. The new configuration is accepted with probability P according to the Metropolis rule

Table 11.1 Simulated annealing (SA): its properties and different modifications

Topic	Authors
Adaptive simulated annealing	Ingber (1993)
Conditional Simulated Annealing	Hazlett (1997), Talukdar et al. (2004)
Convergence of SA	Geman and Geman (19840, Romeo and Sangiovanni-Vincentelli (1984), Gidas (1985), Lundy and Mees (1986), Mitra et al. (1986), Hajek (1988), Charnes and Wolfe (1989), Ferrari et al. (1993), Ingber (2000)
Cooling schedules	Geman and Geman (1984), Mitra et al. (1986), Laarhoven and Aarts (1987), Szu and Hartley (1987), Andresen et al. (1988), Hajek (1988), Nulton and Salamon (1988), Salamon et al. (1988), Matsuba (1989), Ingber (1993), Billings (1994), Nourani and Andresen (1998), Ben-Ameur (2004), Mahdi et al. (2017), Peprah et al. (2017), Khairuddin and Zainuddin (2019)
Critical temperature	Basu and Frazer (1990)
Fast simulated annealing	Szu and Hartley (1987)
General introductions	Hammersley and Handscomb (1964), Binder and Stauffer (1985), Cerny (1985), Laarhaven and Aarts (1987), Aarts and Korst (1989), Rutenbar (1989), Deutsch and Journel (1992), Jensen et al. (1997), Brémaud (1998), Sambridge and Mosegaard (2002), Adewole et al. (2012), Sen and Stoffa (2018)
Gibbs distribution	Geman and Geman (1984)
Heat bath algorithm	Creutz (1983), Rebbi (1984), Rothman (1986)
Hierarchical annealing	Alexander et al. (2009)
Metropolis algorithm	Metropolis et al. (1953)
NP-complete problems	Garey and Johnson (1979)
Optimal simulated annealing	Hajek (1988), Matsuba (1989)
Probabilistic hill climbing algorithms	Romeo and Sangiovanni-Vincentelli (1984), Mitra et al. (1986)
SA-PSO (particle swarm optimization by SA)	Zhang et al. (2017a, b, 2018, 2020)
SA with constant thermodynamic speed	Jakobsen et al. (1987), Nulton and Salamon (1988), Salamon et al. (1988)
Simulated quenching	Ingber (1993)
Thermodynamic properties of SA	Andresen et al. (1988)
Very fast simulated re-annealing	Ingber (1989)

Table 11.2 Applications of simulated annealing in earth sciences

Application	Authors
AVO (amplitude vs. offset) inversion	Gunning and Sams (2018)
Dipmeter analysis	Schneider and Whitman (1990)
Dispersions of particles	Rintoul and Torquato (1997)
Earthquake location	Billings (1994)
Electrical impedance tomography	Pessel and Gibert (2003)
Electromagnetic imaging	Gibert and Virieux (1991)
Fluid distribution in porous media	Knight et al. (1990), Politis et al. (2008b)
Fluid flow	Hazlett (1997)
Geostatistics	Deutsch and Cockerham (1994), Arpat and Caers (2007)
Gravity inversion	Nagihara and Hall (2001)
Image processing	Geman and Geman (1984), Gidas (1989)
Inverse problems	Tarantola and Valette (1982), Sambridge and Mosegaard (2002)
Laboratory robot system	Cabrera et al. (2014)
Lunar seismic data	Khan et al. (2000), Mosegaard and Sambridge (2002)
Magnetotelluric appraisal*	Dosso and Oldenburg (1991)
Multimodal inverse problems	Scales et al. (1992)
Petrophysical inverse problems	Ouenes (1992)
Pore network modeling	Hazlett (1997), Talukdar (2002)
Prestack seismic inversion	Ma (2002)
Reconstruction of cracks	Zhang and Toksöz (2012)
Reconstruction of porous media	Yeong and Torquato (1998a, b), Liang et al. (2000), Manwart et al. (2000), Rozman and Utz (2001), Saccomano et al. (2001), Talukdar (2002), Talukdar and Torsaeter (2002), Talukdar et al. (2002a, b, c, 2004), Čapek et al. (2008), Politis et al. (2008a), Diógenesa et al. (2009)
Reservoir characterization	Ouenes (1992), Ouenes et al. (1994), Goovaerts (1996)
Reservoir modeling	Deutsch (1992), Deutsch and Journel (1994), Sen et al. (1995), Zhang et al. (2017a, b, (2018)
Resistivity sounding	Sen et al. (1993)
Seismic inversion	Vestergaard and Mosegaard (1991), Ma (2002)
Seismic modeling	Jakobsen et al. (1987)
Seismic processing	Rothman (1985, 1986), Vaeudevan et al. (1991)
Seismic reflection traveltimes inversion	Pullammanappalli and Louie (1993)
Seismic traveltime tomography	Ammon and Vidale (1993), Weber (2000)

(continued)

Table 11.2 (continued)

Application	Authors
Seismic velocity analysis	Koren et al. (1991)
Seismic waveform inversion	Sen and Stoffa (1991a, b)
Self-diffusion of tracers	Kainourgiakis et al. (1999)
Shear wave velocity from well logs	Zhang et al. (2020)
Soil thermal conductivity	Stefaniuk et al. (2016)
Transport properties	Kainourgiakis et al. (2000, 2005)
Velocity model estimation	Landa et al. (1989)
Volcano muography (muon tomography, muon radiography)	Vesga-Ramírez et al. (2021)

*For the concept of 'appraisal", see Backus and Gilbert (1968, 1970)

$$P = \begin{cases} 1 \text{ if } E_i \leq E_{i-1} \\ \exp\left[(E_{i-1} - E_i)/T\right] \text{ if } E_i > E_{i-1} \end{cases} \tag{11.1.18}$$

where T plays the role of temperature. In case of rejection, the old configuration is restored. By decreasing T, fields f_i with minimal energy $E_i(\rho)$ will be generated, whose correlation properties will be similar to those of the reference field f_{ref}.

Note that in this example the empirical ACF was fitted by a smooth theoretical model $\exp[-(u/r_0)^n]$. As Saccomano et al. (2001) pointed out, this is by no means necessary, as SA has its capacity to use any shape of the ACF or semivariogram. Deutsch and Journel (1992) used theoretical models of the semivariograms, but (Ouenes 1992; Sen et al.1995; Savioli et al.1996) applied experimental semivariograms "because oscillations, holes and bumps which are difficult to model may be realities of the reservoir that should be taken into account".

(b) *Fluid distribution in a porous rock* (Politis et al. 2008b)

The digitized rock is represented by cubic voxels. Each voxel is labeled by an integer indicating its *phase*. The solid phase is labeled as 0, the fluid phases as 1, 2, 3, ..., n if there are n different pore-fillers. The *saturation* S_i of phase i is the volume fraction of the total pore space occupied by phase i. The distribution of the n fluid phases in the available pore space is determined assuming that it minimizes the total interfacial free energy

$$G_s = \sum_{i=0}^{n-1} \sum_{j>i}^{n} A_{ij}\sigma_{ij}, \tag{11.1.19a}$$

where A_{ij} is the area of the interface between phases i and j, σ_{ij} is the interfacial free energy per unit area between phases i, j. The interfacial free energies obey Young's equation so that the following $n(n-1)/2$ equations are satisfied (van Kats and Egberts 1998):

$$\sigma_{0i} = \sigma_{0j} + \sigma_{ij}\cos\theta_{ij} \tag{11.1.19b}$$

where θ_{ij} is the contact angle of the ij-interface with the solid surface, $i \neq 0, j \neq 0, i < j$.

Suppose n fluid phases are distributed in the pore space. The number of voxels belonging to each phase depends on the saturation S_i. The authors used the *SA algorithm*, adapting the approach of Knight et al. (1990) who dealt with a similar problem.

In each minimization step n voxels, each from a different fluid phase, are randomly selected, and $n(n-1)/2$ swaps are performed. Each trial swap causes a change of G_s by $\Delta G_{s,ij}$ where $i < j$. If at least one $\Delta G_{s,ij} < 0$, the trial swap with the minimum $\Delta G_{s,ij}$ is accepted. If every $\Delta G_{s,ij} > 0$, the swap $i \leftrightarrow j$ is accepted with probability

$$P_{ij} = \frac{\exp\left[-\Delta G_{s,ij}/G_{ref}\right]}{n(n-1)/2}, \tag{11.1.19c}$$

where G_{ref} stands instead of the k_BT parameter ($k_B = 1.380649 \times 10^{-23}$ joules/K is Boltzmann's constant).

After a sufficient number of iterations, the system approaches equilibrium for a specific G_{ref} value. By gradually decreasing G_{ref}, and repeating the process, lower energy levels of G_s will be approached. The "*cooling schedule*" in this example was

$$G_{ref} = f^N G_{ref,N} \tag{11.1.19d}$$

where N is iteration number, and $f\,(0 < f < 1)$ a tunable parameter. Politis et al. (2008b: 216) found that "the value of the parameter f plays a dominant role in the simulation. Small f values decrease G_{ref} quickly resulting in fast simulations, however the risk of trapping by local minima is high. Contrarily, when f approaches unity the simulation becomes lengthy in time but the system escapes more efficiently from metastable configurations".

11.2 Lattice Boltzmann Models

The *Lattice Boltzmann Method* (*LBM*) is a numerical tool for simulating complex fluid flows or other transport processes in rocks (Ghassemi et al. 2010), a *finite-difference method* to solve the *Boltzmann transport equation* on a lattice. It is a simplified *molecular dynamics model* where space, time, and particle-velocities assume discrete values, each lattice node is connected to its neighbors, there can be either 0 or 1 particles at a lattice node, particles move with a given velocity. In successive discrete time instants, each particle moves to a neighboring node, this process is called *streaming*. When more than one particle arrive at the same node from different directions, they *collide* and change their velocities according to *collision rules* conserving particle number (mass), momentum, and energy before and after the collision.

The LBM originated from Boltzmann's *kinetic theory of gases* (Boltzmann 1896, 1898), where the gas consists of a large number of particles in random motion. Transport in this model is described by the equation

$$\frac{\partial f}{\partial t} + \vec{u} \cdot \nabla f = \Omega \tag{11.2.1}$$

where $f(\vec{x}, t)$ is the particle distribution function, $\vec{u}$ is particle velocity, Ω the collision operator. As compared to Boltzmann's *gas dynamics* which gives a full description of the molecular dynamics, in LBM the number of particles is much reduced, and they are confined to the nodes of a lattice. Therefore it is frequently referred to as a *mesoscopic model* (Manwart et al. 2002; Keehm 2003; Aaltosalmi 2005; Ramstad et al. 2010; Pereira 2015a).

The different numerical realizations of LBM are classified according to the *DnQm* convention, where "*Dn*" stands for "*n* dimensions", "*Qm*" stands for "*m* speeds". Here we only discuss *D2Q9*, the 2D model where a particle moves ("streams") in 9 possible directions, including the possibility to stay at rest. These velocities are referred to as *microscopic velocities* and are denoted by e_i, $i = 0, \ldots, 8$. In the *D2Q9* model the 9 velocity vectors are:

i	0	1	2	3	4	5	6	7	8
$\vec{e_i}$	(0, 0)	(1, 0)	(0, 1)	(−1, 0)	(0, −1)	(1, 1)	(−1, 1)	(−1, −1)	(1, −1)

In 3D the D3Q19 model is most common, but D3Q15 is also used, see Fig. 11.1.

In the *D2Q9* model each particle follows a discrete distribution of probabilities $f_i(\vec{x}, t)$, $i = 0 \ldots 8$, of *streaming* in any one particular direction. The *macroscopic fluid*

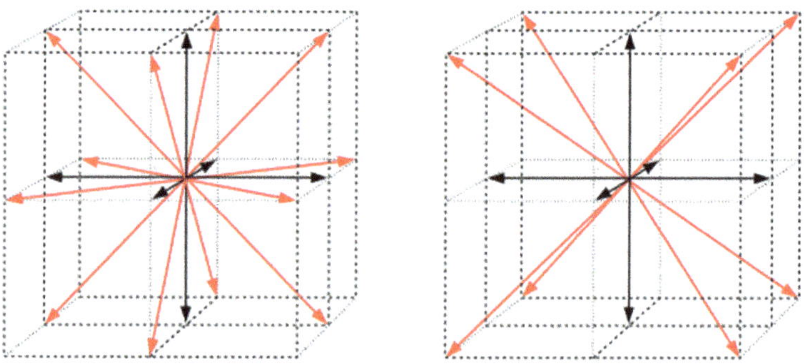

Fig. 11.1 Lattice structures in 2D with allowed directions: the D3Q19 (left) and D3Q15 (right) models. (After Aaltosalmi 2005: Fig. 3.2; reprodudeced with kind permission of *University of Jyväskylä Graduate School*)

density is the sum of microscopic particle distribution functions,

$$\rho(\vec{x}, t) = \sum_{i=0}^{8} f_i(\vec{x}, t), \tag{11.2.2a}$$

the *macroscopic velocity* $\vec{u}(\vec{x}, t)$ is the average of microscopic velocities $\overrightarrow{e_i}$ weighted with the distribution functions f_i,

$$\vec{u}(\vec{x}, t) = \frac{1}{\rho} \sum_{i=0}^{8} c f_i \overrightarrow{e_i}, \tag{11.2.2b}$$

where

$$c = \Delta x / \Delta t \tag{11.2.2c}$$

is the *lattice speed.*

The key steps in LBM are *streaming* and *collision*,

$$\begin{cases} f_i(\vec{x} + c\vec{e}_i \cdot \Delta t, t + \Delta t) - f_i(\vec{x}, t) & (streaming) \\ = -\frac{[f_i(\vec{x},t) - f_i^{eq}(\vec{x},t)]}{\tau} & (collision) \end{cases} \tag{11.2.3}$$

where $f_i^{eq}(\vec{x}, t)$ is the *local equilibrium distribution*, and τ the relaxation time towards local equilibrium. For *single-phase flows*, it suffices to use the simplified *Bhatnagar-Gross-Krook* (*BGK*) collision rules (Bhatnagar et al. 1954) where equilibrium distribution is defined as

$$f_i^{eq}(\vec{x}, t) = w_i \rho + \rho s_i(\vec{u}(\vec{x}, t)) \tag{11.2.4}$$

with

$$s_i(\vec{u}) = w_i \left[3 \frac{\overrightarrow{e_i} \cdot \vec{u}}{c} + \frac{9}{2} \frac{\left(\overrightarrow{e_i} \cdot \vec{u} \right)^2}{c^2} - \frac{3}{2} \frac{\vec{u} \cdot \vec{u}}{c^2} \right] \tag{11.2.5a}$$

and the weights w_i are

$$4/9 \text{ for } i = 0; \;\; 1/9 \text{ for } i = 1, 2, 3, 4; \;\; \text{and } 1/36 \text{ for } i = 5, 6, 7, 8. \tag{11.2.5b}$$

The *BGK* assumption has become so strongly associated with the *BL* Method, that BLM is sometimes referred to as "*BGK method*" (Qian et al. 1992; Flekkey 1993; Qian and Orszag 1993; Mailot and Main 1996; He et al. 1997; Zou and He 1997; Guo et al. 2000, 2002a; Kandhai et al. 1999a, 1999b; Bernsdorf 2008).

The fluid's kinematic viscosity in the *D2Q9* model is related to relaxation time by

$$\nu = \frac{2\tau - 1}{6} \cdot \frac{(\Delta x)^2}{\Delta t}. \tag{11.2.6}$$

Parameters are selected such that the *lattice speed* c be much larger than the maximum fluid velocity v_{max} expected to arise in the simulation. This is expressed by the 'computational' *Mach number*,

$$Ma = v_{max}/c. \tag{11.2.7}$$

Theoretically, it is required that $Ma \ll 1$, in practice, Ma should be, at least $\leq$ 0.1.

The algorithm runs as follows: (1) *Initialize* $\rho, \vec{u}, f_i$ and f_i^{eq}; (2) *Streaming step*: $f_i \to f_i^*$ in the direction of $\overrightarrow{e_i}$; (3) Compute the macroscopic ρ and $\vec{u}$ from f_i^*; (4) Compute f_i^{eq}; (5) *Collision step*: calculate the updated distribution

$$f_i = f_i^* - \frac{1}{\tau}(f_i^* - f_i^{eq}); \tag{11.2.8}$$

(6) Repeat steps 2 to 5.

During *streaming* and *collision*, the boundary nodes need special treatment. In classical fluid dynamics, the interface between solids and fluid is assumed to be *non-slip*, but this requirement would slow down computations. In most LB studies *bounce-back* conditions are used, so that any flux of fluid particles that hits a boundary simply reverses its velocity, that is the direction of the streamed distribution function is simply reversed (He et al. 1996; Gallivan et al. 1997). The *bounce-back boundary condition* is easy to use even in problems with complicated geometry, the small numerical errors what it causes can be corrected by shifting the solid boundary by half a voxel, as done in the *mid-plane bounce-back* version of the BLM (Sukop et al. 2008).

The *BL* (*Boltzmann Lattice*) *Equation can be derived* from the classic Boltzmann Equation (He and Luo 1997a, b; Krafczyk 2001; Bernsdorf 2008) as follows.

Denote by t time, by $\vec{x} \in \mathbb{R}^3$ a point of space, by $\vec{\xi}$ velocity, and let $\vec{F}$ be an external force. The probability distribution of particles, $\left(t, \vec{x}, \overrightarrow{\xi,}\right)$, satisfies the *Boltzmann equation*

$$\frac{\partial f}{\partial t} + \vec{\xi} \cdot \frac{\partial f}{\partial \vec{x}} + \vec{F}\frac{\partial f}{\partial \xi} = \Omega(f), \tag{11.2.9}$$

where $\Omega(f)$ stands for the *collision operator*. In *Statistical Physics* the velocity distribution in equilibrium is the *Maxwell distribution*

$$f^{eq} = \frac{\rho}{\left(2\pi c_s^2\right)^{d/2}} \exp\left[-\frac{\left(\vec{\xi} - \vec{u}\right)^2}{2c_s^2}\right] \tag{11.2.10}$$

where c_s is sound velocity, d dimension of the space, $\vec{u}$ is average velocity of the particles. In the *BGK* (Bhatnagar-Gross-Krook 1954) model it is assumed that the collision operator Ω results in a *local relaxation* towards the Maxwell distribution f^{eq}:

$$\Omega = -\frac{1}{\tau}\left(f - f^{eq}\right), \tag{11.2.11}$$

so that

$$\frac{\partial f}{\partial t} + \vec{\xi}\cdot\frac{\partial f}{\partial \vec{x}} + \vec{F}\frac{\partial f}{\partial \vec{\xi},} = -\frac{1}{\tau}\left(f - f^{eq}\right). \tag{11.2.12}$$

Assume that $|f - f_i| \ll 1$, and approximate the continuous f with a *discrete probability distribution* $\{f_i\}_{i=1}^{N}$ with corresponding discrete velocities $(\vec{e}_i)$. In the absence of external forces the discrete form of (11.2.12) becomes

$$\frac{\partial f_i}{\partial t} + e_{i\alpha}\frac{\partial f}{\partial x_\alpha} = -\frac{1}{\tau}\left(f_i - f_i^{eq}\right) \tag{11.2.13}$$

where $i = 1, \ldots, N$ enumerate the velocities, $\alpha = 1, \ldots, d$ are the Cartesian components. In the numerical implementation of the LB method, Eq. (11.2.13) is discretized using a finite-difference scheme. Let $c_s = 1/\sqrt{3}c$, then on a lattice with lattice vectors $\vec{e}_i = \frac{1}{c}\vec{e}_i$, Eq. (11.2.13) takes the shape

$$\frac{f_i\left(t+\Delta t, \overrightarrow{x'}\right) - f_i(t,\vec{x})}{\Delta t} + c\frac{f_i\left(t+\Delta t, \overrightarrow{x'} + \vec{e}_i\Delta x\right) - f_i\left(t+\Delta t, \overrightarrow{x'}\right)}{\Delta x}$$
$$= -\frac{1}{\tau}\left[f_i\left(t, \overrightarrow{x'}\right) - f_i^{eq}\left(t, \overrightarrow{x'}\right)\right], \tag{11.2.13a}$$

wherefrom, using $\Delta x = c\Delta t$ and multiplying with Δt we obtain the LB equation:

$$f_i\left(t+\Delta t, \overrightarrow{x'} + \vec{e}_i\Delta x\right) = f_i(t,\vec{x}) - \frac{\Delta t}{\tau}\left[f_i\left(t, \overrightarrow{x'}\right) - f_i^{eq}\left(t, \overrightarrow{x'}\right)\right]. \tag{11.2.14}$$

In the 3D case, for small Mach numbers, developing the velocity distribution (11.2.10) into Taylor series yields the equilibrium distribution

$$f_i^q = t_p\rho\left\{1 + \frac{e_{i\alpha}u_\alpha}{c_s^2} + \frac{u_\alpha u_\beta}{2c_s^2}\left(\frac{e_{i\alpha}e_{i\beta}}{c_s^2} - \delta_{\alpha\beta}\right)\right\}, \tag{11.2.15}$$

where t_p is a weight factor depending on the direction of the links: $p = 1$ for diagonal, $p = 2$ for orthogonal, $p = 0$ for the center. For the D3Q19 model, $t_0 = \frac{1}{3}$; $t_1 = \frac{1}{18}$; $t_2 = \frac{1}{36}$. For this model the speed of sound is $c_s = 1/\sqrt{3}$.

As in Eqs. (11.2.2a, 11.2.2b), the *macroscopic fluid density* and *velocity* are derived using the zeroth and first moments of the density distribution:

$$\rho = \sum_i f_i = \sum_i f_i^{eq}; \tag{11.2.15a}$$

$$\rho\vec{u} = \sum_i \vec{e}_i f_i = \sum_i \vec{e}_i f_i^{eq}. \tag{11.2.15b}$$

The *Equation of State*

$$P = \rho c_s^2 = \frac{1}{3}\rho \tag{11.2.16a}$$

yields *pressure, viscosity* is derived from *relaxation time* τ as

$$\nu = \left(\frac{\tau}{3} - \frac{1}{6}\right), \tag{11.2.16b}$$

and the *stress tensor* is computed as

$$S_{\alpha\beta} = -\left(1 - \frac{1}{2\tau}\right)\sum_i e_{i\alpha}e_{i\beta}\left(f_i - f_i^{eq}\right). \tag{11.2.16c}$$

If both the Mach number and Knudsen number are small (these numbers are defined as $Ma = v_{max}/c$, Eq. (11.2.7); and $K_n = \lambda/L_{char}$, Eq. (11.2.16); where L_{char} is a typical length in the geometry where the gas/fluid flows, such as tube radius; and λ is the mean free path of the gas molecules, i.e. average distance traveled by a particle between successive collisions, see Knudsen 1934; Zhang et al. 2015), these macroscopic quantities are solutions of the *Navier–Stokes equation*

$$\rho\left(\frac{\partial\vec{u}}{\partial t} + \vec{u}\nabla u\right) = -\nabla P + \nabla S \tag{11.2.17}$$

where $\hat{S} = (S)_{\alpha\beta 1,2,3}$ is the *strain tensor* defined as

$$S_{\alpha\beta} = \mu\left(\frac{\partial u_\alpha}{\partial x_\beta} + \frac{\partial u_\beta}{\partial x_\alpha}\right) \tag{11.2.18}$$

with $\mu = \rho\nu$ being *dynamic viscosity*.

To check that the *BLM* indeed solves the *NS* equation, and does it with remarkable precision, consider the analytically solvable case of *pipe flow* (*Poiseuille flow*, see Sutera and Skalak 1993), which is a steady, axisymmetric flow inside a long circular pipe of radius *R*, driven by the pressure gradient dp/dx where x is measured along the axis of the pipe. Continuity equation and axial symmetry yield

$$\frac{\partial u_x}{\partial x} = 0, \tag{11.2.19a}$$

so that the axial velocity $u_x(r)$ only depends on the radial coordinate r. Assuming viscosity is constant, the NS equations reduce to

$$\frac{\partial p}{\partial x} = \mu \left\{ \frac{\partial^2 u_x}{\partial r^2} + \frac{1}{r} \frac{\partial u_x}{\partial r} \right\}, \tag{11.2.19b}$$

and

$$\frac{\partial p}{\partial r} = 0. \tag{11.2.19c}$$

By the last equation $\frac{\partial p}{\partial r}$ does not depend on r, and (11.2.19b) integrates to

$$u_x(r) = \frac{r^2}{4\mu}\left(\frac{dp}{dx}\right) + c_1 \log r + c_2 \tag{11.2.19d}$$

To avoid infinite velocity on the axis (where $r = 0$), the integration constant c_1 must be zero. The second constant is determined from the no-slip boundary condition at the pipe wall, which requires $u_x(R) = 0$ wherefrom

$$c_2 = -\frac{R^2}{4\mu}\left(\frac{dp}{dx}\right). \tag{11.2.19e}$$

So, the velocity profile for the Poiseuille flow in the circular pipe is

$$u_x(r) = \frac{1}{4\mu}\left(-\frac{dp}{dx}\right)\left(R^2 - r^2\right). \tag{11.2.19f}$$

This is an excellent and popular *benchmark problem* (Geller et al. 2006; Andrä et al. 2013) for checking the performance and accuracy of any BLM software. We can similarly treat the *planar Poiseuille flow* between two infinitely long parallel plates a distance h apart, driven by a constant pressure gradient $G = -\frac{dp}{dx}$. The NS equations reduce to

$$\frac{d^2 u}{dy^2} = -\frac{G}{\mu} \tag{11.2.20a}$$

subject to the no-slip condition at the walls,

$$u(0) = u(h) = 0. \tag{11.2.20b}$$

The velocity distribution in this case is

$$u(y) = \frac{G}{2\mu} y(h - y). \tag{11.2.20c}$$

Landry (2013: Fig. 26) compares the numerical BL solution with the analytic result (11.2.19f), Bao and Meskas (2011: 7) shows this comparison for the planar case. In the planar case Bao and Meskas (*op.cit.*) applied periodic boundary conditions at the inlet and outlet, had length of the channel $L = 40$, width $h = 32$; pressure difference between inlet/outlet $\Delta p = -0.05$, relaxation time $\tau = 1$, and used for stopping criterion (criterion of steady state)

$$\frac{\sum_{ij} \left| u_{ij}^{n+1} - u_{ij}^{n} \right|}{\sum_{ij} \left| u_{ij}^{n+1} \right|} \leq 5.0 \times 10^{-9}, \tag{11.2.21}$$

where u_{ij}^{n} is the horizontal velocity component in the n-th time step at node (x_i, y_j).

Next, they placed a cylindrical obstacle of circular cross section in the path of the flow (*op. cit.*: Fig. 10) to check the performance of the BLM in the theoretically and experimentally well-studied case of *flow past a circular cylinder* (Rayleigh 1916; Tritton 1959; Batchelor 2000). By theory, the behavior of flow, its change from *laminary* to *turbulent*, only depends on the *Reynolds number*

$$Re = \frac{\rho u_0 d}{\mu}, \tag{11.2.22}$$

where d is the cylinder's diameter. Even in strong turbulence, BLM has proved to be a reliable, satisfactory tool (Derksen and van den Akker 1999; Derksen 2001, 2002, 2007; Lu et al. 2002; Chen et al. 2003; Bao and Meskas 2011; Chikatamarla and Karlin 2013).

We cannot treat here the theoretical aspects of BLM and its mushrooming new versions, new kinds of boundary conditions, parallel implementations on *CPU*s and *GPU*s (*Central* and *Graphical Processing Units*). Hopefully, Table 11.3 will 'guide the perplexed' to the relevant literature. Similarly, an extensive Table 11.4 summarizes the amazingly wide range of applications. The newcomer will find in them clever case histories, careful interpretations, and mathematical details that will shed light on the very problem she tries to solve with BLM. We only discuss here three points, (i) *REV* (Representative Elementary Volume); (ii) simulation of *multiphase flow* (Pereira 2015a), and (iii) the modelling *Knudsen* flow in tight shale (Zhang et al. 2012).

(i) *How large REV is needed for BLM*?

In the *mechanics of composite materials*, the *representative elementary volume* (REV, also called *representative volume element* RVE) is the smallest volume over which a *physical measurement* can be made that will yield a *value representative of the whole*. Rodney Hill (1963) emphasized two important properties of REV: (1) "[it] is entirely typical of the whole mixture on average", and (2) "[it] contains a sufficient

Table 11.3 The Boltzmann Lattice method (BLM): its theory, properties and different modifications

Topic	Authors
Benchmarks	Geller et al. (2006), Andrä et al. (2013)
(BFL) Bouzidi-Firdaouss-Lallemand scheme	Bouzidi et al. (2001)
Bounce-back, and boundary conditions	Noble et al. (1995a, b), Chen et al. (1996), He et al. (1996, 1997), Maier et al. (1996), van Genabeek and Rothman (1996), Gallivan et al. (1997), Zou and He (1997), Kandhai et al. (1999a,b), Manwart et al. (2002), Sukop et al. 2008), d'Humières and Ginzburg (2009), Walsh et al. (2009
Collision rule (BGK rule)	Bhatnagar et al. (1954)
Color gradient method	Rothman and Keller (1988), Gunstensen and Rothman (1991, 1992), Gunstensen et al. (1991), Latva-Kokko and Rothman (improved method) (2005), Tölke et al. (2006 (improved method)), Reis and Phillips (2007), Ahrenholz et al. (2008 (improved method)), Liu et al. (2012), Fauzi et al. (2017)
Curved boundary schemes	Bouzidi et al. (2001), van der Hoef et al. (2005), Pan et al. (2006), Maier and Bernard (2010), Yang et al. (2016), Rao and Schaefer (2020)
Depth-averaged LBM	Dzikowski et al. (2018)
Derivation of LBM	He and Luo (1997a, b), Guo et al. (2000)
Effective viscous drag method	Venturoli and Boek (2006)
Entropic LBM	Chikatamarla and Karlin (2013)
Extrapolation-based boundary scheme	Guo et al. (2002)
Finite differences (FD) based LBM	Guo and Zhao (2005)
Free-energy LBM	Swift et al. (1995, 1996)
General treatments	Arkeryd (1972a, b), McNamara and Zanetti (1988), Doolen (1990), Benzi et al. (1992), Qian et al. (1992), Rothman and Zaleski (1997), Chen and Doolen (1998), Chopard and Droz (1998), Fredrich et al. (1999), Lallemand and Luo (2000), Wolf-Gladrow (2000), Krafczyk (2001), Mohamad (2011), Rivet and Boon (2001), Succi (2001), Sukop and Thorne Jr. (2007), Bao and Meskas (2011), Huang et al. (2015)
Gray-scale LBM	Balasubramaniam et al. (1987), Gao and Sharma (1994), Dardis and McCloskey (1998), Walsh et al. (2009), Zhu and Ma (2013), Li et al. (2014), Pereira (2015a), Ginzburg (2016), Chen et al. (2018a, b)
Higher-order LBM	Shan et al. (2006), Ansumali et al. (2007), Zhang (2010)

(continued)

Table 11.3 (continued)

Topic	Authors
Inter-particle potential LBM	Shan and Chen (1993, 1994), Shan and Doolen (1995), Pereira (2015b)
Kinetic boundary conditions	Ansumali and Karlin (2002)
LB on a single CPU (central processing unit)	Wellein et al. (2006)
LB on GPUs (graphics processing units)	Habich (2008), Tölke (2010), Tölke and Krafczyk (2008), Bailey et al. (2009), Kuznik et al. (2010), Tsuji and Jiang (2016)
Level set simulation	Tan and Zabaras (2006)
Mean-field theory BLM	He et al. (1999), Santos et al. (2003)
Mid-plane bounce-back version of the BLM	Sukop et al. (2008)
Moving contact lines	Brochardwyart and DeGennes (1992), Seppecher (1996), Dhori and Slattery (1997), Pomeau (2002)
Multiple relaxation time schemes	d'Humieres et al. (2002), Dellar (2003), Talon et al. (2012)
Off-lattice Boltzmann method	Zienkiewicz and Codina (1995), Bardow et al. (2006)
Parallel implementation	Aksnes and Elster (2010)
Phase-field LBM	He et al. (1999), Geier et al. (2015)
Pseudopotential model	Shan and Chen (1993, 1994), Sbragaglia et al. (2007), Porter et al. (2012)
Shan-Chen model	Shan and Chen (1993, (1994), Chin et al. (2002)
Shale gas microflow	Zhang et al. (2015)
Spurious currents (non-physical velocities between immiscible fluids)	Connington and Lee (2012), Bao and Schaefer (2013)
Stabilized diffuse-interface LBM	Lee and Liu (2010)
Thermal BLM	He and Doolen (1998)
Unstructured LBM	Rossi et al. (2005), Misztal et al. (2015a, b)
Velocity models	Qian et al. (1992)
Volume of fluid (VOF) method	Hirt and Nichols (1981)

number of inclusions for the apparent properties to be independent of the surface values of traction and displacement, so long as these values are macroscopically uniform.". For numerical computation of effective physical properties from digitized ('voxelized') rock, this definition should be modified. REV is the smallest volume over which *numerical simulation* (in our case BLM) can be made that will yield a value representative of the whole rock on average; and (instead of Hill's second condition) we should require that the effective value computed within the REV should be unaffected by the boundary conditions (such as *periodicity*) that we prescribe on the REV's boundary.

Table 11.4 Applications of the LBM (Lattice Boltzmann Method)

Application	Authors
Anomalous dispersion	Misztal et al. (2015a, b)
Berea sandstone	Prodanovic et al. (2007)
Boussinesq equations	Guo et al. (2002)
Capillary hysteresis	Ahrenholz et al. (2008)
Capillary pressure	Li et al. (2018), Ramstad et al. (2019)
Carbon capture and storage (CCS)	Tsuji and Jiang (2016)
Carbon sequestration	Xie et al. (2017)
Castlegate sandstone	White et al. (2006)
Cellular automaton fluids	Wolfram (1986)
Complex flows	Aidun and Clause (2010)
Complex geometries	Succi et al. (1989)
Compressible flows	Qian and Orszag (1993), Vahala et al. (1998), Yu and Zhao (2000)
Convection	Bertrand et al. (1999), Seta et al. (2006), Chang and Alexander (2007), Huber et al. (2008)
CO_2 flooding	Suekane et al. (2005), Mahmoudi et al. (2014, 2017)
Crystal growth	Miller and Succi (2002), Medvedev and Kassner (2005)
Darcy's law	Balasubramaniam et al. (1987)
Dendritic growth	Rasin et al. (2005), Tan and Zabaras (2006)
Diffusion	McNamara (1990), Maillot and Main (1996)
Displacement of a droplet	Kang et al. (2005)
Dissolution and precipitation	Kang et al. (2003)
Drag force	van der Hoef et al. (2005)
Drop impact on dry surface	Lee and Liu (2010)
Elastic properties	Küntz et al. (1997)
Electric conductivity	Küntz and Mareschal (1997)
Electrokinetic coupling	Wang et al. (2006), Fiorentino et al. (2017)
Electro-osmotic flows	Yang et al. (2014)
Energy dissipation	Pilotti et al. (2002)
EOR (enhanced oil recovery)	Pride et al. (2008), Zhang et al. (2017a, b)
Equations of state	Yuan and Schaefer (2006)
Erosion	Chopard et al. (2000)
Flow in complex geometries	Pan et al. (2006), Chang et al. (2009)
Flow past a circular cylinder	Grucelski and Pozorski (2013)
fluid displacement	Ramstad et al. (2010), Liu et al. (2014a, b), Prodanovic et al. (2007)

(continued)

Table 11.4 (continued)

Application	Authors
Fluid flow	Rothman (1988), Succi et al. (1989), Chen and Doolen (1998), Kandhai et al. (1998, 1999b), Hayashi (2003), Aaltosalmi (2005), Boek and Venturoli (2010), Maier and Bernard (2010), Halisch (2013), Wu et al. (2016)
Fluid flow through fractal media	Chen et al. (1991)
Fluid flow through microfractures	Yang et a. (2018)
Fluid flow with temperature dependent viscosity	Fauzi et al. (2017)
Fluid partitioning	Prodanovic et al. (2006)
Fontainebleau sandstone	Ferréol and Rothman (1995)
Forced convection	Mehrizi et al. (2012)
Formation factor	Sun and Wong (2018)
Fractured media	Meakin and Tartakovsky (2009), Landry (2013), Dzikowski et al. (2018)
Fractured reservoirs	Gunde et al. (2010)
Gas flow in micro- and nanosystems	Zheng et al. (2006)
Heat transfer	Chen and Doolen (1998), Jiaung et al. (2001), Zhang and Chen (2003), Mehrizi et al. (2012)
Heat transfer in micro- and nanosystems	Zheng et al. (2006)
Incompressible axisymmetric flows	Zhang et al. (2016a, b)
Incompressible flows	Guo and Zhao (2002)
Incompressible Navier–Stokes equation	He and Luo (1997c)
Interfacial instabilities	Zhu (2018)
Intermittent flow	de Anna et al. (2013), Kang et al. (2014), Lester et al. (2014)
Knudsen layer	Cornubert et al. (1998), Kim et al. (2008), Aidun and Clausen (2010)
Kozeny constant	Heijs and Lowe (1995)
Lattice-gas automata	McNamara and Zanetti (1988), Higuera and Jiménez (1989)
Lattice gas hydrodynamics	Frisch et al. (1987)
Liesegang patterns	Chopard et al. (1994)
Liquid–gas phase transition	Shan and Chen (1994)
Liquid–vapor thermal flow	Zhang and Chen (2003)
Melting	Bertrand et al. (1999), Mencinger (2004), Huber et al. (2008)

(continued)

Table 11.4 (continued)

Application	Authors
Microscale flow (microfluidics)	Spaid and Phelan (1997), O'Connor et al. (1999), Li and Kwok (2003), Kim et al. (2008), Zhang (2010), Cho et al. (2013)
Middle East carbonates	Natarajan et al. (2014), Khodja et al. (2020)
Miscible fluids	Flekkoy (1993)
Multicomponent fluids	Martys and Chen (1996)
Multicomponent reactive transport	Kang et al. (2006)
Multiphase flow	Gunstensen et al. (1991), Gunstensen and Rothman (1992), Shan and Chen (1993, (1994), Ferréol and Rothman (1995), Shan and Doolen (1995), Swift et al. (1995, 1996), Martys and Chen (1996), Chen and Doolen (1998), He et al. (1999), Chen et al. (2000), Kang et al. (2002, 2005), Briant and Yeomans (2004), Briant et al. (2004), Pan et al. (2004), Turner et al. (2004), Latva-Kokko and Rothman (2005), Geller et al. (2006), Tölke et al. (2006), Reis and Phillips (2007), Schaap et al. (2007), Nordbotten et al. (2008), Sukop et al. (2008), Meakin and Tartakovsky (2009), Lee and Liu (2010), Narváez et al. (2010), Ramstad et al. (2010), Shabro et al. (2010), Silin et al. (2011), Joekar-Niasar et al. (2012), Liu et al. (2012, 2016, 2017), Raeini et al. (2012), Rui et al. (2013), Tölke et al. (2013), Zu and He (2013), Huang et al. (2015), Pereira (2015b), Tsuji and Jiang (2016), Xu and Liu (2018), Bakhshian et al. (2019), McClure et al. (2021)
Multi-physics	Bernsdorf (2008)
Navier–Stokes equation	Frisch et al. (1986), Chen et al. (1992), Qian et al. (1992), Ladd (1994b)
Non-ideal gas	Shan and Chen (1994), Tang et al. (2008)
Non-Newtonian flow	Aharonov and Rothman (1993), Boek et al. (2003), Sullivan et al. (2006), Psihogios et al. (2007)
Particle packings	Vidal et al. (2009)
Particle transport	Chopard et al. (2000), Feng et al. (2007) (in turbulent flow)
Particulate suspension	Ladd (1994a), Aidun and Lu (1995)
Perforation tunnel damage	Crouse et al. (2016)

(continued)

Table 11.4 (continued)

Application	Authors
Permeability	Cancelliere et al. (1990), Martys et al. (1994), Heijs and Lowe (1995), Koponen et al. (1997, 1998), Bosl et al. (1998), Pan et al. (2001), Manwart et al. (2002), Keehm et al. (2003), Pilotti et al. (2002), Guodong et al. (2004), van der Hoef et al. (2005), Fredrich et al. (2006, White et al. (2006), Arab et al. (2009), Narváez et al. (2010), Shabro et al. (2010), Ghassemi and Pak (2011a), Sun et al. (2011a, b), Li et al. (2014), Liu et al. (2014a, b), Wang et al. (2015), Hosa et al. (2016), Sun and Wong (2018), Khodja et al. (2020), Rao and Schaefer (2020)
Phase transition	Nourgaliev et al. (2002)
Poiseuille flow	Kim et al. (2008), Bao and Meskas (2011), Landry (2013)
Poisson equation	Chai and Shi (2008)
Pressure drop	Bernsdorf et al. (2000), Regulski et al. (2015)
Rayleigh–Taylor instability	Shan (1997), He et al. (1999)
Reaction–diffusion equations	Ponce Dawson et al. (1993), Chopard et al. (1994)
Reactive transport	O'Brien et al. (2002), Kang et al. (2006), Meakin and Tartakovsky (2009)
Relative permeability	Civan (2006), Huang and Lu (2009), Silin and Patzek (2009), Gunde et al. (2010), Hao and Cheng (2010), Ramstad et al. (2010, 2012, 2019), Ghassemi and Pak (2011b), Bondino et al. (2013), Dou and Zhou (2013), Arrufat et al. (2014), Hussain et al. (2014), Landry et al. (2014), Natarajan et al. (2014), Goel et al. (2016), Zhang et al. (2016a, b), Mahmoudi et al. (2017), Zhao et al. (2017), Alpak et al. (2018), Li et al. (2018), Shi and Tang (2018), Zhao et al. (2018), Zhu (2018)
Relative permeability upscaling	Apourvari and Arns (2016)
REV (representative elementary volume)	Bear (1972), Bear and Bachmat (1990), White et al. (2006), Khodja et al. (2020), Koestel et al. (2020)
Seepage around solid particles	Kim et al. (2001), Pan et al. (2001, 2006)
Shale transport	Chen et al. (2015), Ning et al. (2015)
Solidification	Boettinger et al. (2002), Semma et al. (2008)),
Stokes flow	Talon et al. (2012)
Streaming potentials	Fiorentino et al. (2016, 2017)
Thermal flow	Parmigiani et al. (2009), Peng et al. (2003)
Tortuosity	Ghassemi and Pak (2011a)
Transfer properties	Keehm et al. (2001, 2003), Keehm (2003), Keehm et al. (2001), Arab et al. (2009)
Transport	Auzerais et al. (1996), Fredrich et al. (2006)

(continued)

Table 11.4 (continued)

Application	Authors
Turbulence	Derksen and van den Akker (1999), Derksen (2001, 2002), Lu et al. (2002), Chen et al. (2003), Derksen et al. (2007), Bao and Meskas (2011), Chikatamarla and Karlin (2013)
Unsteady flows	Pazdniakou and Adler (2013)
Viscoelastic fluid flow	Malaspinas (2009)
Viscous coupling	Yiotis et al. (2007), Huang et al. (2009)
Viscous fingering	Grosfils and Boon (2003), Grosfils et al. (2004), Kang et al. (2004), Dong et al. (2011)
Wave propagation	Chopard et al. (1997)
Wettability	Landry et al. (2014), Zhang et al. (2017a, b), Zhao et al. (2018), Ramstad et al. (2019)
Wetting	Benzi et al. (2006a, b), Huang et al. (2007), Lee and Liu 2008), Wolf et al. (2009), Schmieschek and Harting (2011)

For Earth materials (rocks, soil) the situation is more complex, because REV also have an upper limit. In heterogeneous formations, after a certain size the increasing REV will inevitable include new materials and structures coming from different lithologies. This is beautifully explained in (Fu et al. 2021):

> In order to establish a given porous medium's properties, we are going to have to measure samples of the porous medium. If the sample is too small, the readings tend to oscillate. As we increase the sample size, the oscillations begin to dampen out. Eventually the sample size will become large enough that we begin to get consistent readings. This sample size is referred to as the representative elementary volume. If we continue to increase our sample size, measurement will remain stable until the sample size gets large enough that we begin to include other hydrostratigraphic layers.

REV and its determination have an enormous literature (Bear 1972; Bear and Bachmat 1990; Koestel et al. 2020 and references therein). It has been found that a REV large enough for *porosity* estimation is usually sufficient to determine *tortuosity*, but the linear size of the REV required for *permeability simulation* is an order of magnitude larger (Al-Raoush et al. 2019; Marafini et al. 2020).

White et al. (2006) determined the REV for a synthetic example, and for a *Castlegate sandstone sample*, which is an analogue reservoir rock, fine- to medium-grained, ~0.2 mm grain size, from the Castlegate Formation, Cretaceous Mesa Verde Group, E. Utah (DiGiovanni et al. 2000). Their method was based on monitoring the *kinetic energies* of particles generated by LB simulations on the micron-scale 3D images of the sample. The basic idea had been that REV and the onset of continuum scale can be determined by observing the fluctuations in the *local rate of mechanical dissipation*

$$\mathcal{D}(\boldsymbol{x}) = (2\mu)\varepsilon_{ij}\varepsilon_{ij}^{*}; \quad \varepsilon_{ij} = \frac{1}{2}\left(v_{j,i} + v_{i,j}\right), \tag{11.2.23a,b}$$

where ε_{ij} is the symmetric gradient of the velocity field. As it was shown by (Pilotti et al. 2002) that the evolution of *hydraulic head* is driven by *energy dissipation*, it is reasonable to assume that dissipation is also linked to permeability, and at the REV size where it becomes a *continuous variable*, permeability would similarly behave. White et al. (*op. cit.*) used BL simulation in the D2Q9 model, following the technique of (Zou and He 1997). The pressure drop was kept small to remain in the low Reynold's number regime where Darcy's law applies. The iterations stopped when the rate of change of the velocity field stabilized with a tolerance 10^{-7}. In case of the synthetic example the linear REV size was $\sim 1/18$-th of the full size, for the *Castlegate sandstone*, about one-third of it.

In highly heterogeneous geomaterials (such as *Middle East carbonates*) one has to carefully *preselect* those regions which are sufficiently homogeneous to be used for further, higher resolution scanning, and subsequent BLM processing so that they should yield a meaningful permeability estimate. In a recent study, Khodja et al. (2020) recommend a "*measure of representativeness*" $\mathcal{R}$ to quantify any given part of the rock image as for its relevancy for *BL-based permeability estimation*. They observed that for carbonates there is an empirically well-established correlation (Rezaee et al. 2006) between *permeability* κ, porosity Φ, and *pore-throat radius* r, of the form

$$\kappa \propto \Phi^{\alpha} r^{\beta} \tag{11.2.24}$$

with some constants α, β, so that they assumed the accuracy of κ is related to the accuracy of porosity, and of pore throat size. They defined this measure as

$$\mathcal{R} = \left(\langle(\delta\Phi)^2\rangle + \langle(\delta \mathrm{r})^2\rangle\right)^{1/2}, \tag{11.2.25a}$$

with

$$\delta\Phi = \frac{\Phi - \Phi_{\exp}}{\Phi_{\exp}}; \tag{11.2.25b}$$

$$\delta \mathrm{r} = \frac{\mathrm{r} - \mathrm{r}_{MICP}}{\mathrm{r}_{MICP}}, \tag{11.2.25c}$$

where $\Phi_{\exp}$ is experimentally determined porosity, r_{MICP} is average throat radius found from *Mercury-Injection Capillary-Pressure* (MICP) measurements. Khodja and his group claim that "the best sites to perform the high-resolution scan and the permeability simulation in a core sample are those for which $\mathcal{R} < b$, where b is an upper bound: for instance, $b = 1$ (*op.cit.*: Eq. 10)".

(ii) *A Gray-Scale LBM for Multiphase Flow*

Traditionally the LB method has been applied in a binary, black-and-white scenario, with flow allowed in the pores but prohibited in the solid grains. In many cases, however, under pressure there is a partial flow through the semi-porous grains, and different *gray-scale LB methods* have been developed to simulate *single-phase flow* in such materials (Balasubramaniam et al. 1987; Gao and Sharma 1994; Dardis and McCloskey 1998; Walsh et al. 2009; Zhu and Ma 2013; Li et al. 2014; Pereira 2015a; Ginzburg 2016; Chen et al. 2018). In these models voxels are no longer completely void or completely solid, but somewhere in between. We allow a certain degree of flow in each voxel, by introducing a partial bounce-back rule according to which only a certain fraction of fluid packets is bounced back. We denote the fraction of fluid bounced back at a voxel by n_s, $0 \leq n_s \leq 1$, where each voxel can have a different n_s. The numbers n_s characterize the voxels' *resistance to flow*, as they are between 0 and 1, they can be considered a kind of *gray scale*, in lieau of the traditional binary 0/1 or *black/white* coding. (The use of colors for this purpose, the so-called *color gradient method*, is another possibility: Rothman and Keller 1988; Gunstensen and Rothman 1991, 1992; Gunstensen et al. 1991; Latva-Kokko and Rothman 2005; Tölke et al. 2006; Reis and Phillips 2007; Ahrenholz et al. 2008; Liu et al. 2012; Fauzi et al. 2017). Pereira (2015a) further developed the *Gray Lattice Boltzmannn Method* (GLBM) to simulate multiphase flow (e.g., gas and liquid, or water and oil) in semi-porous media.

To arrive at Pereira's proposed technique we start out from the classic (black-and-white voxels, single fluid) BL algorithm realized with the D3Q19 model, where apart from the zero vector ("stay where you are") there are 18 possible directions of movement, along the vectors $(\pm 1, 0, 0), (0, \pm 1, 0), (0, 0, \pm 1), (\pm 1, \pm 1, 0), (\pm 1, 0, \pm 1), (0, \pm 1, \pm 1)$ (see Fig. 11.1, left). Denote by $f(\boldsymbol{r}, \boldsymbol{u}, t)$ the distribution function of particles which arrive at a lattice node $\boldsymbol{r}$ with velocity $\boldsymbol{u}$ at time t. At every node $\boldsymbol{r}$ and for every velocity $\boldsymbol{e}_i$ we must solve the following equation for the distribution fumction;

$$f_i(\boldsymbol{r} + \boldsymbol{e}_i \Delta t, t + \Delta t) - f_i(\boldsymbol{r}, t) = -\frac{1}{\tau}[f_i(\boldsymbol{r}, t) - f_i^{eq}(\boldsymbol{r}, t)] \tag{11.2.26}$$

compare with (11.2.14).

Here f_i^{eq} is the *equilibrium (Maxwell-) distribution*:

$$f_i^{eq} = w_i \rho \left[1 + \frac{\boldsymbol{e}_i \boldsymbol{u}^{eq}}{c_s^2} + \frac{(\boldsymbol{e}_i \boldsymbol{u}^{eq})^2}{2c_s^4} - \frac{\boldsymbol{u}^{eq} \boldsymbol{u}^{eq}}{2c_s^2} \right]; \tag{11.2.27a}$$

where w_i are the weights of the D3Q19 model

$$w_0 = \frac{1}{3};\ w_1 = \cdots = w_6 = \frac{1}{18};\ w_7 = \cdots = w_{18} = \frac{1}{36} \tag{11.2.27b}$$

see Abas et al. (2016: Table 1);

$$\rho = \sum_i f_i; \tag{11.2.27c}$$

$$c_s = 1/\sqrt{3} \tag{11.2.27d}$$

is sound speed. The *relaxation time* τ is related to *kinematic viscosity* as

$$\nu = c_s^2(\tau - 1/2), \tag{11.2.27e}$$

the *macroscopic velocity* is defined by

$$\rho \boldsymbol{u} = \sum_i f_i \boldsymbol{e}_i. \tag{11.2.27f}$$

Pressure is given by the *Equation of State*

$$P = c_s^2 \rho. \tag{11.2.27g}$$

If a force $\boldsymbol{F}$ is acting on the system, a *forcing term* should be added to to the LB Eq. (11.2.26), this term is defined by (He et al. 1998) as

$$\mathfrak{J}_i = \frac{\boldsymbol{F}(\boldsymbol{e}_i - \boldsymbol{u}^{eq})}{\rho c_s^2} f_i^{eq}. \tag{11.2.28a}$$

The macroscopic velocities should be modified in this case as

$$\rho \boldsymbol{u} = \sum_i f_i \boldsymbol{e}_i + \frac{1}{2}\boldsymbol{F}, \tag{11.2.28b}$$

and

$$\boldsymbol{u}^{eq} = \boldsymbol{u}. \tag{11.2.28c}$$

Classic (black/white) LB solves Eq. (11.2.26) only on void nodes, there is no need to solve it on solid nodes. When the streamed particle arrives at the void/solid boundary, it is bounced back (no-slip condition), that is its direction is reversed. In Gray-Scale LB, the equation is solved at all nodes, and from any given node only a certain fraction of the arriving fluid packet is bounced back. Thus, instead of the governing Eq. (11.2.26) we shall have a slightly modified equation, the *main equation of GLBM* (due to Walsh et al. 2009):

$$\begin{aligned} f_i(\boldsymbol{r} + \boldsymbol{e}_i \Delta t, t + \Delta t) = {} & (1 - n_s) f_i(\boldsymbol{r}, t) - \frac{1}{\tau}(1 - n_s)\left[f_i(\boldsymbol{r}, t) - f_i^{eq}(\boldsymbol{r}, t)\right] \\ & + (1 - n_s)\mathfrak{J}_\mathrm{i} + \mathrm{n_s} \\ & \hat{f}_i(\boldsymbol{r}, t), \end{aligned} \tag{11.2.29a}$$

where the circumflex on $\hat{f}_i(\boldsymbol{r}, t)$ signifies the distribution function should be added in the opposite direction to $\boldsymbol{e}_i$. The macroscopic velocity becomes

$$\rho \boldsymbol{u} = (1 - n_s)\left(\sum_i f_i \boldsymbol{e}_i + \frac{1}{2}\boldsymbol{F}\right). \tag{11.2.29b}$$

Equation (11.2.29a) has been validated both as a numerical benchmark exercise (for the analytically solvable *Darcy-Brinkman equation*, Pereira 2015a), and experimentally, on real samples (Li et al. 2014). The *Darcy-Brinkman equation* (Brinkman 1949; Pereira 2015a, b; Ehlers 2022),

$$\nabla^2 u - \left(\frac{\varphi}{\nu_B}\right) u = \frac{1}{\rho \nu_B} G, \tag{11.2.30}$$

with no-slip conditions at the walls, where φ is *resistance to flow*, ν_B is the *Brinkman viscosity*, G is body force, describes laminar flow through a channel of width H which has a *drag* (non-zero *resistance to flow*). Its solution is known:

$$u(y) = -\frac{G}{\varphi\rho}\left[1 - \frac{\cosh\left[r\left(y - \frac{H}{2}\right)\right]}{\cosh\left(\frac{rH}{2}\right)}\right], \tag{11.2.31a}$$

where

$$r = \sqrt{\frac{\varphi}{\nu_b}}, \tag{11.2.31b}$$

and

$$\varphi = 2n_s. \tag{11.2.31c}$$

Pereira (2015a, b: Fig. 1) compares the analytic solutions of the Darcy-Brinkman equation with its LB simulations.

Pereira (2015a) extended *single-phase Gray-scale BLM* to accomodate *multi-phase flow*. Suppose we have n immiscible phases, each with its own distribution function $f^k(\boldsymbol{r}, \boldsymbol{u}, t)$, $k = 1, \ldots, n$. Let us solve the LB equations at node i for all distribution functions, with possibly different relaxation times, and compute for each phase the macroscopic density, viscosity and momentum flux, as before. Modeling of *immiscibility* between different phases is accomplished by introducing *nearest neighbour inter-particle potentials* (Shan and Chen 1993) to express interactions between the components, similarly to the classic Lennard–Jones (1931) interaction potentials of *Computational Chemistry* acting between near-by molecules. The interaction potential between components appears as a force $\boldsymbol{F}^k$ in the added force term in Eq. (11.2.28a). The equilibrium velocity is also re-defined, because of the interaction forces, as

$$u_k^{eq} = u' = \frac{\sum_k \tau_k^{-1} \rho_k u_k}{\sum_k \tau_k^{-1} \rho_k}. \tag{11.2.32}$$

The fluid–fluid interaction then becomes, for phase k and node $\boldsymbol{r}$:

$$\boldsymbol{F}^k(\boldsymbol{r}) = \rho_k(\boldsymbol{r}) c_s^2 \sum_{k \neq k'} g_{kk'} \sum_i w\left(|\boldsymbol{e}_i|^2\right) \times \rho_{k'}(\boldsymbol{r} + \boldsymbol{e}_i)\boldsymbol{e}_i, \tag{11.2.33a}$$

where $g_{kk'}$ is the interaction potential between dissimilar components, which is assumed to be zero for similar components. The weights w depend on the separation between the interacting nodes:

$$w(1) = \frac{1}{6}; w(2) = 1/12. \tag{11.2.33b}$$

If there is a need to involve in the calculation farther-lying nodes as well to increase numerical accuracy (Porter et al. 2012), the corresponding weights will be found in (Sbragaglia et al. 2007). The Equation of State yielding the *pressure* is

$$P = c_s^2 \sum_k \rho_k + 3 \sum_{kk'} g_{kk'} \rho_k \rho_{k'}. \tag{11.2.33c}$$

We still have to incorporate *voxel resistivities* to flow, these might be different for different phases. So, finally, the governing equation of the *multiphase gray-scale LB model* will be (instead of Eq. 11.2.29a):

$$f_i^k(\boldsymbol{r} + \boldsymbol{e}_i \Delta t, t + \Delta t) = \left(1 - n_s^k\right) f_i^k(\boldsymbol{r}, t) - \frac{1}{\tau_k}\left(1 - n_s^k\right)\left[f_i^k(\boldsymbol{r}, t) - f_i^{k,eq}(\boldsymbol{r}, t)\right]$$
$$+ \left(1 - n_s^k\right)\Im_i^k + n_s^k \hat{f}_i^k(\boldsymbol{r}, t). \tag{11.2.34}$$

The *macroscopic velocity* for each phase is defined as in Eq. (11.2.28b).

In an interesting *Gray-scale LB* simulation of *two-phase channel flow* (Pereira 2015a: Fig. 4) developing viscous fingers were observed, these are characteristic signs of *instability* of the moving front when a less viscous fluid penetrates into a more viscous one (Paterson 1981; Chen 1987).

(iii) *Modeling Knudsen flow*

Gas shales can have extremely low, tens to hundreds of nanodarcy, permeability. Their pores are two magnitude smaller than those of an ordinay sandstone, with throat diameters only 20–30 times the size of methane molecules (Bustin et al. 2008; Loucks et al. 2009; Curtis 2010). Because of the particular flow mechanism in shale gas (*slip flow*, Zhang et al. 2015: Fig. 3), *Darcy's law* or *Kozeny-Carman equation* do not apply, and the boundary conditions and collision rules of the LB apparatus must be substantially modified.

Knudsen (1934), who studied rarefied gas flows in nano-size channels, introduced a dimensionless parameter, later-called "*Knudsen number*" K_n, to quantify collisions between the gas molecules in a given geometry. He defined this number as (Knudsen 1934, Ziarani and Aguilera 2012; Zhang et al. 2015):

$$K_n = \frac{\lambda}{L_{char}}, \tag{11.2.35a}$$

where L_{char} is the characteristic size of the geometry through which the gas flows (such as *tube diameter*), λ is the *mean free* path (MFP) of the gas molecules, that is, the average distance they can travel between successive collisions. For a gas of hard sphere molecules in thermodynamic equilibrium the MFP is given by (Bird 1994; Zhang et al. 2012)

$$\lambda = \frac{1}{\sqrt{2}\pi n_g d^2}, \tag{11.2.35b}$$

where d is diameter, n_g is number density of the gas, $n_g = \delta^{-3}$, δ being *mean molecular spacing*.

The Knudsen number determines the *flow regime* (Zhang et al. 2015). For $0 < K_n \leq 10^{-3}$ we have *continuum flow*, when $10^{-3} < K_n < 10^{-1}$ we have *slip flow*, for $10^1 < K_n < \infty$ there is *free molecular flow*. In the transitional range $10^{-1} \leq K_n \leq 10^1$ the flow is described by the *Burnett equation* (Burnett 1936).

In *shale gas at reservoir conditions* (temperature ~350–450 K, pressure ~25–28 MPa), the *MFP* for methane molecules is between 1.0 and 5.0 nm. For ~50 nm diameter pores $K_n \epsilon \left(10^{-3}, 10^{-2}\right)$, which is *slip flow* regime.

Following (Roy et al. 2003; Javadpour 2009; Zhang et al. 2012, 2015) let us compare the transferred mass fluxes and absolute permeabilities for (*a*) continuum flow, (b) slip flow. The Navier–Stokes equation simplifies to

$$\frac{d^2 v}{dr^2} + \frac{1}{r}\frac{dv}{dr} = -\frac{\Delta P}{\mu L}, \tag{11.2.36a}$$

where v is flow velocity per unit area, m/s; ΔP, Pa, is pressure difference between outlet and inlet; μ is viscosity, Pa s; $r \in (-R_0, R_0)$ is the radial coordinate, m; L is tube length, m. The solution of (11.2.36a) is a parabolic velocity profile

$$v = C_0 - \frac{\Delta P}{4\mu L} r^2, \tag{11.2.36b}$$

the constant C_0 depends on the boundary conditions.

For *micro-scale, continuous flow* the mass flux is (Roy et al. 2003; Javadpour et al. 2009)

$$J_{no\_slip} = -\frac{\rho R_0^2}{8\mu} \cdot \frac{\Delta P}{L}, \tag{11.2.36c}$$

which, by *Darcy's law*, corresponds to an *absolute permeability*

$$K_{no\_slip} = \frac{R_0^2}{8}. \tag{11.2.36d}$$

For *nano-flow* the mass flux can be expressed as

$$J_{slip} = -F \cdot \frac{\rho R_0^2}{8\mu} \cdot \frac{\Delta P}{L}, \tag{11.2.37a}$$

which means an *absolute permeability*

$$K_{no\_\ Slip} = F \cdot \frac{R_0^2}{8}. \tag{11.2.37b}$$

The correction factor is:

$$F = 1 + \sqrt{\frac{8RT}{M}} \times \frac{\mu}{R_0 P_{avg}} \times \left(\frac{2}{\alpha} - 1\right), \tag{11.2.37c}$$

Here, $R = 8.314$ J/mol K is the *gas constant*; T is temperature, K; M is the *molar mass* of the gas, g/mol; $P_{avg} = \frac{1}{2}(P_{inlet} + P_{outlet})$ is average pressure; $\alpha \in (0.65, 0.85)$ is an empirical constant, depending on gas type, smoothness of the wall, etc. Note in Eq. (11.2.37c) that the smaller the tube radius, the larger will be F. Because in the *elastic hard-spheres gas model* viscosity grows with temperature as.

$$\mu = 1.016 \cdot \frac{5}{16\sigma^2} \cdot \left(\frac{k_B m T}{\pi}\right)^{1/2} \tag{11.2.37d}$$

where m is mass; σ diameter, k_B is Boltzmann's constant, the correction factor F, which is *always greater than one, increases with T*.

As Zhang (2010) wrote in his important review "To capture the gas slip at the solid boundary, single phase LB model is widely used, and the slip boundary condition is adopted to implicitly consider the solid–fluid interaction. Tremendous efforts have been devoted to develop accurate and efficient boundary schemes for the LBM." (Some milestones of this exciting research are: Cornubert et al. 1991; Ansumali and Karlin 2002; Succi 2002; Li and Kwok 2003; Shan et al. 2006; Zheng et al. 2006; Ansumali et al. 2007; Kim et al. 2008; Zhang 2010; Chen et al. 2015; Ning et al. 2015; Wang et al. 2016).

The natural candidates to describe slip-flow of the gas at the walls of narrow passages are the classic LB *boundary conditions* describing gas/wall elastic interaction, namely the *bounce-back* rule, the *specular reflection*, and the *Maxwellian diffuse reflection*, or their modifications and suitable combinations.

The *bounce-back condition* prescribes that a particle colliding with the wall is bounced back in the reverse direction,

$$f_i^{bb}(x, t + \delta t) = f_i^{\leftarrow}(x, t), \tag{11.2.38}$$

where $f_i^{\leftarrow}$ denotes the opposite distribution function to f_i with the direction of $\overrightarrow{e_i}$ reversed. Nie et al. (2002) was the first to appy the *bb* condition to simulate slip flow.

Specular reflection mimics elastic collisions between light particles and an immovable, heavy boundary. In this collision only that velocity component of the wall gets reversed which is orthogonal to the wall. Denote this mirrored image of $\overrightarrow{e_i}$ as $\left(\overrightarrow{e_i}\right)^{\perp\leftarrow}$, in *specular reflection* the distribution function changes as

$$f_i^{sr}(x) = f_i^{\perp\leftarrow}(x), \tag{11.2.39}$$

that is only the orthogonal component of the incident vector $\overrightarrow{e_i}$ is reversed. Lim et al. (2002) applied *specular reflection boundary condition* in their study of *pressure driven* and *shear driven* micro-channel flows.

The Maxwellian *diffuse reflection boundary condition* assumes that the impinging particles “forget” the direction of their previous movements and after hitting the wall they are scattered back according to the Maxwell velocity distribution

$$f_i^{dr} = \frac{\sum_{(c_k - u_w)\cdot n < 0} |(\boldsymbol{c}_k - \boldsymbol{u}_w) \cdot \boldsymbol{n}| \overleftarrow{f}_k(x)}{\sum_{(c_k - u_w)\cdot n < 0} |(\boldsymbol{c}_k - \boldsymbol{u}_w) \cdot \boldsymbol{n}| f_k^{eq}(\rho_w, u_w)} \cdot f_k^{eq}(\rho_w, u_w),$$
$$\text{with } (\boldsymbol{c}_k - \boldsymbol{u}_w) \cdot \boldsymbol{n} > 0. \tag{11.2.40}$$

(Ansumali and Karlin 2002: Eqs. 10 and 14). Here $\boldsymbol{n}$ is the inward unit normal vector of the wall, ρ_w, u_w are the density and velocity of the wall. The condition $(\boldsymbol{c}_k - \boldsymbol{u}_w) \cdot \boldsymbol{n} > 0$ assures that all incidental distributions are summed up and then re-distributed over the reflected distributions. Ansumali and Karlin (*op. cit.*) implemented this boundary condition in the D2Q9 LB model and numerically tested it, for the limit $K_n \to 0$, on the *Kramers problem*, for which the analytic solution of the Boltzman equation is known (Cercignani 1975).

These three boundary conditions are usually implemented in LBM in one of the following linear combinations (Zhang et al. 2016a, b):

$$\left.\begin{aligned} f_i(x) &= (1-\sigma) f_i^{sr}(x) + \sigma f_i^{dr}(x) \\ f_i(x) &= (1-r) f_i^{sr}(x) + r f_i^{bb}(x) \\ f_i(x) &= (1-\chi) f_i^{bb}(x) + \chi f_i^{dr}(x) \end{aligned}\right\} \tag{11.2.41a,b,c}$$

Zhang et al. (*op. cit.*: Eqs. 32–34) give application examples for each of them, discuss their relative merits and shortcomings, and recommend values of σ, r, and χ as functions of the *Knudsen number* K_n.

To optimize the performance of these schemes, the relaxation time τ should also be made somehow related to K_n. Nie et al. (2002), who worked on the nano-flow problem between two plates a small distance H apart, proposed a linear relation

$$\tau = H \cdot \frac{K_n}{a} + 0.5. \tag{11.2.42}$$

Here the empirical parameter a is problem-dependent, so that Eq. (11.2.42) is not generally applicable. This is still an active area of research, summarized in Wang et al. (2016: Table 2).

11.3 Renormalization Group (RNG) Models of Phase Transitions

The concept of *Renormalization Group* (RNG) had originally emerged from an effort to control the divergences in *Quantum Field Theory* (Stueckelberg and Peterman 1953; Bogolyubov and Shirkov 1960). Later, the method has proved to be an excellent computational tool in the *Theory of Phase Transitions* (Kadanoff 1966; Kadanoff et al. 1967; Wilson 1971a, b, 1975, 1983; Wilson and Kogut 1974; Goldenfeld 1992; Plascak et al. 1999; Guo et al. 2021). Wilson was awarded the Nobel Prize in Physics in 1982 for his RNG studies of second-order phase transitions. Today, the method is widely applied in diverse fields of science, including Earth Sciences, see Table 11.5. The method has been well treated in technical literature, some excellent reviews, tutorials and special monographs are also referred to in this Table.

In the strict mathematical sense, RNG is not a *group*, but a *semigroup* because its inverse elements cannot be defined (Goldenfeld 1992). To understand this *semi-group property*, consider a typical rock-physical application of RNG to a large-size geomaterial which is near to its point of phase transition, that is self-similar without a characteristic scale (Song et al. 2020). Make a measurement of a certain physical property T at scale l_1, denote the result by T_1. When the scale is increased to l_2, instead of T_1 we shall get some other value T_2, related to T_1 as

$$T_2 = f_1(T_1), \tag{11.3.1}$$

and for an increasing sequence of scales $l_1 < l_2 < \cdots < l_n < \cdots$ we shall similarly find

$$T_{n+1} = f_n(T_n), \quad n = 1, 2, 3, \ldots. \tag{11.3.2}$$

Denote $f_k \circ f_l(T) \equiv f_k(f_l(T))$ and define by

Table 11.5 Some applications of the RNG (renormalization group) method

Application	Authors
Acoustic emissions	Anifrani et al. (1995), Lei et al. (2005)
Conduction	Kirkpatrick (1973), Stinchcombe and Watson (1976), Madden (1983), Karim and Krabbenhoft (2010)
Earthquakes	Allègre et al. (1982, 1995), Olami et al. (1992), Newman et al. (1994, 1995), Sornette and Sammis (1995), Saleur et al. (1996a, b), Borodich (1997)
Flow in fractured rock	Piggott and Elsworth (1992)
Flow in porous media	Özer et al. (2019)
Fracking	Norris et al. (2015)
Fracture	Chelidze (1982)
Fragmentation	Turcotte (1986), Peng et al. (2009)
General literature	Wilson and Kogut (1974), Toulouse and Pfeuty (1975), Wallace and Zia (1978), Wilson (1979), Hu (1982), Amit (1984), Herrmann and Roux (1990), Goldenfeld (1992), Stauffer and Aharony (1992), Barabási and Stanley (1995), Plascak et al. (1999), Sornette (2000), Hunt and Ewing (2009), Guo et al. (2021)
Instability of rock slopes	Chen et al. (2018)
Landscape erosion	Sornette and Zhang (1993), Pastor-Satorras and Rothman (1998a,b), Dodds and Rothman (2000), Antonov and Kakin (2017a, b), Duclut and Delamotte (2017), Antonov et al. (2020)
Microcrack connectivity	Madden (1983), Gomez et al. (1995), Zhao et al. (2016)
Microseismicity	Norris et al. (2015)
Percolation	Shante and Kirkpatrick (1971), Kirkpatrick (1973), Harris et al. (1975), Young and Stinchcombe (1975), Stinchcombe and Watson (1976), Reynolds et al. (1977), Chelidze (1982), Korvin (1992b), Levitan (2022)
Permeability	King (1989), Korvin (1992b), Green and Paterson (2007), Fauzi (2011), Khalili et al. (2013), Hanasoge et al. (2017), Irayani et al. (2019), Wei et al. (2019), Adeyemi et al. (2021)
Radon emission	Sun et al. (2016)
Rock damage	Madden (1983), Anifrani et al. (1995), Lei et al. (2005), Peng et al. (2009), Qin and Pan (2011), Xue et al. (2014), Zhou et al. (2015), Sun et al. (2016), Pan et al. (2018), Pan and Lü (2018), Jiang et al. (2019), Song et al. (2020), Xia et al. (2021)
Seismic waves	Jakobsen and Wu (2016)
Sierpinski gaskets	Gefen et al. (1984), Zhen-quan et al. (1997)
Size distribution	Madden and Williams (1993)
Stick–slip faults	Smalley et al. (1985)

$$f_0(T) \equiv T \text{ for all } T, \tag{11.3.3a, b}$$

then we have the *semigroup property*

$$f_k \circ f_l \equiv f_{k+l}, k, l = 0, 1, 2, \ldots. \tag{11.3.4}$$

The basic assumption of renormalization theory is that for $n \to \infty$ we have $T_n \to T_c$ ("c" stands for "*critical*"), $f_n \to f = f_\infty$, and in the limit, instead of (11.3.2), we shall have

$$T_n \approx T_c = f(T_c) \approx f_n(T_n) \tag{11.3.5}$$

that is T_c is a *fixed point* of the limiting transformation.

Close to the fixed point, the iterative calculation $T_{n+1} = f_n(T_n)$ can behave in two different ways: (a) the calculated value gets farther from the fixed point after each iteration (case of an *unstable fixed point*); or (b) after each iteration it gets closer to it (*stable fixed point*). Stable fixed points are characterized by the property

$$\left| \frac{dT_{n+1}}{dT_n} \right| < 1 \tag{11.3.6}$$

(Atkinson 1978; Piggott and Elsworth 1992).

The simplest applications of RNGs to Rock Physics are related to problems of the sudden breaking through the rock of a hydraulic or electric current (Korvin 1992a, b, 2021; Hunt 2004). Such phenomena are treated in *Percolation Theory*, invented by S. R. Broadbent (Broadbent 1954; Broadbent and Hammersley 1957) who worked on the design of gas masks for use in coal mines. The masks contained porous carbon granules into which the gas could penetrate. Broadbent found that if the pores were large enough and sufficiently *well connected*, the gas could permeate the interior of the granules; but if the pores were too small or *poorly connected*, the gas would not get beyond the granules' surface. There was a *critical porosity* and *pore interconnectedness*, above which the mask worked well and below which it was ineffective.

Thresholds of this sort are typical of *percolation processes*. In the *bond-percolation problem* we assume that a fraction $1 - p$ $(0 < p < 1)$ of the bonds of a regular grid are randomly cut and a fraction p are left uncut. Then there exists a critical fraction p_c (called *percolation threshold*) such that there is no continuous connection along the bonds of the network between the opposite faces for $p < p_c$, and there exists a connection with probability 1 for $p > p_c$. For the 2-dimensional square lattice the percolation threshold is 0.5 (Sykes and Essam 1964). In the more general case the percolation threshold depends on the dimensionality of the network, d, and on its *coordination number* Z (the average number of bonds connected to any node of the network), but *it is independent of the detailed structure of the network.* Table 11.6 lists coordination numbers and percolation thresholds for some common networks. In d-dimensions, the percolation thresholds and coordination numbers

Table 11.6 Regular Lattices with their bond-percolation probabilities (From Korvin 1992a, b, 2021)

Lattice	Dimension	Coordination number Z	p_c
Honeycomb	2	3	0.6527
Square	2	4	0.5
Triangular	2	6	0.3473
Tetrahedral (diamond)	2	4	0.39
Simple cubic	3	6	0.25
Body centered cubic	3	8	0.18
Face centered cubic	3	12	0.12
Hexagonal close packing	3	12	0.12

conform closely to the empirical rule (Vyssotsky et al. 1961)

$$Zp_c = d/(d-1). \tag{11.3.7}$$

Close to the percolation threshold ($p > p_c$) those nodes which are connected with each other by continuous paths form large clusters of average size ξ called the *correlation length*. The correlation length diverges for $p \to p_c, p > p_c$ as

$$p \to p_c, p > p_c \quad \text{as } \xi \propto (p - p_c)^{-\nu}. \tag{11.3.8}$$

For 3-dimensional networks $\nu = 0.83$, independently of their coordination number (Fisch and Harris 1978).

11.3.1 Renormalization Group (RNG) Model of Rock Failure

In my 1992 book on *Fractal Models* (Korvin 1992a: 21–29) I explained in details how to use RNG for finding the crititical percolation probability p_c, and discussed the scaling arguments that yield the exponent ν in Eq. (11.3.8). Rather than repeating this, let us address the analog question of the *critical number of cracks per unit volume* that makes an originally solid rock to fall to pieces. In the *RNG* approach, to study the critical behavior of a large system, we first consider its smaller subsystems, find their behavior, and then we put them together and express in terms of them the behavior of a larger-scale subsystem. Suppose that each of the nth order subsystems can be in one of the states $S_1, S_2, \ldots, S_K$ with probabilities $p_1^{(n)}, p_2^{(n)}, \ldots, p_K^{(n)}$ $\left(\sum_{i=1}^{K} p_i^{(n)} = 1\right)$. Similarly, any ($n$ + *1*)st-order subsystem belongs to state S_i with probability $p_i^{(n+1)}$, $\left(\sum_{i=1}^{K} p_i^{(n+1)} = 1\right)$. The distributions $\left\{p_i^{(n+1)}\right\}$ and $\left\{p_i^{(n)}\right\}$ are related by a system of K nonlinear equations

$$\begin{aligned} p_1^{(n+1)} &= F_1\left[p_1^{(n)}, p_2^{(n)}, \ldots, p_K^{(n)}\right] \\ \cdots \quad &\cdots \qquad \cdots \\ p_K^{(n+1)} &= F_K\left[p_1^{(n)}, p_2^{(n)}, \ldots, p_K^{(n)}\right] \end{aligned} \tag{11.3.9}$$

where the functions F_i depend on the physical properties of the states, and on the rules prescribing how do we put together the $(n+1)$st-order subsystems from the nth order ones (the rules of "*coarse-graining*"). The *critical probabilities* are solutions of Eq. (11.3.9) in the limit $n \to \infty$:

$$\begin{aligned} p_1^{(crit)} &= F_1\left[p_1^{(crit)}, p_2^{(crit)}, \ldots, p_K^{(crit)}\right] \\ \cdots \quad &\cdots \qquad \cdots \\ p_K^{(crit)} &= F_K\left[p_1^{(crit)}, p_2^{(crit)}, \ldots, p_K^{(crit)}\right] \end{aligned} \tag{11.3.10}$$

In the case of *rock damage* (Allègre et al. 1982) there are only two states (*fragile/sound*) with respective probabilities $p^{(n)}, q^{(n)}; p^{(n)} + q^{(n)} = 1$ in the nth step of coarse-graining, that is Eqs. (11.3.9–11.3.10) become and, as

$$p_i^{(n+1)} = F_i\left[p^{(n)}\right] \tag{11.3.11}$$

and, as $n \to \infty$,

$$p_{c,i} = F_i(p_c) \quad i = 1, 2, \ldots, K. \tag{11.3.12}$$

In the model of Allègre et al. (*op. cit.*: Fig. 2a), *coarse-graining* is based on the observation, that *rock failure* is consequence of *fractures*, a fracture is consequence of *microfissures*, a microfissure is consequence of *microcracks*. Each block in the nth step can be *fragile* or *sound* with probabilities $p^{(n)}$ and $1 - p^{(n)}$. Define a cubic cell "*fragile*" if it contains no continuous "*pillar*" connecting any two of its opposite faces. Otherwise, the cell is "*sound*". The probability that a configuration has m fragile corners ($0 \le m \le 8$) and topological multiplicity μ is $\mu \cdot p_{n-1}^m (1 - p_{n-1})^{8-m}$, summing up the probabilities of all possible fragile configurations and letting $n \to \infty$ we find that the critical probability satisfies the algebraic equation

$$p = p^4\left(3p^4 - 8p^3 + 4p^2 + 2\right) \tag{11.3.13}$$

whose non-trivial real solution is $p_c = 0.896$. We note that a different definition (Turcotte 1986) of "fragility" would lead to another equation

$$p = 3p^8 - 32p^7 + 88p^6 - 96p^5 + 86p^4 \tag{11.3.14}$$

whose non-trivial solution yields quite a different (and possibly more realistic!) $p_c = 0.49$.

In this respect, it is instructive to read a recent case-history of Xia et al. (2021) who applied RNG to characterize the *water-induced deterioration process in red-bed soft rocks*. (The red-bed rocks of S. China are composed of mudstone, sandstone, siltstone and shale. They are rich in clay minerals, e.g., montmorillonite and kaolinite, which are sensitive to hydration-induced damage: Zhou et al. 2019; Liu et al. 2021). The authors emphasized that "to apply renormalization group theory in red-bed soft rocks, firstly their microstructures should be analyzed". As the SEM images of 10,000 × magnification (*op. cit.*: Figs. 6.3.3b and d) indicated that the two main types of rock, *light-yellow silty mudstone* and *light-yellow silty mudstone* have different interior structures, Xia et al. applied different RNG models (*op. cit.*: Figs. 4–7). to these rocks, commensurate with their microscopic properties. They defined the *strength reduction coefficient* C of *the red-bed soft rock* as

$$C = (A - B)/A \tag{11.3.15}$$

where A is the *initial stren*gth of the rock; B is the *strength of the rock after saturation*. They assumed, that

$$C = p^*, \tag{11.3.16}$$

where p^* is the *critical failure probability* in the RNG process.

In the RNG flow for *light-yellow silty mudstones* summing up the probabilities yields the recursion

$$p_n = 2p_{n-1}^2(1 - p_{n-1})^2 + 4p_{n-1}^3(1 - p_{n-1}) + p_{n-1}^4, \tag{11.3.17a}$$

letting $n \to \infty$, the critical probability $p_n \approx p^*$ satisfies

$$p_n = 2p_n^2(1 - p_n)^2 + 4p_n^3(1 - p_n) + p_n^4. \tag{11.3.17b}$$

The equation has four roots: 0; 1; −1.6; 0.6, the physically relevant one is $p^* = 0.6$. Similarly, in case of *light-grey silty mudstone*, the limiting equation for the RNG flow becomes

$$p_n = 4p_n^3(1 - p_n) + p_n^4, \tag{11.3.18}$$

which has four roots, (0; 1; −0.4; 0.8), the physical one being $p^* = 0.8$.

11.3.2 *Renormalization Group (RNG) for Failure Time Prediction*

Guo et al. (2021) summarize the three main steps of the RNG method as: (1) Use the renormalization transformation to obtain a *coarse–grained operator* by changing the length scale. This reduces the number of degrees of freedom determining the state variable; (2) Induce the transformation of the model parameters to obtain the *renormalization flow*; (3) Find and analyze the *fixed points* of this transformation, and the behavior of the system near the fixed points.

It is by no means necessary, that the renormalization transformation should be built up on an underlying lattice, as in the previous examples of rock failure. We can easily identify these steps in the *continuous* RNG theory of earthquakes of (Sornette and Sammis 1995; Saleur et al. 1996a, b). Denote by t_f ("time of failure") the time of the impending large quake, introduce the variable

$$x = t_f - t, x \geq 0, \tag{11.3.19}$$

and denote by $F(x)$ a relevant parameter which is singular at $x = 0$. The most general RNG transform for the problem is

$$x' = \phi(x), \tag{11.3.20}$$

and

$$F(x) = g(x) + \frac{1}{\mu} F(\phi(x). \tag{11.3.21}$$

Here $\phi(x)$ is the *renormalization flow map*, μ describes the scaling of F as an effect of the time rescaling, $g(x)$ is the *non-singular part* of $F(x)$. If $F(x)$ is continuous, $\phi(x)$ is differentiable, then the critical points are those time values x for which some finite-order derivative is divergent,

$$\frac{d^k F(x)}{dx^k} = \infty. \tag{11.3.21}$$

Near such a singularity $F(x)$ is a power-law whose exponent provides the scaling law. For a formal solution of (11.3.21), define iteratively

$$f_0(x) \equiv g(x); \tag{11.3.22a}$$

$$f_{n+1} = g(x) + \frac{1}{\mu} f_n(\phi(x)), n = 0, 1, 2 \dots. \tag{11.3.22b}$$

By induction,

$$f_n(x) = \sum_{i=0}^{n} \frac{1}{\mu^i} g\big(\phi^{(i)}(x)\big) \text{ for } n > 0, \tag{11.3.22c}$$

where the superscripts mean composition

$$\phi^{(i+1)}(x) = \phi(\phi^{(i)}(x)). \tag{11.3.22d}$$

Letting $n \to \infty$,

$$\lim_{n\to\infty} f_n(x) = F(x), \tag{11.3.22e}$$

that is

$$F(x) = \sum_{i=0}^{\infty} \frac{1}{\mu^i} g\big(\phi^{(i)}(x)\big). \tag{11.3.23}$$

The behavior of the critical points and the related critical exponents can be found from the form of the RNG flow $\phi(x)$, because the critical points correspond to the unstable fixed points of the flow $\phi(x)$. Consider a fixed point

$$\lambda = F(x) = \left.\frac{d\phi}{dx}\right|_{x=\phi(x)}. \tag{11.3.24}$$

If $|\lambda| > 1$, the i-th term in the series for the k-th derivative of $F(x)$ growds with k as $\propto \left(\frac{\lambda^k}{\mu}\right)^i$ so that it would cause divergence for a sufficiently large k.

If $x = 0$ is a fixed point of the flow, that is $\phi(0) = 0$, and $\phi(x) \approx \lambda x + \mathcal{O}(x^2)$, then the solution of Eq. (11.3.21) near $x = 0$ is

$$F(x) \sim x^m \tag{11.3.25a}$$

where

$$\frac{\lambda^m}{\mu} = 1, \tag{11.3.25b}$$

that is the *real-valued critical exponent* is

$$m = \frac{\log \mu}{\log \lambda}. \tag{11.3.25c}$$

11.3.2.1 Complex Scaling Exponents

The most important contribution of Sornette and co-workers (Sornette and Sammis 1995; Saleur et al. 1996a, b) to RNG theory had been the recognition that in *Complex Analysis* for any integer n the Eq. (11.3.25b) has *infinitely many complex solutions.* Indeed, using the identity $1 = e^{i \cdot 2\pi n}$, the equation

$$\frac{\lambda^m}{\mu} = e^{i \cdot 2\pi m} \tag{11.3.26a}$$

yields a series of *complex exponents*

$$\{m_n = m + in\Omega\}, n = 0, 1, 2, \ldots \tag{11.3.26b}$$

where

$$\Omega = \frac{2\pi}{\log \lambda}. \tag{11.3.26c}$$

As for any real x we have

$$Re\left(x^{m+in\Omega}\right) = x^m \cdot \cos(n\Omega \cdot \log x), \tag{11.3.26d}$$

"*the complex critical exponents describe log-periodic oscillatory corrections decorating the asymptotic power law scaling* $F(x) \propto x^m$" (Saleur et al. 1996a, b: 104). Saleur et al. (*op. cit.*: 104–105) made a reasonable assumption on the functional form of the singular part $g(x)$ of $F(x)$, and showed with a heuristic argument that for practical purposes only the first one or two correcting terms (i.e. with $n = 1, 2$ in Eq. 11.3.26b) need to be taken into account.

To make this plausible, let us Mellin transform $F(x)$:

$$F(x) : \hat{F}(s) = \int_0^\infty F(x) x^{s-1} dx. \tag{11.3.27a}$$

Using the linear approximation

$$\phi(x) \approx \lambda x, \tag{11.3.27b}$$

and denoting by $\hat{g}(s)$ the Mellin transform of $g(x)$, we can write

$$\hat{F}(s) = \hat{g}(s) \frac{\mu\lambda^s}{\mu\lambda^s - 1}, \tag{11.3.27c}$$

so that, by inverse Mellin transform,

$$F(x) = \int_{c-i\infty}^{c+i\infty} \hat{F}(s)x^{-s}ds. \tag{11.3.27d}$$

By Cauchy's *Theorem of Residues* (Ahlfors 1979) $F(x)$ is the sum of the terms corresponding to poles $x^{-s_{pole}}$ with residues $\frac{\hat{g}(s_{pole})}{\log\lambda}$ where the poles satisfy $\mu\lambda^{s_{pole}} = 1$ (that is, $s_{pole} = -m_n$, see Eq. 11.3.26b).

Saleur et al. (*op. cit.*) assume that $g(x)$ has the general form

$$g(x) = \log\left(\sum_{k=0}^{N} a_k x^k\right), \tag{11.3.27e}$$

where all roots $z_j = x_j + iy_j$ of the polynomial are complex, so that

$$\sum_{k=0}^{N} a_k x^k = \prod_j \left[\left(x - x_j\right)^2 + y_j^2\right]. \tag{11.3.27f}$$

The Mellin transform of $g(x)$ is a sum of the transforms of the terms

$$\log\left[\left(x - x_j\right)^2 + y_j^2\right] \sim \log\left(1 + 2x\cos t + x^2\right)$$

that is

$$\int_0^\infty x^{s-1}\log\left(1 + 2x\cos t + x^2\right)dx = \frac{2\pi}{s}\cdot\frac{\cos(st)}{\sin(\pi s)} \tag{11.3.27g}$$

for $|t| < \pi$, and $-1 < Re(s) < 0$. The amplitude of a given pole $s_{pole} = -m + in\Omega$ thus decreases with n as

$$\left|\frac{1}{m - in\Omega}\cdot\frac{\cos(m - in\Omega)t}{\sin(m - in\Omega)\pi}\right|. \tag{11.3.27h}$$

Complex scaling exponents will be further discussed in Sect. 11.4.

11.3.3 Dynamic Renormalization Group Treatment of the Erosion of Tilted Landscapes

In two landmark papers (1998a, b) Pastor-Satorras and Rothman introduced a stochastic differential equation to describe the *erosion of tilted landscapes* (1998a: Eq. 5; 1998b: Eq. 13) and proved that the inclination causes the erosion to be *anisotropic*. By an ingenious application of the *Dynamic Renormalization Group* ("DRG", an RNG operating in *Fourier space*) they determined the roughness exponents (scaling exponents of the surface height-height correlation functions) in the two main structural directions $\boldsymbol{e}_{\parallel}$ and $\boldsymbol{e}_{\perp}$ (parallel and perpendicular to the dip). Lately, their method has been followed up and generalized by (Antonov and Kakin 2017a, b; Duclut and Delamotte 2017; Antonov et al. 2020).

Denote the surface at time t and location $\boldsymbol{x}$ by $h(\boldsymbol{x}, t) = h\big(\boldsymbol{e}_{\parallel}, \boldsymbol{e}_{\perp,}t\big)$ (note that the coordinate-vector $\boldsymbol{e}_h$ is directed downwards). The roughness of the surface for fixed t can be expressed by the *height-height variogram*, defined as

$$C(\boldsymbol{r}) = \left\{ |h(\boldsymbol{x}+\boldsymbol{r}) - h(\boldsymbol{x})|^2\big|_x \right\}^{1/2}. \tag{11.3.28}$$

The surface is not *self-similar*, only *self-affine*, so $C(\boldsymbol{r})$ measured in the two main directions will have different scaling exponents:

$$\left.\begin{array}{l} C_{\parallel}\big(x_{\parallel}\big) \propto \big|x_{\parallel}\big|^{\alpha_{\parallel}} \\ C_{\perp}(x_{\perp}) \propto |x_{\perp}|^{\alpha_{\perp}} \end{array}\right\}, \tag{11.3.29a,b}$$

where in general $\alpha_{\parallel} \neq \alpha_{\perp}$. Equations (11.3.29a,b) can be summarized in a single scaling relation

$$C\big(x_{\parallel}, x_{\perp}\big) \sim b^{\alpha_{\parallel}} C\big(b^{-1}x_{\parallel}, b^{-\zeta_{\parallel}}x_{\perp}\big), \tag{11.3.30a}$$

where $\zeta_{\parallel}$ is the *anisotropy exponent*,

$$\zeta_{\parallel} = \frac{\alpha_{\parallel}}{\alpha_{\perp}}. \tag{11.3.31a}$$

Equivalently with Eq. (11.3.30a)

$$C\big(x_{\parallel}, x_{\perp}\big) \sim b^{\alpha_{\perp}} C\big(b^{-\zeta_{\perp}}x_{\parallel}, b^{-1}x_{\perp}\big), \tag{11.3.30b}$$

where

$$\zeta_{\perp} = 1/\zeta_{\parallel}. \tag{11.3.31b}$$

Pastor-Satorras and Rothman (1998a, b) proposed the following stochastic nonlinear differential equation for the spatio-temporal evolution of the surface $h\big(\boldsymbol{e}_{\parallel}, \boldsymbol{e}_{\perp,}t\big)$:

$$\frac{\partial h}{\partial t} = \nu_{\parallel}\partial_{\parallel}^2 h + \nu_{\perp}\nabla_{\perp}^2 h + \frac{\lambda}{3}\partial_{\parallel}^2\big(h^3\big) + \eta \tag{11.3.32}$$

The equation can be derived in a few heuristic steps.

(i) Assume that the changes in surface shape $\frac{\partial h}{\partial t}$ are caused by a *locally conserved material flow* (the current $\mathcal{J}$) and an additive random noise term η (rain, locally changing resistance to erosion, etc.):

$$\frac{\partial h}{\partial t} = -\nabla \cdot \mathcal{J} + \eta. \tag{11.3.33a}$$

The soil material however is not conserved *globally*, since it disappears at the bottom boundary, and it is also affected by the random term η.

(ii) We postulate that current $\mathcal{I}$ has two components, a locally isotropic *diffusion* continuously smoothing out the surface and erasing irregularities, and a global flow mainly downhill:

$$\mathcal{J} = -\nu \nabla h - \Gamma \nabla_{\parallel} h \tag{11.3.33b}$$

The first term in the RHS of (11.3.33b) describes the isotropic relaxation of the soil by *Fickian diffusion*. The second term corresponds to the flow of soil downhill in the direction $\nabla_{\parallel} h \equiv \partial_{\parallel} h \boldsymbol{e}_{\parallel}$.The factor Γ is *anisotropic diffusivity*. As the eroded material flows downhill, Γ increases with $x_{\parallel}$.

(iii) Parametrize the diffusivity Γ as

$$\Gamma = \Gamma(h). \tag{11.3.33c}$$

Define a function $g(h)$ by with

$$\Gamma(h) = \lambda_0 + g(h) \text{ with } g(0) = 0; \tag{11.3.33d}$$

and let

$$G(h) = \int g(h) dh \tag{11.3.33e}$$

be its integral.

(iv) Substituting Eq. (11.3.33b) into Eq. (11.3.33a), and using the identity (which follows from the *chain rule*)

$$g(h)\partial_{\parallel} h = \frac{dG(h)}{dh} \partial_{\parallel} h = \partial_{\parallel} G(h), \tag{11.3.33f}$$

we get

$$\frac{\partial h}{\partial t} = \nu_{\parallel} \partial_{\parallel}^2 h + \nu_{\perp} \nabla_{\perp}^2 h + \partial_{\parallel}^2 G(h) + \eta, \quad \text{where} \quad \nu_{\perp} = \nu; \; \nu_{\parallel} = \nu + \lambda_0. \tag{11.3.33g}$$

(v) Develop $\Gamma(h)$ in Taylor series in powers of h. As Eq. (11.3.33a) has the joint symmetry $\{h \to -h; \mathcal{J} \to -\mathcal{J}\}$, the series cannot contain odd powers of h. Also, if the surface is d-dimensional, the flux $Q(x_\|)$ cannot grow faster than $Q(x_\|) \propto x_\|^d \sim h^d$ so that in case of a physical surface of dimension $d \leq 3$ we can truncate the Taylor series at the quadratic term, and Eq. (11.3.33g) becomes the *master equation* (11.3.32).

Pastor-Satorras and Rothman (1998a, b) distinguished two types of noise. The *time-dependent noise* $\eta(\boldsymbol{x}, t)$ (e.g. rainfall, etc.) is isotropic, Gaussian with zero mean, its 2-point correlation function is

$$\langle \eta(\boldsymbol{x}_1, t_1)\eta(\boldsymbol{x}_2, t_2)\rangle = 2D_1\delta^{(d)}(\boldsymbol{x}_1 - \boldsymbol{x}_2)\delta(t_1 - t_2). \tag{11.3.34a}$$

We also have a *time-independent noise* (variations in the *erodibility of the landscape*: Czirók et al. 1993; Montgomery 2003), for which

$$\langle \eta(\boldsymbol{x}_1)\eta(\boldsymbol{x}_2)\rangle = 2D_2\delta^{(d)}(\boldsymbol{x}_1 - \boldsymbol{x}_2). \tag{11.3.34b}$$

In Eqs. (11.3.34a, 11.3.34b) the parameters D_1, D_2 measure the strength of noise.

We shall apply the *Dynamic Renormalization Group* (DRG) method (Ma and Mazenko 1975; Medina et al. 1989; Hwa and Kardar 1992) to the landscape model, Eq. (11.3.32), in order to find the *roughness exponents* $\alpha_\|, \alpha_\perp$ (in Eqs. 11.3.29a, b) and *anisotropy exponents* $\zeta_\|, \zeta_\perp$(in Eqs. 11.3.31a, 11.3.31b), as well as a further critical exponent $z_\perp$.

DRG operates in *Fourier space*, where first we integrate over the domain $\Lambda b^{-1} < k < \Lambda$ to filter out the fast modes, corresponding to large momentum $\boldsymbol{k}$. Here Λ is an upper cutoff, b is a rescaling factor. Then, to bring back the system to its original size, we perform *rescaling* through the transform

$$h(x_\|, x_\perp, t) = b^{\alpha_\perp} h(b^{-\zeta_\perp} x_\|, b^{-1} x_\perp, b^{-z_\perp} t) \tag{11.3.35}$$

where $z_\perp$ is the *dynamic critical exponent* describing the scaling of time. The equations of *infinitesimal rescaling* $\{b = e^{dl}; dl \to 0\}$ provide the flow equations of the DRG, and we are interested in their *stable fixed points*. To first order in the coupling constant λ, the flow equations are

$$\frac{d\nu_\|}{dl} = \nu_\|(z_\perp - 2\zeta_\perp + \overline{\lambda}) \tag{11.336a}$$

$$\frac{d\nu_\perp}{dl} = \nu_\perp(z_\perp - 2) \tag{11.3.36b}$$

$$\frac{d\lambda}{dl} = \lambda\left(z_\perp + 2\alpha_\perp - 2\zeta_\perp - \frac{3}{2}\overline{\lambda}\right) \tag{11.3.36c}$$

$$\frac{dD}{dl} = D(2z_\perp - 2\alpha_\perp - \zeta_\perp - d + 1) \tag{11.3.36d}$$

In Eqs. (11.3.36a and 11.3.36c) the term $\overline{\lambda}$ is the *effective coupling constant*, defined as

$$\overline{\lambda} = \frac{\lambda D}{2\nu_\parallel^{3/2}\nu_\perp^{3/2}}\Lambda^{d-4}K_{d-1}, \tag{11.3.36e}$$

where $\varepsilon = 4 - d$; $K_d = S_d/(2\pi)^d$, and S_d is the surface area of the *d-dimensional unit hypersphere*. Because of Eqs. (11.3.36a–11.3.36e) the flow for $\overline{\lambda}$ is described by

$$\frac{d\overline{\lambda}}{dl} = \overline{\lambda}\left(\varepsilon - 3\overline{\lambda}\right). \tag{11.3.36f}$$

Note that out of these flow equations, the last Eq. (11.3.36f) is the only one which does not contain the term $z_\perp$, the *dynamic critical exponent*. In *the statistical theories of surface growth* this exponent measures the *saturation length of correlations* as a function of time (Sornette and Zhang 1993; Barabási and Stanley 1995), its experimental determination would be meaningless because of the enormous time-scale of geomorphologic problems. So, the only practical way to find the fixed points of the flow is from $\frac{d\overline{\lambda}}{dl} = \overline{\lambda}\left(\varepsilon - 3\overline{\lambda}\right)$. For $d < 4$ the stable fixed point of the flow is $\overline{\lambda}^* = \frac{\varepsilon}{3}$, such that to first order in

$$\varepsilon : \alpha_\perp = \frac{5\varepsilon}{12};\ \zeta_\perp = 1 + \frac{\varepsilon}{6}. \tag{11.337a}$$

For 2-D surfaces ($\varepsilon = 2$):

$$\alpha_\perp = \frac{5}{6} \approx 0.83;\ \zeta_\perp = \frac{4}{3};\ \alpha_\parallel = \frac{\alpha_\perp}{\zeta_\perp} = \frac{5}{8} \approx 0.63. \tag{11.337b}$$

Pastor-Satorras and Rothman (1998a, 1998b) compared the result (11.3.37b) with the correlation analysis of the digital map of a submarine canyon off the coast of Oregon (*op cit.*: Figs. 3a, 4). The scaling exponents of the empirical ACFs, $C_\perp \propto x_\perp^{0.78}$ and $C_\parallel \propto x_\parallel^{0.67}$ correspond fairly well to the theoretical $\alpha_\perp = 0.83$; $\alpha_\parallel = 0.63$ predicted by Eq. (11.3.37b).

11.3.4 Permeability Upscaling

In an interesting paper Hanasoge et al. (2017) performed "a statistical study to characterize the relative success of upscaling methods on a large sample of permeability coefficients" and found, as the title of their paper proudly claims "Renormalization group theory outperforms other approaches in statistical comparison between upscaling techniques for porous media". (Not surprisingly, by "Renormalization Group theory" they meant their own version of it, "*Mode-Elimination Renormalization-Group theory*".)

As indicated in Table 11.6, RNG methods were intensively applied to calculate and upscale permeability and estimate its anisotropy (King 1989; Green and Paterson 2007; Fauzi 2011; Khalili et al. 2013; Hanasoge et al. 2017; Irayani et al. 2019; Wei et al. 2019; Adeyemi et al. 2021). King (1989) used it to calculate effective permeability from permeabity distributions; later it was applied to estimate macroscopic permeabilities of petroleum reservoirs (Green and Paterson 2007). Fauzi (2011) pioneered to subject digitized images of thin sections to RNG to estimate permeabilities. In later works many important refinements have been made, and new schemes were introduced, including the exciting appliations where RNG and LBM (Lattice Boltzmann) techhniques are jointly applied to compute the permeability of 3D tomographic rock samples (Khalili et al. 2013; Irayani et al. 2019).

King's 2-dimensional RNG approach to compute effective permeability (King 1989; Adeyemi et al. 2021) is based on the analogy between fluid flow in a pore network, and the flow of electric current in an electric circuit. King mapped a block of cells of different permeabilities into an equivalent resistor network in an iterative manner and finally obtained a single resistor. The effective permeability of a 2 × 2 block was, in this formalism:

$$k_{eff\,(2D)} = \frac{4(k_1+k_3)(k_2+k_4)\left[k_2k_4(k_1+k_3)+k_1k_3(k_2+k_4)\right]}{\left[k_2k_4(k_1+k_3)+k_1k_3(k_2+k_4)\right]\cdot\left[k_1+k_2+k_3+k_4\right]+3(k_1+k_2)(k_3+k_4)(k_1+k_3)(k_2+k_4)} \tag{11.3.38}$$

In 3D, *permeability renormalization* is more complex. The elementary structure now is a 2 × 2 × 2 block. Two ways of renormalization have been suggested in the literature, that of Green and Paterson (2007), and of Karim and Krabbenhoft (2010), see (Irayani et al. 2019: Figs. 1–2). The corresponding equations are as follows.

(a) *The Green – Paterson Equation* (Green and Paterson 2007; Irayani et al. 2019; Adeyemi et al. 2021):

$$\begin{aligned} k_{eff\,(3D)}(k_1,k_2,\ldots,k_8) = \frac{1}{4}\big[&k_{eff\,(2D)}(k_1,k_2,k_3,k_4)+k_{eff\,(2D)}(k_5,k_6,k_7,k_8)\\ &+k_{eff\,(2D)}(k_5,k_6,k_1,k_2)+k_{eff\,(2D)}(k_7,k_8,k_3,k_8)\big]. \end{aligned} \tag{11.3.39}$$

Adeyemi et al. (2021) noted, referring to this equation, that this "is only an approximation, and the exact solution is much more complicated".

(b) *The Karim and Krabbenhoft Equation* (Karim and Krabbenhoft 2010; Irayani et al. 2019; Adeyemi et al. 2021)

If $\alpha_1, \ldots, \alpha_N$ are positive numbers, denote by $\mu_a(\alpha_1, \ldots, \alpha_N)$, $\mu_h(\alpha_1, \ldots, \alpha_N)$, $\mu_g(\alpha_1, \ldots, \alpha_N)$ their *arithmetic*, *harmonic*, and *geometric* mean values, that is

$$\begin{aligned} \mu_a(\alpha_1, \ldots, \alpha_N) &= \frac{1}{N}\sum_{i=1}^{N} \alpha_i;\ \mu_h(\alpha_1, \ldots, \alpha_N) \\ &= \frac{N}{\sum_{i=1}^{N} \alpha_i^{-1}};\ \mu_g(\alpha_1, \ldots, \alpha_N) = \sqrt[N]{\prod_{i=1}^{N} \alpha_i}. \end{aligned} \tag{11.3.40a,b,c}$$

In the method of renormalization introduced by Karim and Krabbenhoft (2010),

$$k_{L,x} = \mu_a[\mu_h(k_1, k_2), \mu_h(k_3, k_4), \mu_h(k_5, k_6), \mu_h(k_7, k_8)], \tag{11.3.41a}$$

$$k_{U,x} = \mu_h[\mu_a(k_1, k_3, k_5, k_7), \mu_a(k_2, k_4, k_6, k_8)]. \tag{11.3.41b}$$

The *effective permeability* in the x direction is the geometric mean of the two previous configurations:

$$\begin{aligned} k_{f,x} &= \mu_g\left[k_{L,x}, k_{U,x}\right] = \left(k_{L,x}.k_{U,x}\right)^{1/2} \\ &= \frac{1}{2}\left[\frac{(k_1+k_3+k_5+k_7)(k_2+k_4+k_6+k_8)}{(k_1+k_2)(k_3+k_4)(k_5+k_6)(k_7+k_8)}\right]^{1/2} \cdot \left(\frac{J}{I}\right)^{1/2}. \end{aligned} \tag{11.3.41c}$$

where

$$I = k_1 + k_2 + k_3 + k_4 + k_5 + k_6 + k_7 + k_8; \tag{11.3.41d}$$

$$J = k_1k_2S_{12} + k_3k_4S_{34} + k_5k_6S_{56} + k_7k_8S_{78}; \tag{11.3.41e}$$

$$S_{ij} = \frac{(k_1+k_2)(k_3+k_4)(k_5+k_6)(k_7+k_8)}{k_i + k_j}. \tag{11.3.41f}$$

The three methods, King's renormalization; renormalization of Green and Paterson; and of Karim and Krabbenhoft can be also used in combination, and—as suggested by Irayani et al. (2019)—in any combinations the permeability of larger blocks can be computed by the *lattice Boltzmann algorithm*. Apparently, *parallel implementations* of such LBM schemes combined with RNG have not been considered as yet.

11.4 Discrete Scale Invariance and Lognormal Periodicity

11.4.1 Mathematical Preliminaries

a. *Scale Invariance and Statistical Self-Similarity*

Consider a stationary random process $S(x)$, denote by $f(x)$ or $\overline{f(x)}$ expected values, define the *autocorrelation function* $R(\xi)$, *normalized autocorrelation function* $\rho(\xi)$, and *semivariogram* (Journel and Huijbregts 1978) $\sigma(\xi)$, respectively, of $S(x)$ as

$$R(\xi) = \left\langle \left(S(x) - \overline{S}\right)\left(S(x+\xi) - \overline{S}\right) \right\rangle;\ \rho(\xi) = \frac{R(\xi)}{R(0)};$$

$$\sigma(\xi) = \frac{1}{2}\left\langle (S(x) - (x+\xi))^2 \right\rangle \qquad (11.4.1.1\text{a,b,c})$$

Simple algebra gives

$$\sigma(\xi) = R(0)[1 - \rho(\xi)] \qquad (11.4.1.2)$$

The process $S(x)$ is *statistically self-similar* if

$$\sigma(r \cdot \xi) = f(r) \cdot \sigma(\xi) \qquad (11.4.1.3)$$

for all positive r and some non-negative function $f(r)$. We say the process is *scaling* (with *exponent* α), if

$$\sigma(\xi) \propto \xi^{2\alpha}. \qquad (11.4.1.4)$$

The exponent α is variously called *Lipschitz exponent*, *Lipschitz-Hölder exponent*, or *Hurst exponent* (Korvin 2022). We shall prove the equivalence of the two definitions (11.4.1.3) and (11.4.1.4), and show that *statistically self-similar processes are scaling ("scale invariant"), and scale invariant processes are statistically self similar.*

For this purpose we shall need *Pexider's Theorem* (1903): *The most general continuous solution of the functional equation*

$$f(xy) = g(x)h(y) \qquad (11.4.1.5)$$

for positive x and y is (apart from the trivial solutions)

$$f(x) = abx^c;\ g(x) = ax^c;\ h(x) = bx^c. \qquad (11.4.1.6)$$

Note that Eq. (11.4.1.5) generalizes Cauchy's (1821) functional equation

$$f(xy) = f(x)f(y) \quad (11.4.1.7a)$$

for positive x, y whose solution is

$$f(x) = x^c \text{ or } f(x) = 0 \text{ (trivial solution)}. \quad (11.4.1.7b)$$

Because of its importance in *scaling problems*, we sketch the proof of *Pexider's Theorem* (Aczel 1966; Korvin 1982: 76). Write out first Eq. (11.4.1.5) for $x = 1, y = t$; then for $x = t, y = 1$:

$$f(t) = g(1)h(t), \text{ i.e. } h(t) = \frac{f(t)}{a} \text{ with } a = g(1); \quad (11.4.1.8a)$$

$$f(t) = g(t)h(1), \text{ i.e. } g(t) = \frac{f(t)}{b} \text{ with } b = h(1); \quad (11.4.1.8b)$$

from these two equations:

$$f(xy) = \frac{f(x)f(y)}{ab}(a, b \neq 0). \quad (11.4.1.9)$$

Let $(t) = f(t)/ab$, then $F(t)$ satisfies $F(xy) = F(x)F(y)$ that is by *Cauchy's functional equation* (11.4.1.7a–11.4.1.7b) $F(t) = t^c$ and we have indeed

$$f(x) = abx^c; g(x) = ax^c; h(x) = bx^c. \quad (11.4.1.6)$$

(For $g(1) = 0$ or $h(1) = 0$ we obtain the trivial solutions: $f(x) = 0, g(x) = 0$, $h(x)$ arbitrary; or $f(x) = 0$, $g(x)$ arbitrary, $h(x) = 0$.)

We are ready to prove the equivalence of statistically self similar processes and scaling processes (Korvin 1982: 255; see also Sornette 1998; Borgnat et al. 2001, 2002; Abed-Pour et al. 2003; Ovdat and Akkermans 2019). If Eq. (11.4.1.4) holds, that is $2\sigma(\xi) = \langle (S(x) - (x + \xi))^2 \rangle \propto \xi^{2\alpha}$, the variations over any scale $r \cdot \xi$ are related to the variations over a scale ξ by

$$\sigma(r\xi) = r^{2\alpha}\sigma(\xi), \quad (11.4.1.10)$$

that is realizations of the function $S(x)$ have the same appearance when subjected to an r^α–fold magnification. Conversely, suppose that $S(x)$ is statistically self similar, that is its semivariogram satisfies $\sigma(r \cdot \xi) = f(r) \cdot \sigma(\xi)$ for all positive values of r. Then, by Pexider's Theorem, $\sigma(\xi)$ must be of the form $\sigma(\xi) \propto \xi^c$ where $c > 0$ because $\sigma(\xi)$ is an increasing function of ξ, and we can take $c = 2\alpha$.

As (Abed-Pour et al. 2003; Camia 2017) pointed out, scale-invariance is a subgroup of a larger transformation group, the *conformal group* (consisting of *translation, rotation, dilatation*, and *special conformal transformation*). *Conformal invariance at phase transition* determines the form of the *correlation functions* which must be *scaling functions* (Polyakov 1970; Belavin et al. 1984; Camia et al. 2015).

b *Stationarity and Statistical Self-Similarity, the Lamperti Transform*

Barenblatt and Zeldovich (1971; see also Sornette 1998) observed that the propagating plane wave $u(x - ct)$ becomes a *self-similar homogeneous function* $U(X/T^c)$ after the transform

$$x = \ln X; t = \ln T, \tag{11.4.1.11}$$

i.e.

$$X = e^x; T = e^t. \tag{11.4.1.12}$$

This shows that performing a *scale transformation* $U(X) \Rightarrow U(X/T^c)$ is the same as a *translation in the logarithmic domain* $u(x) \Rightarrow u(x - ct)$.

A *deterministic signal* $\{X(t)|t \in \mathbb{R}\}$ is *self-similar with exponent H* if for any $\lambda \in \mathbb{R}$

$$X(\lambda t) = \lambda^H X(t). \tag{11.4.1.13}$$

To emphasize the role of exponent H, we shall call $X(t)$ *H-self-similar* or *H-SS* (notation used in Borgnat et al. 2001, 2002).

In case of *stochastic processes* this definition is generalized as

$$X(\lambda t) \triangleq \lambda^H X(t), \tag{11.4.1.14}$$

where the relation "$\triangleq$" means *equality in probabistic sense* (Samorodnitsky and Taqqu 1994), defined by the *equality of any finite-dimensional distributions* (Samorodnitsky and Taqqu 1994), or in a less strict sense, by the *equality of correlation functions* only (Borgnat et al. 2001, 2002).

Because of the relation between *dilations* of stochastic processes and their *translations* in the logarithmic domain (Eqs. 11.4.1.11–11.4.1.12), we expect an analogous correspondence between self-similar and stationary processes. Introduce the *Lamperti transform* (Lamperti 1962; Borgnat et al. 2001, 2002) as follows:

For a random process $\{Y(t)|t \in \mathbb{R}\}$, its *Lamperti-transform* $\left\{X(t)|t \in \mathbb{R}^+\right\}$ is given by

$$X(t) = (\mathcal{L}Y)(t) \triangleq t^H Y(\ln t), t \in \mathbb{R}^+; \tag{11.4.1.15}$$

the inverse Lamperti transform is defined as

$$Y(t) = \left(\mathcal{L}^{-1}X\right)(t) \triangleq e^{-Ht}X\left(e^{t}\right), t \in \mathbb{R}. \tag{11.4.1.16}$$

The Lamperti transformation maps a time-shifted process to a dilated transform of the original process. Let $\mathcal{D}_{\lambda}^{H}$ be the dilation operator, $\mathcal{S}_{\tau}$ the time-shift operator, defined as

$$\left(\mathcal{D}_{\lambda}^{H}X\right)(t) \triangleq \lambda^{-H}X(\lambda t); \ (\mathcal{S}_{\tau})(t) \triangleq Y(t+\tau). \tag{11.4.1.17a, b}$$

Then

$$\left(\mathcal{L}^{-1}\mathcal{D}_{\lambda}^{H}\mathcal{L}Y\right)(t) \triangleq (\mathcal{S}_{\ln\lambda}Y)(t). \tag{11.4.1.18}$$

Using this identity, Lamperti (1962) proved the following.

Theorem *A process $Y(t)$ is stationary if and only if its Lamperti transform $X = \mathcal{L}Y$ is H-SS.*

c *Discrete Scale Invariance (DSI) and Cyclostationarity*

A process $\left\{X(t)|t \in \mathbb{R}^{+}\right\}$ has the property of *discrete scale invariance with scaling exponent H and scale λ* if

$$X(\lambda t) \triangleq \lambda^{H}X(t), t \in \mathbb{R}^{+}. \tag{11.4.1.19}$$

We shall say that the process is (H, λ)-DSI (notation of Borgnat et al. 2001, 2002). By mathematical induction, a (H, λ)-DSI process is also scale invariant for any scaling factor of the form $\{\lambda^{n}|n \in \mathbb{Z}\}$. An example for (H, λ)-DSI processes is the well-known Weierstrass-Mandelbrot function (Berry and Lewis 1980)

$$W(t) = \sum_{n=-\infty}^{\infty} \lambda^{-nH}\left[1 - \cos\left(\lambda^{n}t + \phi_{n}\right)\right] \tag{11.4.1.20}$$

assuming that $\lambda > 1, 0 < H < 1$, and the phases $\{\phi_{n}\}$ are *iid* (*identically disributed independent*) random variables.

A process $\{Y(t)|t \in \mathbb{R}\}$ is *T-cyclostationary* (Gladyshev 1963; Hurd 1969; Gardner and Franks 1975; Borgnat et al. 2001, 2002) if for any value of time $t \in \mathbb{R}$ one has

$$Y(t+T) \triangleq Y(t). \tag{11.4.1.21}$$

It follows from the definition that

$$Y(t+\tau) = Y(t) \tag{11.4.1.22}$$

and

$$Y(t+T)Y(s+T) = Y(t)Y(s), \tag{11.4.1.23}$$

that is the ACF (*autocorrelation function*) $R_{YY}(t, t+\tau)$ is a *periodic function of t with period T* and, consequently, it can be developed into *Fourier series*:

$$R_{YY}(t, t+\tau) = \sum_{n=-\infty}^{\infty} c_n(\tau) e^{i \cdot 2\pi nt/T}. \tag{11.4.1.24}$$

Theorem (Borgnat et al. 2001, 2002). *If* $\{X(t)t \in \mathbb{R}^+\}$ is (H, e^T) *-DSI, then*

$$Y(t) = e^{-Ht} X\left(e^t\right), t \in \mathbb{R} \tag{11.4.1.25}$$

is T-cyclostationary. Conversely, if $\{Y(t)t \in \mathbb{R}\}$ *is T-cyclostationary, then*

$$X(t) = t^H Y(\ln t), t \in \mathbb{R}^+ \tag{11.4.1.26}$$

is (H, e^T)*-DSI.*

Proof Suppose $\{X(t)t \in \mathbb{R}^+\}$ is (H, e^T)*-DSI*, then

$$X\left(e^T t\right) \triangleq e^{HT} X(t), \tag{11.4.1.27}$$

and, consequently,

$$Y(t+T) = e^{-Ht-HT} X\left(e^T e^t\right) \triangleq e^{-Ht} X\left(e^t\right) = Y(t), \tag{11.4.1.28}$$

that is $Y(t)$ is T-cyclostationary.

Conversely, if $Y(t)$ is T-cyclostationary, then

$$Y(T+t) \triangleq Y(t), \tag{11.4.1.29}$$

so that

$$X\left(e^T t\right) = e^{HT} e^{H \ln t} Y(\ln t + T) \triangleq e^{HT} t^H Y(\ln t) = \left(e^T\right)^H X(t), \tag{11.4.1.30}$$

what means $X(t)$ is indeed (H, e^T)*-DSI.*

It immediately follows from this theorem, and from Eq. (11.4.1.24), that the ACF of a (H, λ)-DSI process can be expressed as a series in terms of the *Mellin basis functions* (Bertrand et al. 1996) $\{t^{H+i2\pi n/\ln\lambda}\}_{n=-\infty}^{\infty}$:

$$R_{XX}(t, kt) = k^H t^H \sum_{n=-\infty}^{\infty} c_n(k) t^{H+i2\pi n/\ln\lambda}. \tag{11.4.1.31}$$

Table 11.7 Some applications of discrete scale invariance

Application	Authors
Acoustic emissions	Anifrani et al. (1994, 1995)
Cloud and rain structures	Pecknold et al. (1996), Veneziano (2002)
Cosmic lacunarity	Provenzale et al. (1997)
Crack	Huang et al. (1997)
Crack tip stress singularity	Rice (1988), Rice et al. (1990), Borodich (1993)
DLA (diffusion-limited-aggregation)	Sornette et al. (1996b), Sornette (1998)
Earthquakes	Sornette and Sornette (1989, (1990), Sornette and Sammis (1995), Newman et al. (1995), Johansen et al. (1996, 2000a), Saleur et al. (1996a, b), Bowman and Sammis (1996), Smith and Sammis (1996), Varnes and Bufe (1996), Huang et al. (1996), Sornette (1998), Bak et al. (2002), Kagan (2010), Newberry and Savage (2019)
Fracture dissolution	Szymczak and Ladd (2006)
Infiltration instability	Ortoleva et al. (1987)
Lithosphere	Sornette et al. (1990)
Magneto-resistance	Doucot et al. (1986
River networks	Rinaldo et al. (1993), Roy and Tarafdar (1996)
Ruptures	Anifrani et al. (1994, 1995), Ide and Sornette (2002)
Scattering on fractal structures	Makarov (1994), Sornette and Christian (1992)
Specific heat	Vallejos et al. (1998), Vallejos and Anteneodo (1998)
Titius-Bode laws in the solar system	Graner and Dubrulle (1994)
Turbulence	Novikov (1990), She and Leveque (1994)
Viscous fingering	Roy and Tarafdar (1996), Roy et al. (1999)

If the process X is real-valued, then $c_{-n}(k) = c_n^*(k)$.

11.4.2 *Applications of Discrete Scale Invariance (DSI)*

Discrete Scale Invariance (DSI) has many important applications in Earth Sciences (Table 11.7). We shall only treat two of them: *prediction of the time to failure* (in particular: prediction of earthquakes), and the discrete scaling form of the *Pareto distribution*.

11.4.2.1 Prediction of the Time to Failure

Figure 11.2 shows a spectacular example of phase transition in the *Ising model of ferromagnetism* (Camia 2017). The middle configuration, computed at the point

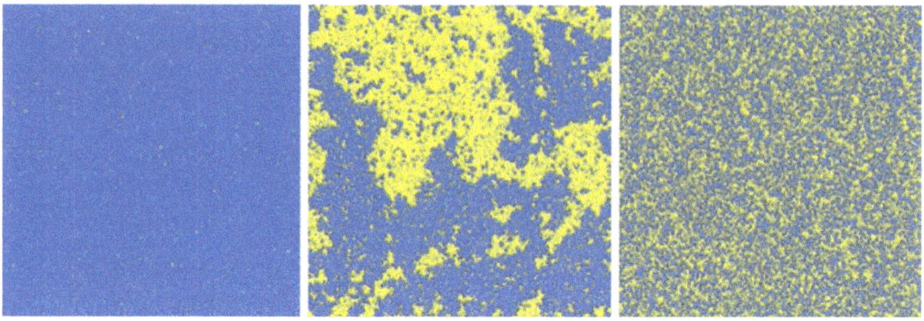

Fig. 11.2 Phase transition in the Ising model (simulation by Wouter Kager). From left to right, at increasing temperatues. Colors denote the two (±) spin states. The middle figure shows the self-similarity of spin distribution at the point of phase transition. (After Camia 2017: Fig. 5; reproduced with kind permissions of Dr. Wouter Kager and Dr. Federico Camia)

of phase transition, is *self-similar*, because the blue and yellow clusters have no characteristic sizes (as they have in case of low temperatures on the left panel, or at high temperatures on the right) so that a scale transformation would not make any difference in their appearance in the statistical sense. The middle image looks like a *fractal*, an imaginary landscape from Mandelbrot's dreamland, with its many islets, lakes, peninsulas and fjords (Mandelbrot 1982; Korvin 1982).

Self-similar objects, or *fractals*, however, do not necessarily have the property of *Continuous Scale Invariance* (*CSI*). Consider, as an example, the prototype of all fractals, the well-known *triadic Cantor set* ("*Cantor bar*", Fig. 11.3) which has the fractal dimension $D = \frac{\log 2}{\log 3}$. To see this, let $N_\lambda(n)$ denote the number of intervals seen at magnification λ at the n-th iteration. The original unit interval corresponds to $\lambda = 1$ by definition. At every iteration, $N_\lambda(n)$ increases by a factor 2, magnification increases by a factor 3. The fractal dimension is

$$D = \lim_{n\to\infty} \frac{\log N_\lambda(n)}{\log \lambda} = \lim_{n\to\infty} \frac{\log 2^n}{\log 3^n} = \frac{\log 2}{\log 3} \approx 0.63 \tag{11.4.2.1}$$

(Korvin 1982: Eq. 2.1.2.5).

Sornette (1998) observed that if, instead of choosing $\lambda = 1, 3, 3^2, \ldots, 3^p, \ldots$, we change λ continuously in the range $1 \leq \lambda < \infty$ then, say, between $\lambda = 3^p$ and

Fig. 11.3 First few step of the construction of a Cantor bar

$\lambda = 3^{p+1}$, the number of intervals $N_\lambda(n)$ will increase by a factor 2 at $\lambda = 3^p$, it remains constant for $3^p < \lambda < 3^{p+1}$, then becomes twice as large again at $\lambda = 3^{p+1}$. The measured fractal dimension $D(\lambda) = \frac{\log N_\lambda(n)}{\log \lambda}$ will decrease for $3^p < \lambda < 3^{p+1}$, then it suddenly jumps up at $\lambda = 3^{p+1}$. For continuous values of λ, $1 < \lambda$, the magnification has a fluctuating form

$$N_\lambda(n) = N_1(n)\lambda^D P\left(\frac{\log \lambda}{\log 3)}\right), \tag{11.4.2.2}$$

where $P(x)$ is a periodic function with period $T = 1$, with Fourier series

$$P\left(\frac{\log \lambda}{\log 3)}\right) = \sum_{-\infty}^{\infty} c_n \exp\left(2n\pi i \frac{\log x}{\log 3}\right) \tag{11.4.2.3}$$

Substituting this to Eq. (11.4.2.2) we see that the fractal dimension D is replaced by infinitely many complex values. We already indicated in Sect. 11.3.2.1 (Eq. 11.3.27h) that the magnitude of the nth term decreases with n. Keeping only the $n = 1$ term in (11.4.2.3), and assume that $c_{-1} = c_1$ to make $N_\lambda(n)$ real, we get from (11.4.2.2):

$$N_\lambda(n) \approx N_1(n)\lambda^D \left[1 + 2\frac{c_1}{c_0} \cos\left(2ni\pi \cdot \frac{\log \lambda}{\log 3}\right)\right] \tag{11.4.2.4}$$

that is the number of intervals $N_\lambda(n)$ as function of magnification λ has a leading power law behavior λ^D *decorated with a log-periodic modulation* with a period of log 3.

We can also derive this result in another way, from the *general theory of scale invariance* (Sect. 11.4.1). An observed physical quantity $\mathcal{O}(x)$ which depends on a parameter x is (continuously) *scale-invariant* under the change $x \Rightarrow \lambda x$ if there exists a number $\mu = \mu(\lambda)$ for which

$$\mathcal{O}(x) = \mu(\lambda) \cdot \mathcal{O}(\lambda x). \tag{11.4.2.5}$$

By *Pexider's theorem* (11.4.1.5–11.4.1.6) $\mathcal{O}(x)$ is of the form

$$\mathcal{O}(x) = Cx^\alpha. \tag{11.4.2.6}$$

From (11.4.2.5) we have $\mu\lambda^\alpha = 1$. Using the identity $e^{i\pi n} = 1$ for an arbitrary integer n, we get

$$\alpha = \left\{-\frac{\log \mu}{\log \lambda} + i\frac{2\pi n}{\log \lambda}\right\}_{n=0}^{\infty}, \tag{11.4.2.7}$$

what is the same as Eq. (11.4.2.3). The case $n = 0$ provides the *power law* describing *continuous scale invariance*, the complex terms correspond to *DSI* (*discrete scale invariance*).

Log-periodic oscillations have been observed and documented in time series of different physical quantities that had been recorded (or numerically simulated) before a forthcoming *natural catastrophe* or *material failure* (Ball and Blumenfield 1990; Bowman and Sammis 1996; Huang et al. 1996; Johansen et al. 1996, 2000a, b, Sahimi and Arbabi 1996; Sornette and Sammis 1995; Saleur et al. 1996a, b; Smith and Sammis 1996; Sornette 1998; Sornette and Sornette 1989, 1990; Varnes and Bufe 1996).

An example for *log-periodic oscillations* is shown in Fig. 11.4. It is related to the Kobe earthquake (an earthquake on January 17, 1995, at 05:46:53 JST in the S. part of Hyōgo Prefecture, Japan; epicenter: 34°35′N 135°04′E; depth: 17.6 km; magnitude 6.9; see Kanamori 1995). The log-periodic fluctuations are especially clearly seen if the control parameter is plotted in logarithmic time scale, as in Fig. 11.4.

Discrete Scale Invariance has found important application in predicting the expected *critical time* t_c (or *critical temperature*, *critical pressure*) for a system undergoing *phase transition*, such as *ordered* $\Rightarrow$ *disordered*; *predictable* $\Rightarrow$ *unpredictable*, etc. (Landau 1970; Ma 1976; Nauenberg 1975; Niemeier and van Leeuwen1976; Goldenfeld 1992; Camia 2017; Ovdat and Akkermans 2019). In phase transitions, there exists an order parameter $\mathcal{O}(|t - t_c|)$ which.

(a) diverges or tends to zero at the critical point:

$$\lim_{t \to t_c} \mathcal{O}(|t - t_c|) = \infty \text{ or } \lim_{t \to t_c} \mathcal{O}(|t - t_c|) = 0 \tag{11.4.2.8}$$

and

(b) is *scaling* as:

$$\mathcal{O}(\lambda|t - t_c|) = \mu(\lambda) \cdot \mathcal{O}(|t - t_c|. \tag{11.4.2.9}$$

As we have seen, the solution to this Pexider-type functional equation is, apart from constant factors, given by

$$\mu(\lambda) = \lambda^\alpha; \mathcal{O}(t) = |t - t_c|^\alpha.$$

It has been found that in some cases scale invariance (11.4.2.9) only holds for *a set of discrete values* of λ. If $\mathcal{O}$ is self-similar for some specific value λ, then—by definition—we have for some exponent

$$\mathcal{O} : \mathcal{O}(\lambda^2 t) = \mathcal{O}(\lambda \cdot \lambda t) = \lambda^(\alpha) \cdot \mathcal{O}(\lambda t) = (\lambda^\alpha)^2 \mathcal{O}(t) = (\lambda^2)^\alpha \mathcal{O}(t),$$

and we find by mathematical induction that $\mathcal{O}$ is also self-similar for any n-th power of λ. Thus, with the notation

$$\tau = |t - t_c| \tag{11.4.2.10}$$

we have

$$\mathcal{O}(\tau) = \tau^{\alpha} \cdot \Theta\left(\frac{\log \tau}{\log \lambda}\right) \tag{11.4.2.11}$$

where Θ is a periodic function of period $T = 1$. Fourier expanding to first order,

$$\mathcal{O}(\tau) = A + B\tau^{\alpha} + C\tau^{\alpha} cos(\omega \cdot \log \tau - \phi), \tag{11.4.2.12}$$

where

$$\omega = 2\pi / \log \lambda. \tag{11.4.2.13}$$

Sornette used this equation in the form

$$E \sim (t_r - t)^m \left[1 + Ccos\left(2\pi \cdot \frac{\log(t_r - t)}{\log \lambda} + \psi\right)\right]; \tag{11.4.2.14}$$

(Sornette 1998: Eq. 27) to predict the point of rupture from the time series of the released *acoustic emission* energy E. Johansen et al. (1996) fitted the artesian water's *chlorine concentration data* (Tsunogai and Wakita 1995) recorded before the Kobe earthquake with a similar equation

$$c(\tau) = A + B(\tau_c - \tau)^m \left[1 + C\cos(\omega \cdot \log((\tau_c - \tau)))\right]; \tag{11.4.2.15}$$

(Johansen et al. 1996: Eq. 1), see Fig. 11.4. In these equations the following are free parameters:

$$\left.\begin{array}{ll} \text{in (6.4.2.10) to (6.4.2.13)} & A, B, C, \alpha, \Phi, t_c, \lambda \\ \text{in (6.4.2.14)} & t_r, m, C, \lambda, \psi \\ \text{in (6.4.2.15)} & A, B, \tau_c, m, C, \omega \end{array}\right\}$$

their determination requires *nonlinear optimization*. The numerical procedures of fitting these equations to the measured data minimize the variance between measured and computed values

$$Var = \frac{1}{N}\sum_{i=1}^{N}\left[y_i - f(t_i)\right]^2, \tag{11.4.2.16}$$

using *downhill simplex minimization* (Nelder and Mead 1965); *simulated annealing* (Sect. 11.1; Metropolis et al. 1953), *genetic algorithm* (Fogel 2006), *tabu search* (Glover 1986), or other techniques (Press et al. 2007). The preferred scaling ratio λ

can also be found by computing and visually inspecting the *Lomb periodogram* of the unevenly sampled data set transformed to *logarithmic time* (Borgnat et al. 2001; VanderPlas 2018).

Abed-Pour et al. (2003) warn us, "log-periodicity per se is not a signature of a critical point. Only within a well-defined model and mechanism can it be used as a potential signature." In the Kobe earthquake study (Johansen et al. 1996) the authors presented a well-defined, plausible model justifying the use of *hydrochemical changes* for *earthquake prediction*:

> We attribute the observed increase of ion concentration to a rise of deep-seated groundwater, which is known to have a higher content of Cl^-, SO_4^{--}, etc., into the artesian layer of the wells. This is due to correlated micro-cracking and fault activation leading to a change of the permeability of the rocks. This follows since there is no principal source of Cl^- and SO_4^{--} in the shallow strata beneath the Kobe area and high ion content are common features of deep-source groundwater rising through fracture zones near the Rokko mountain. (*Op. cit.*: 1401).

In (Johansen et al. 1996) possible other indicators were duly mentioned and carefully excluded: (a) *radon concentration anomaly* before the earthquake had also been observed (Igarashi et al. 1995) but the "data [was] too short and too noisy to attempt a full fit with Eq. 27 [our Eq. 11.4.2.14]" (*ibid.*: 1400); (b) the Benioff strain data set cumulated over the earthquakes which had occurred in the region was "too scarce to really allow for a quantitative analysis" (*ibid.*: 1399). A potentially meaningful indicator, changes in water pressure and temperature (King et al. 1995) "[did] not seem to be useful probes" (*ibid.*: 1401).

It had created large public interest when a group of physicists led by Didier Sornette started to apply the technique to predict *crashes on the stock market*. Any discussion of these attempts would lead far away from the topic of the book, those interested are referred to (Feigenbaum and Freund 1996; Sornette et al. 1996a; Sornette and Johansen 1997, 1998, 2001; Frost and Prechter 1998; Erman 1999; Johansen et al. 1999, 2000a, b; Laloux et al. 1999).

11.4.2.2 Pareto Distributions (Number vs. Size) with Discrete Scaling

Let us consider once more the classic *triadic Cantor bar* (Fig. 11.3). Denote by $N_{(\varepsilon)}^{(n)}$ the *total number of intervals* with length $\geq \varepsilon$ in the *first n* iterations of the object. In the zeroth iteration we have 1 ($= 2^0$) interval of length ≥ 1; in the zeroth and first iterations together we have $(1 + 2)$ intervals with lengths $\geq 1/3$; in the first n iterations we have altogether $(1 + 2 + \cdots 2^n)$ intervals with lengths $\geq 1/3^n$. If $\varepsilon = 3^{-n}$, we have

$$N_{(\varepsilon)}^{(n)} = \left(1 + 2 + \cdots 2^n\right) = \frac{2^{n+1} - 1}{2 - 1} \approx 2 \times 2^n = 2 \times \varepsilon^{-D}, \text{ with} D = \frac{\log 2}{\log 3} \tag{11.4.2.17}$$

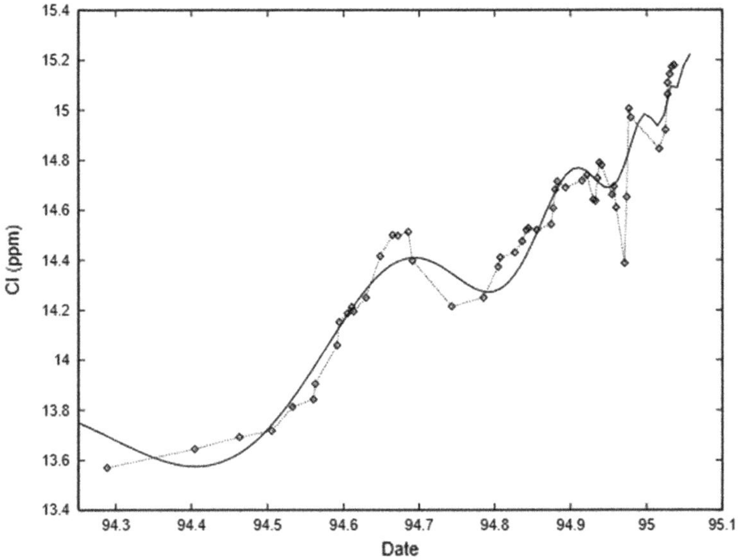

Fig. 11.4 Log-periodic fit of the chlorine concentration before the January 17, 1995 Kobe earthquake. Solid line is best fit with the equation $c(\tau) = A + B(\tau_c - \tau)^m\left[1 + C\cos(\omega \cdot \log(\tau_c - \tau))\right]$ where A, B, m, τ_c, ω are free parameters, τ_c is the time to earthquake. (After Johansen et al. 2000a, b: Fig. 1 left top panel; reproduced by kind permission of Prof. Didier Sornette and *Springer Nature*.)

If $\varepsilon \to 0$ continuously, rather than in discrete steps $\varepsilon = 3^{-n}$, an argument similar to what has resulted in Eq. (11.4.2.4) would lead to an expression of the form

$$N_{(\varepsilon)}^{(n)} \propto \varepsilon^{-D}\left[1 + \text{log periodic corrections}\right] \tag{11.4.2.18}$$

The oscillatory nature of the *number vs. size distribution* has been observed in several growth processes (Huang et al. 1997; Roy et al. 1999: Figs. 1, 2; Szymczak and Ladd 2006: Figs. 1, 2). Sornette (1998) noted: "if an unstable eddy in turbulent flow typically breaks up into two or three smaller eddies, but not into 10 or 20 eddies, then one can suspect the existence of a preferable scale factor, hence the log-periodic oscillations". In a similar vein, Newberry and Savage (2019) observed: "many self-similar systems are scale invariant only in discrete steps. A blood vessel tends to branch into two smaller vessels, a fluid vortex into two or three smaller vortices, and the Sierpinski triangle is self-similar only by powers of two … It is a weaker condition than the continuous scale invariance underlying the Pareto distribution. Whereas strict scale invariance implies a power law and vice versa, discrete-scale invariance allows log-periodic modulations in the frequencies of observations". (Korvin 1982: Figs. 3.17–3.18 on pp. 217–218 also pointed out the "waviness" of empirical Pareto distributions.)

In the classic, *continuous Pareto distribution* (Korvin 1982: 215 – 218; Arnold 2015) the observable variables have a minimal value $x_{min} > 0$, and the *probability density* of measuring a continuous value x is

$$p(x) = a x_{min}^{a} x^{-(a+1)}. \tag{11.4.2.19}$$

The probability of observing some value y greater than x is

$$Prob(y \geq x) = P(x) = \int_{x}^{\infty} p(\xi) d\xi = \left(\frac{x}{x_{min}}\right)^{-a}, \tag{11.4.2.20}$$

that is, it is a *continuous power law*. A simple example of *discrete Pareto distribution* (Newberry and Savage 2019: Fig. 1b) is provided by a uniformly random draw from the values at the nodes of a tree where each node with value x has one *parent* with value λx and n *children* with value x/λ. (The model is self-similar because a similar tree is obtained after the transform $x \Rightarrow \lambda x$.) Denote the smallest observable value by $x_m = x_{min}$, then we see that the probability $p(x)$ to find a value x at a randomly selected node is

$$p(x) = \begin{cases} p(x_m) \cdot n^{-k} & \text{if } x \text{ is of the form } \quad x = \lambda^k x_m \\ 0 & \text{if } x \text{ is not of the form } x = \lambda^k x_m \end{cases} ; k \geq 0 \text{ is an interger} \tag{11.4.2.21}$$

Taking the base-λ logarithm of $x = \lambda^k x_m$ we get $k = \log_\lambda\left(\frac{x}{x_m}\right)$, so that

$$n^{-k} = n^{-\log_\lambda\left(\frac{x}{x_m}\right)} = n^{-\log_n\left(\frac{x}{x_m}\right)/\log_n \lambda} = \left(\frac{x}{x_m}\right)^{-1/\log_n \lambda} = \left(\frac{x}{x_m}\right)^{-\alpha}, \tag{11.4.2.22}$$

where we used the notation

$$\alpha = 1/\log_n \lambda. \tag{11.4.2.23a}$$

With this notation we have

$$n = \lambda^\alpha, \tag{11.4.2.23b}$$

and

$$p(x) = p(x_m) \cdot \left(\frac{x}{x_m}\right)^{-\alpha} \tag{11.4.2.23c}$$

for allowable values $x = \lambda^k x_m$. As the probabilities must sum to 1, we have

$$1 = \sum_{x=\lambda^k x_m} p(x) \approx \sum_{k=}^{\infty} \left(x_m \lambda^k\right)^{-\alpha} = x_m^{-\alpha} \left(1 - \lambda^{-\alpha}\right)^{-1} \tag{11.4.2.24}$$

(The $\approx$ sign means that instead of $\sum_{k=0}^{k_{\max}}$, $k_{\max} \gg 1$ we continued the sum up to $k = \infty$, and used the same finite minimal observable x_m in all terms.) Thus, instead of (11.4.2.21) we can write

$$p(x) = \begin{cases} \left(1 - \lambda^{-\alpha}\right)\left(\frac{x}{x_m}\right)^{-\alpha} & \text{if } x \text{ is of the form } \quad x = \lambda^k x_m \\ 0 & \text{if } x \text{ is not of the form } x = \lambda^k x_m \end{cases}; k \geq 0 \text{ is an integer} \tag{11.4.2.25}$$

The *cumulative distribution function* (Newburry and Savage 2019: Fig. 1a) is a *stair-step function displaying logperiodic behavio*r, which is equal to the Pareto CDF,

$$P(x) = \int_x^{\infty} p(\xi) d\xi = \left(\frac{x}{x_{\min}}\right)^{-a} \tag{11.4.2.20}$$

at allowed values of x.

Suppose we make $N \gg 1$ independent observations of a variable x which follows a power-like probability distribution of the form

$$Prob\left(x = x_j\right) = p\left(x_j\right) = Cx_j^{-(\alpha+1)}. \tag{11.4.2.26}$$

The *likelihood* of getting a set of values $\left\{x_1, \ldots, x_j\right\}$ is

$$\mathcal{L} = \prod_{j=1}^{N} p\left(x_j\right). \tag{11.4.2.27}$$

The *logarithmic likelihood*, in case of the distribution (11.4.2.26), is

$$\log \mathcal{L} = N \cdot \log C - (\alpha + 1) \sum_{j=1}^{n} \log x_j.$$

To find the MLE (*Maximum-Likelihood Estimate*) of the exponent α, we maximize the log-likelihood (which is a monotonic function of likelihood) with respect to α, by putting

$$\frac{\partial \mathcal{L}}{\partial \alpha} = 0 = N \cdot \frac{\partial \log C}{\partial \alpha} - \sum_{j=1}^{n} \log x_j,$$

i.e.

$$\frac{1}{C}\frac{\partial C}{\partial \alpha} = \frac{1}{N}\sum_{j=1}^{n} \log x_j = \langle \log x_j \rangle. \tag{11.4.2.28}$$

For the distribution (11.4.2.25) the normalizing constant is

$$C = \left[\sum_{k=0}^{\infty} \left(x_m \lambda^k\right)^{-\alpha}\right]^{-1} = x_m^{\alpha}\left(1 - \lambda^{-\alpha}\right), \tag{11.4.2.29}$$

and—after elementary calculations—Eq. (11.4.2.28) yields the *MLE* estimate for the exponent as

$$\hat{\alpha}_{discrete} = \log_\lambda\left[1 + \left\langle \log_\lambda\left(\frac{x_j}{x_m}\right)\right\rangle^{-1}\right]. \tag{11.4.2.30}$$

For the classical Pareto distribution (Muniruzzaman 1957; Clauset et al. 2009)

$$C = \left[\int_{x_{min}}^{\infty} x^{-(\alpha+1)} dx\right]^{-1} = \alpha x_m^{\alpha} \tag{11.4.2.30}$$

and the MLE optimization (11.4.2.28) would yield

$$\hat{\alpha}_{Pareto} = \left\langle \log\left(\frac{x_j}{x_m}\right)\right\rangle^{-1}. \tag{11.4.2.31}$$

As Newberry and Savage (2019) pointed out,

$$\lim_{\lambda \searrow 1} \hat{\alpha}_{discrete} = \hat{\alpha}_{Pareto}, \tag{11.4.2.32}$$

and, *unless* $\lambda = 1 + \varepsilon$ with $0 \leq \varepsilon \ll 1$, *the estimates* $\hat{\alpha}_{discrete}$ and $\hat{\alpha}_{Pareto}$ *are significantly different*. These authors also discussed the role, and proper selection of λ.

Earthquake magnitudes also follow an approximately power-law distribution. Bak et al. (2002) used *regression* on an extensive dataset and determined the exponent of this law—called the "*Gutenberg-Richter b value*"—as $b = 0.95 \pm 0.01$. The Pareto MLE (Eq. 11.4.2.31) yields $b = \hat{\alpha}_{Pareto} = 1.054 \pm 0.007$ for the same dataset. Newberry and Savage (2019: Table 1, Fig. 3) used their discrete model with $x_m = 2$, $\lambda = 10^{0.1} = 1.26$ and found $b = \hat{\alpha}_{discrete} = 0.943 \pm 0.006$ from Eq. (11.4.2.30), which is close to the regression result 0.95 ± 0.01.

References

Aaltosalmi U (2005) Fluid flow in porous media with the Lattice-Boltzmann method. PhD dissertation, University oF Jyväskylä, Finnland

Aarts E, Korst J (1989) Simulated annealing and Boltzmann machines: a stochastic approach to combinatorial optimization and neural computing. Wiley, Chichester

Abas A, Gan ZL, Ishak MHH, Abdullah MZ, Khor SF (2016) Lattice Boltzmann method of different BGA orientations on I-type dispensing method. PLoS ONE 11(7):e0159357

Abed-Pour N, Aghamohammadi A, Khorrami M, Reza Rahimi Tabar M (2003) Discrete scale invariance and its logarithmic extension. Nuclear Phys B 655(3):342–352

Aczel J (1966) Lectures on functional equations and their applications. Academic Press, New York and London

Adewole AP, Otubamowo K, Egunjobi TO (2012) A comparative study of simulated annealing and genetic algorithm for solving the travelling salesman problem. Int J Appl Inf Syst 4(4):6–12

Adeyemi B, Ghanbarian B, Winter CL, King PR (2021) Determining effective permeability at reservoir scale: numerical simulations and theoretical modeling. arXiv:2109.13708v1 [physics.geo-ph]

Aharonov E, Rothman DH (1993) Non-Newtonian flow (through porous media): a lattice Boltzmann method. Geophys Res Lett 20:679–682

Ahlfors L (1979) Complex analysis. McGraw Hill, New York

Ahrenholz B, Tölke J, Lehmann P, Peters A, Kaestner A, Krafczyk M, Durner W (2008) Prediction of capillary hysteresis in a porous material using lattice-Boltzmann methods and comparison to experimental data and a morphological pore network model. Adv Water Resour 31:1151–1173

Aidun C, Lu Y (1995) Lattice Boltzmann simulation of solid particles suspended in a fluid. J Stat Phys 81:49–59

Aidun CK, Clausen JR (2010) Lattice-Boltzmann method for complex flows. Annual Rev Fluid Mech 4:439–472

Aksnes EO, Elster AC (2010) Porous rock simulations and lattice Boltzmann on GPUs. In: Chapman B, Desprez F, Joubert GR, Lichnewsky A, Peters F, Priol T (eds) Parallel computing: from multi-cores and GPU's to petascale. Advances in parallel computing, vol 19. IOS Press, Amsterdam, pp 536–545

Alexander SK, Fieguth P, Ioannidis MA, Vrscay ER (2009) Hierarchical annealing for synthesis of binary images. Math Geosci 41:357–378

Allègre C, Le Mouel J, Provost A (1982) Scaling rules in rock fracture and possible implications for earthquake prediction. Nature 297:47–49

Allègre CJ, Le Mouel JL, Chau HD, Narteau C (1995) Scaling organization of fracture tectonics (SOFT) and earthquake mechanism. Phys Earth Planet Inter 92:215–233

Alpak FO, Berg S, Zacharoudiou I (2018) Prediction of fluid topology and relative permeability in imbibition in sandstone rock by direct numerical simulation. Adv Water Resour 122:49–59

Al-Raoush R, Hannun J, Jarrar Z, Alshibli K, Jung J (2019) Grouping minimum REV of porosity and tortuosity based on descriptors of sand grains. In: Conference on proceedings of the third EAGE WIPIC workshop: reservoir management in carbonates, Nov 2019, pp 1–5

Ammon CJ, Vidale JE (1993) Tomography without rays. Bull Seismol Soc Am 83(2):509–528

Amit DJ (1984) Field theory, the renormalization group, and critical phenomena. World Scientific, Singapore

Andrä H, Combaret N, Dvorkin J, Glatt E, Han J, Kabel M, Keehm Y, Krzikalla F, Lee M, Madonna C, Marsh M, Mukerji T, Saenger EH, Sain R, Saxena N, Ricker S, Wiegmann A, Zhan X (2013) Digital rock physics benchmarks—part II: computing effective properties. Comput Geosci 50:33–43

Andresen B, Hoffmann KH, Mosegaard K, Nulton J, Pedersen JM, Salamon P (1988) On lumped models for thermodynamic properties of simulated annealing problems. J Phys France 49:1485–1492

Anifrani J-C, Johansen AA, Le Floc'h C, Ouillon G, Sornette D, Vanneste C, Souillard B (1994) New approaches for exploiting acoustic emission. In: Proceedings of the 6th European conference on non-destructive testing, 24–28 October 1994, Nice: Presentation N72
Anifrani J-C, Le Floc'h C, Sornette D, Souillard B (1995) Universal log-periodic correction to renormalization group scaling for rupture stress prediction from acoustic emissions. J Phys I France 5(6):631–638
Ansumali S, Karlin IV (2002) Kinetic boundary conditions in the lattice Boltzmann method. Phys Rev E 66(2):026311
Ansumali S, Karlin IV, Arcidiacono S, Abbas A, Prasianakis N (2007) Hydrodynamics beyond Navier–Stokes: exact solution to the Lattice Boltzmann hierarchy. Phys Rev Lett 98:124502
Antonov NV, Kakin PI (2017a) Scaling in landscape erosion: renormalization group analysis of a model with infinitely many couplings. Theor Math Phys 190(2C):193–203 (in Russian)
Antonov NV, Kakin PI (2017b) Scaling in erosion of landscapes: renormalization group analysis of a model with turbulent mixing. J Phys a: Math Theor 50:085002
Antonov NV, Kakin PI, Lebedev NM (2020) Static approach to renormalization group analysis of stochastic models with spatially quenched noise. J Stat Phys 178(2):392–419
Apourvari SN, Arns CH (2016) Image-based relative permeability upscaling from the pore scale. Adv Water Resour 95:161–175
Arab MR, Semma E, Pateyron B, El Ganaoui M (2009) Determination of physical properties of porous materials by a lattice Boltzmann approach. FDMP Fluid Dyn Mat Process J 5(2):161–175
Arkeryd L (1972a) On the Boltzmann equation part I: Existence. Arch Ration Mech Anal 45:1–16
Arkeryd L (1972b) On the Boltzmann equation part II: The full initial value problem. Arch Ration Mech Anal 45:17–34
Arnold BC (2015) Pareto distributions, 2nd edn. Chapman and Hall/CRC, Boca Raton, FL
Arpat GB, Caers J (2007) Conditional simulation with patterns. Math Geol 39(2):542–547
Arrufat T, Bondino I, Zaleski S, Lagrée B, Keskes N (2014) Developments on relative permeability computation in 3D rock images. SPE 172025
Atkinson KE (1978) An introduction to numerical analysis, 2nd edn. Wiley, New York
Auzerais FM, Dunsmuir J, Ferreol BB, Martys N, Olson J, Ramakrishnan TS, Rothman DH, Schwartz LM (1996) Transport in sandstone: a study based on three dimensional microtomography. Geophys Res 23:705–708
Backus GE, Gilbert F (1968) Numerical applications of a formalism for geophysical inverse problems. Geophys JR Astr SocW:247–276
Backus GE, Gilbert F (1970) Uniqueness in the inversion of inaccurate gross earth data. Phil Trans R Soc Lond a, Math Phys Sci 266:123–192
Bailey P, Myre J, Walsh SDC, Lilja DJ, Saar MO (2009) Accelerating Lattice Boltzmann fluid flow simulations using graphics processors. In: International conference on parallel processing, Vienna, Austria, pp 550–557
Bak P, Christensen K, Danon L, Scanlon T (2002) Unified scaling law for earthquakes. Phys Rev Lett 88(17):178501
Bakhshian S, Hosseini SA, Shokri N (2019) Pore-scale characteristics of multiphase flow in heterogeneous porous media using the lattice Boltzmann method. Sci Rep 9(1):3377
Balasubramaniam K, Hayot F, Saam WF (1987) Darcy's law from lattice-gas hydrodynamics. Phys Rev A 36:2248–2253
Ball RC, Blumenfeld R (1990) Universal scaling of the stress field at the vicinity of a wedge crack in two dimensions and oscillatory self-similar corrections to scaling. Phys Rev Lett 65(14):1784
Bao YB, Meskas J (2011) Lattice Boltzmann method for fluid simulations. Courant Institute of Mathematical Sciences, New York University, Technical Report, Department of Mathematics, p 44
Bao J, Schaefer L (2013) Lattice Boltzmann equation model for multi-component multi-phase flow with high density ratios. Appl Math Model 37(4):1860–1871
Barabási A-L, Stanley HE (1995) Fractal concepts in surface growth. Cambridge University Press, New York

Bardow A, Karlin IV, Gusev AA (2006) General characteristic-based algorithm for off-lattice Boltzmann simulations. Europhys Lett 75:434–440
Barenblatt GI, Zeldovich YB (1971) Intermediate asymptotics in mathematical physics. Russian Math Surv 26:45–61
Basu A, Neil Frazer L (1990) Rapid determination of the critical temperature in simulated annealing inversion. Science 4975:1409–1412
Batchelor GK (2000) An introduction to fluid dynamics. Cambridge University Press, Cambridge
Bear J (1972) Dynamics of fluids in porous media. American Elsevier, New York
Bear J, Bachmat Y (1990) Introduction to modelling of transport phenomena in porous media. Kluwer Academic Publishers, Dordrecht
Belavin AA, Polyakov AM, Zamolodchikov AB (1984) Infinite conformal symmetry in two-dimensional quantum field theory. Nucl Phys B 241(2):333–380
Ben-Ameur W (2004) Computing the initial temperature of simulated annealing. Comput Optim Appl 29:369–385
Benzi R, Succi S, Vergassola M (1992) The lattice Boltzmann equation: theory and applications. Phys Rep 222(3):145–197
Benzi R, Biferale L, Sbragaglia M, Succi S, Toschi F (2006a) Mesoscopic modeling of a two-phase flow in the presence of boundaries: the contact angle. Phys Rev E 74(2):021509
Benzi R, Biferale L, Sbragaglia M, Succi S, Toschi F (2006b) Mesoscopic modelling of heterogeneous boundary conditions for microchannel flows. J Fluid Mech 548:257
Bernsdorf JM (2008) Simulation of complex flows and multi-physics with the Lattice-Boltzmann method. PhD dissertation, Universiteit van Amsterdam
Bernsdorf J, Brenner G, Durs F (2000) Numerical analysis of the pressure drop in porous media with lattice Boltzmann (BGK) automata. Comput Phys Commun 129(1–3):247–255
Berry MW, Lewis ZV (1980) On the Weierstrass-Mandelbrot fractal function. Proc R Soc Lond A 370:459–484
Bertrand J, Bertrand P, Ovarlez JP (1996) The Mellin transform. In: Poularikas AD (ed) The transforms and applications handbook. CRC Press, Boca Raton, FL
Bertrand O, Binet B, Combeau H, Couturier S, Delannoy Y, Gobin D, Lacroix M, Le Quéré P, Médale M, Mencinger J, Sadat H, Vieira G (1999) Melting driven by natural convection, a comparison exercise: first results. Int J Therm Sci 38:5–26
Bhatnagar P, Gross E, Krook M (1954) A model for collisional processes in gases I: Small amplitude processes in charged and neutral one-component system. Phys Rev A 94:511–524
Billings SD (1994) Simulated annealing for earthquake location. Geophys J Int 118:680–692
Binder K, Stauffer D (1985) A simple introduction to Monte Carlo simulations and some specialized topics. In: Binder K (ed) Applications of the Monte Carlo method in statistical physics. Springer-Verlag, Berlin, pp 1–36
Bird GA (1994) Molecular gas dynamics and the direct simulation of gas flows. Clarendon Press, Oxford
Boek ES, Venturoli M (2010) Lattice-Boltzmann studies of fluid flow in porous media with realistic rock geometries. Comput Math Appl 59(7):2305–2314
Boek ES, Chin J, Coveney PV (2003) Lattice Boltzmann simulation of the flow of non-Newtonian fluids in porous media. Int J Mod Phys B 17:99–102
Boettinger WJ, Warren JA, Beckermann C, Karma A (2002) Phase-field simulation of solidification. Annu Rev Mater Res 32:163–194
Bogolyubov NN, Shirkov DV (1960) Introduction to the theory of quantized fields. Interscience Publishers, New York
Boltzmann L (1896) Vorlesungen über Gastheorie, vol I. J.A. Barth, Leipzig
Boltzmann L (1898) Vorlesungen über Gastheorie, vol II. J.A. Barth, Leipzig
Bondino I, Hamon G, Kallel W, Kachuma D (2013) Relative permeabilities from simulation in 3D rock models and equivalent pore networks: critical review and way forward. Petrophysics 54(6, SI):538

Borgnat P, Flandrin P, Amblard P-O (2001) Stochastic discrete scale invariance and Lamperti transformation. In: IEEE workshop statistics signal processing, pp 66–69
Borgnat P, Flandrin P, Amblard PO (2002) Stochastic discrete scale invariance. IEEE Signal Process Lett 9(6):181–184
Borodich FM (1993) The Hertz frictional problem contact between nonlinear elastic anisotropic bodies (the similarity approach). Int J Solids Struct 30:1513–1526
Borodich FM (1997) Renormalization schemes for earthquake prediction. Geophys J Int'l 131:171–178
Bosl WJ, Dvorkin J, Nur A (1998) A numerical study of pore structure and permeability using a Lattice Boltzmann simulation. Geophys Res Lett 25:1475–1478
Bouzidi M, Firdaouss M, Lallemand P (2001) Momentum transfer of a Boltzmann-lattice fluid with boundaries. Phys Fluids 13(11):3452–3459
Bowman DD, Sammis CG (1996) An observational determination of the critical region before the 1983 M = 6.7 Coalinga earthqake (Abst.). EOS Trans Am Geophys U 77:F486
Brémaud P (1998) Markov chains: Gibbs fields, Monte Carlo simulation, and queues. Texts in applied mathematics, vol 31. Springer, New York
Briant AJ, Yeomans JM (2004) Lattice Boltzmann simulations of contact line motion. II. Binary fluids. Phys Rev E 69(3):031603
Briant AJ, Wagner AJ, Yeomans JM (2004) Lattice Boltzmann simulations of contact line motion. I. Liquid-gas systems. Phys Rev E 69(3):031602
Brinkman HC (1949) A calculation of the viscous force exerted by a flowing fluid on a dense swarm of particles. Appl Sci Res A 1:27–34
Broadbent SR (1954) Discussion on symposium on Monte Carlo methods. J R Stat Soc B 16:68
Broadbent SR, Hammersley JM (1957) Percolation processes I: crystals and mazes. Proc Camb Philos Soc 53:629–641
Brochardwyart F, DeGennes P (1992) Dynamics of partial wetting. Adv Coll Interface Sci 39:1
Burnett D (1936) The distribution of molecular velocities and the mean motion in a non-uniform gas. Proc Lond Math Soc 40:382
Bustin RM, Bustin AM, Cui X, Ross DJK, Pathi VS (2008) Impact of shale properties on pore structure and storage characteristics. In: SPE Shale gas production conference: SPE-119892
Cabrera C, Fine-Morris M, Pokross M, Kish K, Michalczyk S, Cahn M, Klei H, Russo MF (2014) Dynamically optimizing experiment schedules of a laboratory robot system with simulated annealing. J Lab Autom 19:517–527
Camia F (2017) Phase transitions, Euclidean fields and self-similar random fractals. Niew Arhief voor Wiskunde 5/18(1):49–54
Camia F, Garban C, Newman CM (2015) Planar Ising magnetization field I. Uniqueness of the critical scaling limit. Ann Probab 43:528–571
Cancelliere A, Chang C, Foti E, Rothman D, Succi S (1990) The permeability of a random medium: comparison of simulation with theory. Phys Fluids A 2(12):2085–2088
Čapek PH, Brabec V, Zikánová AL, Kočiřík M (2008) Stochastic reconstruction of particulate media using simulated annealing: improving pore connectivity. Transp Porous Media 76(2):179–198
Cauchy AL (1821) Cours d'Analyse de l'Ecole Royale Polytechnique. Chez Debure frères, Paris
Cercignani C (1975) Theory and application of the Boltzmann equation. Scottish Academic Press, London
Cerny V (1982) A thermodynamical approach to the travelling salesman problem: an efficient simulation algorithm. Report, Comenius University, Bratislava, Czechoslovakia
Cerny V (1985) A thermodynamical approach to the traveling salesman problem: an efficient simulated algorithm. J Optim Theory Appl 45:41–51
Chai Z, Shi B (2008) A novel Lattice Boltzmann model for the Poisson equation. Appl Math Model 32:2050–2058
Chang C, Liu C-H, Lin C-A (2009) Boundary conditions for lattice Boltzmann simulations with complex geometry flows. Comput Math Appl 58(5):940–949

Chang Q, Alexander JID (2007) Study of Marangoni-natural convection in a two-layer liquid system with density inversion using a lattice Boltzmann model. Phys Fluids 19:102107
Charnes A, Wolfe M (1989) Extended Pincus theorems and convergence of simulated annealing. Int J Syst Sci 20(S):1521–1533
Chelidze TL (1982) Percolation and fracture. Phys Earth Planet Inter 28(2):93–101
Chen JD (1987) Radial viscous fingering patterns in Hele-Shaw cells. Exp Fluids 5:363–371
Chen S, Doolen GD (1998) Lattice Boltzmann method for fluid flows. Ann Rev Fluid Mech 30:329–364
Chen S, Diemer K, Doolen G, Eggert K, Fu C, Gutman S, Travis B (1991) Lattice gas automata for flow through porous media. Phys D 4:72–84
Chen H, Chen S, Matthaeus WH (1992) Recovery of the Navier-Stokes equations using a lattice-gas Boltzmann method. Phys Rev A 45:R5339–R5342
Chen S, Martínez D, Mei R (1996) On boundary conditions in lattice Boltzmann methods. J Phys Fluids 8:2527–2536
Chen HY, Jasnow D, Vinals J (2000) Interface and contact line motion in a two phase fluid under shear flow. Phys Rev Lett 85(8):1686
Chen H, Kandasamy S, Orszag S, Shock R, Succi S, Yakhot V (2003) Extended Boltzmann kinetic equation for turbulent flows. Science 301(5633):633–636
Chen C, Li L, Mei R, Klausner JF (2018a) Chapman-Enskog analyses on the gray lattice Boltzmann equation method for fluid flow in porous media. J Stat Phys 171(3):493–520
Chen H, Qin S, Xue L, Yang B, Zhan K (2018b) A physical model predicting instability of rock slopes with locked segments along a potential slip surface. Eng Geol 242:34–43
Chen L, Zhang L, Kang Q, Viswanathan HS, Yao J, Tao W (2015) Nanoscale simulation of shale transport properties using the lattice Boltzmann method: permeability and diffusivity. Sci Rep 5:8089
Chikatamarla S, Karlin I (2013) Entropic lattice Boltzmann method for turbulent flow simulations: boundary condition. Phys A 392(9):1925–1930
Chin J, Boek ES, Coveney PV (2002) Lattice Boltzmann simulation of the flow of binary immiscible fluids with different viscosities using the Shan-Chen microscopic interaction model. Proc R Soc Lond A 360:547–558
Cho H, Jeong N, Sung HJ (2013) Permeability of microscale fibrous porous media using the lattice Boltzmann method. Int J Heat Fluid Flow 44:435–443
Chopard B, Droz M (1998) Cellular automata modelling of physical systems. Cambridge University Press, Cambridge
Chopard B, Luthi PO, Droz M (1994) Reaction-diffusion cellular automata model for the formation of Liesegang patterns. Phys Rev Lett 72(9):13841387
Chopard B, Luthi PO, Wagen JF (1997) Lattice Boltzmann method for wave propagation in urban microcells. IEE Proc Microw Antennas Propag 144(4):251–255
Chopard B, Masselot A, Dupuis A (2000) A lattice gas model for erosion and particles transport in a fluid. Comput Phys Commun 129(1–3):167–176
Civan F (2006) Viscosity-temperature correlation for crude oils using an Arrhenius-type asymptotic exponential function. Pet Sci Technol 24:6
Clauset A, Shalizi CR, Newman ME (2009) Power-law distributions in empirical data. SIAM Rev 51:661–703
Connington K, Lee T (2012) A review of spurious currents in the Lattice Boltzmann method for multiphase flows. J Mech Sci Technol 26(12):3857–3863
Cornubert R, d'Humires D, Levermore D (1998) A Knudsen layer theory for lattice gases. Phys D 47:241–259
Creutz M (1983) Microcanonical Monte Carlo simulation. Phys Rev Lett 50(19):1411–1414
Crouse B, Freed DM, Koliha N, Balasubramanian G, Satti R, Bale D, Zuklic S (2016) A Lattice-Boltzmann based method applied to digital rock characterization of perforation tunnel damage. In: SCA2016-058: 1/6

Curtis ME (2010) Structural characterization of gas shales on the micro-and nano-scales. In: SPE Canada unconventional resources conference: SPE-137693
Czirók A, Somfai E, Vicsek T (1993) Experimental evidence for self-affine roughening in a micromodel of geomorphological evolution. Phys Rev Lett 71:2154
Dardis O, McCloskey J (1998) Lattice Boltzmann scheme with real numbered solid density for the simulation of flow in porous media. Phys Rev E 57:4834–4837
de Anna P, Le Borgne T, Dentz M, Tartakovsky AM, Bolster D, Davy P (2013) Flow intermittency, dispersion, and correlated continuous time random walks in porous media. Phys Rev Lett 11:184502
Dellar PJ (2003) Incompressible limits of lattice Boltzmann equations using multiple relaxation time. J Comput Phys 190(2):351–370
Derksen J (2001) Assessment of large eddy simulations for agitated flows. Chem Eng Res Des 79(8):824–830
Derksen JJ (2002) Lattice-Boltzmann based large-eddy simulations applied to industrial processes. In: Sloot PMA et al (eds) Lecture notes in computer science, vol 2329, pp 713–722
Derksen JJ (2007) The Lattice-Boltzmann method for multiphase fluid flow simulations and Euler-Lagrange large-eddy simulations. In: Marchisio DL, Fox RO (eds) Multiphase reacting flows: modelling and simulation. Springer, Cham
Derksen JJ, van den Akker HEA (1999) Large-eddy simulations on the flow driven by a Rushton turbine. AIChE J 45:209–221
Derksen JJ, Kontomaris K, McLaughlin JB, van den Akker HEA (2007) Large-eddy simulations of single-phase flow dynamics and mixing in an industrial crystallizer. Chem Eng Res Des 85(2):169–179
Deutsch CV (1992) Annealing techniques applied to reservoir modeling and the integration of geological and engineering (well test) data. PhD dissertation, Stanford University, Stanford CA
Deutsch CV, Cockerham PW (1994) Practical considerations in the application of simulated annealing to stochastic simulation. Math Geol 26:67–82
Deutsch CV, Journel AG (1992) GSLIB-geostatistical software library and user's guide. New York University Press, New York
Deutsch CV, Journel AG (1994) The application of simulated annealing to stochastic reservoir modeling. Soc Pet Eng Adv Technol Ser 1994:222
Dhori PK, Slattery JC (1997) Common line motion I: implications of entropy inequality. J Nonnewton Fluid Mech 71(3):197–213
d'Humieres D, Ginzburg I, Krafczyk M, Lallemand P, Luo LS (2002) Multiple-relaxation-time Lattice Boltzmann Models in three dimensions. Philos Trans Math Phys Eng Sci 360:437
DiGiovanni AA, Fredrich JT, Holcomb DJ, Olsson WA (2000) Micromechanics of compaction in an analogue reservoir sandstone. In: Proceedings of the 4th U.S. Rock Mech. Symposium, pp 1153–1160
Diógenesa AN, dos Santos LOE, Fernandes CP, Moreira AC, Apolloni CR (2009) Porous media microstructure reconstruction using pixel-based and object-based simulated annealing – comparison with other reconstruction methods. Engenharia Térmica (Thermal Eng) 8(02):35–41
Dodds PS, Rothman DH (2000) Scaling, universality, and geomorphology. Annu Rev Earth Planet Sci 28(1):571–610
Doolen GD (ed) (1990) Lattice gas methods for partial differential equations. Addison-Wesley, Redwood City CA
Dosso SE, Oldenburg DW (1991) Numerical applications of a formalism for geophysical inverse problems. Geophys J Intern 106:379–385
Dou Z, Zhou ZF (2013) Numerical study of non-uniqueness of the factors influencing relative permeability in heterogeneous porous media by lattice Boltzmann method. Int J Heat Fluid Flow 42:23–32
Doucot B, Wang W, Chaussy J, Pannetier B, Rammal R (1986) First observation of the universal periodic corrections to scaling: magnetoresistance of normal-metal self-similar networks. Phys Rev Lett 57:1235–1238

Duclut C, Delamotte B (2017) Nonuniversality in the erosion of tilted landscapes. Phys Rev E 96(1):012149

Dzikowski M, Jasinski L, Dabrowski M (2018) Depth-averaged Lattice Boltzmann and finite element methods for single-phase flows in fractures with obstacles. Comput Math Appl 75(10):3453–3470

Ehlers W (2022) Darcy, Forchheimer, Brinkman and Richards: classical hydromechanical equations and their significance in the light of the TPM. Arch Appl Mech 92:619–639

Erman WT (1999) Log spirals in the stock market. In: Technical analysis of stock commodities, February, 16–34

Fauzi U (2011) An estimation of rock permeability and its anisotropy from thin sections using a renormalization group approach. Energy Sourc Part a. Recov Utilization Environ Effects 33:539–548

Fauzi U, Latief FDE, Akbar RM, Qohar UNA (2017) Lattice-Boltzmann color gradient method for fluid flow simulation in porous media with viscosity as a function of temperature. In: 4th international workshop on rock physics, 29 May–2 June 2017, Trondheim, Norway

Feigenbaum JA, Freund PGO (1996) Discrete scale invariance in stock markets before crashes. Int J Mod Phys 10(27):3737–3745

Feng YT, Han K, Owen DRJ (2007) Coupled lattice Boltzmann method and discrete element modeling of particle transport in turbulent fluid flows: computational issues. Int J Numer Meth Eng 72:1111–1134

Ferrari PA, Frigessi A, Schonmann RH (1993) Convergence of some partially parallel Gibbs samplers with annealing. Ann Appl Probab 3(1):137–153

Ferréol BD, Rothman H (1995) Lattice-Boltzmann simulations of flow through Fontainebleau sandstone. Transp Porous Media 20(1–2):3–20

Fiorentino EA, Toussaint R, Jouniaux L (2016) Lattice Boltzmann modelling of: variations with salinity in monophasic conditions. Geophys Suppl Mon Notices R Astron Soc 205(1):648–664

Fiorentino EA, Toussaint R, Jouniaux L (2017) Two-phase Lattice Boltzmann modelling of streaming potentials: influence of the air–water interface on the electrokinetic coupling. Geophys J Int 208(2):1139–1156

Fisch R, Harris AB (1978) Critical behavior of random resistor networks near the percolation threshold. Phys Rev B 18:416

Flekkoy E (1993) Lattic BGK models for miscible fluids. Phys Rev E 47(6):4247–4257

Fogel D (2006) Evolutionary computation: toward a new philosophy of machine intelligence, 3rd edn. IEEE Press, Piscataway, NJ

Fredrich JT, Noble DR, O'Connor RM, Lindquist WB (1999) Development, implementation, and experimental validation of the Lattice-Boltzmann method for modeling three-dimensional complex flows. Sandia National Laboratory Report: SAND99-0369

Fredrich JT, Di Giovanni AA, Noble DR (2006) Predicting macroscopic transport properties using microscale image data. J Geophys Res 111:B03201

Frost AJ, Prechter R (1998) Elliott wave principle: key to market behavior, 7th edn. New Classics Library, Gainesville, Georgia

Fu J, Thomas HR, Li C (2021) Tortuosity of porous media: image analysis and physical simulation. Earth Sci Rev 212:103439

Gallivan M, Noble D, Georgiadis J, Buckius R (1997) An evaluation of the bounce-back boundary condition for lattice-Boltzmann simulation. Int J Numer Meth Fluids 25(3):249–263

Gao Y, Sharma MM (1994) A LGA model for fluid flow in heterogeneous porous media. Transp Porous Media 17:1–17

Gardner W, Franks L (1975) Characterization of cyclostationary random signal processes. IEEE Trans Inform Theory IT-21(1):4–14

Garey MR, Johnson DS (1979) Computers and intractability: a guide to the theory of NP-completeness. Freeman, San Francisco

Gefen Y, Aharony A, Shapir Y, Mandelbrot BB (1984) Phase transitions on fractals. II. Sierpinski gaskets. J Phys A: Math Gen 17(2):435–444

Geier M, Fakhari A, Lee T (2015) Conservative phase-field lattice Boltzmann model for interface tracking equation. Phys Rev E 91(6):063309
Geller S, Krafczyk M, Tölke J, Turek S, Hron J (2006) Benchmark computations based on lattice-Boltzmann, finite element and finite volume methods for laminar flows. Comput Fluids 35(8–9):888–897
Geman S, Geman D (1984) Stochastic relaxation, Gibbs distribution and the Bayesian restoration in images. IEEE Trans Patt Anal Mach Int 6(6):721–741
Ghassemi A, Pak A (2011a) Pore scale study of permeability and tortuosity for flow through particulate media using Lattice Boltzmann method. Int J Numer Anal Meth Geomech 35(8):886–901
Ghassemi A, Pak (2011b) Numerical study of factors influencing relative permeabilities of two immiscible fluids flowing through porous media using lattice Boltzmann method. J Pet Sci Eng 77(1):135–145
Ghassemi A, Pak A, Shahir H (2010) Numerical study of the coupled hydro-mechanical effects in dynamic compaction of saturated granular soils. Comput Geotech 37(1–2):10–24
Gibert D, Virieux J (1991) Electromagnetic imaging and simulated annealing. J Geophys Res Solid Earth 96:8057–8067
Gidas B (1985) Non-stationary Markov chains and convergence of the annealing algorithm. J Statist Phys 3:73–131
Gidas B (1989) A renormalization group approach to image processing problems. IEEE Trans Pattern Anal Mach Intell 11(2):164–180
Ginzburg I (2016) Comment on "An improved gray lattice Boltzmann model for simulating fluid flow in multi-scale porous media": intrinsic links between LBE Brinkman schemes". Adv Water Resour 8:241–249
Gladyshev E (1963) Periodically and almost periodically correlated random processes with continuous time parameter. Theory Prob. Appl. 8:173–177
Glover F (1986) Future paths for integer programming and links to artificial intelligence. Comput Oper Res 13(5):533–549
Goel G, Abidoye LK, Chahar BR, Das DB (2016) Scale dependency of dynamic relative permeability–saturation curves in relation with fluid viscosity and dynamic capillary pressure effect. Environ Fluid Mech 16:945–963
Goldenfeld N (1992) Lectures on phase transitions and the renormalization group. CRC Press, Boca Raton
Gomez JB, Pacheco AF, Segui-Santonja AJ (1995) A model for crack connectivity in rocks, a discussion. Math Geol 2:23–39
Goovaerts P (1996) Stochastic simulation of categorical variables using a classification algorithm and simulated annealing. Math Geol 28:909–921
Graner F, Dubrulle B (1994) Titius-Bode laws in the solar system: I. Scale invariance explains everything. Astron Astrophys 282:262–268; II. Build your own law from disk models. Astron Astrophys 282:269–276
Green CP, Paterson L (2007) Analytical three-dimensional renormalization for calculating effective permeabilities. Transp Porous Media 68:237–248
Grosfils P, Boon JP (2003) Viscous fingering in miscible, immiscible and reactive fluids. Int J Mod Phys B 17(1–2):15–20
Grosfils P, Boon JP, Chin J, Boek ES (2004) Structural and dynamical characterization of Hele-Shaw viscous fingering. Proc R Soc Lond A 362(1821):1723–1734
Grucelski A, Pozorski J (2013) Lattice Boltzmann simulations of flow past a circular cylinder and in simple porous media. Comput Fluids 71:406–416
Gunde A, Babadagli T, Mitra S (2010) Lattice-Boltzmann method to estimate relative permeabilities for matrix-fracture interaction in naturally fractured reservoirs. In: SPE eastern regional meeting meeting, Morgantown, West Virginia, October 13–15
Gunning J, Sams M (2018) Joint facies and rock properties Bayesian amplitude-versus-offset inversion using Markov random fields. Geophys Prospect 66:904–919

Gunstensen AK, Rothman DH (1991) Lattice Boltzmann model of immiscible fluids. Phys Rev A 43:4320–4327

Gunstensen AK, Rothman DH (1992) Lattice-Boltzmann studies of immiscible two-phase flow through porous media. J Geophys Res 98(B4):6431–6441

Gunstensen AK, Rothman DH, Zaleski S, Zanetti G (1991) Lattice Boltzmann model of immiscible fluids. Phys Rev A 43:4320

Guo L, Chen YangQuan, Shi S, West BJ (2021) Renormalization group and fractional calculus methods in a complex world: a review. Fract Calc Appl Anal 24(1):5–53

Guo Z, Shi B, Wang N (2000) Lattice BGK model for incompressible Navier-Stokes equation. J Comput Phys 165:288–306

Guo Z, Zhao TS (2002) Lattice Boltzmann model for incompressible flows through porous media. Phys Rev E 66:036304

Guo Z, Shi B, Zheng C (2002) A coupled lattice BGK model for the Boussinesq equations. Int J Num Methods 39:325–342

Guodong J, Tad W, Patzek S, Silin DB (2004) Direct prediction of the absolute permeability of unconsolidated and consolidated reservoir rock. In: SPE 90084

Habich J (2008) Performance evaluation of numeric compute kernels on nVIDIA GPUs8. MSc thesis. Friedrich-Alexander-Universität

Hajek B (1988) Cooling schedules for optimal annealing. Math Oper Res 13(2):311–329

Halisch M (2013) Application and assessment of the Lattice Boltzmann method for fluid flow modeling in porous rocks. PhD dissertation. Technische Universität Berlin

Hammersley JM, Handscomb DC (1964) Monte Carlo methods. Methuen & Co., London; Wiley, New York, London

Hanasoge S, Agarwal U, Tandon K, Koelman JVA (2017) Renormalization group theory outperforms other approaches in statistical comparison between upscaling techniques for porous media. Phys Rev E 96(3):033313

Hao L, Cheng P (2010) Pore-scale simulations on relative permeabilities of porous media by lattice Boltzmann method. Int J Heat Mass Transf 53(9–10):1908–1913

Harris AB, Lubensky TC, Holcomb WK, Dasgupta C (1975) Renormalization group approach to percolation problems. Phys Rev Lett 35:327

Hayashi H (2003) Lattice Boltzmann method and its application to flow analysis in porous media. R&D Rev Toyota CRDL 38(1):17–25

Hazlett RD (1997) Statistical characterization and stochastic modeling of pore networks in relation to fluid flow. Math Geol 29:801–822

He X, Luo L-S (1997a) A priori derivation of the lattice Boltzmann equation. Phys Rev E 55(6):R6333–R6336

He X, Luo LS (1997b) Theory of the lattice Boltzmann method: from the Boltzmann equation to the lattice Boltzmann equation. Phys Rev E 56:6811

He X, Zou Q, Luo L-S, Dembo SM (1996) Some progress in the lattice Boltzmann method. Part I, Non-uniform mesh grids. J Comp Phys 129:357–363

He X, Zou Q, Luo LS, Dembo M (1997) Analytic solutions of simple flows and analysis of nonslip boundary conditions for the lattice Boltzmann BGK model. J Stat Phys 87:115–136

He X, Luo L-Sh (1997) Lattice Boltzmann Model for the Incompressible Navier-Stokes Equation J. Stat. Phys. 88(3/4):927–944

He X, Chen S, Doolen G (1998) A novel thermal model for the lattice Boltzmann method in the incompressible limit. J Comput Phys 146:282–300

He X, Chen S, Zhang R (1999) A lattice Boltzmann scheme for incompressible multiphase flow and its application in simulation of Rayleigh-Taylor instability. J Comput Phys 152(2):642–663

Heijs AWJ, Lowe CP (1995) Numerical evaluation of the permeability and the Kozeny constant for two types of porous media. Phys Rev E 51(5):4,346–4,352

Herrmann HJ, Roux S (1990) Statistical models for the fracture of disordered media. North-Holland, Amsterdam

Hosa A, Curtis A, Wood R (2016) Calibrating lattice Boltzmann flow simulations and estimating uncertainty in the permeability of complex porous media. Adv Water Res 94:60–74
Higuera FJ, Jiménez J (1989) Boltzmann approach to lattice gas simulations. Europhys Lett 9(7):663–668
Hill R (1963) Elastic properties of reinforced solids: some theoretical principles. J Mech Phys Solids 11(5):357–372
Hu B (1982) Introduction to real-space renormalization group methods in critical and chaotic problems. Phys Rep 91:233
Huang H Jr, Thorne DT, Schaap MG, Sukop MC (2007) Proposed approximation for contact angles in Shan-and-Chen-type multicomponent multiphase lattice Boltzmann models. Phys Rev E 76:066701
Huang H, Lu XY (2009) Relative permeabilities and coupling effects in steady-state gas-liquid flow in porous media: a lattice Boltzmann study. Phys Fluids 21(9):092104
Huang H, Li Z, Liu S, Lu XY (2009) Shan-and-Chen-type multiphase lattice Boltzmann study of viscous coupling effects for two-phase flow in porous media. Int J Numer Methods Fluid 61(3):341–354
Huang H, Sukop M, Lu X-Y (2015) Multiphase Lattice Boltzmann method, theory and application. Wiley, Chichester, West Sussex, UK
Huang Y, Saleur H, Sammis CG, Sornette D (1996) Precursors, aftershocks, criticality and self-organized criticality. Europhys Lett 41:43–48
Huang Y, Ouillon G, Saleur H, Sornette D (1997) Spontaneous generation of discrete scale invariance in growth models. Phys Rev E 55:6433–6447
Huber C, Parmigiani A, Chopard B, Manga M, Bachmann O (2008) Lattice Boltzmann model for melting with natural convection. Int J Heat Fluid Flow 29:1469–1480
Hunt AG (2004) Continuum percolation theory and Archie's law. Geoph Res Lett 3:L19503
Hunt A, Ewing R (2009) Percolation theory for flow in porous media. Lecture notes in physics. Springer, Cham
Hurd H (1969) An investigation of periodically correlated stochastic processes. PhD dissertation, Dept. Elect. Eng., Duke Univ., Durham, NC
Hussain F, Pinczewski WV, Cinar Y, Arns JY, Arns CH, Turner ML (2014) Computation of relative permeability from imaged fluid distributions at the pore scale. Transp Porous Media 104(1):91–107
Hwa T, Kardar M (1992) Avalanches, hydrodynamics, and discharge events in models of sandpiles. Phys Rev A 45:7002
Igarashi G, Saeki S, Takahata N, Sumikawa K, Tasaka S, Sasaki Y, Takahashi M, Sano Y (1995) Groundwater radon anomaly before the Kobe earthquake in Japan. Science 269:60–61
Ide KD, Sornette D (2002) Oscillatory finite-time singularities in finance, population and rupture. Physica A 307(1-2):63–106
Ingber L (1989) Very fast simulated re-annealing. Math Comput Model 12:967–973
Ingber L (1993) Simulated annealing: practice versus theory. Math Comput Model 18(11):29–57
Ingber L (2000) Adaptive simulated annealing (ASA): lessons learned. arXiv:cs/0001018
Irayani Z, Fauzi U, Viridi S, Latief FDE (2019) Calculation of anisotropy permeability from 3D tomographic images using renormalization group approaches and lattice Boltzmann method. J Pet Explor Prod Technol 9:889–897
Jakobsen M, Wu RS (2016) Renormalized scattering series for frequency-domain waveform modelling of strong velocity contrasts. Geophys J Int 20:880–899
Jakobsen MO, Mosegaard K, Pedersen JM (1987) Global model optimisation in reflection seismology by simulated annealing. In: Proceedings of the mathematical geophysics fifth international seminar on model optimization in exploration Geophysics, Berlin
Javadpour F (2009) Nanopores and apparent permeability of gas flow in mudrocks (shales and siltstone). J Can Petr Technol 48(8):16–21
Jensen JL, Lake LW, Corbett PWM, Goggin DJ (1997) Statistics for petroleum engineers and geoscientists. Prentice Hall PTR, New Jersey

Jiang Q, Li J, Luo Z, Xu X, Assefa E, Deng H (2019) Study on the time-lag failure of sandstone with different degrees of unloading damage. Period Polytech Civil Eng 63(1):206–214
Jiaung W-S, Ho J-R, Kuo C-P (2001) Lattice-Boltzmann method for the heat conduction problem with phase change. Numer Heat Transfer, Part B 39:167–187
Joekar-Niasar V, Van Dijke MIJ, Hassanizadeh SM (2012) Pore-scale modeling of multiphase flow and transport: achievements and perspectives. Transp Porous Media 94(2):461–464
Johansen A, Sornette D, Wakita H, Tsunogai U, Newman WI, Saleur H (1996) Discrete scaling in earthquake precursory phenomena: evidence in the Kobe earthquake. Jpn J Phys I France 6:1391–1402
Johansen A, Sornette D, Ledoit O (1999) Predicting financial crashes using discrete scale invariance. arXiv preprint cond-mat/9903321
Johansen A, Ledoit O, Sornette D (2000a) Crashes as critical points. Int J Theor Appl Finance 3(1)
Johansen A, Saleur H, Sornette D (2000b) New evidence of earthquake precursory phenomena in the 17 January 1995 Kobe earthquake. Jpn Eur Phys J B 15:551–555
Journel AG, Huijbregts CJ (1978) Mining geostatistics. Academic Press, London
Kadanoff LP (1966) Scaling laws for Ising models near T_c. Physics 2(6):263–272
Kadanoff LP, Götze W, Hamblen D, Hecht R, Lewis EAS, Palciauskas VV, Rayl M, Swift J, Aspnes D, Kane J (1967) Static phenomena near critical points: theory and experiment. Rev Mod Phys 39(2):395–431
Kagan YY (2010) Earthquake size distribution: power-law with exponent $\beta \equiv 1/2$? Tectonophysics 490(1–2):103–114
Kainourgiakis ME, Kikkinides ES, Stubos AK, Kanellopoulos NK (1999) Simulation of self-diffusion of point-like and finite-size tracers in stochastically reconstructed Vycor porous glasses. J Chem Phys 111(6):2735–2743
Kainourgiakis ME, Kikkinides ES, Steriotis TA, Stubos AK, Tzevelekos KP, Kanellopoulos NK (2000) Structural and transport properties of alumina porous membranes from process-based and statistical reconstruction techniques. J Colloid Interface Sci 231(1):158–167
Kainourgiakis ME, Kikkinides ES, Galani A, Charalambopoulou GC, Stubos AK (2005) Digitally reconstructed porous media: transport and sorption properties. Transp Porous Media 58(1–2):43–62
Kanamori H (1995) The Kobe (Hyogo-ken Nanbu), Japan, earthquake of January 16, 1995. Seismol Res Lett 66(2):6–10
Kandhai D, Vidal DJ-E, Hoekstra AG, Hoefsloot H, Iedema P, Sloot PMA (1998) A comparison between lattice-Boltzmann and finite-element simulations of fluid flow in static mixer reactors. Int J Mod Phys 9:1123–1128
Kandhai D, Koponen A, Hoekstra A, Kataja M, Timonen J, Sloot P (1999a) Implementation aspects of 3D lattice-BGK: boundaries, accuracy and a new fast relaxation method. J Comp Phys 150:482–501
Kandhai D, Vidal DJ-E, Hoekstra AG, Hoefsloot H, Iedema P, Sloot PMA (1999b) Lattice-Boltzmann and finite-element simulations of fluid flow in a SMRX static mixer reactor. Int J Numer Methods Fluids 31:1019–1033
Kang PK, de Anna P, Nunes JP, Bijeljic B, Blunt MJ, Juanes R (2014) Pore-scale intermittent velocity structure underpinning anomalous transport through 3-d porous media. Geophys Res Lett 41(17):6184–6190
Kang Q, Zhang D, Chen S (2002) Displacement of a two-dimensional immiscible droplet in a channel. Phys Fluids 14(9):3203–3214
Kang Q, Zhang D, Chen S (2003) Simulation of dissolution and precipitation in porous media. J Geophys Res 108:2505
Kang Q, Zhang D, Chen S (2004) Immiscible displacement in a channel: simulations of fingering in two dimension. Adv Water Resour 27(1):13–22
Kang Q, Zhang D, Chen S (2005) Displacement of a three-dimensional immiscible droplet in a duct. J Fluid Mech 545:41–66

Kang Q, Lichtner PC, Zhang D (2006) Lattice Boltzmann pore-scale model for multicomponent reactive transport in porous media. J Geophys Res 111:B05203
Karim M, Krabbenhoft K (2010) New renormalization schemes for conductivity upscaling in heterogeneous media. Transp Porous Media 85:677–690
Keehm Y (2003) Computational rock physics: transport properties in porous media and applications. PhD dissertation, Stanford University
Keehm Y, Mukerji T, Nur A (2001) Computational rock physics at the pore scale: transport properties and diagenesis in realistic pore geometries. Lead Edge 20:180–183
Keehm Y, Mukerji T, Prasad M, Nur A (2003) Permeability prediction from thin sections using the Lattice-Boltzmann flow simulation. In: SEG annual meeting.
Khairuddin R, Zainuddin ZM (2019) A comparison of simulated annealing cooling strategies for redesigning a warehouse network problem. J Phys Conf Ser (IOP Publishing) 012078
Khalili AD, Arns J-Y, Hussain F, Cinar Y, Pinczewski WV, Arns CH (2013) Permeability upscaling for carbonates from the pore scale by use of multiscale X-ray-CT images. SPE Res Eval Eng 16:353–368
Khan A, Mosegaard K, Rasmussen KL (2000) A new seismic velocity model for the Moon from a Monte Carlo inversion of the Apollo Lunar seismic data. Geophys Res Lett 27:1591–1594
Khodja Mohamed R, Li J, Hussaini SR, Ali AZ, Al-Mukainah HS, Jangda ZZ (2020) Consistent prediction of absolute permeability in carbonates without upscaling. Oil Gas Sci Technol - Rev IFP Energies Nouvelles 75:44
Kim J, Lee J, Lee K (2001) Nonlinear correction to Darcy's law for a flow through periodic arrays of elliptic cylinders. Phys A 293:13–20
Kim SH, Pitsch H, Boyd ID (2008) Slip velocity and Knudsen layer in the lattice Boltzmann method for microscale flows. Phys Rev E 77:026704
King P (1989) The use of renormalization for calculating effective permeability. Transp Porous Media 4:37–58
King CY, Koizumi N, Kitagawa Y (1995) Hydrogeochemical anomalies and the 1995 Kobe earthquake. Science 269(5220):38–39
Kirkpatrick S (1973) Percolation and conduction. Rev Mod Phys 45(4):574–588
Kirkpatrick SC, Gelatt D, Vecchi MP (1982) Optimisation by simulated annealing. IBM Research Report, RC 9355, New York
Kirkpatrick S, Gelatt CD Jr, Vecchi MP (1983) Optimization by simulated annealing. Science 220(4598):671–680
Knight R, Chapman A, Knoll M (1990) Numerical modeling of microscopic fluid distribution in porous media. J Appl Phys 68(3):994–1001
Knopp K (1954) Theory and application of infinite series. Blackie and Son Limited, London
Knudsen M (1934) The kinetic theory of gases: some modern aspects. Methuen & Co, London
Koestel J, Larsbo M, Jarvis N (2020) Scale and REV analyses for porosity and pore connectivity measures in undisturbed soil. Geoderma 366:114206
Koponen A, Kataja M, Timonen J (1997) Permeability and effective porosity of porous media. Phys Rev E 6(3):3319–3325
Koponen A, Kandhai D, Hellen E, Alava M, Hoekstra A, Kataja M, Niskanen K (1998) Permeability of three-dimensional random fiber webs. Phys Rev Lett 80(4):716–719
Koren Z, Mosegaard K, Landa E, Thore P, Tarantola A (1991) Monte Carlo estimation and resolution analysis of seismic background velocities. J Geophys Res 96(20):289–299
Korvin G (1982) Axiomatic characterization of the general mixture rule. Geoexploration 19(4):267–276
Korvin G (1992a) Fractal models in the earth sciences. Elsevier, Amsterdam
Korvin G (1992b) A percolation model for the permeability of kaolinite-bearing sandstone. Geophys Trans 37(2–3):177–209
Korvin G (2021) Statistical rock physics. In: Daya Sagar B, Cheng Q, McKinley J, Agterberg F (eds) Encyclopedia of mathematical geosciences. Encyclopedia of earth sciences series. Springer, Cham

Korvin G (2022) Rescaled range analysis. In: Daya Sagar BS, Cheng Q, McKinley J, Agterberg F (eds) Encyclopedia of mathematical geosciences. Encyclopedia of earth sciences series. Springer, Cham

Krafczyk M (2001) Gitter-Boltzmann-Methoden: Von der Theorie zur Anwendung. Habilitationsschrift, TU München

Küntz M, Mareschal JC (1997) Numerical estimation of the effective conductivity of heterogeneous media with a 2D cellular automata fluid. Geophys Res Lett 24:2865–2868

Küntz M, Lavallée P, Mareschal JC (1997) Determination of elastic properties of very heterogeneous media with cellular automata. J Geophy Res 102:7647–7658

Kuznik F, Obrecht C, Rusaouen G, Roux JJ (2010) LBM based flow simulation using GPU computing processor. Comput Math Appl 59(7):2380–2392. IDAIG

van Laarhaven PJM, Aarts EHL (1987) Simulated annealing; theory and applications. D. Reidel Publishing, Dordrecht

Ladd AJC (1994a) Numerical simulation of particulate suspensions via a discretized Boltzmann equation. J Fluid Mech 271:285–309

Ladd AJC (1994b) Numerical simulations of particulate suspensions via a discretized Boltzmann equation: Part2. Numerical results. J Fluid Mech 271:311–339

Lallemand P, Luo LS (2000) Theory of the lattice Boltzmann method: dispersion, dissipation, isotropy, Galilean invariance, and stability. Phys Rev E 61(6):6546

Laloux L, Potters M, Cont R, Aguilar JP, Bouchaud JP (1999) Are financial crashes predictable? Europhys Lett 45:1–5

Lamperti J (1962) Semi-stable stochastic processes. Trans Am Math Soc 104(1):62–78

Landa E, Beydoun WB, Tarantola A (1989) Reference velocity model estimation from prestack waveforms; coherency optimization by simulated annealing. Geophysics 54(8):984–990

Landry CJ (2013) Pore-scale imaging and Lattice Boltzmann modeling of sngle- and multi-phase flow in fractured and mixed-wet permeable media. PhD dissertation, The Pennsylvania State University

Landry CJ, Karpyn ZT, Ayala O (2014) Relative permeability of homogenous-wet and mixed-wet porous media as determined by pore-scale lattice Boltzmann modeling. Water Resour Res 5:3672–3689

Latva-Kokko, Rothman DH (2005) Diffusion properties of gradient-based Lattice Boltzmann models of immiscible fluid. Phys Rev E 71:056702

Lawler EL, Lenstra JK, Rinnooy Kan AHG, Shmoys DB (eds) (1985) The traveling salesman problem. Wiley, Chichester

Lei X, Nishizawa O, Moura A, Satoh T (2005) Hierarchical fracture process in brittle rocks by means of high-speed monitoring of AE hypocenter. J Acoust Emiss 23:102–112

Lester DR, Metcalfe G, Trefry MG (2014) Anomalous transport and chaotic advection in homogeneous porous media. Phys Rev E 90:063012

Levin DA, Peres Y (2017) Markov chains and mixing times, 2nd edn. American Mathematical Society, Providence

Levitan A (2022) Renormalization group approach to percolation in hierarchical lattices. arXiv: 2202.09436 [cond-mat.stat-mech]

Li B, Kwok DY (2003) Discrete Boltzmann equation for microfluidics. Phys Rev Lett 90:124502

Li RR, Yang YS, Pan J, Pereira GG, Taylor JA, Clennell B, Zou C (2014) Lattice Boltzmann modelling of permeability in porous materials with partially percolating voxels. Phys Rev E 90(3):033301

Li Z, Galindo-Torres S, Yan G, Scheuermann A, Li L (2018) A lattice Boltzmann investigation of steady-state fluid distribution, capillary pressure and relative permeability of a porous medium: Effects of fluid and geometrical properties. Adv Water Resour 116:153–166

Liang Z, Ioannidis MA, Chatzis I (2000) Reconstruction of 3D porous media using simulated annealing. In: Proceedings of the XIII international conference on computational methods in water resources, Alberta, Canada, pp 25–29

Lim C, Shu C, Niu X, Chew Y (2002) Application of lattice Boltzmann method to simulate microchannel flows. Phys Fluids 14(7):2299–2308

Liu G-Y, Chen Y-L, Du X, Azzam R (2021) A fractional viscoplastic model to predict the time-dependent displacement of deeply buried tunnels in swelling rock. Comput Geotech 129:103901

Liu H, Valocchi AJ, Kang Q (2012) Three-dimensional lattice Boltzmann model for immiscible two-phase flow simulations. Phys Rev E 85:046309; Erratum: Phys Rev E 85:069901

Liu H, Valocchi AJ, Werth C, Kang Q, Oostrom M (2014a) Pore-scale simulation of liquid CO_2 displacement of water using a two-phase lattice Boltzmann model. Adv Water Resour 73:144–158

Liu J, Pereira GG, Regenauer-Lieb K (2014b) From characterisation of pore-structures to simulations of pore-scale fluid flow and the upscaling of permeability using micro-tomography: a case study of heterogeneous carbonates. J Geochem Expl 144:84–96

Liu H, Kang Q, Leonardi CR, Schmieschek SMP, Narvaez Salazar AE, Jones BD, Williams JR, Valocchi AJ, Harting JDR (2016) Multiphase lattice Boltzmann simulations for porous media applications : a review. Comput Geosci 20(4):777–805

Liu H, Kang Q, Leonardi CR, Schmieschek S, Narváez A, Jones BD, Williams JR, Valocchi AJ, Harting J (2017) Multiphase lattice Boltzmann simulations for porous media applications. Comput Geosci 20(4):777

Loucks RG, Reed RM, Ruppel SC, Jarvie DM (2009) Morphology, genesis, and distribution of nanometer-scale pores in siliceous mudstones of the Mississippian Barnett Shale. J Sediment Res 79(12):848–861

Lu ZY, Liao Y, Qian DY, McLaughlin JB, Derksen JJ, Kontomaris K (2002) Large eddy simulations of a stirred tank using the lattice-Boltzmann method on a nonuniform grid. J Comput Phys 181:675–704

Lundy M, Mees A (1986) Convergence of an annealing algorithm. Math Program 3:111–124

Ma Sh-K, Mazenko GF (1975) Critical dynamics of ferromagnets in $6-\varepsilon$ dimensions: General discussion and detailed calculation. Phys Rev B 11:4077

Ma SK (1976) Modern theory of critical phenomena. Benjamin, New York

Ma X-Q (2002) Simultaneous inversion of prestack seismic data for rock properties using simulated annealing. Geophysics 67:1877–1885

MacCluer CR (2000) The many proofs and applications of Perron's theorem. SIAM Rev 42(3):487–498

Madden TR (1983) Microcrack connectivity in rocks: a renormalization group approach to the critical phenomena of conduction and failure in crystalline rocks. J Geophys Res 88:585–592

Madden TR, Williams E (1993) Role of size distributions on physical properties: real size renormalization group. J Geophys Res 98:15,951–15,965

Mahdi W, Medjahed SA, Ouali M (2017) Performance analysis of simulated annealing cooling schedules in the context of dense image matching. Comput Sist 2:493–501

Mahmoudi S, Hashemia A, Kord Sh (2014) Lattice Boltzmann relative permeability for CO_2-brine system in a 2D porous medium. Sci Iran C 21(6):2116–2125

Mahmoudi S, Mohammadzadeh O, Hashemi A, Kord S (2017) Pore-scale numerical modeling of relative permeability curves for CO_2–oil fluid system with an application in immiscible CO_2 flooding. J Pet Explor Prod Technol 7(1):235–249

Maier R, Bernard R (2010) Lattice-Boltzmann accuracy in pore-scale flow simulation. J Comput Phys 2292:233–255

Maier RS, Bernard RS, Grunau DW (1996) Boundary conditions for the lattice Boltzmann method. Phys Fluids 8(7):1788–1801

Maillot B, Main IG (1996) A lattice BGK model for the diffusion of pore pressure, including anisotropy, heterogeneity and gravity effects. Geophys Res Lett 23:13–16

Makarov KA (1994) Asymptotic expansions for Fourier transform of singular self-affine measures. J Math Anal Appl 187:259–286

Malaspinas O (2009) Lattice Boltzmann method for the simulation of viscoelastic fluid flows. PhD thesis, Ecole Polythecnique Federale de Lausanne

Mandelbrot BB (1982) The fractal geometry of nature. W.H. Freeman, San Francisco
Manwart C, Torquato S, Hilfer R (2000) Stochastic reconstruction of sandstones. Phys Rev E 62(1 Pt.B):893–899
Manwart C, Aaltosalmi U, Koponen A, Hilfer R, Timonen J (2002) Lattice-Boltzmann and finite-difference simulations for the permeability for three-dimensional porous media. Phys Rev E 66(1):016702
Marafini E, La Rocca M, Fiori A, Battiato I, Prestininzi P (2020) Suitability of 2D modelling to evaluate flow properties in 3D porous media. Transp Porous Media 134:315–329
Martys NS, Chen H (1996) Simulation of multicomponent fluids in complex three dimensional geometries by the lattice Boltzmann method. Phys Rev E 53:743–750
Martys NS, Torquato S, Bentz DP (1994) Universal scaling of fluid permeability for sphere packings. Phys Rev E 50(1):403–409
Matsuba I (1989) Optimal simulated annealing method based on stochastic-dynamic programming. Phys Rev A 39:2635–2642
McClure JE, Li Z, Berrill M, Ramstad T (2021) The LBPM software package for simulating multiphase flow on digital images of porous rocks. Comput Geosci 25:871–895
McNamara GR, Zanetti G (1988) Use of the Boltzmann equation to simulate lattice-gas automata. Phys Rev Lett 61(20):2332
McNamara G (1990) Diffusion in a lattice gas. Europhys Lett 12(4):329–334
Meakin P, Tartakovsky MA (2009) Modeling and simulation of pore-scale multiphase fluid flow and reactive transport in fractured and porous media. Rev Geophys 47:RG3002
Medina E, Hwa T, Kardar M, Zhang YC (1989) Burgers equation with correlated noise: renormalization-group analysis and applications to directed polymers and interface growth. Phys Rev A 39(6):3053
Medvedev D, Kassner K (2005) Lattice Boltzmann scheme for crystal growth in external flows. Phys Rev E 7:056703
Mehrizi AA, Sedighi K, Afrouzi, HH, Latif AA (2012) Lattice Boltzmann simulation of forced convection in vented cavity filled by porous medium with obstruction. World Appl Sci J 16(Spec Issue Appl Math):31–36
Mencinger J (2004) Numerical simulation of melting in two-dimensional cavity using adaptative grid. J Comp Phys 198:243–264
Metropolis N, Rosenbluth AW, Rosenbluth MN, Teller AH, Teller E (1953) Equation of state calculations by fast computing machines. J Chem Phys 21(6):1087–1092
Miller W, Succi S (2002) A lattice Boltzmann model for anisotropic crystal growth from melt. J Stat Phys 112:173–186
Misztal MK, Hernandez-Garcia A, Matin R, Müter D, Jha D, Sørensen HO, Mathiesen J (2015a) Simulating anomalous dispersion in porous media using the unstructured lattice Boltzmann method. Front Phys 3:50
Misztal MK, Hernandez-Garcia A, Matin R, Sørensen HO, Mathiesen J (2015b) Detailed analysis of the lattice Boltzmann method on unstructured. J Comput Phys 297:349
Mitra D, Romeo F, Sangiovanni-Vincentelli A (1986) Convergence and finite-time behavior of simulated annealing. Adv Appl Probab 18(3):747–771
Mohamad A (2011) Lattice Boltzmann method. Springer-Verlag Limited, London
Montgomery DR (2003) Predicting landscape-scale erosion rates using digital elevation models. CR Geosci 335(16):1121–1130
Mosegaard K, Sambridge M (2002) Monte Carlo analysis of inverse problems. Inverse Prob 18(3):R29–R54
Muniruzzaman ANM (1957) On measures of location and dispersion and tests of hypotheses in a Pareto distribution. Bull Calcutta Stat Assoc 7:115–123
Nagihara S, Hall SA (2001) Three-dimensional gravity inversion using simulated annealing: constraints on the diapiric roots of allochthonous salt structures. Geophysics 66:1438–1449

Narváez AT, Zauner FR, Hilfer R, Harting J (2010) Quantitative analysis of numerical estimates for the permeability of porous media from lattice-Boltzmann simulations. J Stat Mech: Theory Exp 11:P11026

Natarajan D, Verma NK, Abdul Salam T, Al-Sammak I, Koronfol S, Dernaika M, Teh J (2014) Computation of curves in middle eastern carbonates using digital rock physics. In: International petroleum technology conference: IPTC 18211

Nauenberg M (1975) Scaling representations for critical phenomena. J Phys A 8:925

Nelder JA, Mead R (1965) A simplex method for function minimization. Comput J 7(4):308–313

Newberry M, Savage VM (2019) Self-similar processes follow a power law in discrete logarithmic space. Phys Rev Lett 122:158303

Newman W, Gabrielov A, Durand T, Phoenix SL, Turcotte DL (1994) An exact renormalization model for earthquakes and material failure, statics and dynamics. Phys D 77:200–216

Newman WI, Turcotte DL, Gabrielov AM (1995) Log-periodic behavior of a hierarchical failure model with applications to precursory seismic activation. Phys Rev E 52(5):4827–4835

Nie X, Doolen GD, Chen S (2002) Lattice-Boltzmann simulations of fluid flows in mems. J Stat Phys 107(1–2):279–289

Ning Y, He S, Chen T, Jiang Y, Qin G (2015) Simulation of shale gas transport in 3D complex nanoscale-pore structures using the Lattice Boltzmann method. In: SPE Asia pacific unconventional resources conference and exhibition.

Noble D, Georgiadis J, Buckius R (1995a) Direct assessment of lattice Boltzmann hydrodynamics and boundary conditions for recirculating flow. J Stat Phys 81(1):17–33

Noble D, Chen JG, Buckius R (1995b) A consistent hydrodynamic boundary condition for the lattice-Boltzmann method. Phys Fluids 7(1):203–209

Nordbotten J, Celia M, Dahle H, Hassanizadeh S (2008) On the definition of macroscale pressure for multiphase flow in porous media. Water Resour Res 44; W06S02

Norris JQ, Turcotte DL, Rundle JB (2015) Anisotropy in fracking: a percolation model for observed microseismicity. Pure Appl Geophys 172:7–21

Nourani Y, Andresen B (1998) A comparison of simulated annealing cooling strategies. J Phys a: Math Gen 31(41):8373

Nourgaliev RR, Dinh TN, Sehgal BR (2002) On lattice Boltzmann modeling of phase transition in an isothermal non-ideal fluid. Nucl Eng Des 211(2–3):153–171

Novikov EA (1990) The effect of intermittency on statistical characteristics of turbulence and scale similarity of breakdown coefficients. Phys Fluids A 2:814–820

Nulton JD, Salamon P (1988) Statistical mechanics of combinatorial optimization. Phys Rev A Gen Phys 37:1351–1356

O'Brien G, Bean C, Frank McDermott (2002) A comparison of published experimental data with a coupled lattice Boltzmann-analytic advection–diffusion method for reactive transport in porous media. J Hydrol 268:143–157

Olami Z, Feder HJS, Christensen K (1992) Self-organized criticality in a continuous, nonconservative cellular automaton modeling earthquakes. Phys Rev Lett 68(8):1244–1247

Ortoleva P, Chadam J, Merino E, Sen A (1987) Geochemical self-organisation II: The reactive-infiltration instability. Am J Sci 28:1008–1040

Ouenes A (1992) Application of simulated annealing to reservoir characterization and petrophysics inverse problems. PhD dissertation, New Mexico Tech, Socorro, NM

Ouenes A, Bhagavan S, Bunge PH, Travis BJ (1994) Application of simulated annealing and other global optimization methods to reservoir description: myths and realities. In: Paper SPE 28415

Ovdat O, Akkermans E (2019) Breaking of continuous scale invariance to discrete scale invariance: a universal quantum phase transition. Mesoscale Nanoscale Phys 209–238

Özer UC, King PR, Vvedensky DD (2019) Path integral renormalization of flow through random porous media. arXiv: 1911.11218

Pan C, Hilpert M, Miller C (2001) Pore-scale modeling of saturated permeabilities in random sphere packings. Phys Rev E 64(6):066702

Pan C, Hilpert M, Miller CT (2004) Lattice-Boltzmann simulation of two-phase flow in porous media. Water Resour Res 40:W01501
Pan C, Luo L-S, Miller CT (2006) An evaluation of lattice Boltzmann schemes for porous medium flow simulation. Comput Fluids 35(8–9):898–909
Pan XH, Lü QA (2018) Quantitative strain energy indicator for predicting the failure of laboratory-scale rock samples: application to shale rock. Rock Mech Rock Eng 51:2689–2707
Pan XH, Xiong QQ, Wu ZJ (2018) A new method for determining the homogeneity index m of Weibull distribution by linking the peak strain with the crack damage strain of the uniaxial compression test. Int J Geomech 18(6):04018034
Parmigiani A, Huber C, Chopard B, Latt J, Bachmann O (2009) Application of the multi distribution function lattice Boltzmann approach to thermal flows. Eur Phys J 171:37–43
Pastor-Satorras R, Rothman DH (1998a) Stochastic equation for the erosion of inclined topography. Phys Rev Lett 80(19):4349
Pastor-Satorras R, Rothman DH (1998b) Scaling of a slope: the erosion of tilted landscapes. J Stat Phys 93:477–500
Paterson L (1981) Radial fingering in a Hele-Shaw cell. J Fluid Mech 113:513–529
Pazdniakou K, Adler PM (2013) Dynamic permeability of porous media by the lattice Boltzmann method Advances in Water Resources 62B:292–302
Pecknold S, Lovejoy S, Schertzer D (1996) The morphology and texture of anisotropic multifractals using generalized scale invariance. In: Molchanov SA, Woyczynski WA (eds) Stochastic models in geosystems. IMA Volumes in mathematics and its applications, vol 85, pp 269–312
Peng S, Zhi-weia X, Hou-quana Z, Yong-nian H (2009) Evolution of blast-induced rock damage and fragmentation prediction. Procedia Earth Planet Sci 1:585–591
Peng Y, Shu C, Chew YT (2003) Simplified thermal lattice Boltzmann model for incompressible thermal flow. Phys Rev E 68:046308
Peprah AK, Appiah SK, Amponsah SK (2017) An optimal cooling schedule using a simulated annealing based approach. Appl Math 8:1195
Pereira GG (2015a) A multiphase gray-scale lattice Boltzmann model. In: Eleventh international conference on CFD in the minerals and process industries. CSIRO, Melbourne, Australia, 7–9 December 2015
Pereira GG (2015b) Lattice Boltzmann simulations applied to understanding the stability of multiphase fluid interfaces. In: Progress in applied FD. Sintef Academic Press, Oslo, Norway, pp 231–237
Pessel M, Gibert D (2003) Multiscale electrical impedance tomography. J Geophys Res 108:2054
Pexider JV (1903) Notiz über Funktionaltheoreme. Monatsh Math Phys 14:293–301
Piggott AR, Elsworth D (1992) Analytical models for flow through obstructed domains. J Geophys Res Solid Earth 97(B2):2085–2093
Pilotti M, Succi S, Menduni G (2002) Energy dissipation and permeability in porous media. Europhys Lett 60:72–78
Pincus M (1970) A Monte Carlo method for the approximate solution of certain types of constrained optimization problems. Oper Res 18:1225–1228
Plascak JA, Figueiredo W, Grandi BCS (1999) Phenomenological renormalization group methods. Braz J Phys 29(3):579–610
Politis MG, Kikkinides ES, Kainourgiakis ME, Stubos AK (2008a) A hybrid process-based and stochastic reconstruction method of porous media. Microporous Mesoporous Mater 110(1):92–99
Politis M, Kainourgiakis ME, Kikkinides ES, Stubos AK (2008b) Application of simulated annealing on the study of multiphase systems. In: Tan CM (ed) Simulated annealing. I-Tech Education and Publishing, Vienna, pp 207–226
Polyakov AM (1970) Conformal symmetry of critical fluctuations. JETP Lett 12:381–383
Pomeau Y (2002) Recent progress in the moving contact line problem: a review. CR Mec 330(3):207–222

Ponce Dawson S, Chen S, Doolen GD (1993) Lattice Boltzmann computations for reaction – diffusion equations. J Chem Phys 98(2):1514–1523
Porter ML, Coon ET, Kang Q, Moulton J, Carey JW (2012) Multicomponent inter-particle potential lattice Boltzmann model for fluids with large viscosity ratios. Phys Rev E 86:036701
Press WH, Teukolsky SA, Vetterling WT, Flannery BP (2007) Numerical recipes: the art of scientific computing, 3rd edn. Cambridge University Press, New York
Pride SR, Flekkøy EG, Aursjø O (2008) Seismic stimulation for enhanced oil recovery. Geophysics 73(5):O23–O35
Prodanovic M, Lindquist WB, Seright RS (2006) Porous structure and fluid partitioning in polyethylene cores from 3D X-ray microtomographic imaging. J Colloid Interface Sci 298:282–297
Prodanovic M, Lindquist WB, Seright RS (2007) 3D image-based characterization of fluid displacement in a Berea core. Adv Water Resour 30:214–226
Provenzale A, Spiegel EA, Thieberger R (1997) Cosmic lacunarity. Chaos 7(1):82–88
Psihogios J, Kainourgiakis ME, Yiotis AG, Papaioannou A Th, Stubos AK (2007) A Lattice Boltzmann study of non-Newtonian flow in digitally reconstructed porous domains. Transp Porous Med 70:279–292
Pullammanappalli SK, Louie JN (1993) Inversion of seismic reflection traveltimes using a nonlinear optimization scheme. Geophysics 58:1607–1620
Qian YH, d'Humiéres D, Lallemand P (1992) Lattice-BGK models for Navier-Stokes equation. Europhys Lett 17(6):479–484
Qian Y, Orszag S (1993) Lattice BGK models for the Navier-Stokes equation: nonlinear deviation in compressible regimes. Europhys Lett 21:255–259
Qin SQ, Pan XH (2011) Stress and strain instability criteria for crustal rocks under shear condition. Chin J Geophys 54(7):1767–1771
Raeini AQ, Blunt MJ, Bijeljic B (2012) Direct simulations of two-phase flow on micro-ct image of porous media and upscaling of pore-scale forces. J Comput Phys 231:5653
Ramstad T, Oren PE, Bakke S (2010) Simulation of two phase flow in reservior rocks using a lattice Boltzmann method. SPE J 15(4):917–927
Ramstad T, Idowu N, Nardi C, Øren PE (2012) Relative permeability calculations from two-phase flow simulations directly on digital images of porous rocks. Transp Porous Media 94(2):487–504
Ramstad T, Kristoffersen A, Ebeltoft E (2019) Uncertainty span for relative permeability and capillary pressure by varying wettability and spatiality flow directions utilizing pore scale modelling. In: Annual symposium of society of core analysis, SCA2019-003
Rao P, Schaefer L (2020) Permeability estimation on tomographic images using curved boundary schemes in the lattice Boltzmann method. Adv Water Resour 143:103685
Rasin I, Miller W, Succi S (2005) Phase-field lattice kinetic scheme for the numerical simulation of dendritic growth. Phys Rev E 72:066705
Rayleigh L (1916) I. On the flow of compressible fluid past an obstacle. The London, Edinburgh, and Dublin Philos Mag J Sci 32(187):1–6
Rebbi C (1984) Monte Carlo calculations in lattice gauge theory. In: Binder K (ed) Applications of the Monte Carlo method in statistical physics. Springer-Verlag, New York, pp 277–298
Regulski W, Szumbarski J, Łaniewski-Wołłk Ł, Gumowski K, Skibiński J, Wichrowski M, Wejrzanowski T (2015) Pressure drop in flow across ceramic foams—a numerical and experimental study. Chem Eng Sci 137:320–337
Reis T, Phillips TN (2007) Lattice Boltzmann model for simulating immiscible two-phase flows. J Phys a: Math Theor 4:4033
Reynolds PS, Klein W, Stanley HE (1977) A real space renormalization group for site and bond percolation. J Phys c: Solid State Phys 10:1167
Rezaee MR, Jafari A, Kazemzadeh E (2006) Relationships between permeability, porosity and pore throat size in carbonate rocks using regression analysis and neural networks. J Geophys Eng 3:370–376
Rice JR (1988) Elastic fracture mechanics concepts for interfacial cracks. ASME J Appl Mech 55(1):98–103

Rice JR, Suo Z, Wang J-S (1990) Mechanics and thermodynamics of brittle interface failure in bimaterial systems. In: Ruhle M, Evans AG, Ashby MF, Hirth JP (eds) Metal-ceramic interfaces. Acta-scripta metallurgica proceedings, vol 4, pp 269–294
Rinaldo A, Rodriguez Iturbe I, Rigon R, Ijjasz Vasquez E, Bras RL (1993) Self-organized fractal river networks. Phys Rev Lett 70:822–826
Rintoul MD, Torquato S (1997) Reconstruction of the structure of dispersions. J Colloid Interface Sci 186(2):467–476
Rivet J-P, Boon JP (2001) Lattice gas hydrodynamics. Cambridge nonlinear science series, vol 11. Cambridge University Press, Cambridge
Romeo F, Sangiovanni-Vincentelli AL (1984) Probabilistic hill climbing algorithms: properties and applications. Report, EECS Department, University of California, Berkeley: UCB/ERL M84/34
Rossi N, Ubertini S, Bella G, Succi S (2005) Unstructured lattice Boltzmann method in three dimensions. Int J Numer Meth Fluids 49(6):619–633
Rothman DH (1985) Nonlinear inversion, statistical mechanics, and residual statics estimation. Geophysics 50:2784–2796
Rothman DH (1986) Automatic estimation of large residual static corrections. Geophysics 51:332–346
Rothman DH, Keller JM (1988) Immiscible cellular-automaton fluids. J Stat Phys 52:1119–1127
Rothman DH (1988) Cellular automaton fluids: a model for flow in porous media. Geophys 53:509–518
Rothman DH, Zaleski S (1997) Lattice-gas cellular automata: simple models of complex hydrodynamics. Cambridge University Press, Cambridge
Roy A, Roy S, Bhattacharyya AJ, Banerjee S, Tarafdar S (1999) Discrete scale invariance in viscous fingering patterns. Eur Phys J B-Condens Matter Compl Syst 12:1–3
Roy S, Tarafdar S (1996) Patterns in the variable Hele-Shaw cell for different viscosity ratios: similarity to river network geometry. Phys Rev E 54(6):6495
Roy S, Raju R, Chuang HF, Cruden BA, Meyyappan M (2003) Modeling gas flow through microchannels and nanopores. J Appl Phys 93(8):4870–4879
Rozman MG, Utz M (2001) Efficient reconstruction of multiphase morphologies from correlation functions. Phys Rev E 63(6):066701
Rui S, Jianjun L, Dahui Q (2013) Numerical simulation of two phase flow in reconstructed pore network based on lattice Boltzmann method. Int J Comput Sci Issues 10(1):193–200
Rutenbar RB (1989) Simulated annealing algorithms: an overview. IEEE Circuits Devices Mag (January):19–26
Saccomano AF, Savioli GB, Bidner MS (2001) Stochastic modeling of rock heterogeneities applying new autocorrelation estimators and simulated annealing. In: SPE 69654
Sahimi M, Arbabi S (1996) Scaling laws for fracture of heterogeneous materials and rock. Phys Rev Lett 77:3689
Salamon P, Nulton JD, Harland JR, Pedersen J, Ruppeiner G, Liao L (1988) Simulated annealing with constant thermodynamic speed. Comput Phys Commun 49(3):423–428
Saleur H, Sammis CG, Sornette D (1996a) Renormalization group theory of earthquakes. Nonlinear Process Geophys 3(2):102–109
Saleur H, Sammis CG, Sornette D (1996b) Discrete scale invariance, complex fractal dimensions and log-periodic corrections in earthquakes. J Geophys Res 101:17661–17677
Samorodnitsky G, Taqqu M (1994) Stable non-Gaussian random processes. Chapman & Hall, London
Sambridge M, Mosegaard K (2002) Monte Carlo methods in geophysical inverse problems. Rev Geophys 40:3
Santos LOE, Facin PC, Philippi PC (2003) Lattice-Boltzmann model based on field mediators for immiscible fluids. Phys Rev E 68:056302
Savioli GB, Falcigno ED, Bidner MS, Lake LW (1996) Applications of simulated annealing on actual but atypical permeability data. In: SPE Paper 35345

Sbragaglia M, Benzi R, Biferale L, Succi S, Sugiyama S, Toschi F (2007) Generalized lattice Boltzmann method with multirange pseudopotential. Phys Rev E 75:026702

Scales JA, Smith ML, Fischer TL (1992) Global optimization methods for multimodal inverse problems. J Comput Phys 103(2):258–268

Schaap MG, Porter ML, Christensen B, Wildenschild D (2007) Comparison of pressure-saturation characteristics derived from computed tomography and lattice Boltzmann simulations. Water Resour Res 43:W12S06

Schmieschek S, Harting J (2011) Contact angle determination in multicomponent lattice Boltzmann simulations. Commun Comput Phys 9(5):1165

Schneider WA, Whitman WW (1990) Dipmeter analysis by a Monte Carlo technique. Geophysics 55:320–326

Searcóid MÓ (2006) Metric Sspaces. Springer undergraduate mathematics series. Springer-Verlag, Berlin, New York

Semma E, El Ganaoui M, Bennacer R, Mohamad AA (2008) Investigation of flows in solidification by using the lattice Boltzmann method. Int J Therm Sci 47:201–208

Sen MK, Stoffa PL (1991a) Simulated annealing, genetic algorithms and seismic waveform inversion. In: SEG technical program expanded abstracts, pp 945–947

Sen MK, Stoffa PL (1991b) Nonlinear one-dimensional seismic waveform inversion using simulated annealing. Geophysics 56:1624–1638

Sen MK, Stoffa PL (2018) Global optimization methods in geophysical inversión, revised edition. Cambridge University Press

Sen MK, Bhattacharya BB, Stoffa PL (1993) Nonlinear inversion of resistivity sounding data. Geophysics 58(4):496–507

Sen MK, Datta-Gupta A, Stoffa PL, Lake LW, Pope GA (1995) Stochastic reservoir modeling using simulated annealing and genetic algorithms. SPE Form Eval 10(1):49

Seppecher P (1996) Moving contact lines in the Cahn-Hilliard theory. Int J Eng Sci 34(9):977–992

Seta T, Takegoshi E, Okui K (2006) Lattice Boltzmann simulation of natural convection in porous media. Math Comput Simul 72:195–200

Shabro V, Prodanović M, Arns CH, Bryant SL, Torres-Verdin C, Knackstedt MA (2010) Pore scale modeling of two-phase flow. In: Carrera J (ed) XVIII international conference on water resources CMWR 2010, CIMNE Barcelona

Shan X (1997) Simulation of Rayleigh-Bénard convection using a lattice Boltzmann method. Phys Rev E 55:2780–2788

Shan X, Chen H (1993) Lattice Boltzmann model for simulating multiple phases and components. Phys Rev E 4:1815–1819

Shan X, Chen H (1994) Simulation of nonideal gases and liquid-gas phase transitions by the lattice Boltzmann equation. Phys Rev E 49(4):2941

Shan X, Doolen G (1995) Multicomponent lattice-Boltzmann model with interparticle interaction. J Stat Phys 81:379–393

Shan X, Yuan X, Chen H (2006) Kinetic theory representation of hydrodynamics: a way beyond the Navier-Stokes equation. J Fluid MecH 550:413–441

Shante VKS, Kirkpatrick S (1971) An introduction to percolation theory. Adv Phys 20(85):325–357

She Z-S, Leveque E (1994) Universal scaling laws in fully developed turbulence. Phys Rev Lett 72:336

Shi Y, Tang GH (2018) Relative permeability of two-phase flow in three-dimensional porous media using the lattice Boltzmann method. Int J Heat Fluid Flow 73:101–113

Silin D, Patzek T (2009) Predicting relative-permeability curves directly from rock images. In: SPE 124974

Silin D, Tomutsa L, Benson SM, Patzek TW (2011) Microtomography and pore-scale modeling of two-phase fluid distribution. Transp Porous Media 86:495–515

Smalley RF, Turcotte DL, Solla SA (1985) A renormalization group approach to the stick-slip behaviour of faults. J Geophys Res 90:1894–1900

Smith SW, Sammis CG (1996) Discrete hierarchic cellular model for fore-shocks (Abst.). EOS Trans Am Geophys U 77:F48

Song Z, Ji H, Liu Z, Sun L (2020) Study on the critical stress threshold of weakly cemented sandstone damage based on the renormalization group method. Int J Coal Sci Technol 7:693–703

Sornette D (1998) Discrete-scale invariance and complex dimensions. Phys Rep 297(5):239–270

Sornette D (2000) Critical phenomena in natural sciences: chaos, fractals, self-organization and disorder: concepts and tools. Springer Series in synergetics. Springer, Berlin

Sornette A, Sornette D (1989) Self-organized criticality and earthquakes. Europhys Lett 9:197

Sornette A, Sornette D (1990) Earthquake rupture as a critical point: consequences for telluric precursors. Tectonophysics 179(3–4):327–334

Sornette D, Johansen A (1997) Large financial crashes. Phys A 245(3–4):411–422

Sornette D, Johansen A (1998) A hierarchical model of financial crashes. Phys A 261(3–4):581–598

Sornette D, Johansen A (2001) Significance of log-periodic precursors to financial crashes. Quant Finance 1(4):452

Sornette D, Sammis CG (1995) Complex critical exponents from renormalization group theory of earthquakes: implications for earthquake predictions. J Phys I France 607–619

Sornette D, Christian V (1992) Dynamics and memory effects in rupture of thermal fuse networks. Phys Rev Lett 68(5):612–615

Sornette D, Zhang YC (1993) Non-linear Langevin model of geomorphic erosion processes. Geophys J Int 113(2):382–386

Sornette D, Davy Ph, Sornette A (1990) Structuration of the lithosphere in plate tectonics as a self-organized critical phenomenon. J Geophys Res 95:17353

Sornette D, Johansen A, Bouchaud J-P (1996a) Stock market crashes, precursors and replicas. J Phys I France 6:167–175

Sornette D, Johansen A, Arneodo A, Muzy J-F, Saleur H (1996b) Complex fractal dimensions describe the hierarchical structure of diffusion-limited-aggregate clusters. Phys Rev Lett 76:251

Spaid MAA, Phelan FR Jr (1997) Lattice Boltzmann methods for modeling microscale flow in fibrous porous media. Phys Fluids 9:2468–2474

Stauffer D, Aharony A (1992) Introduction to percolation theory. Taylor and Francis, London

Stefaniuk D, Adrian R, Dariusz Ł (2016) Recovery of microstructure properties: random variability of soil solid thermal conductivity. Studia Geotech Mech 38(1):99–107

Stinchcombe RB, Watson BP (1976) Renormalization group approach for percolation conductivity. J Phys c: Solid State Phys 9:3221

Stueckelberg E, Petermann A (1953) La renormalisation des constants dans la théorie de quanta. Helv Phys Acta 26:499–520

Succi S (2001) The Lattice Boltzmann equation for fluid dynamics and beyond. Oxford University Press, Oxford

Succi S (2002) Mesoscopic modeling of slip motion at fluid-solid interfaces with heterogeneous catalysis. Phys Rev Letts 89(6):064502

Succi S, Foti E, Higuera F (1989) Three-dimensional flows in complex geometries with the lattice Boltzmann method. Europhys Lett 10(5):433–438

Suekane T, Soukawa S, Iwatani S, Tsushima S, Hirai S (2005) Behavior of supercritical CO_2 injected into porous media containing water. Energy 30:2370–2382

Sukop MC, Thorne Jr DT (2007) Lattice Boltzmann modeling, an introduction for geoscientists and engineers. Springer, Berlin, Heidelberg

Sukop MC, Huang HB, Lin CL, Deo MD, Oh K, Miller JD (2008) Distribution of multiphase fluids in porous media: Comparison between lattice Boltzmann modeling and micro-x-ray tomography. Phys Rev E 77:026710

Sullivan SP, Gladden LF, Johns ML (2006) Simulation of power-law fluid flow through porous media using lattice Boltzmann techniques. J Non-Newtonian Fluid Mech 133:91–98

Sun Q, Zhao C, Lu H (2016) Radon emission evolution and rock failure. Acta Geod Geophys 51:583–595

Sun WC, Wong T-F (2018) Prediction of permeability and formation factor of sandstone with hybrid lattice Boltzmann/finite element simulation on microtomographic images. Int J Rock Mech Min Sci 106:269–277
Sun WC, Andrade JE, Rudnicki JW (2011a) Multiscale method for characterization of porous microstructures and their impact on macroscopic effective permeability. Int J Num Methods Eng. 88:1260–1279
Sun WC, Andrade JE, Rudnicki JW, Richhubl P (2011b) Connecting microstructural attri-butes and permeability from 3D tomographic images of in situ shear-enhanced compaction bands using multiscale computations. J Geophys Res 38:L10302
Sutera SP, Skalak R (1993) The history of Poiseuille's law. Annu Rev Fluid Mech 25:1–19
Swift MR, Osborn WR, Yeomans JM (1995) Lattice Boltzmann simulation of nonideal fluids. Phys Rev Lett 75(5):830
Swift MR, Orlandini E, Osborn WR, Yeomans JM (1996) Lattice Boltzmann simulations of liquid-gas and binary fluid systems. Phys Rev E 54:5041
Sykes MF, Essam JW (1964) Exact critical percolation probabilities for site and bond problems in two dimensions. J Math Phys 5(8):1117–1127
Szu H, Hartley R (1987) Fast simulated annealing. Phys Lett A 122(3/4):157–162
Szymczak P, Ladd AJC (2006) A network model of channel competition in fracture dissolution. Geophys Res Lett 33:L05401
Talon L, Bauer D, Gland N, Youssef S, Auradou H, Ginzburg I (2012) Assessment of the two relaxation time lattice-Boltzmann scheme to simulate Stokes flow in porous media. Water Resour Res 376:48
Talukdar MS (2002). Ekofisk Chalk: core measurement, stochastic reconstruction, network modeling and simulation. PhD dissertation, Department of Petroleum Engineering and Applied Geophysics, Norwegian University of Science and Technology, Trondheim, Norway
Talukdar MS, Torsaeter O (2002) Reconstruction of chalk pore networks from 2D backscatter electron micrographs using a simulated annealing technique. J Pet Sci Eng 33:265–282
Talukdar MS, Talukdar MS, Torsaeter O, Ioannidis MA, Howard JJ (2002a) Stochastic reconstruction, 3D characterization and network modeling of chalk. J Pet Sci Eng 35(1–2):1–21
Talukdar MS, Torsaeter O, Ioannidis MA, Howard JJ (2002b) Stochastic reconstruction of chalk from 2D images. Transp Porous Media 48:101–123
Talukdar MS, Torsaeter O, Ioannidis MA (2002c) Stochastic reconstruction of particulate media from two dimensional images. J Colloid Interface Sci 248(2):419–428
Talukdar MS, Torsaeter O, Howard JJ (2004) Stochastic reconstruction of chalk samples containing vuggy porosity using a conditional simulated annealing technique. Transp Porous Media 57:1–15
Tan L, Zabaras N (2006) A level set simulation of dendritic solidification with combined features of front-tracking and fixed-domain methods. J Comp Phys 211:36–63
Tang GH, Zhang YH, Emerson DR (2008) Lattice Boltzmann models for nonequilibrium gas flows. Phys Rev E 77:046701
Tarantola A, Valette B (1982) Inverse problems = quest for information. J Geophys 50:150–170
Tölke J (2010) Implementation of a lattice Boltzmann kernel using the compute unified device architecture developed by nVIDIA. Comput vis Sci 13(1):29–39
Tölke J, Krafczyk M (2008) TeraFLOP computing on a desktop PC with GPUs for 3D CFD. Int J Comput Fluid Dyn 22(7):443–456
Tölke J, Freudinger S, Krafczyk M (2006) An adaptive scheme using hierarchical grids for lattice Boltzmann multi-phase flow simulations. Comput Fluids 35:820–830
Tölke J, Prisco GD, Mu Y (2013) A lattice Boltzmann method for immiscible two-phase Stokes flow with a local collision operator. Comput Math Appl 65:864–881
Toulouse Gérard et Pierre Pfeuty (1975) Introduction au groupe de renormalisation et à ses applications. Presses Universitaires de Grenoble, Grenoble
Tritton D (1959) Experiments on the flow past a circular cylinder at low Reynolds numbers. J Fluid Mech 6:547–567

Tsuji T, Jiang F (2016) Two-phase porous flow simulation for natural sandstone on GPU supercomputer. TSUBAME e-Sci J 14:02–07
Tsunogai U, Wakita H (1995) Precursory chemical changes in ground water: Kobe earthquake, Japan. Science 269(5220):61–63
Turcotte DL (1986) Fractals and fragmentation. J Geophys Res B91:1921–1926
Turner ML, Knufing L, Arns CH, Sakellariou A, Senden TJ, Sheppard AP, Sok RM, Limaye A, Pinczewski WV, Knackstedt MA (2004) Three-dimensional imaging of multiphase flow in porous media. Phys A Stat Mech Appl 339:166–172
Vaeudevan K, Wilson WG, Laidlaw WG (1991) Simulated annealing statics computation using an order-based energy function. Geophysics 56:1831–1839
Vahala G, Pavlo P, Vahala L, Martys NS (1998) Thermal lattice-Boltzmann models for compressible flows. Int J Modern Phys C 9(8):1247–1261
Vallejos RO, Anteneodo C (1998) Thermodynamical fingerprints of fractal spectra. Phys Rev E 58:4134
Vallejos RO, Mendes RS, da Silva LR, Tsallis C (1998) Connection between energy-spectrum self-similarity and specific heat log-periodicity. Phys Rev E 58:1346
van der Hoef MA, Beetstra R, Kuipers J (2005) Lattice-Boltzmann simulations of low-Reynolds-number flow past mono-and bidisperse arrays of spheres: results for the permeability and drag force. J Fluid Mech 528:233–254
VanderPlas JT (2018) Understanding the Lomb-Scargle periodogram. Astrophys J Suppl Ser 236:16
van Genabeek O, Rothman DH (1996) Macroscopic manifestations of microscopic flows through porous media. Annu Rev Earth Planet Sci 24:63–87
Van Kats FM, Egberts PJP (1998) Spreading dynamics modeled by lattice-Boltzmann techniques. J Colloid Interface Sci 205(1):166–177
Varnes DJ, Bufe CG (1996) The cyclic and fractal seismic series preceding an 4.8 earthquake on 1980 February 14 near the Virgin Islands. Geophys J Int 124:149–158
Veneziano D (2002) Iterated random pulse processes and their spectral properties. Fractals 10(01):1–11
Venturoli M, Boek ES (2006) Two-dimensional LB simulations of single phase flow in a pseudo two-dimensional micromodel. Physica A 362:23–29
Vesga-Ramírez A, Sanabria-Gómez JD, Sierra-Porta D, Arana-Salinas L, Asorey H, Kudryavtsev VA, Calderón-Ardila R, Núñez LA (2021) Simulated annealing for volcano muography. J S Am Earth Sci 109:103248
Vestergaard PD, Mosegaard K (1991) Inversion of post-stack seismic data using simulated annealing. Geophys Prospect 39(5):613–624
Vidal D, Ridgway C, Pianet G, Schoelkopf J, Roy R, Bertrand F (2009) Effect of particle size distribution and packing compression on fluid permeability as predicted by lattice-Boltzmann simulations. Comput Chem Eng 33(1):256–266
Vyssotsky VA, Gordon SB, Frisch HL, Hammersley JM (1961) Critical percolation probabilities (bond problem). Phys Rev 123:1566
Wallace DJ, Zia RKP (1978) The renormalization group approach to scaling in physics. Rep Prog Phys 41:1
Walsh SDC, Burwinkle H, Saar MO (2009) A new partial bounce-back LB method for fluid flow through heterogeneous media. Comput Geosci 3:1186–1193
Wang J, Chen L, Kang Q, Rahman SS (2016) The lattice Boltzmann method for isothermal micro-gaseous flow and its application in shale gas flow: a review. Int J Heat Mass Transf 95:94–108
Wang J, Ju Y, Huang Y, Zheng J, Zheng Z (2015) Study of the influence of porous structure on the permeability of rock using Lattice Boltzmann method. Proc Eng 102:1835–1841
Wang M, Wang J, Chen S, Pan N (2006) Electrokinetic pumping effects of charged porous media in microchannels using the lattice Poisson-Boltzmann method. J Colloid Interface Sci 304(1):246–253
Weber Z (2000) Seismic traveltime tomography: a simulated annealing approach. Phys Earth Planet Inter 11:149–159

Wei S, Shen J, Yang W, Li Z, Di S, Ma C (2019) Application of the renormalization group approach for permeability estimation in digital rocks. J Pet Sci Eng 179:631–644

Wellein G, Zeiser T, Donath S, Hager G (2006) On the single processor performance of simple lattice Boltzmann kernels. Comput Fluids 35:8–9

White JA, Borja RI, Fredrich JT (2006) Calculating the effective permeability of sandstone with multiscale lattice Boltzmann/finite element simulations. Acta Geotech 1:195–209

Wilson KG (1971a) Re-normalization group and critical phenomena. I. Re-normalization group and the Kadanoff scaling picture. Phys Rev B 4(9):3174–3183

Wilson KG (1971b) Re-normalization group and critical phenomena. II. Phase-space cell analysis of critical behavior. Phys Rev B 4(9):3184–3205

Wilson KG (1975) The renormalization group: critical phenomena and the Kondo problem. Rev Mod Phys 47(4):773–840

Wilson KG (1979) Problems in physics with many scales of length. Sci Am 241:158–179

Wilson KG (1983) The renormalization group and critical phenomena. Rev Mod Phys 55(3):583–600

Wilson KG, Kogut J (1974) The renormalization group and the ε expansion. Phys Rev C 12:75–200

Wolf FG, dos Santos LOE, Philippi PC (2009) Modeling and simulation of the fluid–solid interaction in wetting. J Stat Mech Theory Exp 06:P06008

Wolf-Gladrow DA (2000) Lattice-gas automata and Lattice Boltzmann models. Lecture notes in mathematics, vol 1725. Springer, Cham

Wolfram S (1986) Cellular automaton fluids 1: basic theory. J Stat Phys 45:471–526

Wu ZS, Dong PC, Lei G, Yang S, Cao N (2016) Lattice Boltzmann simulation of fluid flow in complex porous media based on CT image. J Ind Intell Inf 4(1):65–68

Xia C, Zhou C, Zhu F, Liu Z, Cui G (2021) The critical indicator of red-bed soft rocks in deterioration process induced by water basing on renormalization group theory. Appl Sci 11:7968

Xie JF, He S, Zu YQ, Lamy-Chappuis B, Yardley BWD (2017) Relative permeabilities of supercritical CO_2 and brine in carbon sequestration by a two-phase lattice Boltzmann method. Heat Mass Transf 53:2637–2649

Xue L, Qin SQ, Sun Q, Wang YY, Lee LM, Li WC (2014) A study on crack damage stress thresholds of different rock types based on uniaxial compression tests. Rock Mech Rock Eng 47(4):1183–1195

Yang X, Shi B, Chai Z, Guo Z (2014) A coupled Lattice Boltzmann method to solve Nernst-Planck model for simulating electro-osmotic flows. J Sci Comput 61(1):222–238

Yang X, Mehmani Y, Perkins WA, Pasquali A, Schönherr M, Kim K, Perego M, Parks ML, Trask N, Balhoff MT, Richmond MC (2016) Intercomparison of 3D pore-scale flow and solute transport simulation methods. Adv Water Resour 95:176–189

Yang Y, Liu Z, Yao J, Zhang L, Ma J, Hejazi SH, Luquot L, Ngarta TD (2018) Flow simulation of artificially induced microfractures using digital rock and lattice Boltzmann methods. Energies 11(8):2145

Yeong CLY, Torquato S (1998a) Reconstructing Random Media. Phys Rev E 57(1):495–506

Yeong CLY, Torquato S (1998b) Reconstructing random media. II. Three-dimensional media from two-dimensional cuts. Phys Rev E 58(1):224–233

Yiotis AG, Psihogios J, Kainourgiakis ME, Papaioannou A, Stubos AK (2007) A lattice Boltzmann study of viscous coupling effects in immiscible two-phase flow in porous media. Colloids and Surfaces a: Physicochem Eng Aspects 300:35–49

Young AP, Stinchcombe RB (1975) A renormalization group theory for percolation problems. J Phys c: Solid State Phys 8:1535

Yu H, Zhao K (2000) Lattice Boltzmann method for compressible flows with high Mach numbers. Phys Rev E 61(4):3867–3870

Yuan P, Schaefer L (2006) Equations of state in a lattice Boltzmann model. Phys Fluids 18(4):042101

Zhang J (2010) Lattice Boltzmann method for microfluidics: models and applications. Microfluid Nanofluid 10:1–28

Zhang R, Chen H (2003) Lattice Boltzmann method for simulations of liquid-vapor thermal flows. Phys Rev E 67:066711
Zhang B, Liu C, Guo Z, Lu N, Liu X (2017a) Probabilistic reservoir-properties estimation for anisotropic shales using statistical rock physics and seismic data. SEG Techn Program Expand Abst 2017:3179–3183
Zhang B, Liu C, Guo Z, Liu X, Liu Y (2018) Probabilistic reservoir parameters inversion for anisotropic shale using a statistical rock physics model. Chin J Geophys Chin Ed 61(6):2601–2617
Zhang B, Jin S, Liu C, Guo Z, Liu X (2020) Prediction of shear wave velocity based on a statistical rock-physics model and Bayesian theory. J Petrol Sci Eng 195:107710
Zhang Y, Nafi Toksöz M (2012) Impact of the cracks lost in the imaging process on computing linear elastic properties from 3D microtomographic images of Berea sandstone. Geophysics 77(2):R95–R104
Zhang WM, Meng G, Wei XY (2012) A review on slip models for gas microflows. Microfluid Nanofluid 13(6):845–882
Zhang X-L, Xiao L-Z, Guo L, Xie Q-M (2015) Investigation of shale gas microflow with the Lattice Boltzmann method. Pet Sci 12:96–103
Zhang D, Papadikis K, Gu S (2016a) A lattice Boltzmann study on the impact of the geometrical properties of porous media on the steady state relative permeabilities on two-phase immiscible flows. Adv Water Resour 95:61–79
Zhang L, Yang S, Zeng Z, Yin L, Zhao Y, Chew JW (2016b) Consistent lattice Boltzmann methods for incompressible axisymmetric flows. Phys Rev E 94(2):023302
Zhang L, Yao J, Zhao JL, Li A, Sun H, Wan Y, Su Y (2017b) The influence of wettability and shut-in time on oil recovery through microscale simulation based on an ideal model. J Nat Gas Sci Eng 48:178–185
Zhao H, Ning Z, Kang Q, Chen L, Zhao T (2017) Relative permeability of two immiscible fluids flowing through porous media determined by lattice Boltzmann method. Int Commun Heat Mass Transfer 8:53–61
Zhao J, Kang Q, Yao J, Viswanathan H, Pawar R, Zhang L, Sun H (2018) The effect of wettability heterogeneity on relative permeability of two-phase flow in porous media: A lattice Boltzmann study. Water Resour Res 54(2):1295–1311
Zhao Y-P, Chen J, Yuan Q, Cheng C (2016) Microcrack connectivity in rocks: a real-space renormalization group approach for 3D anisotropic bond percolation. J Stat Mech: Theory Exp 2016:013205
Zheng Y, Reese JM, Scanlon TJ, Lockerby DA (2006) Scaled Navier-Stokes-Fourier equations for gas flow and heat transfer phenomena in micro- and nanosystems. In: Proceedings of ASME ICNMM2006, June 19–21, Limerick, Ireland: 96066
Zhen-quan L, Zhan-ru Y, Yong Q (1997) Renormalization group approach to the bond percolation on Sierpinski carpets. Acta Phys Sin (overseas Edition) 6(4):257–265
Zhou CY, Li BT, Zhang XH, Liu Z (2015) Study on percolation threshold of red-layer soft rock failure process based on the renormalization group method. J Eng Geol 23(5):965–970
Zhou CY, Yang X, Liang YH, Du ZC, Liu Z, Huang W, Ming WH (2019) Classification of red-bed rock mass structures and slope failure modes in South China. Geosciences 9(6):273
Zhu J, Ma J (2013) An improved gray lattice Boltzmann model for simulating fluid flow in multi-scale porous media. Adv Water Resour 56:61–76
Zhu Z (2018) Pore-scale analysis of interfacial instabilities and impact of heterogeneity on relative permeability by Lattice Boltzmann method. Master's Thesis, Louisiana State University
Ziarani AS, Aguilera R (2012) Knudsen's permeability correction for tight porous media. Transp Porous Media 91(1):239–260
Zienkiewicz OC, Codina R (1995) A general algorithm for compressible and incompressible flow–part I: the split, characteristic-based scheme. Int J Numer Meth Fluids 20:869–885
Zou Q, He X (1997) On pressure and velocity boundary conditions for the lattice Boltzmann BGK model. Phys Fluids 9(6):1591–1598

Zu Y, He S (2013) Phase-field-based lattice Boltzmann model for incompressible binary fluid systems with density and viscosity contrasts. Phys Rev E 87(4):043301

Author Index

G. Korvin, *Statistical Rock Physics*, Earth and Environmental Sciences Library,
https://doi.org/10.1007/978-3-031-46700-4

C

D

E

M

O

P

T

X

Y

Z

Subject Index

G. Korvin, *Statistical Rock Physics*, Earth and Environmental Sciences Library,
https://doi.org/10.1007/978-3-031-46700-4

D

F

G

S

T

U

GPSR Compliance
The European Union's (EU) General Product Safety Regulation (GPSR) is a set of rules that requires consumer products to be safe and our obligations to ensure this.

If you have any concerns about our products, you can contact us on

ProductSafety@springernature.com

In case Publisher is established outside the EU, the EU authorized representative is:

Springer Nature Customer Service Center GmbH
Europaplatz 3
69115 Heidelberg, Germany

www.ingramcontent.com/pod-product-compliance
Ingram Content Group UK Ltd.
Pitfield, Milton Keynes, MK11 3LW, UK
UKHW022217051026
15624UKWH00001B/207
* 9 7 8 3 0 3 1 4 6 6 9 9 1 *